MW00710091

Contemporary Theory and Pragmatic Approaches in Fuzzy Computing Utilization

Toly Chen
Feng Chia University, Taiwan

Information Science
REFERENCE

Managing Director: Lindsay Johnston
Senior Editorial Director: Heather A. Probst
Book Production Manager: Sean Woznicki
Development Manager: Joel Gamon
Assistant Acquisitions Editor: Kayla Wolfe
Typesetter: Jennifer Romanchak
Cover Design: Nick Newcomer

Published in the United States of America by
Information Science Reference (an imprint of IGI Global)
701 E. Chocolate Avenue
Hershey PA 17033
Tel: 717-533-8845
Fax: 717-533-8661
E-mail: cust@igi-global.com
Web site: http://www.igi-global.com

Library of Congress Cataloging-in-Publication Data

Contemporary theory and pragmatic approaches in fuzzy computing utilization / Toly Chen, editor.
p. cm.
Summary: "This book presents the most innovative systematic and practical facets of fuzzy computing technologies to students, scholars, and academicians, as well as practitioners, engineers, and professionals"--Provided by publisher.
Includes bibliographical references and index.
ISBN 978-1-4666-1870-1 (hardcover) -- ISBN 978-1-4666-1871-8 (ebook) -- ISBN 978-1-4666-1872-5 (print & perpetual access) 1. Fuzzy logic. I. Chen, Toly, 1969-
QA9.64.C667 2013
003'.3633--dc23
2012005393

British Cataloguing in Publication Data
A Cataloguing in Publication record for this book is available from the British Library.

Table of Contents

Detailed Table of Contents

Preface xv

Acknowledgment xxiii

Chapter 1
Fuzzy C-Means in High Dimensional Spaces 1
Roland Winkler, German Aerospace Center, Germany
Frank Klawonn, Ostfalia University, Germany
Rudolf Kruse, Otto-von-Guericke University, Germany

High dimensions have a devastating effect on the FCM algorithm and similar algorithms. One effect is that the prototypes run into the centre of gravity of the entire data set. The objective function must have a local minimum in the centre of gravity that causes FCM's behaviour. In this paper, examine this problem. This paper answers the following questions: How many dimensions are necessary to cause an ill behaviour of FCM? How does the number of prototypes influence the behaviour? Why has the objective function a local minimum in the centre of gravity? How must FCM be initialised to avoid the local minima in the centre of gravity? To understand the behaviour of the FCM algorithm and answer the above questions, the authors examine the values of the objective function and develop three test environments that consist of artificially generated data sets to provide a controlled environment. The paper concludes that FCM can only be applied successfully in high dimensions if the prototypes are initialized very close to the cluster centres.

Chapter 2
Fuzzy Cluster Validation Based on Fuzzy PCA-Guided Procedure 17
K. Honda, Osaka Prefecture University, Japan
A. Notsu, Osaka Prefecture University, Japan
T. Matsui, Osaka Prefecture University, Japan
H. Ichihashi, Osaka Prefecture University, Japan

Cluster validation is an important issue in fuzzy clustering research and many validity measures, most of which are motivated by intuitive justification considering geometrical features, have been developed. This paper proposes a new validation approach, which evaluates the validity degree of cluster partitions from the view point of the optimality of objective functions in FCM-type clustering. This approach makes it possible to evaluate the validity degree of robust cluster partitions, in which geometrical features are not available because of their possibilistic natures.

This paper proposes a fuzzy clustering model for fuzzy data with outliers. The model is based on Wasserstein distance between interval valued data, which is generalized to fuzzy data. In addition, Keller's approach is used to identify outliers and reduce their influences. The authors also define a transformation to change the distance to the Euclidean distance. With the help of this approach, the problem of fuzzy clustering of fuzzy data is reduced to fuzzy clustering of crisp data. In order to show the performance of the proposed clustering algorithm, two simulation experiments are discussed.

This paper presents a new fuzzy modeling method that can be classified as a grid partitioning method, in which the domain space is partitioned by the fuzzy equalization method one dimension at a time, followed by the computation of rule weights according to the max-min composition. Five datasets were selected for testing. Among them, three datasets are high-dimensional; for these datasets only selected features are used to control the model size. An enumerative method is used to determine the best combination of fuzzy terms for each variable. The performance of each fuzzy model is evaluated in terms of average test error, average false positive, average false negative, training error, and CPU time taken to build model. The results indicate that this method is best, because it produces the lowest average test errors and take less time to build fuzzy models. The average test errors vary greatly with model sizes. Generally large models produce lower test errors than small models regardless of the fuzzy modeling method used. However, the relationship is not monotonic. Therefore, effort must be made to determine which model is the best for a given dataset and a chosen fuzzy modeling method.

This article discusses the basic features of information provided in terms of possibilistic uncertainty. It points out the entailment principle, a tool that allows one to infer less specific from a given piece of information. The problem of fusing multiple pieces of possibilistic information is and the basic features of probabilistic information are described. The authors detail a procedure for transforming information between possibilistic and probabilistic representations, and using this to form the basis for a technique for fusing multiple pieces of uncertain information, some of which is possibilistic and some probabilistic. A procedure is provided for addressing the problems that arise when the information to be fused has some conflicts.

Accurately forecasting the foreign exchange rate is important for export-oriented enterprises. For this purpose, a fuzzy and neural approach is applied in this study. In the fuzzy and neural approach, multiple experts construct fuzzy linear regression (FLR) equations from various viewpoints to forecast the foreign exchange rate. Each FLR equation can be converted into two equivalent nonlinear programming problems to be solved. To aggregate these fuzzy foreign exchange rate forecasts, a two-step aggregation mechanism is applied. At the first step, fuzzy intersection is applied to aggregate the fuzzy forecasts into a polygon-shaped fuzzy number to improve the precision. A back propagation network is then constructed to defuzzify the polygon-shaped fuzzy number and generate a representative/crisp value to enhance accuracy. To evaluate the effectiveness of the fuzzy and neural approach, a practical case of forecasting the foreign exchange rate in Taiwan is used. According to the experimental results, the fuzzy and neural approach improved both the precision and accuracy of the foreign exchange rate forecasting by 79% and 81%, respectively.

In this paper, a novel stock trading framework based on a neuro-fuzzy associative memory (FAM) architecture is proposed. The architecture incorporates the approximate analogical reasoning schema (AARS) to resolve the problem of discontinuous (staircase) response and inefficient memory utilization with uniform quantization in the associative memory structure. The resultant structure is conceptually clearer and more computationally efficient than the Compositional Rule Inference (CRI) and Truth Value Restriction (TVR) fuzzy inference schemes. The local generalization characteristic of the associative memory structure is preserved by the FAM-AARS architecture. The prediction and trading framework exploits the price percentage oscillator (PPO) for input preprocessing and trading decision making. Numerical experiments conducted on real-life stock data confirm the validity of the design and the performance of the proposed architecture.

As an emerging human-computer interaction (HCI) technology, recognition of human hand gesture is considered a very powerful means for human intention reading. To construct a system with a reliable and robust hand gesture recognition algorithm, it is necessary to resolve several major difficulties of hand gesture recognition, such as inter-person variation, intra-person variation, and false positive error caused by meaningless hand gestures. This paper proposes a learning algorithm and also a classification technique, based on multivariate fuzzy decision tree (MFDT). Efficient control of a fuzzified decision boundary in the MFDT leads to reduction of intra-person variation, while proper selection of a user dependent (UD) recognition model contributes to minimization of inter-person variation. The proposed method is tested first by using two benchmark data sets in UCI Machine Learning Repository and then by a hand gesture data set obtained from 10 people for 15 days. The experimental results show a discernibly enhanced classification performance as well as user adaptation capability of the proposed algorithm.

Thanks to the rapid advancement of human-computer interaction technologies it is becoming easier for the elderly and/or people with disabilities to operate various electrical systems. Operation of home appliances by using a set of predefined hand gestures is an example. However, hand gesture recognition may fail when the predefined command gestures are similar to some ordinary but meaningless behaviors of the user. This paper uses a gesture spotting method to recognize a designated gesture from other similar gestures. A fuzzy garbage model is proposed to provide a variable reference value to determine whether the user's gesture is the command gesture or not. Further, the authors propose two-stage user adaptation to enhance recognition performance: that is, off-line (batch) adaptation for inter-person variation and on-line (incremental) adaptation for intra-person variation. For implementation of the two-stage adaptation method, a genetic algorithm (GA) and the steepest descent method are adopted for each stage. Experimental results were obtained for 5 different users with left and up command gestures.

In this paper, the authors propose two indirect adaptive fuzzy control schemes for a class of uncertain continuous-time single-input single-output (SISO) nonlinear dynamic systems with known and unknown control direction. Within these schemes, fuzzy systems are used to approximate unknown nonlinear functions and the Nussbaum gain technique is used to deal with the unknown control direction. This paper first presents a singularity-free indirect adaptive control algorithm for nonlinear systems with known control direction, and then this control algorithm is generalized for the case of unknown control direction. The proposed adaptive controllers are free from singularity, allow initialization to zero of all adjustable parameters of the used fuzzy systems, and guarantee asymptotic convergence of the tracking error to zero. Simulations performed on a nonlinear system are given to show the feasibility of the proposed adaptive control schemes.

This paper examines the size reduction of the fuzzy rule base without compromising the control characteristics of a fuzzy logic controller (FLC). A 49-rule FLC is approximated by a 4-rule simplest FLC using compensating factors. This approximated 4-rule FLC is implemented to control the shunt active power filter (APF), which is used for harmonic mitigation in source current. The proposed control methodology is less complex and computationally efficient due to significant reduction in the size of rule base. As a result, computational time and memory requirement are also reduced significantly. The control performance and harmonic compensation capability of proposed approximated 4-rule FLC based shunt APF is compared with the conventional PI controller and 49-rule FLC under randomly varying nonlinear loads. The simulation results presented under transient and steady state conditions show that dynamic performance of approximated simplest FLC is better than conventional PI controller and comparable with 49-rule FLC, while maintaining harmonic compensation within limits. Due to its effectiveness and reduced complexity, the proposed approximation methodology emerges out to be a suitable alternative for large rule FLC.

Building a linear fitting model for a given interval-valued data set is challenging since the minimization of the residue function leads to a huge combinatorial problem. To overcome such a difficulty, this article proposes a new semidefinite programming-based method for implementing linear fitting to interval-valued data. First, the fitting model is cast to a problem of quadratically constrained quadratic programming (QCQP), and then two formulae are derived to develop the lower bound on the optimal value of the nonconvex QCQP by semidefinite relaxation and Lagrangian relaxation. In many cases, this method can solve the fitting problem by giving the exact optimal solution. Even though the lower bound is not the optimal value, it is still a good approximation of the global optimal solution. Experimental studies on different fitting problems of different scales demonstrate the good performance and stability of our method. Furthermore, the proposed method performs very well in solving relatively large-scale interval-fitting problems.

This paper proposes a novel fuzzy modeling approach for identification of dynamic systems. A fuzzy model, recurrent interval type-2 fuzzy neural network (RIT2FNN), is constructed by using a recurrent neural network which recurrent weights, mean and standard deviation of the membership functions are updated. The complete back propagation (BP) algorithm tuning equations used to tune the antecedent and consequent parameters for the interval type-2 fuzzy neural networks (IT2FNNs) are developed to handle the training data corrupted by noise or rule uncertainties for nonlinear system identification involving external disturbances. Only by using the current inputs and most recent outputs of the input layers, the system can be completely identified based on RIT2FNNs. In order to show that the interval IT2FNNs can handle the measurement uncertainties, training data are corrupted by white Gaussian noise with signal-to-noise ratio (SNR) 20 dB. Simulation results are obtained for the identification of nonlinear system, which yield more improved performance than those using recurrent type-1 fuzzy neural networks (RT1FNNs).

This paper extends the technique for order preference by similarity to ideal solution (TOPSIS) for solving multi-attribute group decision making (MAGDM) problems under Atanassov intuitionistic fuzzy set (IFS) environments. In this methodology, weights of attributes and ratings of alternatives on attributes are extracted from fuzziness inherent in decision data and making process and described using Atanassov IFSs. An Euclidean distance measure is developed to calculate the differences between alternatives for each decision maker and an Atanassov IFS positive ideal solution (IFSPIS) as well as an Atanassov IFS negative ideal-solution (IFSNIS). Degrees of relative closeness to the Atanassov IFSPIS for all alternatives with respect to each decision maker in the group are calculated. Then all decision makers in the group may be regarded as "attributes" and a corresponding classical MADM problem is generated and hereby solved by the TOPSIS. The proposed methodology is validated and compared with other similar methods. A numerical example is examined to demonstrate the implementation process of the methodology proposed in this paper.

Supply chain management is a new and evolving paradigm for enterprises to cope with international competition and to improve global logistics efficiency. The suppliers' performances affect not only supply chain execution results but also the profit capability and business survivability. However, suppliers' performance assessment always involves a large dimension of supplier behaviors. Information on supplier behaviors is often difficult to be accurately demonstrated as quantitative data. For this reason, the study employs a 2-tuple linguistic variable to perform the initial evaluation and final assessment while keeping track of both linguistic information and data, which can avoid a tied result. Additionally, the modified linguistic ordered weighted averaging (M-LOWA) operator with maximum entropy is used to derive the maximum aggregation value under the current business strategy to reflect on the criteria. The focal company can then rapidly rely on the assessment results to represent the performance of suppliers and provide integrated information to decision makers. This study draws the complete framework for the issue of supplier performance assessment without limitations on categories of variables and scales.

Preface

Real-life problems in industries or service sectors are usually filled with a lot of uncertainties, sometimes because of the unpredictability of what lies ahead, sometimes it is in part due to the subjectivity of the decision-maker, and sometimes it are technical limitations. For example, in building models for decision making, many coefficients are based on forecasts, and therefore cannot be certain. Some constraints can also be relaxed to a certain degree. These considerations may involve subjective judgments that are beyond the scope of parametric analyses or probabilistic methods. The concept of fuzzy sets proposed by Zadeh in 1985 provides us with an effective tool to solve these problems. Fuzzy logic is a form of many-valued logic. Contrary to the traditional binary logic theory, fuzzy logic variables have a truth value that ranges between 0 and 1. Reasoning in fuzzy logic is similar to human reasoning. It allows for approximate values and inferences as well as incomplete or ambiguous data. In addition, compared to nonlinear or stochastic methods, problems are generally easier to solve using fuzzy methods, since they are more flexible, and better reflect and incorporate human judgment. These features allow fuzzy techniques to be widely applied, including decision making, problem solving, system learning, clustering, system control, simulation, optimization, and others.

From a theoretical point of view, fuzzy control systems have advanced rapidly since the late 1980s. The rapid advances in computer technology have resulted in the proposal of many advanced computing techniques such as machine intelligence, artificial neural networks, data mining, soft computing, computing with words/expressions, semantic web, ubiquitous computing, and others. The combination of fuzzy sets with these new computing techniques has received a lot of attention. For example, Figure 1 shows the numbers of relevant articles on fuzzy neural networks in the past ten years. On average, more than one thousand papers are published and indexed each year.

In 2011, with the full support of IGI Global, we started a new journal on fuzzy systems – International journal of Fuzzy System Applications, which focuses on:

- Fuzzy clustering
- Fuzzy data analysis
- Fuzzy decision support systems
- Fuzzy evolutionary computing
- Fuzzy expert systems
- Fuzzy mathematical programming
- Fuzzy modeling and fuzzy control of biotechnological processes
- Fuzzy neural systems or neuro-fuzzy systems
- Fuzzy pattern recognition
- Fuzzy process control
- Fuzzy reasoning system

Figure 1. The number of published articles on fuzzy neural networks from 2002 to 2011 (data source: Engineering Village©)

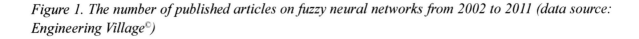

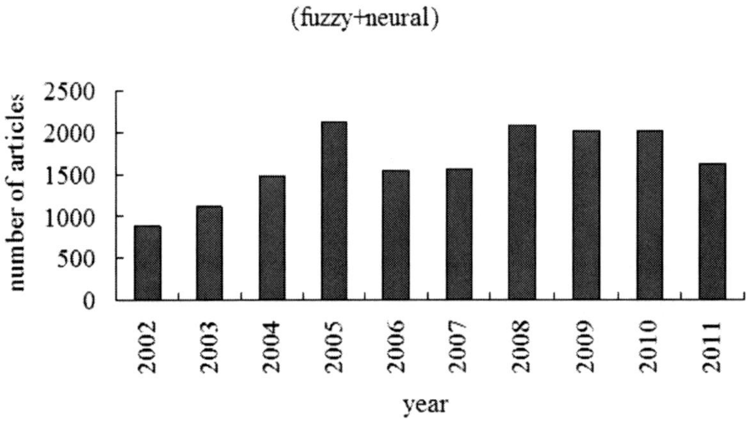

- Fuzzy-rule based system
- Fuzzy system applications in computer vision
- Fuzzy system applications in electronic commerce
- Fuzzy system applications in human-machine interface
- Fuzzy system applications in robotics
- Fuzzy system applications in system and control engineering

This book collected fourteen representative articles from this journal. Most articles included in this book are reprinted in their original form. New typesetting was employed only for those articles whose original form was not of sufficiently high quality. In certain articles, we made some minor adjustments to avoid confusion possibly caused if the articles are not entirely read.

With the popularity of the Internet and inexpensive data storage devices, massive amounts of data are stored everywhere. These huge amounts of data lead to difficulties when they need to be analyzed. Data clustering is an important task in fields such as data mining, statistical data analysis, machine learning, pattern recognition, image analysis, information retrieval, and bioinformatics, and thus encounters the same difficulties of these fields. Clustering is the task of assigning a set of objects to certain groups (called clusters) such that the objects in the same cluster are more similar to each other than to those in the other clusters. In the first chapter, Winkler et al. discusses the clustering of high dimensional data, which is especially meaningful for solving the difficulties in high dimensional data analyses. Fuzzy c-means (FCM) is an effective clustering approach, and works quite well in low dimensions. FCM and a number of its variants have been proven to be useful for data summarization. In addition, the clustering results using these FCM approaches are flexible in interpretation, since an object can be simultaneously assigned to different categories and to different degrees. However, one of the problems of FCM is that the prototypes tend to run into the center of gravity of the complete data set. Winkler et al. showed this property with some interesting examples. They also provided some clues to answer the following questions: What is the minimum number of dimensions necessary to cause unacceptable behavior by FCM? How does the number of prototypes influence unacceptable behavior? Why does the objective function have a local minima in the center of gravity, and how must FCM be initialized to avoid the local minima in the center of gravity? The answers to these questions allow FCM to be effectively applied to the clustering of high dimensional data, such as sales data, shop floor data, and others.

In order to objectively identify clusters in a data set, it is usually necessary to define a measure of similarity or proximity to establish a rule for assigning patterns to the domain of a particular cluster center. As expected, the measure of similarity is problem dependent, and thus most FCM algorithms adopt the iterative optimization scheme. However, FCM only considers the distances from the data to the cluster centers and their memberships. If two distinct clusters have a common mean, then the performance of FCM is poor. In addition, FCM variants are well designed for finding local optimal solutions, but may derive several different local solutions in the multi-starting strategy. Since the optimal cluster number is not known a priori, a validation measure for selecting the optimal cluster partition from multiple solutions is needed. In the next chapter, Honda et al. discusses the issue of clustering validation for FCM. Cluster validation is an important issue in fuzzy clustering research. Many validity measures have been developed, most of which are motivated by intuitive justification based on geometrical features, Honda et al. proposes a new validation approach, which evaluates the validity degree of cluster partitions from the point of view of the optimality of objective functions in FCM-type clustering. Their approach makes it possible to evaluate the validity of robust cluster-partitions, in which geometric features are not available because of their possibilistic nature.

In the third chapter, Zarandi and Razaee propose a fuzzy cluster model based on a new distance measure for classifying the fuzzy data. Their model can be used in noisy environments. The authors also added some weighting factors to reduce the effects of outliers. In short, their approach presents the results of successfully solving the problems of transforming the fuzzy data into the crisp data. Furthermore, for determining the optimal number of clusters, there is no need to define a cluster validity index for fuzzy data. The ones existing in literature for crisp data can be applied.

A fuzzy system can be divided into three parts: input, processing, and output. Typical fuzzy systems include the general fuzzy system, the Takagi-Sugeno-Kang (TSK) fuzzy system, the Mamdani fuzzy system, and the adapted network-based fuzzy inference system (ANFIS). Inputs to a fuzzy system usually have to undergo the fuzzification step. In this step, each input dimension is fuzzily partitioned. That is to say, it is segmented into several intervals with equal or unequal widths. These intervals overlap each other according to pre-defined membership functions. The partition method has a great influence on the performance of the fuzzy system. In Chapter 4, with the focus on this issue, Liao proposes a grid partitioning method. Here the domain space is partitioned by the fuzzy equalization method, one dimension at a time, followed by the computation of rule weights based on the max-min composition. Liao also proposed an enumeration method to determine the best combination of fuzzy terms for each variable. After applying it to five datasets (weld recognition, welding flaw identification, Haberman's survival, Pima diabetes, and Wisconsin breast cancer datasets), and comparing it with some existing approaches (FCM variants, ant colony optimization (ACO)-based fuzzy partitioning, Wang and Mendel (WM) method, fuzzy subtractive (FS), ANFIS, FS-ANFIS, and fuzzy equalization), the proposed methodology produced the lowest average test errors and required less time to build the fuzzy models.

Information used in decision making generally comes from multiple sources. The fusion of multi-source information is an important issue, but not an easy one, especially not when the information provided has some uncertainty. There are however many different viewpoints and methods for dealing with uncertainty, including probabilistic, possibilistic, fuzzy, grey, chaotic, and other ways. Each method has its own merits and disadvantages, and is suitable for dealing with certain topics. The results obtained by these methods tend to differ by their basic nature. If we split a problem into different parts and then treat each part using a different approach, then the question of how to integrate the results from these approaches becomes important. For example, as a fuzzy clustering approach, FCM can be generalized many ways, including generalizing the memberships to include possibilities. In addition, if there is a discrepancy in these results, then the question remains of how to solve the problem. In Chapter 5, Yager investigated the fusion of possibilistic and probabilistic information. He discussed the basic features of information provided in terms of possibilistic uncertainty, and pointed out the entailment principle, "a tool that allows one to infer less specific from a given piece of information". He then provided a procedure for addressing the problems that arise when the information to be fused has some conflicts.

In recent years, many advanced evolutionary algorithms have been proposed, such as the ant colony optimization, swarm intelligence, genetic algorithms, genetic programming, artificial neural networks, immunological computing, morphic computing, and others. As a result of the pioneering studies of Pedrycz, fuzzy collaborative intelligence approaches have been shown to have the potential for improving the accuracy of forecasting. In addition, seeing a problem from different perspectives ensures that no parts are ignored when solving the problem. Collaborative forecasting is not derived from academic discussion, but arises from practical applications. However, in fuzzy collaborative systems, fuzzy collaborative clustering is one of the more often discussed issues. It relies on information granules rather than on data type to optimize the process. Fuzzy information granulation and granular computing are important concepts in fuzzy set and rough set theories. Fuzzy set, rough set and their combinations seem to be efficient tools for granular computing – the approaches to generating fuzzy information granules from data. Why use a system that seems complex, time consuming, and requires the collaboration of a number of domain experts? Although the existing methods can provide the same forecast in a more realistic manner and in a shorter time, the accuracy of that method of forecasting is often far from perfect, mainly due to unpredictable changes in observation. In Chapter 6, Chen proposed a fuzzy and neural approach for forecasting the foreign exchange rate. The foreign exchange rate between two currencies specifies how much one currency is worth in terms of the other. Accurately forecasting the foreign exchange rate is very important for export-oriented enterprises. Unfavorable foreign exchange rates result in the increase of raw material costs and the decrease of gross margin for these enterprises. In the fuzzy and neural approach, a group of domain experts are asked to configure their fuzzy linear regression equations to forecast the foreign exchange rate based on their views. A collaboration mechanism is therefore established to develop the views. To facilitate this collaboration process and to derive a single representative value from these forecasts, Chen used the fuzzy intersection and back propagation network approach. He used the historical data of the foreign exchange rate from NTD to USD to evaluate the effectiveness of the fuzzy and neural approach.

Forecasting the stock market is another concern. There are so many stock market gurus and experts out there suggesting ways to forecast the stock market. Basically, making the right buying/selling action at the right time is the only way to get rich in stock market. There are two viewpoints when it comes to forecasting the stock market of a future period. One is the input-output relationship viewpoint. It determines the factors that are influential in the stock market, and then applies different approaches such as multiple linear regression (MLR) or artificial neural network (ANN), to model the relationship between the stock market and these factors in order to forecast the stock market. The other viewpoint, the time-series viewpoint, is to treat fluctuations in the stock market as a type of time series. Theoretically there are many approaches, e.g. moving average (MA), weighted moving average (WMA), exponential smoothing (ES), MLR, ANN, auto-regressive integrated moving average (ARIMA), and others that can be applied to forecast the stock market. Generally speaking, ANN is suitable for modeling a short-term nonlinear pattern of the stock market, while traditional approaches such as MA, WMA, ES, and MLR have good performances when the trend in the stock market is stable. In Chapter 7, Quek et al. proposed a novel stock trading framework based on a neuro-fuzzy associative memory (FAM) architecture. They incorporated the approximate analogical reasoning schema (AARS) to resolve the problem of discontinuous (staircase) response and inefficient memory utilization with uniform quantization in the associative memory structure. Their experimental results showed that a more advanced prediction allows the stock to be switched earlier to the correct position, which in turn leads to a substantial increase in wealth.

Fuzzy systems are increasingly common in daily life, and there have been many innovative applications based on the needs of people, such as a fuzzy washing machine, fuzzy air conditioner, the antilock braking system, and so on. These applications allow us to interact with the objects around us in more intelligent ways. For example, we may want to communicate with computers using gestures instead of, screens, and keyboards. Several new concepts have emerged lately, one of them being ambient intelligence (AmI). Ambient Intelligence is the vision of a future in which environments support the people inhabiting them. In this vision the traditional computer input and output media disappear, and instead processors and sensors are integrated in everyday objects. We can communicate with our clothes, household devices, and furniture, which in turn may communicate with each other and with the devices and furniture of other people. In this vision the environment is sensitive to the needs of its inhabitants, and capable of anticipating their needs and behavior. As an interesting investigation in this field, in Chapter 8, Jeon et al. recognizes hand gestures using a multivariate fuzzy decision tree and user adaptation. Recognizing the human hand gestures, that are usually culture-specified and can convey very different meanings in different social or cultural settings, is very important for understanding human intention. To reliably recognize hand gestures, it is necessary to resolve several major difficulties such as inter-person variation, intra-person variation, and false positive error caused by meaningless hand gestures. In their viewpoint, the efficient control of fuzzified decision boundary leads to the reduction of intra-person variation, while the proper selection of a user dependent recognition model contributes to the minimization of inter-person variation.

In Chapter 9, Yang et al. provides another way to achieve the same objective. With the rapid advances in human-computer interaction technologies, it is becoming easier for the elderly and/or people with disabilities to operate a variety of electrical systems, such as using gestures to operate home appliances. The gesture-based machine control system, allows you to wave your arm or hand in such a way that it is picked up by sensors, which is then translated into the movement of a device. This device can be a robotic surgical device, equipment control, an energy delivery system, or any other device. Gesture recognition is a prerequisite for gesture-based machine control. Gesture recognition is a field in the area of computer science and language technology that focuses on interpreting human gestures via computer hardware and software. Gestures can originate from any motion or state of your body but the face or hand are most commonly used. The current focus is on emotion recognition from the face and on hand gesture recognition. However, hand gesture recognition may fail when a predefined command gesture is similar to a common but meaningless behavior of the user. Thus, a gesture spotting procedure was developed to distinguish designated gestures from other similar gestures. In particular, a fuzzy garbage model was established to provide a variable reference value to determine whether the user's gesture is a command gesture or not.

Fuzzy systems have been applied to a wide variety of fields ranging from control, signal processing, communication, to circuit design optimization and manufacturing. In system control, for example, Canon developed an autofocusing camera that uses a fuzzy control system with 12 inputs, 13 rules, and an output to determine the position of the lens. An industrial air conditioner designed by Mitsubishi uses a fuzzy control system with 25 heating rules and 25 cooling rules. Maytag invented an intelligent dishwasher that uses a fuzzy controller. The operation procedure of a fuzzy controller can be divided into three stages. First, the input stage maps the sensor or other inputs, such as switches, thumbwheels, and so on, to the appropriate membership functions and truth values. Subsequently, the processing stage invokes each appropriate rule and generates a result for each rule, then combines the results of the rules. Finally, the output stage converts the combined result back into a specific control output value. In Chapter 10, Laboid et al. proposes two indirect adaptive fuzzy control schemes for a class of uncertain continuous single-input single-output (SISO) nonlinear dynamic systems with a known and an unknown control direction, respectively. In these schemes, fuzzy systems are used to approximate unknown nonlinear functions, and the Nussbaum gain technique is applied to deal with the unknown control direction. The effectiveness of the proposed methodology was verified through simulation experiments. The actual trajectories were found to be quite close to the desired ones.

Sometimes, to be able to control a system more precisely, we need to build a considerable number of fuzzy rules. This of course reduces the efficiency of the system, and results in a dilemma. In order to solve this problem, some researchers tried to achieve the effects of many rules while using fewer rules. For example, Singh et al. explored how relatively fewer rules can achieve similar control performance. In Chapter 11, a 49-rule fuzzy logic controller (FLC) used to control the shunt active power filter is approximated by a 4-rule one using compensating factors. Traditionally conventional PI controllers are used to regulate the dc link voltage of the shunt APF. They provide efficient harmonic compensation but suffer from poor dynamic response. This drawback is overcome by the 49-rule FLC. As the number of rules increases the control action becomes more efficient due to the smooth transition from one membership function to another. However, this is achieved at the cost of increased complexity. On the other hand, the 4-rule FLC is less complex and computationally more efficient due to the significant reduction in rule base size. In addition, computational time and memory requirement are also significantly reduced. The simulation results showed that the dynamic performance of the approximated simplest FLC is comparable with that of the 49-rule FLC, under transient and steady state conditions. Therefore, the 4-rule FLC is a suitable alternative for a large-rule FLC. In addition, the proposed scheme has several important features – it is system independent, is less complex in design, takes less memory, reduces computational time and provides efficient control action that satisfies the demand of effective compensation and provides a better dynamic response.

Li and Yu discussed another approximation problem. In Chapter 12, they proposed a semi-definite programming-based method for implementing linear fitting to interval-valued data. The central concern of fuzzy regression is linear regression analysis involving fuzzy data. Tanaka was the first to propose a fuzzy linear regression model by minimizing the index of fuzziness of the system. Diamond then introduced a metric on the set of fuzzy numbers and used this metric to define a least-sum-of-squares criterion function in the usual sense. In the field of symbolic data analysis, at the interval level, a similar topic has also been discussed extensively. Especially, building a linear fitting model for a given interval-valued data set is challenging since the minimization of the residue function leads to a huge combinatorial problem. Diamond first cast the fitting model to a problem of quadratically constrained quadratic programming (QCQP), and then derived two formulae to develop the lower bound on the optimal value of the non-convex QCQP by semi-definite relaxation and Lagrangian relaxation. In many cases, their method solves the fitting problem by giving the exact optimal solution. Even though the lower bound is not the optimal value, their solution is still a good approximation of the global optimal solution. According to the numerical results, their method performs very well in solving relatively large-scale interval-fitting problems.

In Chapter 13, Lin, Chang and Lee proposes a novel fuzzy modeling approach for identification of dynamic systems. They construct a recurrent interval type-2 fuzzy neural network (RIT2FNN) by using a recurrent neural network which recurrent weights, mean, and standard deviation of the membership functions are updated. The complete back propagation (BP) algorithm tuning equations used to tune the antecedent and consequent parameters for the interval type-2 fuzzy neural networks (IT2FNNs) are developed to handle the training data corrupted by noise or rule uncertainties for nonlinear system identification involving external disturbances. By using the current inputs and most recent outputs of the input layers, the dynamic system can be completely identified based on RIT2FNNs. According to the simulation results, the proposed methodology yielded more improved performance than those using recurrent type-1 fuzzy neural networks (RT1FNNs).

Nowadays, the electromagnetic compatibility (EMC) analysis of integrated circuit (IC) design has become an important issue due to the rapid increasing of electromagnetic interference (EMI) and the tendency of adopting tremendous technologies such as higher operating frequency, higher dissipation and lower supply voltage. The next paper written by Lin et al. demonstrates the extraction capability of a single model to predict the electromagnetic emission of a digital circuit by using a fuzzy logic system. This paper also evaluates the overall variation of the output response in the circuit and automatically extracts suitable values for the critical parameters of the ICEM.

The incorporation of fuzzy sets into the traditional mathematical programming models leads to the fuzzy versions, such as fuzzy linear programming (FLP), fuzzy integer programming (FIP), fuzzy goal programming (FGP), and fuzzy nonlinear programming (FNP). The previous studies on fuzzy mathematical programming are largely limited in the range of LP or multiple-objective LP, but FNP is rarely involved. In many practical problems, however, these are many kinds of nonlinearity and uncertainty, and they cannot be described or solved by traditional crisp or linear mathematical programming models. Therefore, the modeling and optimization methods for nonlinear programming under a fuzzy environment are not only important in theory, but also have a great application value for a wide range of practical problems. Chapter 15 is by Al-Refaie and Li. They propose a goal programming model to optimize the performance of the injection molding process by considering three important quality responses: number of defects, cycle time, and weight of the products. The injection molding process has become increasingly important because it provides high-quality products, short product cycles, and is light weight. The process consists of first constructing a weighted additive goal programming model. The three quality responses and process factors are described by appropriate membership functions. Then, Taguchi's orthogonal array is utilized to provide the experimental layout. A linear optimization based on the weighted additive goal programming model is built to minimize the deviations of the product/process targets from their corresponding imprecise fuzzy values specified by the process engineer's preferences. The result shows that a good setting can be obtained for the controllable factors of the plastic injection modeling process by integrating the Taguchi method with their proposed fuzzy model.

Competitiveness is the ability and performance of a firm, sub-sector or country to sell and supply goods and/or services in a given market. Competitiveness engineering is a systematic procedure, and includes a series of activities to assess and enhance competitiveness, with competitiveness assessment being one of the major tasks. There have been many relevant references in this field, but most of them focused on exploring the factors affecting competitiveness (such as cost, quality, customer satisfaction, technical competence, etc.) and ways to improve competitiveness (such as balanced scorecard, blue ocean strategy, lean production, green supply chain, learning organization, etc.). However, how to assess competitiveness in a quantitative way is rarely discussed. Chapter 16 discusses the fuzzy approach for assessing the long-term competitiveness of a semiconductor product proposed by Chen. Initially, the fuzzy linear regression approach is applied to estimate the unit cost of a semiconductor product. Then the cost region for the semiconductor product to be competitive is specified. The mid-term cost competitiveness of the semiconductor product is assessed by comparing the estimated unit cost and the competitive region. To obtain the long-term cost competitiveness, the mid-term competitiveness trend is taken into consideration. A practical example with data collected from a real semiconductor-manufacturing factory is used to demonstrate the applicability of the proposed methodology.

Chapter 17 by Li and Nan extends the technique for order preference by similarity to ideal solution (TOPSIS) for solving multi-attribute group decision making (MAGDM) problems under Atanassov intuitionistic fuzzy set (IFS) environments. The basic principle of TOPSIS is that the chosen alternative should have the shortest distance from the ideal solution and the farthest distance from the negative-ideal solution. This principle can also be used to demonstrate that any alternative which has the shortest distance from the ideal solution is also guaranteed to have the longest distance from the negative-ideal solution, while there will be a compromise solution at the point where the level of satisfaction of both criteria are the same. However, the highest ranked alternative by TOPSIS is the best in terms of the ranking index, which does not mean that it is always the closest to the ideal solution. Many studies have combined fuzzy logic and TOPSIS for applications under an uncertain environment. In Li and Nan's method, the weights of the attributes and the ratings of the alternatives for the attributes are extracted from the fuzziness inherent in the decision data and the decision making process, and are described using Atanassov IFSs. An Euclidean distance measure is then developed to calculate the differences between the alternatives for each decision maker and the Atanassov IFS positive ideal solution (IFSPIS) as well as the Atanassov IFS negative ideal solution (IFSNIS). The degree of relative closeness to the Atanassov IFSPIS are then calculated for all alternatives with respect to each decision maker in the group. Then all decision makers in the group may be regarded as "attributes" and a corresponding classical MAGDM problem is generated and then solved by the TOPSIS.

In the field of supply chain management, Wang, Chang, and Wang develop a comprehensive framework to determine supplier behavior in the last chapter. They employ a 2-tuple linguistic variable to perform the initial evaluation and final assessment while keeping track of both fuzzy linguistic information and data from suppliers. Their study also draws the complete framework for the issue of supplier performance assessment without limitations on categories of variables and scales.

The purpose of this book is twofold. First, it is intended to provide audiences with an extensive exploration of fuzzy technologies, computing, and systems, while providing pragmatic examples of application, making this book valuable to practitioners and professionals in various industries. Second, it is expected to play a useful role in higher education, as a rich source of supplementary readings in relevant courses and seminars. We hope that this book will serve its purpose well.

Tin-Chih Toly Chen
Taichung, Taiwan

Yi-Chi Wang
Taichung, Taiwan

March, 2012

Acknowledgment

The production of the book would not have been possible without the invaluable help from Yi-Chi Wang (Feng Chia University). Without his continued support this book would never have been published.

Toly Chen
Feng Chia University, Taiwan

Chapter 1
Fuzzy C-Means in High Dimensional Spaces

Roland Winkler
German Aerospace Center, Germany

Frank Klawonn
Ostfalia University, Germany

Rudolf Kruse
Otto-von-Guericke University, Germany

ABSTRACT

High dimensions have a devastating effect on the FCM algorithm and similar algorithms. One effect is that the prototypes run into the centre of gravity of the entire data set. The objective function must have a local minimum in the centre of gravity that causes FCM's behaviour. In this paper, examine this problem. This paper answers the following questions: How many dimensions are necessary to cause an ill behaviour of FCM? How does the number of prototypes influence the behaviour? Why has the objective function a local minimum in the centre of gravity? How must FCM be initialised to avoid the local minima in the centre of gravity? To understand the behaviour of the FCM algorithm and answer the above questions, the authors examine the values of the objective function and develop three test environments that consist of artificially generated data sets to provide a controlled environment. The paper concludes that FCM can only be applied successfully in high dimensions if the prototypes are initialized very close to the cluster centres.

INTRODUCTION

Clustering high dimensional data has many interesting applications. For example clustering similar music files, semantic web applications, image recognition or biochemical problems. Many tools today are not designed to handle hundreds of dimensions, or in this case, it might be better to call it degrees of freedom. Many clustering approaches work quite well in low dimensions, especially the fuzzy c-means algorithm (FCM) (Dunn, 1973; Bezdek, 1981; Höppner et al., 1999; Kruse et al., 2007) seems to fail in high dimensions. Hence FCM is so useful in cases where data object arrangements that are not crisp, this paper is dedicated to give some insight into the behaviour of FCM in high dimensions.

DOI: 10.4018/978-1-4666-1870-1.ch001

One of the problems of FCM is, that the prototypes tend to run into the centre of gravity of the complete data set, almost independently of the initialisation of the prototypes. Figure 1 shows on the right hand side a 50 dimensional artificial data set, projected on 2 dimensions, containing 100 clusters, 100 data objects each. The lines represent the way the prototypes took from their initial position. The left hand side shows FCM in 2 dimensions with 4 clusters where it works quite well. It appears that the structural problem of FCM is independent from the data set, this question is addressed later in the paper. The 4 questions presented in the abstract will be answered in this paper:

- How many dimensions are at least necessary to cause an ill behaviour of FCM?
- How does the number of prototypes influence ill behaviour?
- Why has the objective function a local minima in the centre of gravity?
- How must FCM be initialised to avoid the local minima in the centre of gravity?

These questions are independent of the actual data set FCM is applied to. The idea is to use test environments that are optimal for FCM, so that it is likely that FCM fails on all data sets if it fails on the optimal ones. Four data sets D_1 through D_4 are used which have different properties and are introduced Q_1. is answered using D_1, Q_2 is answered using D_2, Q_3 by analysing the objective function and for Q_4 the data sets D_3 and D_4 are used. The answers to the questions are found by observing the objective function of FCM.

Related Work

The curse of dimensionality was often addressed in literature. Most of the time this term appears is, to invent a clustering algorithm that does not suffer from it by design. For example in Hinneburg and Keim (1999) a method is proposed to subdivide the input space by creating a non-axis parallel grid. This subdivides the input space in such a way, that the grid boarders do not cross areas of high density but rather follow empty space areas. Also Steinbach et al. (2004) gives a good overview about a collection of methods that address clustering in high dimensional spaces.

In this paper, outliers are not addressed. However, since there are many methods to detect them in lower dimensions, they tend to fail in higher

Figure 1. Fuzzy c Means in 2 dimensions (left) and 50 dimensions (right)

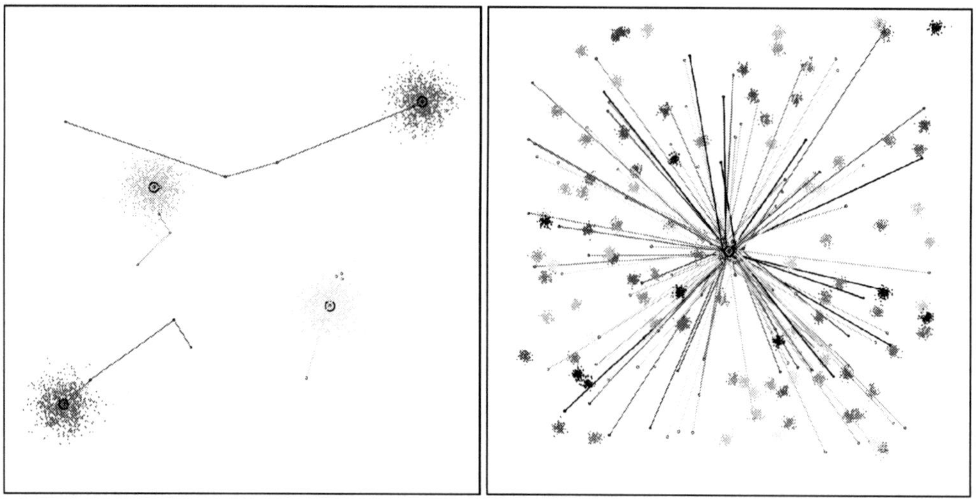

dimensions. Aggarwal and Yu (2001) provides an interesting inside in detecting outliers in high dimensional spaces. They find outliers by intelligently choosing projections into lower dimensional spaces.

In high dimensional spaces, it is very likely that clusters roughly form subspaces. In this case, approaches like FCM and alike do not work well because it assumes a full dimensional cluster. In some cases, not even subspaces are formed, even though the data is not fully dimensional like a piece of a spheres surface. Gionis et al. (2005) provides a method to find lower dimensional manifolds. Dimension reduction in general is a good idea if possible. Similar to the last paper but with a completely different approach, Verleysen (2003) provides an overview to map high dimensional data into lower dimensions using artificial neural networks.

The most theoretical approach however was done in Beyer et al. (1999), that addresses the question when the nearest neighbour is meaningful in high dimensional spaces. This paper is dedicated to the nearest neighbour problem in a high dimensional input space and cannot be applied directly on clustering problems. But it gives a good insight into the effects of high dimensional spaces and was the starting point for the analysis of the effects of many dimensions on FCM as it is presented in this paper. Also very interesting to read is Donoho (2000) which provides a general view on high dimensionality without concentrating on a particular data analysis method.

THE TEST DATA SETS FOR FCM

The FCM Algorithm

Although fuzzy c-means (FCM) is well known, some mathematical details are needed in the next section. Let $\mathbf{X} \subset \mathrm{R}$ be a finite set of data objects of the vector space R^n and $|X| = m$. The clusters are represented by a set of prototypes $\mathbf{Y} = \{y_1, ..., y_c\} \subset \mathrm{R}^n$. Let

$$Y_2 = \text{prototypes}(\text{HKM}(X_2, c))\,\mathrm{R}$$

be the fuzzifier and $\mathbf{U} \in \mathrm{R}^{c \times m}$ be the partition matrix with $u_{ij} \in [0,1]$ and $\forall j{:}\sum_{i=1}^{c} u_{ij} = 1$. And finally, let $d{:}\, \mathrm{R}^n \times \mathrm{R}^n \rightarrow \mathrm{R}$ be the Euclidean distance function with its abbreviation $d_{ij} = d(y_i, x_j)$.

Fuzzy c-means clustering is based on an objective function $\mathbf{J}$ that is to be minimised:

$$\mathbf{J}(\mathbf{X}, \mathbf{U}, \mathbf{Y}) = \sum_{i=1}^{c} \sum_{j=1}^{m} u_{ij}^{\omega} d_{ij}^2 \tag{1}$$

The minimisation of J is done by iteratively updating the members of $\mathbf{U}$ and $\mathbf{Y}$ and is computed using a Lagrange extension to ensure the constraints $\sum_{i=1}^{c} u_{ij} = 1$. The iteration steps are denoted by a time variable $t \in \mathrm{N}$ denoting $t = 0$ as the initialisation step:

$$u_{ij}^{t+1} = \frac{\left(d_{ij}^{t}\right)^{\frac{2}{1-\omega}}}{\sum_{k=1}^{c}\left(\left(d_{kj}^{t}\right)^{\frac{2}{1-\omega}}\right)}. \tag{2}$$

$$y_i^{t+1} = \frac{\sum_{j=1}^{n}\left(u_{ij}^{t}\right)^{\omega} \cdot x_j}{\sum_{j=1}^{n}\left(u_{ij}^{t}\right)^{\omega}}. \tag{3}$$

For noise clustering, an additional cluster is specified which is represented by a virtual prototype y_0 which has no location in V. Instead, it has a constant distance $0 < d_{\text{noise}} \in \mathrm{R}$ to all data objects: $\forall j{:} d_{0j} = d(y_0, x_j) = d_{\text{noise}}$ which is called noise distance. y_0 is not represented as a member of V, which means, it is not updated during the iteration process. However, the membership values are updated as if y_0 was a normal prototype. The clustering is done iteratively until it converges.

Requirements for the Test Data Sets

The first step towards a formal description in this paper is done by setting the test environment. At first, a motivation for the later used test environments is given, followed by their definition. The goal is to exploit structural problems with FCM that are independent of the data set FCM is applied on. Since it is very hard to do an analysis on all possible data sets, we decided to use a data set that is optimal for FCM. If FCM fails on a optimal data set, it is safe to conclude that it will fail on any data set under the same conditions. We consider a data set optimal for FCM if the clusters are very well separated and the data objects of one cluster are tightly packed. In other words, the maximal distance among data objects of of the same cluster is much lower than the distance to a data object of a different cluster. In this situation, FCM is not really necessary because hard k-means (HKM) would be better suited due to the lack of fuzzy areas, but this is for testing FCM under optimal conditions. Since FCM finds circular shaped clusters, the clusters should have roughly the shape of a hypersphere.

We are interested in the performance of FCM in high dimensions. Let $V \subset \mathcal{R}^n$ be the n-dimensional data set. Let there be c clusters with $c \leq n$. The cluster centres induce a $c-1$ dimensional linear subspace $W \subset V$. W still contains n-dimensional vectors but there is a linear transformation from W into a $c-1$ dimensional vector space $\bar{W}$. For the ill behaviour of FCM, there is no difference whether the algorithm is applied in W instead of $\bar{W}$.

The data examples and the tests form a unity that is hard to separate. The goal is to understand why the prototypes run into the centre of gravity of a data set. Because that fact is already known, in an optimal data set, the clusters should be placed as far away from the centre of gravity as possible. This requirement is added to the list of requirements, because if the prototypes still run into the centre of gravity, even if the clusters are maximal distant from it, it might reveal the reason for the ill behaviour of the FCM algorithm.

Defining the Test Data Sets

The first test data set D_1 consists of $c = n$ clusters. The clusters consist of only 1 data object each. Obviously, it is not possible to pack the data objects for each cluster more tight since there is only one. To achieve an optimal separation, the clusters are located at the scaled unit vectors of R^n so they form a simplex. The unit vectors are scaled in such a way, that they have a distance of 1 to the centre of gravity. The scaling ensures better comparability among tests at different dimensions, it leads to an objection function value of $c^{2-\omega}$ if all prototypes are located in the centre of gravity.

If the fuzzifier ω is set to 2, the objective function value is 1 for all dimensions. The separation of the clusters is optimal, because all pairwise distances among cluster centres are identical. One disadvantage of D_1 however is, that the number of the clusters depends on the number of dimensions. At this point it is not sure that the number of prototypes is the reason for FCM failure in high dimensions.

D_1 is defined as the set of data objects with $\gamma = \sqrt{\frac{n}{n-1}}$ and the n-dimensional unit vector e_i:

$$D_1 = \{x_i \mid x_i = \gamma e_i, i = 1, ..., n\} \tag{4}$$

In the left part of Figure 2, D_1 is presented in the 2-dimensional case. Obviously, the data objects x_1 and x_2 create an 1-dimensional subspace. The prototypes start in the centre of gravity and end in the data objects. The dotted lines show the way of the prototypes from the center of gravity to the data objects.

The second data set D_2 is designed to overcome the disadvantage of D_1. D_2 consists of $c > n$ clusters, again with one data object each. In principle, what is needed is a data set that has a constant number of clusters, for all dimensions strictly less than c. Still, it is required to fulfill the constraints that the clusters should be best separated and maximal distant from the centre of gravity. Therefore, D_2 consists of c clusters that are located on an n-dimensional hypersphere surface in such a way, that the minimal distance between each pair of clusters is maximised. It seems to be a hard problem to find an analytical solution for this problem, hence a heuristic approach is presented in the next subsection. A pseudo-code version is shown below where *HKM* (dataset, number of prototypes) denotes the HKM algorithm with randomly initialized prototypes. D_2 has the advantage over D_1 that it has independent numbers of dimensions and clusters. Unfortunately, it has the disadvantage to be heuristic and that the prototypes may not be as well separated as in D_1.

$$X_2 = \left\{ \frac{x}{\|x\|} : x \sim N(\vec{0}, 1) \right\} \tag{5}$$

$$Y_2 = \text{prototypes}(\text{HKM}(X_2, c)) \tag{6}$$

$$D_2 = \left\{ \frac{y}{\|y\|} : y \in Y_2 \right\} \tag{7}$$

In the right part of Figure 2, D_2 is shown for 2 dimensions and 10 prototypes. The data objects are located at a circle around the origin of the coordinate system. The separation among the data objects is not perfect due to the heuristic approach like HKM. As a consequence, the centre of gravity is not in the coordinate system origin. However, the prototypes start in the centre of gravity and end in the data objects. The dotted lines show the way of the prototypes from the center of gravity to the data objects. Each dot represents one iteration step for calculating the objective function values in later graphs.

Figure 2. Visualization of the data sets D_1 (left) and D_2 (right) in 2 dimensions and in case of D_2 for 10 prototypes

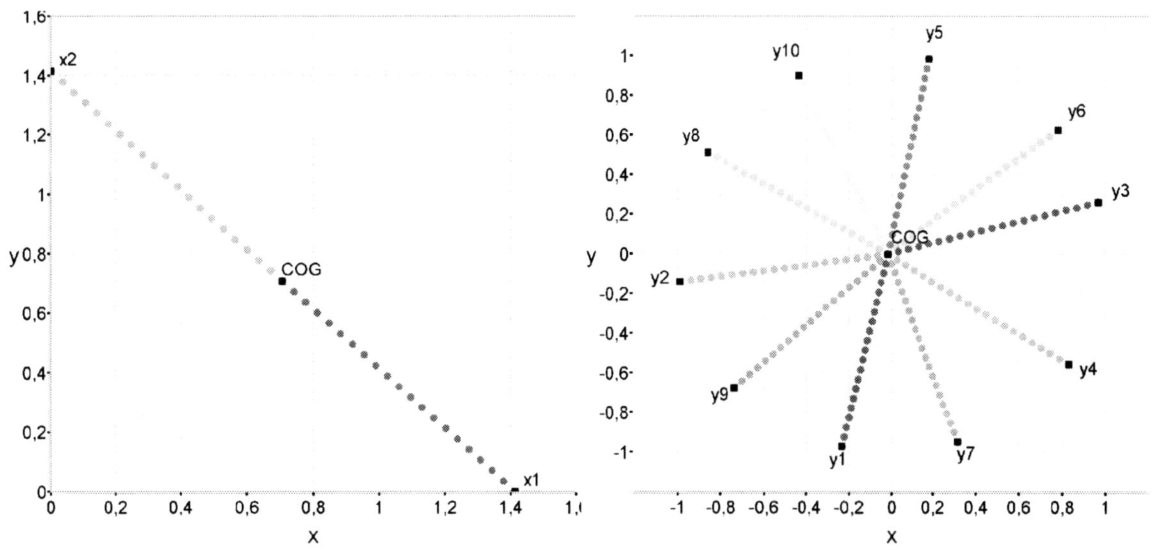

The third data D_3 set is introduced to verify if the first two data sets have anything to do with reality. An example of the third data set is already illustrated in the introduction to present the problem of FCM in Figure 1. D_3consists of c cluster centres that are uniformly distributed in all dimensions. Around each cluster centre, a predefined number of data objects are scattered. The scattered data objects are normal distributed with a small variance to ensure a good separation of clusters. Let σ^2 be the variance (for all clusters) and X_3 be the set of of coordinates that are used as cluster centres:

$$X_4 = \{\mu_1,...,\mu_c\} \sim N(\vec{0},1) \tag{8}$$

$$\mathrm{D}_3 = \bigcup_{i=1}^{c}(\{x_{i1},...,x_{im}\}) \sim N(\mu_i\sigma^2) \tag{9}$$

The fourth data set is almost identical to the third. The only difference is the distribution of the cluster centres X_4, which are standard normal distributed instead of uniform.

$$X_4 = \{\mu_1,...,\mu_c\} \sim N(\vec{0},1) \tag{10}$$

$$D_4 = \bigcup_{i=1}^{c}\left(\{x_{i1},..., x_{im}\} \sim N(\mu_i,\sigma^2)\right) \tag{11}$$

Generating Points on a Hypersphere with Maximal Pairwise Distance

For convenience, let S_n be the set of points in a hypersphere of n dimensions with centre 0 and radius 1 and let $\overline{S_n}$ be the surface of S_n:

$$S_n = \{x:\| x \| \leq 1\} \tag{12}$$

$$\overline{S_n} = \{x:\| x \| = 1\} \tag{13}$$

Generating a set D_2 of $|D_2| = c$ data objects on $\overline{S_n}$ with optimal separation is not trivial. An optimal separation is reached, if for each data object the distance to the closest other data object is maximised. A heuristic approach is used that consists of two steps. The first step generates a uniformly distributed data objects on $\overline{S_n}$. The second step applies HKM on that data set which leads to the final data set of D_2.

The first problem is, how to define a random variable with uniform distribution on $\overline{S_n}$. A simple and direct approach is, to generate a uniform distribution on the hypersphere S_n and projecting the data objects on the surface $\overline{S_n}$. Generating a uniform distribution on S_n is also not trivial. The straightforward approach of generating a uniform distribution on the enclosing hypercube and removing all data objects outside the hypersphere does not work well in high dimensions. That is because the volume of the hypersphere in relation to the volume of the hypercube approaches zero for many dimensions. The volume of S_n with radius r is given by

$$V_n(r) = \frac{\pi^{\frac{n}{2}}}{\Gamma(\frac{n}{2}+1)} \tag{14}$$

where Γ is the Gamma function[1]. The Gamma function in the denominator grows much faster than the exponential value in the numerator: $\lim_{n\to\infty} V_n(r) = 0$. For a large amount of dimensions, almost all data objects must be deleted, hence the computation time becomes too high for practical uses.

Instead, the uniform distribution on the hypersphere surface is generated by an n-dimensional standard normal distribution $x_i \in \mathcal{R}^n$ that is projected on the hypersphere surface. The n-dimensional normal distribution is generated by joining n independent 1-dimensional standard normal distributions $\mathcal{N}(0,1)$. Let R_2 be an n-dimensional random variable following a standard normal distribution: $R_2 \sim \mathcal{N}(0,E)$. Since the covariance matrix of R_2 is the identity matrix, the distribution is rotation-symmetric. Let $\overline{R_2}$ be a random variable by projecting R_2 on $\overline{S_n}$ and has the probability density function $f_{\overline{R_2}}$. From the rotation symmetry follows, that $f_{\overline{R_2}}(x)=f_{\overline{R_2}}(y)$ for all $x, y \in \overline{S_n}$. For the next step, let $X_2 \sim \overline{R_2}$ be a set of m data objects, uniformly distributed on $\overline{S_n}$.

The HKM algorithms tends to generate data set partitions with approximately equal numbers of data objects. Because the data objects are uniformly distributed on $\overline{S_n}$, the surface area should be partitioned in pieces of approximately equal sizes. The HKM algorithm is initialised with c prototypes which are located at random picked data objects of X_2. HKM does not suffer from the same dimensional problems as FCM which makes it possible to apply it in this situation. Due to the dimensional structure of the hypersphere, the prototypes end up somewhere inside the hypersphere. Projecting them on the surface does not change the partitioning of the data objects with respect to the closest prototype, hence does not change the area partitioning as well.

The test data set D_2 is defined as the set of prototype locations after the projection.

To test the result, some properties are listed in Table 1 which indicates that the separation is indeed quite good. At least it is possible to see that there are no elements in D_2 that are close together. In the table, several numbers are presented. The number of dimensions, the number of clusters and the number of data objects in X_2 are shown in the first three columns, HKM is initialised with. The other values are calculated after performing the clustering and projecting the prototypes on the hypersphere surface. Data objects per cluster shows the minimal and maximal number of data objects that are allocated to a cluster. The minimal angle of each pair of data objects shows the minimal angular separation. There is no pair of clusters that come close together, even in 50 dimensions with 500 clusters, the clusters are separated of at least one radius of the hypersphere. For comparison, Table 2 contains the separation properties of D_1 for several dimensions.

Table 1. Separation properties of D_2

Dim	Clusters	Data Objects	per Cluster		Angle [ ]	Distance	
			(Min	Max)	(Min)	(Min	Max)
2	100	100000	811	3256	1.46	0.0253	2.0
5	100	100000	1800	2235	38.8	0.665	1.99
10	100	100000	1795	2237	54.1	0.910	1.95
20	100	100000	1831	2151	60.9	1.01	1.84
50	100	100000	1826	2162	74.0	1.20	1.74
100	100	100000	1853	2137	77.3	1.25	1.63
50	50	100000	3772	4192	80.3	1.29	1.60
50	100	100000	1839	2120	72.5	1.18	1.72
50	200	100000	901	1114	70.0	1.15	1.75
50	500	100000	309	480	63.0	1.04	1.77

Table 2. Separation properties of D_1

Dim	Clusters	DataObjects	per Cluster		Angle [ ]	Distance	
			(Min	Max)	(Min)	(Min	Max)
2	2	2	1	1	180.0	2.0	2.0
5	5	5	1	1	104.4	1.58	1.58
10	10	10	1	1	96.4	1.49	1.49
20	20	20	1	1	93.0	1.45	1.45
50	50	50	1	1	91.2	1.43	1.43
100	100	100	1	1	90.6	1.42	1.42

THE OBJECTIVE FUNCTION IN HIGH DIMENSIONAL SPACES

The initialisation of FCM provides some random effects on the clustering results. Multiple initialisations are helpful to recognise local minima and to increase the probability of finding a global minimum. Since this is an analysis of the structural behaviour of FCM, random effects are not wanted. In fact, the question is, why the prototypes run into the centre of gravity, almost independently of the initialisation. Therefore, FCM is not applied on D_1 to D_3 directly. Instead the prototypes are all initialised in the COG and then moved step by step towards the clusters. Let $\alpha \in [0,1]$, $\mathbf{J}(D,\alpha)=\sum_{i=1}^{c}\sum_{j=1}^{m} u_{ij}^{\omega}\left(d_{ij}(\alpha)\right)^2$ the i-th cluster and $cog(D) \in \mathcal{R}^n$ the centre of gravity of a data set D. Than the location of prototype i is defined by a function $y_i : [0,1] \rightarrow \mathcal{R}^n$ with

$$y_i(\alpha) = \alpha \cdot x_i + (1-\alpha) \cdot cog(D) \qquad (15)$$

which implies a distance function of $d_{ij}(\alpha) = d(y_i(\alpha) - x_j)$. The objective function than can be plotted by using

$$\omega \qquad (16)$$

or equivalently,

$$\mathbf{J}(D,\alpha) = \sum_{i=1}^{c}\sum_{j=1}^{m} \left(\frac{\left(d_{ij}(\alpha)\right)^{\frac{2}{1-\omega}}}{\sum_{k=1}^{c}\left((d_{kj}(\alpha))^{\frac{2}{1-\omega}} \right)} \right)^{\omega} \left(dij(\alpha)\right)^2 \qquad (17)$$

The fuzzifier plays an important role for the ill behaviour of FCM. But a fixed fuzzifier only shifts the problem towards higher dimensions. Also, if it is left as a variable, further argumentation becomes very complicated without educational benefits. Therefore, the influence of the fuzzifier is discussed in a separate subsection. For the following four subsections, the fuzzifier is set to $\omega = 2$.

How Many Dimensions are at Least Necessary to Cause an Ill Behaviour of FCM?

The first question is answered by observing the objective function for the data set D_1: $\mathbf{J}(D_1,\alpha)$ with α=0 for several dimensions: see the left part of Figure 3. As already explained above, in the first data set, the c prototypes form a $c-1$ dimensional subspace. In the figure, the dimension of the data set is presented, which is equal to c in D_1. In all examples, the value α is drawn on the x-axis while the (sometimes normalized) value of the objective function is presented on the y-

Figure 3. D_1 for Dimensions 2 to 10 (left) and 10 to 500 dimensions (right)

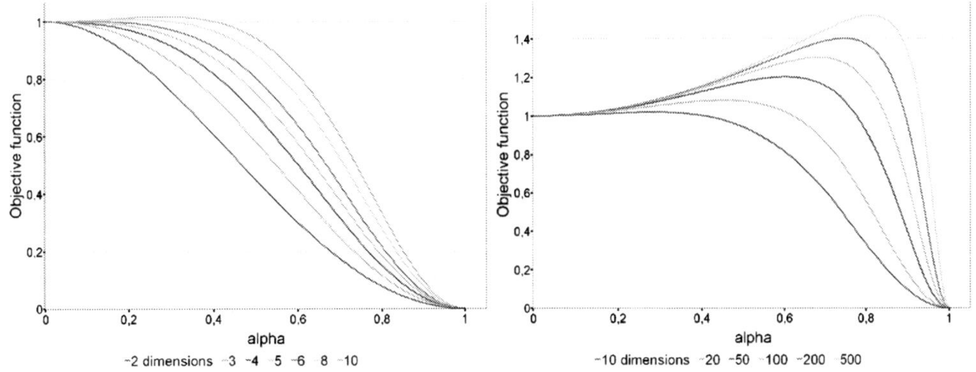

axis. For an ill behaviour, it is necessary to form a local minimum in the centre of gravity. Or in other words, to have a local maximum between the centre of gravity and the cluster centres. In Figure 1 the first part of the objective function is enlarged to show the effect of the local maximum. As it can be seen, under D_1 circumstances, a local maximum exists at approximately $\alpha = 0.1$ for $c = 6$, i.e. for a 5-dimensional data set. At 5 dimensions, the effect is very small and even for 10 dimensions, the valley is very small.

If (in general) a prototype sits inside a cluster, this cluster has almost no influence on the other clusters. Hence, from the point of view of the other clusters, it can be considered to be removed from the data set. In a 10-dimensional data set, its very likely, that at least a few clusters are found due to a lucky initialisation of the prototypes. The remaining data objects form a data set of lower dimensionality which in turn reduces the effect of the local minimum in the centre of gravity. The right part of Figure 4 shows the effect of very

Figure 4. The enlarged first section of D1 for Dimensions 2 to 10

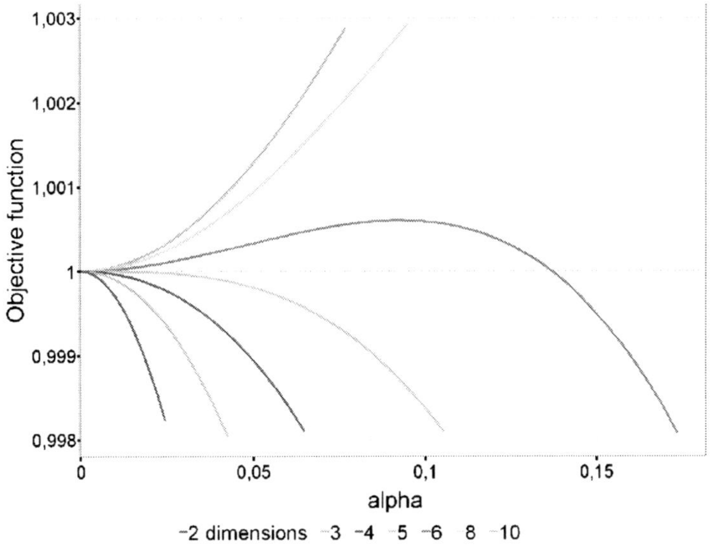

high dimensions. The important thing is, that the local maximum shifts further away from the centre of gravity for higher dimensions. Additionally, the height of the local maximum increases. This reduces the chance of a lucky initialization considerably.

How Does the Number of Prototypes Influences the Ill Behaviour?

In D_1, the number of prototypes equals the number of dimensions plus one. So the number of prototypes strictly depends on the number of dimensions. This rises the question, whether effects on the objective function is caused by the dimensions or by the number of prototypes? To answer this question, data set D_2 is used because it provides a set of well separated cluster centres, independent of the number of dimensions. Observe the left part of Figure 5, it provides the curve of the objective function for D_2, where the prototypes start in the centre of gravity for $d_{ij}^2(\alpha) = \left\| y_i(\alpha) - x_j \right\|^2 = \sum_{k=1}^{c} (y_{ik} - x_{jk})^2$ and end up at the cluster centres at $\alpha = 1$. In this figure, the number of dimensions is fixed to 20 and the number of prototypes ranges from 2 to 500. The objective function shows a very similar behaviour as for D_1, but the conclusion that the dimensions do not cause the local minimum in the centre of gravity is wrong.

Observe the right part of Figure 5, which presents the same scenario for 4 instead of 20 dimensions. In fact, there is no local minimum in the centre of gravity, this only occurs for dimensions of 5 or higher. However, the objective function has a very long, almost flat area for low α values, which means it is only optimized close to a good solution. Figure 6 shows the effect of the dimension for a fixed number of prototypes. In this figure, it is visible that the number of dimensions influences the occurrence and height of local maxima, but the location changes only very slightly. In fact, it is possible to conclude from these tests with D_2, that the number of prototypes influences the location of the local maximum, while the number of dimensions influences its height.

In Figure 6, it is also visible that the objective function is not influenced by the number of dimensions if it is higher than the number of prototypes as it is shown for dimensions above 50 (all lines are almost identical). This empirically supports the earlier stated argument that the prototypes form a lower dimensional subspace.

Figure 5. D_2 for different number of prototypes and 20 dimensions (left) as well as 4 dimensions (right)

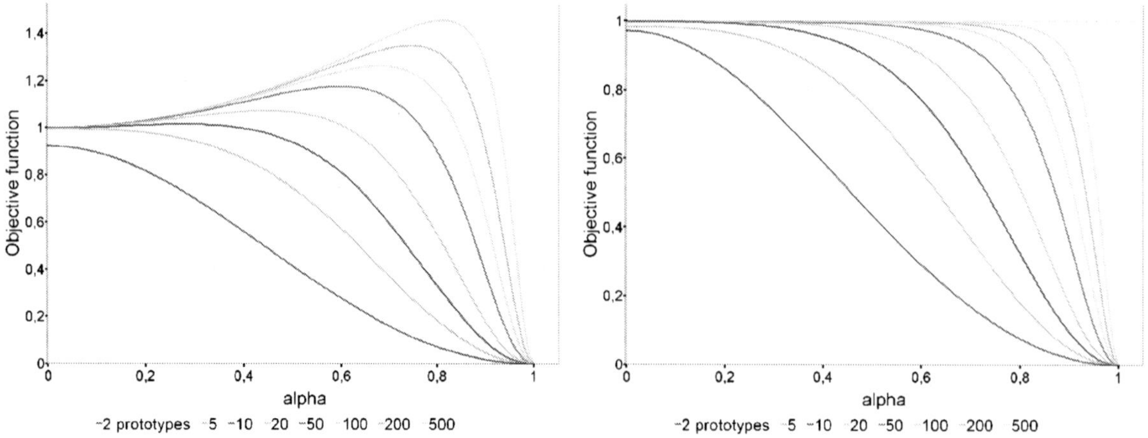

Why Has the Objective Function a Local Minimum in the Centre of Gravity?

The last two questions were answered using especially designed data sets and artificially applying FCM to them. So far, there is no explanation given for the effects that have been discussed. D_1 is very useful for analysing the objective function because it leaves no degree of freedom. To answer the question, why there is a local minimum, the formula of the objective function is taken apart. To make things a little easier, the scaling parameter in D_1 is skipped. Instead, the following definition is applied for c dimensions or c prototypes respectively:

$$D = \{x_i \mid x_i = e_i, i = 1, \ldots, c\} \tag{18}$$

which implies a centre of gravity of

$$cog_D = \begin{pmatrix} \frac{1}{c} \\ \vdots \\ \frac{1}{c} \end{pmatrix}. \tag{19}$$

The prototypes are defined depending on $\alpha \in [0,1]$:

$$y_i(\alpha) = (1-\alpha)cog_D + \alpha x_i = \begin{pmatrix} (1-\alpha)\frac{1}{c} \\ \vdots \\ (1-\alpha)\frac{1}{c} + \alpha \\ \vdots \\ (1-\alpha)\frac{1}{c} \end{pmatrix} \tag{20}$$

From this definitions, the squared euclidean distance from prototype y_i to data object x_j is defined by

$$\begin{aligned} d_{ij}^2(\alpha) &= \|y_i(\alpha) - x_j\|^2 = \sum_{k=1}^{c} (y_{ik} - x_{jk})^2 \\ &= \begin{cases} \overbrace{(c-2)\left((1-\alpha)\frac{1}{c}\right)^2}^{k\neq i, k\neq j} + \overbrace{\left((1-\alpha)\frac{1}{c}+\alpha\right)^2}^{k=i} + \overbrace{\left((1-\alpha)\frac{1}{c}-1\right)^2}^{k=j} & i \neq j \\ \underbrace{(c-1)\left((1-\alpha)\frac{1}{c}\right)^2}_{i=j\neq k} + \underbrace{\left((1-\alpha)\frac{1}{c}+\alpha-1\right)^2}_{i=j=k} & i = j \end{cases} \\ &= \begin{cases} (\alpha+1) - (1-\alpha)^2\frac{1}{c} & i \neq j \\ \frac{(c-1)(2c-1)}{c^2}(1-\alpha)^2 & i = j \end{cases}. \end{aligned} \tag{21}$$

Figure 6. D_2 for different number of dimensions and 50 prototypes

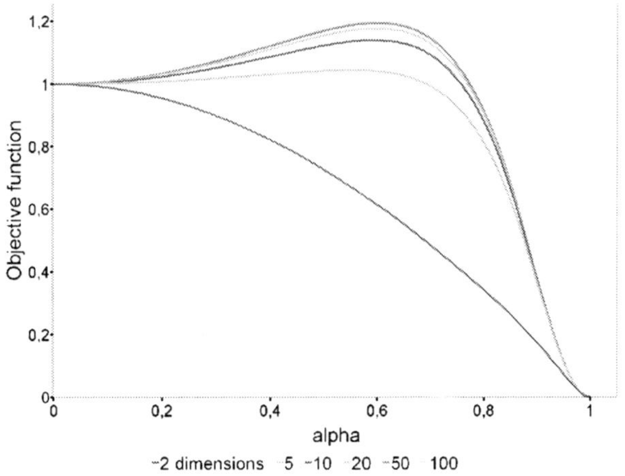

To make the situation a little simpler, the fuzzifier ω is set to 2. So the objective function $\mathbf{J}(D,\alpha)$ can be reduced as follows:

$$\mathbf{J}(D,\alpha) = \sum_{i=1}^{c}\sum_{j=1}^{c}\left(\frac{\frac{1}{d_{ij}^2(\alpha)}}{\sum_{k=1}^{c}\frac{1}{d_{kj}^2(\alpha)}}\right)^2 d_{ij}^2(\alpha)$$

$$= \sum_{j=1}^{c}\sum_{i=1}^{c}\frac{\frac{1}{d_{ij}^2(\alpha)}}{\left(\sum_{k=1}^{c}\frac{1}{d_{kj}^2(\alpha)}\right)^2}. \quad (22)$$

Because of the highly symmetric setting of the prototypes and data objects, each prototype gives the same contribution to the membership value. Therefore, the outer sum can be removed.

$$\mathbf{J}(D,\alpha) = c\sum_{i=1}^{c}\frac{\frac{1}{d_{i1}^2(\alpha)}}{\left(\sum_{k=1}^{c}\frac{1}{d_{k1}^2(\alpha)}\right)^2}$$

$$\frac{1}{c}$$

$$= c\frac{1}{\sum_{k=1}^{c}\frac{1}{d_{k1}^2(\alpha)}}. \quad (23)$$

By inserting Equation (21) for the distance values, we obtain

$$\mathbf{J}(D,\alpha) = \frac{c}{(c-1)\frac{c}{(\alpha^2+1)c-(1-\alpha)^2}+\frac{c^2}{(c-1)(2c-1)(1-\alpha)^2}}$$

$$= \frac{1}{\underbrace{\frac{1}{\alpha^2+\frac{2\alpha}{c-1}+1}}_{\mathbf{A}}+\underbrace{\frac{c}{(c-1)(2c-1)(1-\alpha)^2}}_{\mathbf{B}}} \quad (24)$$

Term **A** represents the influence of the data objects a prototype does not belong to, while term **B** represents the influence of the remaining, "good" data object. Because it is below the fractional, the higher the sum of **A** and **B** is, the smaller the objective function value becomes. In other words, the term that holds the largest value has the most influence on the objective function. One notices, that the value of **A** does not change much; for α=0 it is 1 and for α =1, it is $2+\frac{2}{c-1}$, which is less than 4 for $c \leq 2$. The transition between these two extremes is almost linear, since the effect of the square α does not have a large effect for values between 0 and 1. It is important to notice here, that the value of **A** is almost independent of c which means, that the "bad" data objects have roughly the same influence on a prototype, independently of the number of dimensions.

B on the other hand, can be characterized by $\frac{1}{c}(1-\alpha)^{-2}$. For α values close to 1, **B** becomes very large and this is the reason for the objective function to approach zero. But the factor $\frac{1}{c}$ shifts this influence towards the higher values of α, or in other words, closer to the optimal solution. So the number of dimensions and the number of "bad" data objects, reduces the influence of the "good" data objects for a prototype.

In combination, leaving the area around the centre of gravity, increases the distance to most data objects which cannot be compensated by approaching some data objects. For a reasonably high number of dimensions in combination with a high number of prototypes, the influence of the "good" data objects becomes insignificantly small compared to the majority of all data objects.

How Must FCM be Initialised to Avoid the Local Minimum in the Centre of Gravity?

This question can only be approached in a very common way since the success of an initialization depends on the data set that should be analysed. As a substitute for a real data set, data sets D_3 and D_4 are used, which provide randomly placed cluster, with data objects scattered around their centres. In Figure 7, the sequences of objective function values are presented. They differ mostly in the fact that at $\alpha = 1$, they do not hit the objective function value at 0 because the data objects are scattered around the cluster centres rather than consist of one data object only. Other than that, the objective function values show the same (bad) characteristic as in D_1 and D_2. With this in mind, it is safe to assume that the problem, described so far is not just academic, but occurs in many real high dimensional data sets as well.

The question for this subsection is, how to avoid that prototypes are running into to centre of gravity. The answer is quite simple, the initialization has to make sure, that all prototypes are initialized behind the "hill" in the objective function plot. Solving the objective function Equation for its maximum depending on α is very difficult. Therefore, a numeric approach is used and the result is presented in Figure 8. The plots present the location (in terms of an α -value) of the local maximum in relation to the number of prototypes in the data set. For each number of prototypes, D_3 and D_4 are generated for the range of dimensions from 5 to 100. The colored background area is bounded by the 100 dimensional line while the other lines represent the local maximum at different dimensions between 5 and 100. Despite the fact that D_3 and D_4 are different in cluster centre distribution, the maximum membership value graphs are very similar. This similarity supports the assumption that the presented problems are symptomatic for all high dimensional data sets, regardless of the data distribution. For a data set with a given number of prototypes, each prototype should be initialised closer than $(1-\alpha)\|cog-clustercentre\|$ in order to find the cluster as presented in the graph. As can be seen, most of the lines are close to the above border of the coloured area defined by the 100 dimensional line. For higher dimensions, the location of this line does not change very much. For the lower dimensions, the location of the maximum is not well defined, hence the lines are not as straight as for higher dimensions.

Figure 7. D_3 (left) and D_4 (right) for 100 prototypes and a series of different dimensions

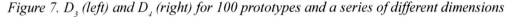

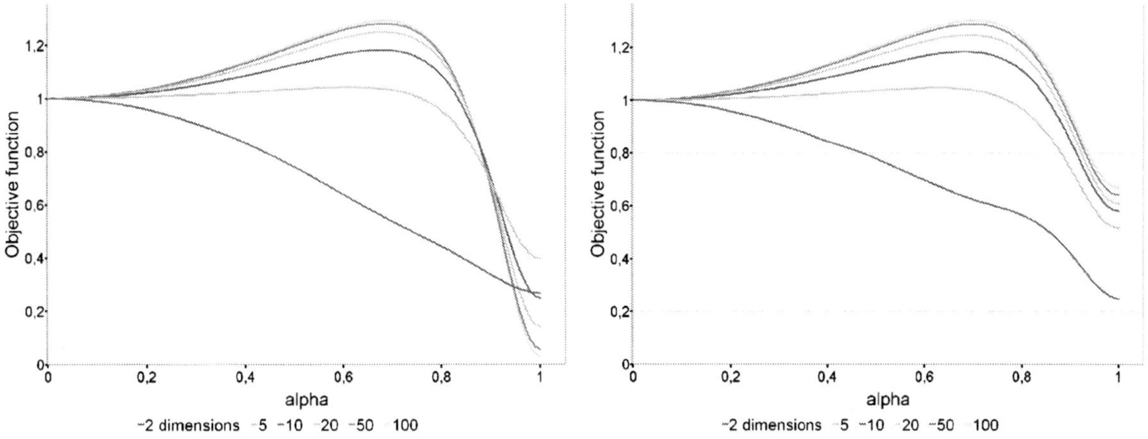

A little example should illustrate the full magnitude of the problem, which is presented in Figure 8. For 20 dimensions and 20 prototypes, the membership function maximum occurs at $\alpha = 0.4$. In other words, the prototype has to be initialised in 0.6 d with d is the distance from the centre of gravity to the cluster. The centre of gravity is usually in the centre of the data space, hence it is save to assume that the data space forms a hypercube with a side length of twice the distance from the centre of gravity to the cluster centre. Therefore, the prototype has to be initialised in a distance 0.3 a with a is the hypercube side length. For the sake of argument, let $a = 1$. The volume of a hypersphere of radius 0.3 in 20 dimensions is $9 \cdot 10^{-13}$ which means a prototype has to be initialised in a fraction of $9 \cdot 10^{-13}$ of the full data space, which is almost impossible for random initialisations. In other words, the position of the clusters have to be known in advance in order to use FCM in high dimensional spaces.

The Effect of the Fuzzifier

So far, the influence of the fuzzifier was neglected. The reason is, that the fuzzifier is used to adjust the fuzziness of the clustering result. However, it is possible to take advantage of the effect the fuzzifier provides to overcome problems with FCM in high dimensions. Effectively, if the fuzzifier approaches 1, the maximum in the membership function shifts towards the centre of gravity. In Figure 9-left, the effect of the fuzzifier is presented for D_2 with 50 dimensions and 200 prototypes. The value of the fuzzifier influences the scale of the objective function. Therefore, the objective function value of each experiment is scaled by the value of the objective function at $\alpha = 0$:

$$\omega = \frac{2+n}{n} \tag{25}$$

For large fuzzifier values, shifts very near $\alpha = 1$. But if the fuzzifier is chosen very close to 1, the result becomes almost crisp and it cannot be considered to be fuzzy. However, if the result is just needed to be clustered, regardless the fuzziness, a dimension-dependent fuzzifier is a good choice. Equation (13) where n is the number of dimensions is an example for a dimension-dependent fuzzifier where the maximum in the objective function does not occur. Its effect is presented in the right part of Figure 9, it is again scaled as in the left part of Figure 9 because the fuzzifier would produce extreme different membership values.

Figure 8. Locations of the local maximum in the objective function for D_3 (left) and D_4 (right)

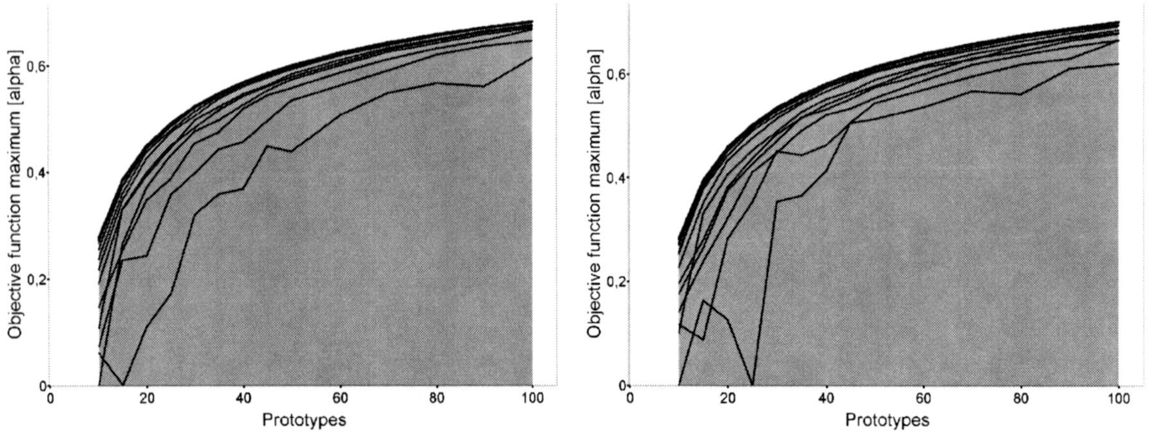

Figure 9. D_2 with scaled objective functions for different fuzzifiers at 200 prototypes in 50 dimensions (left) and the scaled objective function values with different dimensions and according to Equation (13) adjusted fuzzifier (right)

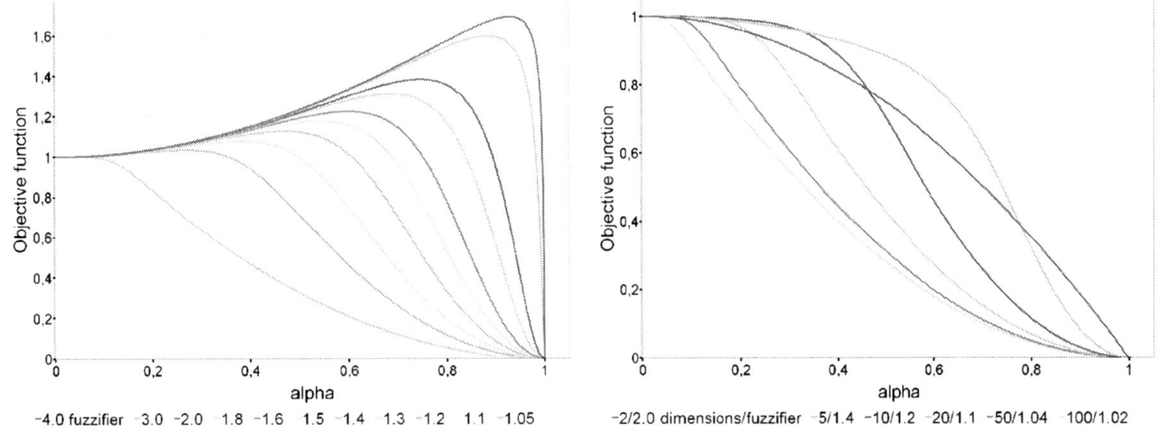

CONCLUSION

In high dimensional data spaces, pairwise distances of data objects become more similar than in low dimensions, hence it is very hard to define regions of overlapping clusters. Therefore, the argument to use FCM is questionable, not only because of its weaknesses, but also because the fuzzy data representation does not fit high dimensional data sets. This effect is analysed in Beyer et al. (1999). Beyer states that in high dimensional spaces, all distances become similar. The effect on fuzzy clustering is, the membership matrix tends to hold values of roughly $\frac{1}{c}$.

In this paper, we presented tests that show that FCM works well until 5 dimensions. For higher dimensions, FCM has to be started with well initialized prototypes. In very high dimensions of more than around 20, the position of the clusters have to be known in advance to use FCM. In other words, FCM with random initialisation cannot be used to find the cluster centres. That is a problem in particular because FCM is often used to initialise other clustering algorithms.

As part of FCM, the fuzzifier can be adjusted to represent the fuzziness of the data set. This can be exploited to create a more crisp behaving algorithm that generates better results in high dimensions. But even this approach has its limits because FCM approaches crisp clustering which reduces its benefits in contrast to HKM. Consequently, the argument that FCM is used to represent fuzzy cluster boundaries does not hold any more. However, FCM with a fuzzifier near 1 is still a little more stable than HKM in high dimensions.

Our future work will do a similar analysis as it is done in this paper for other, FCM related algorithms.

REFERENCES

Aggarwal, C. C., & Yu, P. S. (2001). Outlier detection for high dimensional data. In *Proceedings of the 2001 ACM SIGMOD international conference on Management of data* (pp. 37-46).

Beyer, K., Goldstein, J., Ramakrishnan, R., & Shaft, U. (1999). When is nearest neighbor meaningful? In *ICDT '99: Proceedings of the 7th International Conference on Database Theory* (LNCS 1540, pp. 217-235).

Bezdek, J. C. (1981). *Pattern Recognition with Fuzzy Objective Function Algorithms*. New York: Plenum Press.

Donoho, D. (2000, August). *High-dimensional Data Analysis: the Curses and Blessings of Dimensionality*. Paper presented at the conference "Math Challenges of the 21st Century" of the American Mathematics Society, Los Angeles.

Dunn, J. (1973). A fuzzy relative of the isodata process and its use in detecting compact well-separated clusters. *Cybernetics and Systems: An International Journal*, *3*(3), 32–57. doi:10.1080/01969727308546046

Gionis, A., Hinneburg, A., Papadimitriou, S., & Tsaparas, P. (2005). Dimension induced clustering. In *Proceedings of the Eleventh ACM SIGKDD International Conference on Knowledge Discovery in Data Mining* (pp. 51-60).

Hinneburg, A., & Keim, D. A. (1999). Optimal grid-clustering: Towards breaking the curse of dimensionality in high-dimensional clustering. In *Proceedings of 25th International Conference on Very Large Data Bases* (pp. 506-517).

Höppner, F., Klawonn, F., Kruse, R., & Runkler, T. (1999). *Fuzzy Cluster Analysis*. Chichester, UK: John Wiley & Sons.

Kruse, R., Döring, C., & Lesot, M.-J. (2007). Advances in fuzzy clustering and its applications. In *Fundamentals of Fuzzy Clustering* (pp. 3–30). London: John Wiley & Sons.

Steinbach, M., Ertös, L., & Kumar, V. (2004). The challenges of clustering high dimensional data. In Wille, L. T. (Ed.), *New Directions in Statistical Physics: Econophysics, Bioinformatics, and Pattern Recognition* (pp. 273–307). Berlin: Springer.

Verleysen, M. (2003). Learning high-dimensional data. In Ablameyko, S., Gori, M., Goras, L., & Piuri, V. (Eds.), *Limitations and future trends in neural computation* (pp. 141–162). IOS Press.

ENDNOTE

1 Which is defined for $z \in \mathbb{C}$ by $\Gamma(z) = \int_0^\infty t^{z-1} e^{-t} dt$ and can be reduced to $\Gamma(n+1) = n!$ for $n \in \mathbb{N}$

This work was previously published in the International Journal of Fuzzy System Applications, Volume 1, Issue 1, edited by Toly Chen, pp. 1-16, copyright 2010 by IGI Publishing (an imprint of IGI Global).

Chapter 2
Fuzzy Cluster Validation Based on Fuzzy PCA-Guided Procedure

K. Honda
Osaka Prefecture University, Japan

T. Matsui
Osaka Prefecture University, Japan

A. Notsu
Osaka Prefecture University, Japan

H. Ichihashi
Osaka Prefecture University, Japan

ABSTRACT

Cluster validation is an important issue in fuzzy clustering research and many validity measures, most of which are motivated by intuitive justification considering geometrical features, have been developed. This paper proposes a new validation approach, which evaluates the validity degree of cluster partitions from the view point of the optimality of objective functions in FCM-type clustering. This approach makes it possible to evaluate the validity degree of robust cluster partitions, in which geometrical features are not available because of their possibilistic natures.

INTRODUCTION

Fuzzy clustering (or fuzzy cluster analysis) is a most active research area in the fuzzy systems field. The fuzzy c-means (FCM) algorithm (Bezdek, 1981) and its variants (Höppner et al., 1999; Miyamoto et al., 2008) have been proved to be useful for data summarization. In the FCM-type clustering models, the objective function is given as the fuzzy membership-weighted inner-cluster errors between data points and cluster prototypes. Although the algorithms are designed well for finding local optimal solutions based on the iterative optimization scheme, they often derive several different local solutions in the multi-starting strategy. Additionally, the optimal cluster number is not known a priori. Therefore, we need the validation measure for selecting the optimal cluster partition from multiple solutions.

Many validation measures have been proposed, some of which were designed for finding compact and separate clusters from the view point of intuitive geometrical features. Xie-Beni index (Xie & Beni, 1987) and other indices based

DOI: 10.4018/978-1-4666-1870-1.ch002

on similar concepts (Dunn, 1974, Fukuyama & Sugeno, 1989) measures the cluster separateness using distances among cluster centers. Another approach considers cluster overlapping without using prototypes (Bezdek, 1981; Kim et al., 2003). Although these measures have been intuitively justified, it is not necessarily guaranteed that the validation measures really suit the evaluation of 'local optima' of objective functions. So, Rankler (2007) considered the pareto optimal solutions in multi-objective problems of objective functions and validation measures. In this paper, the cluster validity is discussed considering only the optimality of objective functions.

Another topic considered in this paper is noise rejection mechanism in the FCM-type clustering. Noise fuzzy clustering (Davé, 1991) uses an additional "noise cluster" so that noise samples are dumped into it. Masulli and Rovetta (2006) proposed a technique for soft transition from the conventional FCM constraint to a robust situation where noise samples are rejected (or ignored). Although noise rejection mechanisms are useful for obtaining cluster structures for contaminated data, they also create difficulties when we apply conventional validity measures designed for fuzzy clustering.

In this paper, a new cluster validation approach is considered based on principal component analysis (PCA)-guided procedures. The PCA-guided *k*-means (Ding & He, 2004a) is a deterministic method for finding the optimal *k*-means partition, in which a relaxed cluster indicator is estimated from a rotated principal component score matrix, though the rotation matrix cannot be explicitly estimated. Honda et al. (2010) introduced the noise rejection mechanism and proposed fuzzy PCA-guided robust *k*-means (FPR *k*-means). Because these PCA-guided procedures can find the optimal cluster structure considering only the optimality of objective functions, we can validate candidate cluster partitions if we estimate the rotation matrix for reconstructing the cluster indicators. In the proposed validation approach, a candidate for the pseudo-optimal indicator is first reconstructed by Procrustean transformation (Procrustean rotation) and a fair deviation from the pseudo-optimal solution is calculated for the current partition. Then, the cluster partition having least deviation is selected.

The remainder of the article is organized as follows. The next section describes some related works. The following section outlines the proposed validation approach. Experimental results are then shown. The final section outlines a summary and the conclusions.

RELATED WORK

FCM-Type Fuzzy Clustering and Cluster Validation

Let $\mathbf{x}_i$, $i=1,...n$ be n samples characterized by m dimensional observation. The FCM objective function for partitioning samples into C fuzzy clusters is defined as the generalized sum of within-cluster errors.

$$L_{fcm} = \sum_{c=1}^{C}\sum_{i=1}^{n} u_{ci}^{\theta} \left\| \mathbf{x}_i - \mathbf{b}_c \right\|^2 \qquad (1)$$

where $u_{ci} \in [0, 1]$ represents the membership degree of sample i to cluster c. Usually, the sum of u_{ci} with respect to c is constrained to be 1 and the constraint is called the "probabilistic constraint". θ is the fuzzification index called "fuzzifier" that tunes the degree of fuzziness of memberships. $\mathbf{b}_c$ is the prototypical centroid vector of cluster c. When $\theta=1$, the FCM model is reduced to *k*-means (or hard *c*-means) (MacQueen, 1967), in which $u_{ci} \in \{0, 1\}$ is given based on the nearest prototype assignment.

In order to determine the optimal fuzzy partition in the iterative optimization scheme, we must select the optimal cluster number and the suitable initialization. A simple way for finding "good partition" is to evaluate the degree of cluster overlapping. Bezdek et al. proposed the partition coefficient and partition entropy (Bezdek, 1981) that are designed for finding the cluster partition, which is most similar to the crisp partition of $u_{ci} \in \{0, 1\}$. Kim et al. (2003) extended the idea to the degree of inter-cluster proximity where the proximity degree of cluster prototypes is estimated only using memberships. However, it is also the case that non-overlapping partition is not necessarily the optimal one.

Another intuitive approach considers geometrical features of cluster prototypes. It is natural to assume that cluster centers should be mutually distant in well-separated partitions. Xie-Beni index (Xie & Beni, 1987) measures the compactness and separateness of cluster partitions and used the following criterion:

$$V_{XB} = \frac{\sum_{c=1}^{C}\sum_{i=1}^{n} u_{ci} \left\|\mathbf{x}_i - \mathbf{b}_c\right\|^2}{n\left(\min_{c \neq j} \left\|\mathbf{b}_c - \mathbf{b}_j\right\|^2\right)}. \tag{2}$$

We also have several other indices based on similar concepts (Dunn, 1974; Fukuyama & Sugeno, 1989). Although these measures have been intuitively justified, it is not evident that the local optima search in FCM-type clustering is directly connected to such validation criteria. Additionally, these validation criteria cannot be applied to the modified FCM-type objective functions such as noise rejection model (Davé, 1991; Masulli & Rovetta, 2006) or possibilistic one (Krishnapuram & Keller, 1993).

Noise Rejection in Fuzzy Clustering

In noise fuzzy clustering (Davé, 1991), an additional noise cluster is considered where all samples have same distance (penalty weight) to the noise cluster. If a sample has larger distances to all other clusters than the noise cluster, the sample is dumped into the noise cluster because of the sum-to-one condition including the noise cluster whose index is "*":

$$\sum_{c=1}^{C} u_{ci} + u_{*i} = 1. \tag{3}$$

Masulli and Rovetta (2006) considered a soft transition from the sum-to-one condition to free membership where each membership can take arbitrary value from [0,1] without constraint. In noise rejection version, additional constraint of $\sum_{c=1}^{C} u_{ci} \leq 1$ is considered in order to derive exclusive partitions. While the original model was constructed for a deterministic annealing (DA)-based approach or entropy regularization-based FCM (Miyamoto et al. 2008), several variants have also been proposed. In the standard fuzzification-based model (Honda et al., 2006), the objective function is defined in the same way with noise fuzzy clustering context.

$$L_{fcm}^{st} = \sum_{c=1}^{C}\sum_{i=1}^{n} u_{ci}^{\theta} \left\|\mathbf{x}_i - \mathbf{b}_c\right\|^2 + \gamma \sum_{c=1}^{C}\sum_{i=1}^{n} \left(1 - u_{ci}\right)^{\theta} \tag{4}$$

where γ is the noise penalty weight. A suitable value of γ is given by

$$\gamma = \frac{\sum_{c=1}^{C}\sum_{i=1}^{n} u_{ci}^{\theta} \left\|\mathbf{x}_i - \mathbf{b}_c\right\|^2}{\sum_{c=1}^{C}\sum_{i=1}^{n} u_{ci}^{\theta}}. \tag{5}$$

Memberships are updated by the following formula:

$$u_{ci} = \frac{\varphi_{ci}}{\kappa_i}. \tag{6}$$

where φ_{ci} is a free membership of sample *i* to cluster *c* drawn from [0,1] without constraint as:

$$\varphi_{ci} = \frac{1}{1 + \left(\frac{\|\mathbf{x}_i - \mathbf{b}_c\|^2}{\gamma} \right)^{1/(\theta-1)}}. \quad (7)$$

κ_i is a normalization weight given by

$$\kappa_i = \sum_{l=1}^{C} \varphi_{li}, \quad \text{if} \sum_{l=1}^{C} \varphi_{li} > 1, \quad (8)$$

$$\kappa_i = \left(\sum_{l=1}^{C} \varphi_{li}^{\alpha} \right)^{\frac{1}{\alpha}}, \quad \text{if} \left(\sum_{l=1}^{C} \varphi_{li}^{\alpha} \right)^{\frac{1}{\alpha}} < 1, \quad (9)$$

$$\kappa_i = 1, \quad \text{otherwise}. \quad (10)$$

Equation (8) is responsible for keeping the exclusive constraint of $\sum_{c=1}^{C} u_{ci} \leq 1$. Equation (9) determines the lower boundary of the sum of memberships. α is a running parameter for gradually transposing the conventional FCM model to noise rejecting one. When α =1, this model is equivalent to the conventional FCM. When α =0, the model is reduced to a fully noise rejecting one, i.e., noise samples can take $u_{ci} = 0$ for all clusters. By gradually decreasing α from 1 to 0, the soft transition of clustering model is implemented.

PCA-Guided *k*-Means and Noise Rejection

Although the goal of *k*-means clustering is unsupervised classification, the derived cluster indicator has a close connection with principal component analysis (PCA). Ding and He (2004a) introduced a deterministic procedure for *k*-means clustering based on a PCA-guided manner. Honda et al. (2010) applied the noise rejection mechanism to the PCA-guided *k*-mean procedure based on the noise fuzzy clustering scheme. The objective function to be minimized is defined as follows:

$$L_{rkm} = \gamma \sum_{i=1}^{n} \left(1 - w_i\right)^{\theta} + \sum_{c=1}^{C} \sum_{i \in G_c} w_i^{\theta} \left\|\mathbf{x}_i - \mathbf{b}_c\right\|^2, \quad (11)$$

where the goal is to assign samples to one of C clusters G_c, c=1,...,C. w_i is the responsibility degree of sample i to *k*-means process and the model is reduced to the conventional *k*-means when all w_i is 1. If sample i is regarded as a noise, w_i becomes small and the sample is ignored in *k*-means process.

The prototypical cluster centroid $\mathbf{b}_c$ is derived as

$$\mathbf{b}_c = \frac{\sum_{i \in G_c} w_i^{\theta} \mathbf{x}_i}{\sum_{i \in G_c} w_i^{\theta}}. \quad (12)$$

Substituting Equation (12) to Equation (11), the objective function can be reduced to the following centroid-less formulation:

$$L_{rkm} = \gamma \sum_{i=1}^{n} \left(1 - w_i\right)^{\theta} + \sum_{i=1}^{n} w_i^{\theta} \left\|\mathbf{x}_i\right\|^2 - \sum_{c=1}^{C} \frac{\sum_{i,j \in G_c} w_i^{\theta} \mathbf{x}_i^{\mathrm{T}} \mathbf{x}_j w_j^{\theta}}{\sum_{j \in G_c} w_j^{\theta}} \quad . \quad (13)$$

Assume that w_i are fixed and that the goal is to estimate a cluster indicator matrix $H_C = \left(\mathbf{h}_1, \cdots, \mathbf{h}_C\right)$,

$$h_{ci} = \begin{cases} \frac{\left(w_i^{\theta}\right)^{1/2}}{\left|\sum_{j \in G_c} w_j^{\theta}\right|^{1/2}} & : i \in G_c \\ 0 & : \text{otherwise} \end{cases} \quad (14)$$

Here, $\mathbf{h}_c$ is normalized so that $H_C^{\mathrm{T}} H_C = I$, and h_{ci} becomes large when sample i belongs to cluster c and is not regarded as a noise. In the PCA-guided procedures, a rotated indicator matrix $Q_C = (\mathbf{q}_1, \cdots, \mathbf{q}_C) = H_C T$ instead of the original H_c is considered where T is an unknown rotation matrix. It has been shown that a continuous (relaxed) solution for Q_{C-1} is derived as the eigenvectors corresponding to the $C-1$ largest eigenvalues of $W^{\theta/2}Y^{\mathrm{T}}YW^{\theta/2}$ where $Y = \{y_{ij}\} = \{x_{ij} - \bar{x}_j\}$ is a normalized data matrix so that $\sum_{i=1}^{n} w_i^{\theta} y_{ij} = 0$ and $W = \mathrm{diag}(w_1, \cdots, w_n)$. In order to avoid redundancies, $\mathbf{q}_C$ is given by

$$\mathbf{q}_C = \left(\frac{\left(w_1^{\theta}\right)^{1/2}}{\left(\sum_{i=1}^{n} w_i^{\theta}\right)^{1/2}}, \cdots, \frac{\left(w_n^{\theta}\right)^{1/2}}{\left(\sum_{i=1}^{n} w_i^{\theta}\right)^{1/2}} \right). \tag{15}$$

Here, $\mathbf{q}_c$, $c = 1, \cdots, C-1$ are identified with the fuzzy principal component score vectors in fuzzy PCA (Yabuuchi & Watada, 1997) using a generalized membership weight w_i^{θ} instead of w_i. In this way, a rotated cluster indicator matrix Q_C can be derived in a fuzzy PCA-guided procedure while the rotation matrix is not explicitly determined.

Next, the responsibility of each sample with respect to the *k*-means process is estimated with fixed *k*-means cluster assignment. Using the rotated cluster indicator Q_C, the centroid-less objective function with normalized data is rewritten as

$$\begin{aligned} L_{rkm} &= \gamma \sum_{i=1}^{n} (1 - w_i)^{\theta} + \sum_{i=1}^{n} w_i^{\theta} \|\mathbf{y}_i\|^2 \\ &- \sum_{i=1}^{n} w_i^{\theta} \sum_{j=1}^{n} \sum_{c=1}^{C} q_{ci} q_{cj} \mathbf{y}_i^{\mathrm{T}} \mathbf{y}_j \left(\frac{w_j^{\theta}}{w_i^{\theta}} \right)^{\frac{1}{2}}, \end{aligned} \tag{16}$$

and w_i is estimated as

$$w_i = \left[1 + \left(\frac{d_i}{\gamma} \right)^{\frac{1}{\theta - 1}} \right]^{-1}, \tag{17}$$

where d_i is calculated with fixed weight ratio w_j / w_i as

$$d_i = \|\mathbf{y}_i\|^2 - \sum_{j=1}^{n} \sum_{c=1}^{C} q_{ci} q_{cj} \mathbf{y}_i^{\mathrm{T}} \mathbf{y}_j \left(\frac{w_j^{\theta}}{w_i^{\theta}} \right)^{\frac{1}{2}}. \tag{18}$$

The above two phases of cluster indicator estimation and responsibility estimation are repeated until convergence.

When cluster boundaries are non-linear, the kernel method (Bishop, 2006) with a certain kernel function can also be used.

After mapping to a high dimensional data space $\mathbf{x}_i \to \varphi(\mathbf{x}_i)$, the objective function is rewritten as:

$$\begin{aligned} L_{krkm} &= \gamma \sum_{i=1}^{n} (1 - w_i)^{\theta} + \sum_{i=1}^{n} w_i^{\theta} g_{ii} \\ &- \mathrm{tr}\left(Q_C^{\mathrm{T}} W^{\theta/2} G W^{\theta/2} Q_C \right), \end{aligned} \tag{19}$$

where $G = \{g_{ij}\}$ is the kernel matrix whose element is $g_{ij} = \varphi(\mathbf{x}_i)^{\mathrm{T}} \varphi(\mathbf{x}_j)$. In the kernel method, a suitable kernel function such as Gaussian kernel or polynomial kernel is used instead of the exact form of function $\varphi(\mathbf{x}_i)$.

Following the normalization of the kernel matrix, the optimal cluster indicator Q_{C-1} is derived as the eigenvector corresponding to the $C-1$ largest eigenvalue of $W^{\theta/2} G W^{\theta/2}$.

The noise criterion d_i is also calculated as

$$d_i = g_{ii} - \sum_{j=1}^{n} \sum_{c=1}^{C} q_{ci} q_{cj} g_{ij} \left(\frac{w_j^{\theta}}{w_i^{\theta}} \right)^{\frac{1}{2}}. \tag{20}$$

In order to capture the cluster structure in the rotated cluster indicator matrix, some visual assessment techniques (Ding & He, 2004b; Honda et al., 2010) are used instead of explicitly estimating the rotation matrix because the matrix estimation is not easy in many cases.

PCA-GUIDED CLUSTER VALIDATION

Cluster Indicator Reconstruction by Procrustean Rotation

Although it is not easy to estimate the rotation matrix for reconstructing the original indicator matrix in PCA-guided *k*-means models, we can calculate a suitable rotation matrix when we have the optimal solution as the target matrix.

Procrustean transformation (Procrustean rotation) is a popular technique used in factor analysis (Lawley & Maxwell, 1963) for intuitively interpreting the derived factor loadings that rotates factors to a target structure specified prior to the analysis based on a certain hypothesis or a priori knowledge. Assume that B is a target matrix and A is the factor loading matrix to be rotated. The goal of the rotation is to estimate a suitable rotation matrix T so that the sum of the squared error $\mathrm{tr}\left(E^{\mathrm{T}}E\right)$, $E = AT - B$ is minimized. A solution is derived as

$$T = A^{T}B\left(B^{T}AA^{T}B\right)^{-\frac{1}{2}}, \tag{21}$$

and $\left(B^{T}AA^{T}B\right)^{-1/2}$ is calculated from the eigen decomposition $B^{\mathrm{T}}AA^{\mathrm{T}}B = PDP^{\mathrm{T}}$ as

$$\left(B^{\mathrm{T}}AA^{\mathrm{T}}B\right)^{-\frac{1}{2}} = PD^{-\frac{1}{2}}P^{\mathrm{T}}, \tag{22}$$

where P is the square matrix whose columns are the eigenvectors and D is the diagonal matrix whose diagonal elements are the corresponding eigenvalues.

In order to evaluate the validity of a candidate FCM-type cluster partition $U_C = \{u_{ci}\}$, Procrustean rotation of a PCA-guided solution Q_C is considered. Here, we can expect that a similar partition matrix with U_C is reconstructed by Procrustean rotation of Q_C with the target matrix being U_C if the candidate solution is the optimal one. On the other hand, there is no plausible rotation matrix if the candidate solution is not optimal. Therefore, a PCA-guided cluster validity criterion is defined as the sum of squared errors:

$$V_{PCA} = \frac{1}{C}\,\mathrm{tr}\left(E^{\mathrm{T}}E\right), \tag{23}$$

where $E = Q_C T - U_C$. Here, in order to avoid a tendency for the error to monotonically increase with the number of clusters, the error measure is divided by the number of clusters C.

Procedure of PCA-Guided Cluster Validation

There may be a simple question "Why is the *k*-means solution used for validating fuzzy partitions?" because the PCA-guided solution is a relaxed optimal solution of *k*-means clustering, i.e., crisp (hard) clustering model. The PCA-guided *k*-means models use the normalization constraint of $\mathbf{h}_c^{\mathrm{T}}\mathbf{h}_c = 1$ instead of $\sum_{c=1}^{C} u_{ci} = 1$ that is used in *k*-means (and FCM). The constraint is rather related to possibilistic partition as is pointed out in (Pal et al., 2005; Honda et al., 2007), i.e., h_{ci} does not necessarily represent the probability of sample i belonging to cluster c but represents the relative typicality of sample i in cluster c. Therefore, h_{ci} is not a crisp assignment but a kind of "fuzzy or possibilistic solution". On the other hand, $H_C^{\mathrm{T}} H_C = I$ implies an exclusive partition where each sample can belong to only a single cluster with a high degree. So, this constraint makes a possibilistic partition to a fuzzy partition or noise rejecting partition following $\sum_{c=1}^{C} u_{ci} \leq 1$. In this sense, the proposed validity measure is more suitable for fuzzy partitions or noise rejecting partitions than crisp *k*-means partitions.

The procedure for calculating the cluster validation measure is as follows:

Step 1: In order to relate the *c*-means solution $U_C = \{u_{ci}\}$ to its corresponding PCA-guided solution Q_C, the target matrix $\tilde{U}_C = \{\tilde{u}_{ci}\}$ is defined as $\tilde{u}_{ci} = \frac{\left(u_{ci}^{\theta}\right)^{1/2}}{\left(\sum_{j=1}^{n} u_{cj}^{\theta}\right)^{1/2}}$. (24)

Step 2: A plausible rotation matrix T is calculated by Procrustean transformation, in which the target matrix B and the original matrix A are given by $\tilde{U}_C$ and Q_C, respectively.

Step 3: The PCA-guided cluster validity criterion (V_{PCA}) is calculated as the sum of squared errors $V_{PCA} = \frac{1}{C} \mathrm{tr}\left(E^{\mathrm{T}} E\right)$, (25)

where $E = Q_C T - U_C$.

The optimal *c*-means solution is determined by finding the solution with the minimum V_{PCA} value.

NUMERICAL EXPERIMENTS

Selection of Optimal Initialization

A numerical experiment was performed with an artificially generated data set composed of 200 samples. The data set forms three spherical clusters following Gaussian distributions in 2-D data space, one of which has larger variance than other two data masses. In this first example, the availability of the proposed validation measure for selecting the optimal initial partition is demonstrated with a fixed cluster number. The proposed measure was calculated with the kernel method using the Gaussian kernel where $\lambda = 5.0$,

$$g_{ij} = \exp\left(-\lambda \left\|\mathbf{x}_i - \mathbf{x}_j\right\|^2\right). \quad (26)$$

When the cluster number is fixed as $C = 3$, the FCM algorithm with $\theta = 2.0$ derived two different types of cluster partitions in 100 trials with various random initial partitions. Figures 1 and 2 show the derived two partitions. The first pattern shown in Figure 1 successfully recognized two small clusters located on the left side. On the other hand, the second pattern shown in Figure 2 illegally separated the larger cluster of the right side and jointed the two small clusters. The objective function values L_{fcm} for the two cluster partitions were almost similar (or slightly better for the first pattern).

Here, the proposed validation index V_{PCA} had a plausible value for the first pattern but a much larger value for the second one, i.e., the index successfully worked for selecting the optimal initial partition. On the other hand, the Xie-Beni index selected the second cluster partition to be optimal. It is because the geometrical feature of inter-cluster separateness was under-estimated in the situation where cluster variances are unbalanced.

In this way, the proposed validation measure works well for selecting the optimal initial partition while the conventional measures supported by intuitive geometrical justification may fail to select the optimal one.

Figure 1. The first pattern of derived cluster partitions with C=3 ($V_{PCA} = 0.027$, $V_{XB} = 0.564$, $L_{fcm} = 40.71$*)*

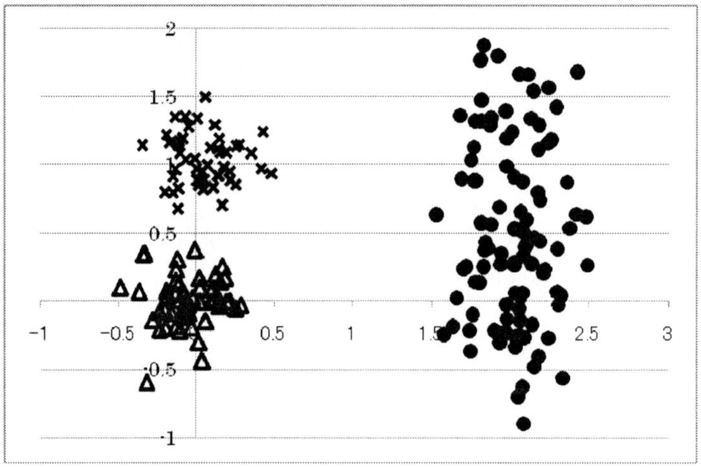

Selection of Cluster Number

Next, the validation measures were used for selecting the optimal cluster number with the same data set. Table 1 compares the values of validation measures for various cluster numbers. The value for each cluster number was selected the minimum one derived in 100 trials with random initialization.

The proposed validation measure V_{PCA} could also select the optimal cluster number of *C=3* while the Xie-Beni index V_{XB} chose *C=2*. It is because the small inter-cluster separation of the left two clusters concealed the cluster structure. However, the proposed method considering the optimality of the objective function could capture the cluster structure well. By the way, the objective function L_{fcm} has the monotonically decreasing tendency with respect to *C*. So the proposed method does not search just the smallest value of L_{fcm} but search the cluster structure although the method considers the optimality of the objective function.

These results support the availability of the proposed index for selecting the optimal cluster partition.

Application to Noise Rejection Models

Finally, the availability of the proposed measure is tested in an application to noise rejection models. The data set used in this experiment includes 4 clusters that are buried in uniformly distributed noise. In this experiment, two noise rejection models of the noise fuzzy clustering and the soft transition-based model were applied.

Table 2 compares the values of the validity measure, in which "N/A" means that some clusters coincided, i.e., the possibilistic nature of soft transition-based method located multiple cluster centers on a same place because the cluster number was too large. The table indicates that the proposed index V_{PCA} selected *C=5* for noise fuzzy clustering model while *C=4* for soft transition-based model. Figures 3 and 4 show the selected partitions where small dots represent noise samples ignored in *c*-means clustering. Noise clustering divided the left large cluster into two spherical-shape clusters because of its probabilistic nature that implies a tendency to produce exclusive partitions. On the other hand, the soft transition-based model preferred to assign cluster

Figure 2. The second pattern of derived cluster partitions with C=3 ($V_{PCA} = 0.597$, $V_{XB} = 0.452$, $L_{fcm} = 41.00$ *)*

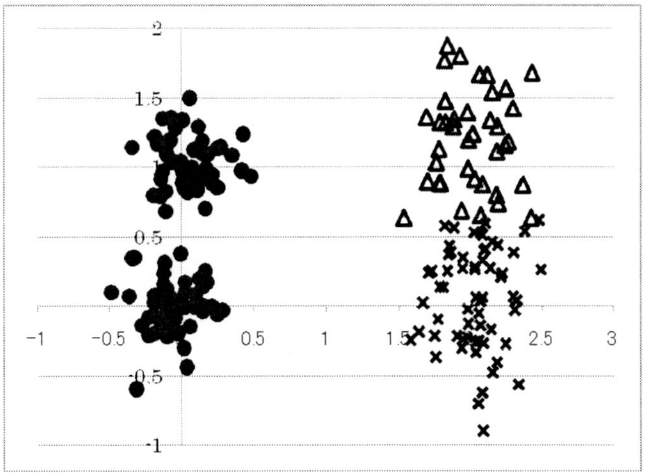

centers to "centers of mass" because of its possibilistic nature, i.e., cluster centers should be assigned to each center of mass independently. The proposed validation measure works well for both noise rejection models.

The right side of Table 2 shows the values of Xie-Beni index and indicates that the index selected the minimum number of C. It is because we can extract a few well-separated small clusters by ignoring many other samples whereas some significant cluster structures are buried in the removed noise samples.

These results imply that the proposed validation approach, which is based on the optimality of the objective function, is available for the modified objective functions (some FCM variants) although the conventional validation measures cannot be applied in a straightforward manner.

CONCLUSION

This paper proposed a fuzzy cluster validation approach, in which the optimal c-means partition is selected considering only the optimality of the objective function. It was demonstrated in several experiments that the proposed approach is still applicable to unbalanced situation although the conventional indices, which are based on the

Table 1. Comparison of validation measures for data set without noise

C	V_PCA	V_XB	L_fcm
2	0.724	***0.179***	70.652
3 (1st)	***0.027***	0.564	40.713
3 (2nd)	0.597	0.452	41.000
4	0.052	0.328	17.845
5	0.048	0.657	11.926
6	1.364	1.234	10.037
7	0.306	2.313	8.856

Table 2. Comparison of validation measures for artificial data with noise

C	V_PCA		V_XB	
	NFC	ST	NFC	ST
3	0.073	0.066	***0.148***	***0.080***
4	0.069	***0.033***	0.266	0.142
5	***0.052***	0.104	0.870	1.411
6	0.318	N/A	4.276	N/A
7	0.392	N/A	2.625	N/A

Figure 3. Cluster partition selected by noise fuzzy clustering model (C=5, $V_{PCA} = 0.052$*)*

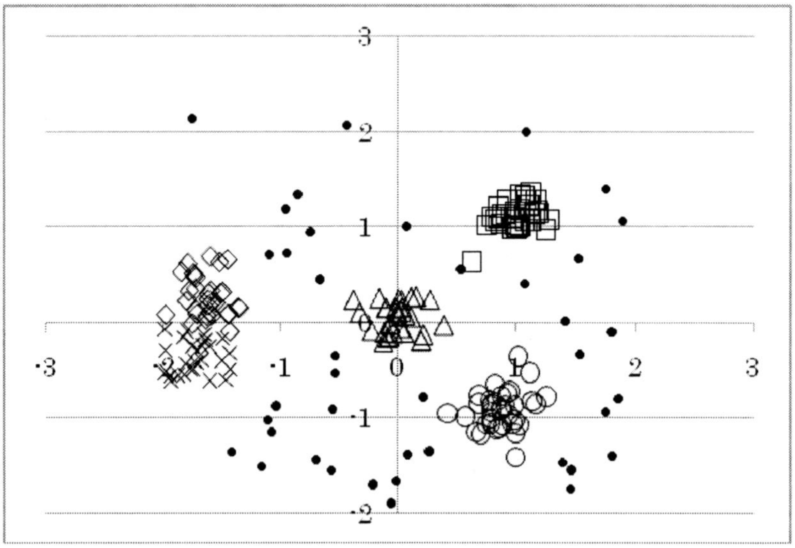

Figure 4. Cluster partition selected by soft transition-based model (C=4, $V_{PCA} = 0.033$*)*

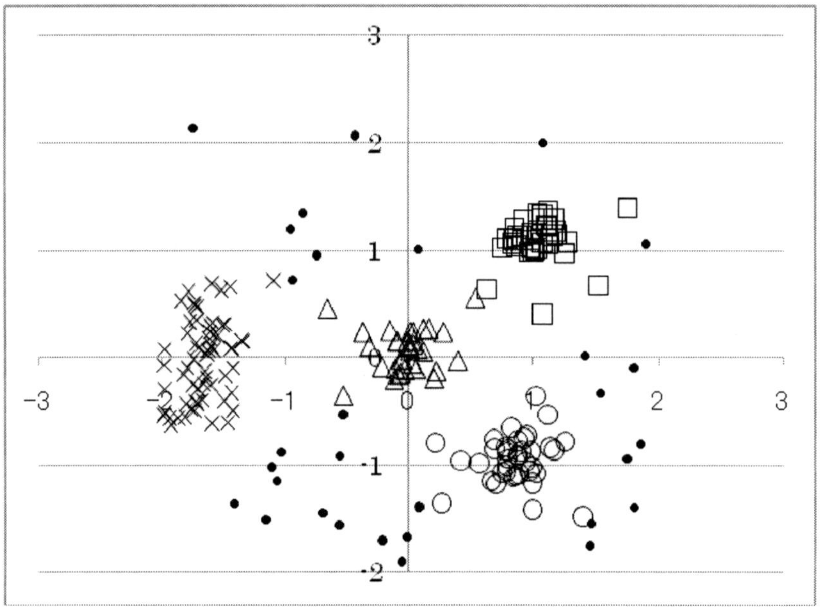

geometrical intuitive justification, may fail to find correct cluster structures.

The proposed approach is also applicable to some extended c-means objective functions. In numerical experiments, the proposed validation measure was applied to noise rejection models including both of noise fuzzy clustering with probabilistic constraint and soft transition-based model with possibilistic constraint.

Future work may include application to other modified objective functions such as fully possibilistic model (Krishnapuram & Keller, 1993) and the models for handling missing values (Hathaway & Bezdek, 2001; Honda & Ichihashi, 2004). The comparison with other clustering models based on probabilistic mixture density models (Bishop, 2006; Honda & Ichihashi, 2005) is also remained in future work.

ACKNOWLEDGMENT

This work was supported in part by the Ministry of Internal Affairs and Communications, Japan, under the Strategic Information and Communications R&D Promotion Programme (SCOPE).

REFERENCES

Bezdek, J. C. (1981). *Pattern Recognition with Fuzzy Objective Function Algorithms*. New York: Plenum Press.

Bishop, C. M. (2006). *Pattern Recognition and Machine Learning*. Berlin: Springer Verlag.

Davé, R. N. (1991). Characterization and detection of noise in clustering. *Pattern Recognition Letters*, *12*(11), 657–664. doi:10.1016/0167-8655(91)90002-4

Ding, C., & He, X. (2004a). *K*-means clustering via principal component analysis. In *Proceedings of the Twenty-first International Conference on Machine learning* (pp. 225-232).

Ding, C., & He, X. (2004b). Linearized cluster assignment via spectral ordering. In *Proceedings of the Twenty-first International Conference on Machine learning* (pp. 233-240).

Dunn, J. C. (1974). Well separated clusters and optimal fuzzy partitions. *Journal of Cybernetics*, *4*(1), 95–104. doi:10.1080/01969727408546059

Fukuyama, Y., & Sugeno, M. (1989). A new method of choosing the number of clusters for the fuzzy *c*-means method. In *Proceedings of the 5th Fuzzy System Symposium* (pp. 247-250).

Hathaway, R. J., & Bezdek, J. C. (2001). Fuzzy *c*-means clustering of incomplete data. *IEEE Transactions on Systems, Man, and Cybernetics. Part B, Cybernetics*, *31*(5), 735–744. doi:10.1109/3477.956035

Honda, K., & Ichihashi, H. (2004). Linear fuzzy clustering techniques with missing values and their application to local principal component analysis. *IEEE Transactions on Fuzzy Systems*, *12*(2), 183–193. doi:10.1109/TFUZZ.2004.825073

Honda, K., & Ichihashi, H. (2005). Regularized linear fuzzy clustering and probabilistic PCA mixture models. *IEEE Transactions on Fuzzy Systems*, *13*(4), 508–516. doi:10.1109/TFUZZ.2004.840104

Honda, K., Ichihashi, H., Masulli, F., & Rovetta, S. (2007). Linear fuzzy clustering with selection of variables using graded possibilistic approach. *IEEE Transactions on Fuzzy Systems*, *15*(5), 878–889. doi:10.1109/TFUZZ.2006.889946

Honda, K., Ichihashi, H., Notsu, A., Masulli, F., & Rovetta, S. (2006). Several formulations for graded possibilistic approach to fuzzy clustering. In S. Greco, et al. (Eds.), *Rough Sets and Current Trends in Computing (RSCTC) 2006* (LNCS 4259, pp. 939-948). Berlin: Springer Verlag.

Honda, K., Notsu, A., & Ichihashi, H. (2010). Fuzzy pca-guided robust *k*-means clustering. *IEEE Transactions on Fuzzy Systems*, *18*(1), 67–79. doi:10.1109/TFUZZ.2009.2036603

Höppner, F., Klawonn, F., Kruse, R., & Runkler, T. (1999). *Fuzzy Cluster Analysis*. Chichester, UK: John Wiley & Sons.

Kim, D.-W., Lee, K. H., & Lee, D. (2003). Fuzzy cluster validation index based on inter-cluster proximity. *Pattern Recognition Letters*, *24*(15), 2561–2574. doi:10.1016/S0167-8655(03)00101-6

Krishnapuram, R., & Keller, J. M. (1993). A possibilistic approach to clustering. *IEEE Transactions on Fuzzy Systems*, *1*(2), 98–110. doi:10.1109/91.227387

Lawley, D. N., & Maxwell, A. E. (1963). *Factor Analysis as a Statistical Method.* London: Butterworth.

MacQueen, J. B. (1967). Some methods of classification and analysis of multivariate observations. In *Proceedings of the Fifth Berkeley Symposium on Mathematical Statistics and Probability* (Vol. 1, pp. 281-297).

Masulli, F., & Rovetta, S. (2006). Soft transition from probabilistic to possibilistic fuzzy clustering. *IEEE Transactions on Fuzzy Systems*, *14*(4), 516–527. doi:10.1109/TFUZZ.2006.876740

Miyamoto, S., Ichihashi, H., & Honda, K. (2008). *Algorithms for Fuzzy Clustering.* Berlin: Springer Verlag.

Pal, N. R., Pal, K., Keller, J. M., & Bezdek, J. C. (2005). A possibilistic fuzzy *c*-means clustering algorithm. *IEEE Transactions on Fuzzy Systems*, *13*(4), 508–516. doi:10.1109/TFUZZ.2004.840099

Runkler, T. A. (2007). Pareto optimality of cluster objective and validity functions. In *Proceedings of 2007 IEEE International Conference on Fuzzy Systems* (pp. 79-84).

Xie, X. L., & Beni, G. (1987). A validity measure or fuzzy clustering. *IEEE Transactions on Pattern Analysis and Machine Intelligence*, *13*(8), 841–847. doi:10.1109/34.85677

Yabuuchi, Y., & Watada, J. (1997). Fuzzy principal component analysis and its application. *Biomedical Fuzzy and Human Sciences*, *3*, 83–92.

This work was previously published in the International Journal of Fuzzy System Applications, Volume 1, Issue 1, edited by Toly Chen, pp. 49-60, copyright 2010 by IGI Publishing (an imprint of IGI Global).

Chapter 3
A Fuzzy Clustering Model for Fuzzy Data with Outliers

M. H. Fazel Zarandi
Amirkabir University of Technology, Iran

Zahra S. Razaee
Amirkabir University of Technology, Iran

ABSTRACT

This paper proposes a fuzzy clustering model for fuzzy data with outliers. The model is based on Wasserstein distance between interval valued data, which is generalized to fuzzy data. In addition, Keller's approach is used to identify outliers and reduce their influences. The authors also define a transformation to change the distance to the Euclidean distance. With the help of this approach, the problem of fuzzy clustering of fuzzy data is reduced to fuzzy clustering of crisp data. In order to show the performance of the proposed clustering algorithm, two simulation experiments are discussed.

INTRODUCTION

Clustering is a division of a given set of objects into subgroups or clusters, so that objects in the same cluster are as similar as possible, and objects in different clusters are as dissimilar as possible. From a machine learning perspective, clustering is an unsupervised learning of a hidden data concept (Berkhin, 2002). In conventional (hard) clustering analysis, each datum belongs to exactly one cluster, whereas in fuzzy clustering, data points can belong to more than one cluster, and associated with each datum is a set of membership degrees. Fuzzy data are imprecise data obtained from measurements, human judgements or linguistic assessments. In cluster analysis, when there is simultaneous uncertainty in the partition and data, a fuzzy clustering model for fuzzy data should be applied (D'Urso & Giordani, 2006).

In recent literature, there are several works regarding the fuzzy clustering of fuzzy data. Hathaway et al. (1996) and Pedrycz et al. (1998) introduced models that convert parametric or non-parametric linguistic variables into generalized coordinates before performing fuzzy c-means clustering. Yang and Ko (1996) presented a fuzzy k-numbers clustering model that uses a squared distance between each pair of fuzzy numbers.

DOI: 10.4018/978-1-4666-1870-1.ch003

Yang and Liu (1999) extended the Yang and Ko (1996) work and proposed a fuzzy k-means clustering model for conical fuzzy vectors. Yang et al. (2004) proposed a fuzzy K-means clustering model for handling both symbolic and fuzzy data. Hung and Yang (2005) proposed an alternative fuzzy k-numbers clustering model which is based on exponential-type distance measure. D'Urso and Giordani (2006) proposed a weighted fuzzy c-means clustering model which considers fuzzy data with a symmetric LR membership function.

In this paper, we first propose a new distance measure for comparison of fuzzy data. On account of the fact that all the α-cuts of fuzzy data are intervals, we obtain the distance between two fuzzy data from the distances between their α-cuts. To this purpose, a special case of Wasserstein distance is utilized. The choice of α-cuts is motivated by the fact that, fuzzy data with different shapes can be used. After introducing our distance, we use it for fuzzy clustering of fuzzy data. Moreover, with the help of Keller's (2000) approach, an additional weighting factor is added for each datum to identify outliers and reduce their effects. In other approach, by definition of a transformation, triangular fuzzy data are changed to crisp data. With this novel approach, after applying the transformation, any fuzzy clustering model for crisp data can be used. Furthermore, for determining the optimal number of clusters, there is no need to define a cluster validity index for fuzzy data. The ones existing in literature for crisp data can be applied.

The rest of the paper is organized as follows. First, the concept of LR-type fuzzy data is introduced. Some related works regarding metrics for fuzzy data are then reviewed. We propose a distance measure for ; $L(1)= 0$ or ($L(x) > 0$, $\forall x$ and $L(+\infty) = 0$)(Zimmerman, 2001). Then, a fuzzy number $\tilde{A}$ is of LR-type if for $c, l > 0; r > 0$ in R,

$$\mu_{\tilde{A}}(x) = \begin{cases} L(\dfrac{c-x}{l}) & \text{for } x \le c, \\ R(\dfrac{x-c}{r}) & \text{for } x \ge c, \end{cases} \qquad (1)$$

where, c, l, r are the center, left and right spreads of $\tilde{A}$, respectively. Symbolically we can write $\tilde{A} = (c, l, r)$.

In LR-type fuzzy numbers, the triangular fuzzy numbers (TFNs) are most commonly used. An LR-type fuzzy number $\tilde{A}$ is called triangular fuzzy number if $L(x) = R(x) = 1 \mid x$, characterized by the following membership function:

$$\mu_{\tilde{A}}(x) = \begin{cases} 1-(\dfrac{c-x}{l}) & \text{for } x \le c, \\ 1-(\dfrac{x-c}{r}) & \text{for } x \ge c. \end{cases} \qquad (2)$$

RELATED WORKS

In the recent literature, there are some distance measures for fuzzy data. We review some of them in this section.

Definition 1. *Considering two crisp sets A, $B \subset R^K$ and a distance $d(x, y)$ where, $x \in A$ and $y \in B$, the Hausdorff distance is defined as follows:*

$$d_H(A,B) = \max\left\{\sup_{x\in A}\inf_{y\in B} d(x,y), \sup_{y\in B}\inf_{x\in A} d(x,y)\right\}$$

According to the concept of α-cuts, the Hausdorff metric d_H can be generalized to fuzzy numbers $\tilde{F}$, $\tilde{G}$, where ($\tilde{F}$ or $\tilde{G}$): R → [0, 1]:

$$d_\rho(\tilde{F},\tilde{G}) = \begin{cases} \left[\int_0^1 (d_H(F_\alpha,G_\alpha))^\rho d\alpha\right]^{1/\rho} & \text{if } \rho[1,\infty) \\ \sup_{\alpha\in[0,1]} d_H(F_\alpha,G_\alpha) & \text{if } \rho = \infty \end{cases} \quad (4)$$

where, the crisp set $F_\alpha \equiv \{x \in R^K\colon F(x) \geq \alpha\}$, $a \in [0,1]$, is called the α-cut of $\tilde{F}$ (Näther, 2000).

Tran and Duckstein (2002) proposed the following distance between two intervals:

$$\begin{aligned} d_{TD}(A,B) &= \int_{-\frac{1}{2}}^{\frac{1}{2}}\int_{-\frac{1}{2}}^{\frac{1}{2}} \left\{\left[\left(\frac{a+b}{2}\right)+x(b-a)\right] - \left[\left(\frac{u+v}{2}\right)+y(v-u)\right]\right\}^2 dx\,dy \\ &= \left[\left(\frac{a+b}{2}\right)-\left(\frac{u+v}{2}\right)\right]^2 \\ &\quad + \frac{1}{3}\left[\left(\frac{b-a}{2}\right)+\left(\frac{v+u}{2}\right)^2\right]. \end{aligned} \quad (5)$$

Then, they used it to formulate their distance measure for fuzzy numbers, but d_{TD} does not satisfy the reflexivity property (Irpino & Verde, 2008):

$$\begin{aligned} d_{TD}(A,A) &= \left[\left(\frac{a+b}{2}\right)-\left(\frac{a+b}{2}\right)\right]^2 \\ &\quad + \frac{1}{3}\left[\left(\frac{b-a}{2}\right)^2+\left(\frac{b-a}{2}\right)^2\right] \\ &\quad + \frac{2}{3}\left(\frac{b-a}{2}\right)^2 \geq 0. \end{aligned} \quad (6)$$

A squared Euclidean distance between a pair of LR-type fuzzy data $\tilde{A}_1 = (c_1, l_1, r_1)$ and $\tilde{A}_2 = (c_2, l_2, r_2)$, where c denotes the center and l, r indicate, respectively, the left and right spread, is defined by Yang and Ko (1996):

$$\begin{aligned} d_{YK}^2(\lambda,\rho) &= (c_1-c_2)^2 + [(c_1-\lambda l_1)-\lambda l_2]^2 \\ &\quad + [(c_1+\rho r_1)-(c_2+\rho r_2)^2 \end{aligned} \quad (7)$$

where $\lambda = \int_0^1 L^{-1}(t)dt$, $\rho = \int_0^1 R^{-1}(t)dt$ are parameters that summarize the shape of the left and right tails of the membership function and L,R are decreasing shape functions which were defined previously.

THE PROPOSED DISTANCE FOR FUZZY DATA

In this section, we first present a new distance measure for interval-valued data, and then it is used to formulate the distance measure for fuzzy data. Let $I_i = [a_i, b_i]$, be an interval for $i = 1, 2$. We can parameterize I_i as follows:

$$I_i(t) = a_i + t(b_i\ a_i)\ 0 \leq t \leq 1. \quad (8)$$

If we represent I_i by means of its midpoint $m_i = \frac{a_i+b_i}{2}$ and radius $\delta_i = \frac{b_i-a_i}{2}$, Equation (8) can be rewritten as follows:

$$I_i(t) = m_i + (2t\ 1)\delta_i\ \ 0 \leq t \leq 1. \quad (9)$$

The distance measure between I_1 and I_2 can be defined as follows:

$$\begin{aligned} d^2(I_1,I_2) &= \int_0^1 [I_1(t)-I_2(t)]^2 dt \\ &= \int_0^1 [(m_1-m_2)+(\delta_1-\delta_2)(2t-1)]^2 dt \\ &= (m_1-m_2)2 + \frac{1}{3}(\delta_1-\delta_2)^2. \end{aligned} \quad (10)$$

This distance takes into account all the points in both intervals. Irpino and Verde (2008) has derived Equation (10) from another point of view, using the Wasserstein distance. To be more specific, let F_1 and F_2 be distribution functions, the Wasserstein L_2 metric is defined as follows (Gibbs & Su, 2002):

$$d_{Wass}(F_1, F_2) = \left\{\int_0^1 (F_1^{-1}(t) - F_2^{-1}(t))^2 dt\right\}^{1/2} \tag{11}$$

where F_1^{-1} and F_2^{-1} are the quantile functions of the two distributions. If we assume F_i for i=1,2 to be the uniform distribution function on $[a_i, b_i]$, then $F_i^{-1}(t)$ is the same as the parametric representation $I_i(t)$ in Equation (8). Thus, the Wasserstein distance coincides with the distance defined in Equation (10).

Now we are ready to construct a distance between fuzzy data. According to α-cuts, the Wasserstein distance d_{wass} can be generalized to fuzzy numbers $\tilde{A}_1$ and $\tilde{A}_2$:

$$d(\tilde{A}_1, \tilde{A}_2) = \left\{\int_0^1 d_{Wass}^2((\tilde{A}_1)_\alpha, (\tilde{A}_2)_\alpha) d\alpha\right\}^{\frac{1}{2}}. \tag{12}$$

We calculate this distance for triangular fuzzy numbers. Let $\tilde{A}_i = (c_i, 1_i, r_i)$, $i = 1,2$ be triangular fuzzy numbers and $(\tilde{A}_i)_\alpha = [1_i\alpha + (c_i - 1_i) - r_i\alpha + (c_i + r_i)]$, the midpoint and the radius of $(\tilde{A}_i)_\alpha$ are as follows:

$$m_{(\tilde{A}_i)\alpha} = c_i + \frac{1}{2}(1-\alpha)(r_i - l_i). \tag{13}$$

$$\delta_{(\tilde{A}i)\alpha} = \frac{1}{2}(1-\alpha)(r_i + l_i). \tag{14}$$

Then we have:

$$\begin{aligned}
&d^2(\tilde{A}_1, \tilde{A}_2) \\
&= \int_0^1 d_{Wass}^2((\tilde{A}_1)_\alpha, ((\tilde{A}_2)_\alpha) d\alpha \\
&= \int_0^1 \left\{[m_{(\tilde{A}_1)_\alpha} - m_{(\tilde{A}_2)_\alpha}]^2 + \frac{1}{3}[\delta_{(\tilde{A}_1)_\alpha} - \delta_{(\tilde{A}_2)_\alpha}]^2\right\} d\alpha \\
&= \int_0^1 \left\{\left((c_1 - c_2) + \frac{1}{2}(1-\alpha)[(r_1 - r_2) - (l_1 - l_2)]\right)^2\right. \\
&\left. + \frac{1}{12}(1-\alpha)^2[(r_1 - r_2) + (l_1 - l_2)]^2\right\} d\alpha \\
&= (c_1 - c_2)^2 + \frac{1}{9}[(l_1 - l_2)^2 + (r_1 - r_2)^2 - (l_1 - l_2) \\
&\times (r_1 - r_2)] - \frac{1}{2}(c_1 - c_2)[(l_1 - l_2) - (r_1 - r_2)].
\end{aligned} \tag{15}$$

We can use the distance (15) to define a distance between any two vectors of fuzzy numbers, by considering the sum of squared distances between individual elements. See Equation (20) ahead for more details. In the next section, this distance is used for fuzzy clustering of fuzzy data.

FUZZY CLUSTERING OF FUZZY DATA WITH OUTLIERS

In this section we propose two approaches. In the first approach, based on our distance, we propose a fuzzy clustering model for fuzzy data, by modifying Keller's (2000) algorithm. In the second approach, by defining a transformation, we reduce the problem of fuzzy clustering of fuzzy data to fuzzy clustering of crisp data. With the help of the second approach, any fuzzy clustering algorithms for crisp data can be used for fuzzy clustering of fuzzy data. For the sake of comparison with the first approach, we again use Keller's algorithm. Before describing the approaches, let us introduce some notations.

Let $U = \{u_{ik} : i = 1,...,c; k = 1,...,n\}$ be the $c \times n$ membership matrix, where c is the number of clusters, n the number of data vectors and $u_{ik} \in [0,1]$ the membership degree of the k-th object to the i-th cluster. We consider each data point, denoted as $\tilde{x}_k$, and each cluster prototype, denoted as $\tilde{v}_i$, to be a p-dimensional vector of triangular fuzzy data. This is in contrast to the Keller's approach where data elements and cluster prototypes are crisp.

To be more specific, let $\tilde{x}_{kj}$ denote the j-th component of $\tilde{x}_k$, the k-th data point. Then, $\tilde{x}_{kj}$ can be represented as a 3-vector collecting its center, left spread and right spread. In symbols, we have

$$\widetilde{X}_{kj} := [c_{\widetilde{X}_{kj}} \;\; l_{\widetilde{X}_{kj}} \;\; r_{\widetilde{X}_{kj}}]^T \in \mathbb{R}^3 \tag{16}$$

$$\widetilde{X}_k := \left[\tilde{x}_{k1}^T, \tilde{x}_{k2}^T ... \tilde{x}_{kP}^T\right]^T \in \mathbb{R}^{3p} \tag{17}$$

for $k = 1,...,n$. In other words, we may view each data point, $\tilde{x}_k$, either as a p-dimensional vector of fuzzy elements $\widetilde{X}_{kj}$ or as a $3p$-dimensional vector of real numbers. Both viewpoints are helpful and will be used interchangeably in what follows. A similar representation will be used for cluster prototypes, $\tilde{v}_i$ That is,

$$\tilde{v}_{ij} := [c_{\tilde{v}_{ij}} \;\; l_{\tilde{v}_{ij}} \;\; r_{\tilde{v}_{ij}}]^T \in \mathbb{R}^3 \tag{18}$$

$$\tilde{v}_i := \left[\tilde{v}_{i1}^T \;\; \tilde{v}_{i2}^T \;\; ... \;\; \tilde{v}_{iP}^T\right]^T \in \mathbb{R}^{3p} \tag{19}$$

for $i = 1,...,c$.

As mentioned earlier, we consider the following (squared) distance between fuzzy vectors $\tilde{x}_k$ and $\tilde{v}_i$,

$$d^2(\tilde{v}_i, \tilde{x}_k) = \sum_{j=1}^{p} d^2(\tilde{v}_{ij}, \tilde{x}_{kj}) \tag{20}$$

where $d^2(\tilde{v}_{ij}, \tilde{x}_{kj})$ is the (squared) distance (15) between fuzzy numbers $\tilde{v}_{ij}$ and $\tilde{x}_{kj}$.

Approach I

Following Keller, we minimize the objective function:

$$J(U, \widetilde{V}; \widetilde{X}) = \sum_{i=1}^{c}\sum_{k=1}^{n} u_{ik}^m \cdot \frac{1}{w_k^q} \cdot d^2(\tilde{v}_i, \tilde{x}_k) \tag{21}$$

subject to the constraints

$$\sum_{k=1}^{n} w_k = w \tag{22}$$

$$\sum_{i=1}^{c} u_{ik} = 1 \tag{23}$$

where, m is the degree of fuzziness and $d^2(\tilde{v}_i, \tilde{x}_k)$ is as defined in Equation (20).

The factor w_k represents the weight of the k-th datum and ω is a constant real valued parameter. According to Keller, the introduction of these weight factors helps in identifying outliers and reducing their effects. With constant parameter q, the influence of the outlier weight factors can be controlled. For this purpose, outliers are assigned a large weight $\frac{1}{w_k^q}$, so is small in this case.

The necessary conditions for minimizing the objective function are as follows:

$$c_{\tilde{v}_{ij}} = \frac{\sum_{k=1}^{n} u_{ik}^m \cdot \frac{1}{w_k^q}\left[2c_{\tilde{x}_{kj}} + \frac{1}{2}[(l_{\tilde{v}_{ij}} - l_{\tilde{x}_{kj}}) - (r_{\tilde{v}_{ij}} - r_{\tilde{x}_{kj}})]\right]}{2\sum_{k=1}^{n} u_{ik}^m \cdot \frac{1}{w_k^q}} \tag{24}$$

$$l_{\tilde{v}_{ij}} = \frac{\sum_{k=1}^{n} u_{ik}^{m} \cdot \frac{1}{w_k^q}\left[\frac{2}{9} l_{\tilde{x}_{kj}} + \frac{1}{9}(r_{\tilde{v}_{ij}} - r_{\tilde{x}_{kj}}) + \frac{1}{2}(c_{\tilde{v}_{ij}} - c_{\tilde{x}_{kj}})\right]}{\frac{2}{9}\sum_{k=1}^{n} u_{ik}^{m} \cdot \frac{1}{w_k^q}} \tag{25}$$

$$r_{\tilde{v}_{ij}} = \frac{\sum_{k=1}^{n} u_{ik}^{m} \cdot \frac{1}{w_k^q}\left[\frac{2}{9} r_{\tilde{x}_{kj}} + \frac{1}{9}(l_{\tilde{v}_{ij}} - l_{\tilde{x}_{kj}}) - \frac{1}{2}(c_{\tilde{v}_{ij}} - c_{\tilde{x}_{kj}})\right]}{\frac{2}{9}\sum_{k=1}^{n} u_{ik}^{m} \cdot \frac{1}{w_k^q}} \tag{26}$$

$$w_k = \frac{(\sum_{i=1}^{c} u_{ik}^{m} \cdot d^2(\tilde{v}_i, \tilde{x}_k))^{\frac{1}{q+1}}}{\sum_{i=1}^{n}(\sum_{i=1}^{c} u_{il}^{m} \cdot d^2(\tilde{v}_i, \tilde{x}_l))^{\frac{1}{q+1}}} \cdot w \tag{27}$$

$$u_{ik} = \frac{1}{\sum_{r=1}^{c}\left(\frac{d^2(\tilde{v}_i \tilde{x}_k)}{d^2(\tilde{v}_r \tilde{x}_k)}\right)^{\frac{1}{m-1}}}. \tag{28}$$

As it is observed, the membership degrees are left unchanged, while the cluster centers take into account the weights; points with high representativeness are more effective than outliers. On the basis of the necessary conditions, we can construct an iterative algorithm as follows:

Algorithm 1

Step 1: Fix the degree of fuzziness (m), the number of clusters (c), ω and q. Choose an initial fuzzy c-partition $U^{(0)}$. Also, choose initial spreads and weights for each datum subject to Equation (22). Set t=0.

Step 2: Calculate $\widetilde{V}^{(t)} = \left(c_{\tilde{V}}^{(t)}, l_{\tilde{V}}^{(t)}, r_{\tilde{V}}^{(t)}\right)$ using $U^{(t)}$, spreads, weights and Eqs.(24) through (26).

Step 3: Update $w_k^{(t)}, k = 1, ..., n$ using Equation (27) and update $U^{(t)}$ by $U^{(t+1)}$ using $\widetilde{V}^{(t)} = \left(c_{\tilde{V}}^{(t)}, l_{\tilde{V}}^{(t)}, r_{\tilde{V}}^{(t)}\right)$ and Equation (28).

Step 4: If, where ε is a non-negative small number fixed by the researcher, the algorithm has converged. Otherwise, set $t = t + 1$ and go to step 2.

Approach II

This approach is based on a different view of the distance (20). With some linear algebra, one can reduce this distance to the usual $3p$-dimensional Euclidean distance. For any N-vector, say $y = (y_1, y_2, ..., y_n) \in R^N$, let $\| y \|_2 := (\sum_{i=1}^{N} y_i^2)^{1/2}$ denote its Euclidean norm.

Consider two triangular fuzzy numbers $\widetilde{A}_i = (c_i, l_i, r_i), i = 1, 2$ Letting $c = c_1 - c_2$; $l = l_1 - l_2$; $r = r_1 - r_2$ and $z = [c, l, t]^T$, Equation (15) can be rewritten as:

$$d^2(\widetilde{A}_1, \widetilde{A}_2) = c^2 + \frac{1}{9}l^2 + \frac{1}{9}r^2 - \frac{1}{9}lr - \frac{1}{2}cl + \frac{1}{2}cr \tag{29}$$

or equivalently as:

$$d^2(\widetilde{A}_1, \widetilde{A}_2) = [c, l, r]\begin{bmatrix} 1 & -\frac{1}{4} & \frac{1}{4} \\ -\frac{1}{4} & \frac{1}{9} & -\frac{1}{18} \\ \frac{1}{4} & -\frac{1}{18} & \frac{1}{9} \end{bmatrix}\begin{bmatrix} c \\ l \\ r \end{bmatrix} \tag{30}$$

Let us denote the matrix above as Q. The eigenvalues of Q are:

$$\lambda_1 = \frac{7 + \sqrt{43}}{12}, \lambda_2 = \frac{1}{18}, \lambda_3 = \frac{7 - \sqrt{43}}{12}.$$

Since Q is a real symmetric matrix, it is diagonalizable by orthogonal matrices. That is, there is an orthogonal 3×3 matrix U (whose columns are orthonormal eigenvectors of Q) for which we have

$$Q = U \underbrace{\begin{pmatrix} \lambda_1 & 0 & 0 \\ 0 & \lambda_2 & 0 \\ 0 & 0 & \lambda_3 \end{pmatrix}}_{\Lambda} U^T \tag{31}$$

Let T be the (symmetric) square root of Q, i.e.

$$T := Q^{1/2} = U\Lambda^{1/2}U^T \tag{32}$$

Then, we may write

$$\begin{aligned} d(\widetilde{A}_1, \widetilde{A}_2) &= \sqrt{z^T Q z} = \sqrt{z^T Q^{1/2} Q^{1/2} z} \\ &= \sqrt{z^T T^T T z} = \|Tz\|_2 \end{aligned} \tag{33}$$

Now, recalling definitions (16) and (18) of $\tilde{x}_{kj}$ and $\tilde{v}_{ij}$, consider the following transformations:

$$\hat{x}_{kj} = T\tilde{x}_{kj}, \hat{v}_{ij} = T\tilde{v}_{ij} \tag{34}$$

where $\tilde{x}_{kj}$ and $\tilde{v}_{ij}$ are treated as 3-vectors. Furthermore, let us stack $\{\tilde{x}_{kj}\}_j$ and $\{\tilde{v}_{ij}\}_j$ into -dimensional vectors as usual, i.e.,

$$\hat{x}_k := \begin{bmatrix} \hat{x}_{k1}^T & \hat{x}_{k2}^T & \dots & \hat{x}_{kP}^T \end{bmatrix}^T \in \mathbb{R}^{3P}$$

$$\hat{v}_i := \begin{bmatrix} \hat{v}_{i1}^T & \hat{v}_{i2}^T & \dots & \hat{v}_{iP}^T \end{bmatrix}^T \in \mathbb{R}^{3P}.$$

Combining (20), (33) and the definitions of $\hat{x}_k$ and $\hat{v}_i$, we obtain:

$$\begin{aligned} d(\tilde{x}_k, \tilde{v}_i) &:= \sqrt{\sum_{j=1}^{p} d^2(\tilde{x}_{kj}, \tilde{v}_{ij})} \\ &= \sqrt{\sum_{j=1}^{p} \left\|\hat{x}_{kj} - \hat{v}_{ij}\right\|_2^2} \\ &= \left\|\hat{x}_k - \hat{v}_i\right\|_2 \end{aligned} \tag{35}$$

Equation (35) shows that the distance between fuzzy vectors $\tilde{x}_k$ and $\tilde{v}_i$ is the same as the Euclidean distance between the transformed vectors $\hat{x}_k$ and $\hat{v}_i$. In other words, we have reduced the problem of fuzzy clustering of fuzzy data to fuzzy clustering of crisp data. Thus, after applying transformations (34), any fuzzy clustering algorithm for crisp data can be used.

In particular, we can directly apply Keller's algorithm to $\{\hat{x}_k\}$ by minimizing the objective function:

$$J(U, \widehat{V}; \widehat{X}) = \sum_{i=1}^{c}\sum_{k=1}^{n} u_{ik}^m \cdot \frac{1}{w_k^q} \left\|\hat{x}_k - \hat{v}_i\right\|_2^2 \tag{36}$$

under the same constraints (22) and (23) on $\{w_k\}$ and $\{u_{ik}\}$

Necessary conditions for minimizing (36) are as follows:

$$\hat{v}_i = \frac{\sum_{k=1}^{n} u_{ik}^m \cdot \frac{1}{w_k^q} \hat{x}_k}{\sum_{k=1}^{n} u_{ik}^m \cdot \frac{1}{w_k^q}} \tag{37}$$

$$w_k = \frac{\left(\sum_{i=1}^{c} u_{ik}^m \cdot d^2(\hat{v}_i, \hat{x}_k)\right)^{\frac{1}{q+1}}}{\sum_{l=1}^{n}\left(\sum_{i=1}^{c} u_{il}^m \cdot d^2(\hat{v}_i, \hat{x}_l)\right)^{\frac{1}{q+1}}} \cdot w \tag{38}$$

$$u_{ik} = \frac{1}{\sum_{i=1}^{c}\left(\frac{d^2(\hat{v}_i, \hat{x}_k)}{d^2(\hat{v}_r, \hat{x}_k)}\right)^{\frac{1}{m-1}}} \quad (39)$$

After iterations, equation (37) provides cluster prototypes in the transformed domain. To retrieve the fuzzy prototypes, one should apply the inverse transformation T^{-1}, i.e. $\tilde{v}_{ij} = T^{-1}\hat{v}_{ij}$, for $i = 1, \ldots, c$ and $j = 1, \ldots, p$.

SIMULATION EXPERIMENTS

In order to show how well our method works, two simulation experiments are conducted; one in an environment without outliers and the other one in presence of outliers. We almost obtained the same results with both approaches.

Clustering Fuzzy Data Without Outliers

We now discuss the results of a simulation study carried out in order to compare the performance of our model with existing models able to handle fuzzy data. These models are proposed by D'Urso and Giordani (2006) by Yang et al. (2004), by Hathaway et al. (1996) and by Yang and Liu (1999). In order to compare the models, 2160 fuzzy data sets were randomly generated. After running several models for different values of q and ω, we chose $q = 1$, $\omega = 200$. The other parameters for clustering algorithm were set as follows: Number of objects ($n = 10, 50, 100$), number of variables ($k = 2, 8, 16$) and the weighting exponent ($m = 2, 3$). We constructed the data sets in such a way that $c = 2$ patterns can be found all over the simulation. To this purpose, the centers corresponding to the first $n/2$ objects were generated from the uniform distribution in [0, 1], and those corresponding to the latter $n/2$ from the uniform distribution in $[0 + \theta,\ 1 + \theta]$. All the spreads were generated from the uniform distribution in [0, 1] (case α). On the other hand, in case β, all the centers were generated from the uniform distribution in [0,1], while the spreads corresponding to the first $n/2$ objects were generated from the uniform distribution in [0, 1], and those corresponding to the latter $n/2$ from the uniform distribution in $[0 + \theta,\ 1 + \theta]$. θ *was* set to 1.5 and 0.75. In case of $\theta = 1.5$, the clusters are separated, whereas they are overlapped when θ is set to 0.75. Moreover, three sizes of centers with respect to the ones of the spreads were considered by defining a parameter h having the values $\frac{1}{2,1,2}$. This parameter means that the size of the spreads is h times that of the centers.

In Table 1 and Table 2, the percentage of well-classified objects by the models is given. This is done by fixing one parameter at a time and averaging over the rest. So, the left columns of Table 1 and Table 2 display the fixed parameters. Inasmuch as the cluster membership functions were known in advance, it is presumed that an object is assigned to a cluster correctly if the membership degree was the highest among all ($u = 0.5$). In addition, membership degrees higher than 0.75 and 0.9 are reported so that the strength of our model can be evaluated. It can be seen that our model works better in most of the conditions. As a case in point, when $\theta = 1.5$ and in conditions $n = 10, 50, 100$, $k = 2, 8, 16$, α, $m = 2, 3$, our model had better performance for $u = 0.5, 0.75, 0.9$. When $\theta = 0.75$ and in conditions $n = 10, 50, 100$, $k = 2, 8, 16$, $m = 2, 3$, our model worked better, whereas in case β the models proposed by Hathaway et al. (1996) and Yang and Liu (1999), had the best performance. As reported in the Table 3 and Table 4, when $\theta = 1.5$, the average percentage of well-classified objects for our model is 97.39(u = 0.5), 71.38(u = 0.75) and 33.69(u = 0.9). The model proposed by D'Urso and Giordini (2006) had the second highest performance after our model with 93.25(u = 0.5), 60.81(u = 0.75), 27.93(u = 0.9). When θ passed from 1.5 to 0.75, the average performance of all models got worse. In this case, the average percentage of well-classified objects for our model is 91.12(u = 0.5), 37.70(u = 0.75) and 11.09(u = 0.9) and for the model proposed by D'Urso and Giordini (2006) is 89.00(u = 0.5), 32.56(u = 0.75) and 7.35(u = 0.9). As mentioned earlier, the simulation study showed that our model had much better results than the other existing models.

Clustering Fuzzy Data with Outliers

In order to evaluate how our model is able to detect the prototypes in case of possible presence of observations that can be seen as outliers, we added some outliers to cases α and β, mentioned above. After running several models for different values of q and ω, we chose $q = 2$, $\omega = 200$. The other parameters for clustering algorithm were set as follows: Number of objects (n = 100, 200, 300), where n/10 of them are outliers and the rest of the them are inliers, number of variables k = (2, 8, 16) and the weighting exponent (m = 2). The modified cases α and β are as follows:

- **Case α:** The centers corresponding to the first 1/2 of inliers were generated from the uniform distribution in [0, 1], and those corresponding to the rest of the inliers from the uniform distribution in [1.5, 2.5]. The number of outliers is $\frac{n}{10}$. The centers of outliers were generated from Normal distribution with mean=-2 and variance=2. The left and the right spreads were generated from the uniform distribution in [0, 1].
- **Case β:** The left and the right spreads corresponding to the first 1/2 of inliers were generated from the uniform distribution in [0, 1], and those corresponding to the rest of the inliers from the uniform distribution in [1.5, 2.5]. The number of outliers is n/10. The left and the right spreads of outliers were generated from Normal distribution with mean = 5 and variance = 2. All the centers were generated from the uniform distribution in

Table 1. Percentages of well-classified objects with membership higher than u = 0.5, u = 0.75, u = 0.9 (Θ = 1.5)

	Our model			D'Urso and Giordani (2006)			Hathaway et. al. (1996) Yang and Liu (1999)			Yang et. al. (2004)		
	u = 0.5	u = 0.75	u = 0.9	u = 0.5	u = 0.75	u = 0.9	u = 0.5	u = 0.75	u = 0.9	u = 0.5	u = 0.75	u = 0.9
n = 10	97.53	74.94	38.14	94.11	66.22	31.78	94.78	56.61	19.11	84.83	54.94	24.39
n = 50	97.86	70.14	32.03	94.18	58.47	26.14	94.34	48.26	15.32	85.97	47.81	19.74
n = 100	97.68	69.12	31.16	91.45	57.74	25.87	92.88	47.21	14.82	85.57	46.27	18.90
k = 2	93.33	73.17	38.02	84.36	64.74	31.08	85.06	53.29	18.16	82.18	55.59	24.33
k = 8	98.95	70.86	31.80	96.11	50.95	27.15	98.26	49.17	15.67	86.50	47.80	19.59
k = 16	99.68	70.21	31.11	99.27	58.73	25.56	98.69	49.71	15.42	87.69	45.64	19.11
α	100.00	98.83	52.63	100.00	97.30	51.32	93.38	50.91	16.47	100.00	87.29	39.82
β	94.78	43.09	15.33	86.49	24.32	4.54	94.64	50.54	16.37	70.91	12.06	2.30
m = 2	97.13	81.64	62.27	93.12	71.75	53.56	94.34	69.79	32.62	85.50	60.51	40.90
m = 3	97.49	60.48	4.43	93.37	49.87	2.30	93.66	31.06	0.21	85.41	38.84	1.13
h = 1/2	91.94	55.20	29.02	91.96	54.79	25.44	91.96	51.31	21.95	76.28	54.23	27.59
h = 1	99.72	70.65	29.88	100.00	74.77	28.33	99.91	50.65	5.29	82.29	52.24	25.01
h = 2	100.00	88.96	42.06	87.78	55.87	30.01	90.10	50.21	22.01	97.79	42.55	10.43

Table 2. Percentages of well-classified objects with membership higher than u = 0.5, u = 0.75, u = 0.9 (Θ = 0.75)

	Our model			D'Urso and Giordani (2006)			Hathaway et. al. (1996) Yang and Liu (1999)			Yang et. al. (2004)		
	u = 0.5	u = 0.75	u = 0.9	u = 0.5	u = 0.75	u = 0.9	u = 0.5	u = 0.75	u = 0.9	u = 0.5	u = 0.75	u = 0.9
n = 10	93.86	47.72	17.47	80.78	30.80	12.39	88.72	32.58	5.06	81.80	37.30	10.31
n = 50	91.84	35.28	8.25	89.43	29.17	5.47	97.67	30.61	1.73	78.30	28.37	4.71
n = 100	92.28	32.06	7.19	87.78	29.86	4.58	96.27	19.92	1.61	98.20	27.63	4.76
k = 2	82.82	44.91	19.49	81.40	41.08	18.78	78.59	32.27	7.67	74.43	35.50	11.97
k = 6	91.25	35.11	6.06	88.95	29.18	4.99	91.11	21.31	1.12	62.85	27.07	4.41
k = 16	97.17	32.48	6.58	90.61	27.71	3.29	94.78	19.43	0.61	84.29	29.22	5.88
α	98.52	90.97	18.29	96.97	22.20	12.57	96.33	21.77	3.96	88.80	51.97	11.39
β	81.32	14.01	3.15	79.02	10.07	2.12	88.86	21.88	1.34	81.80	7.38	1.81
m = 2	90.58	93.62	20.41	89.14	58.50	14.44	97.88	41.08	6.18	90.64	58.71	12.05
m = 3	91.94	10.16	1.95	88.80	9.77	0.25	97.35	3.58	0.08	90.34	11.54	0.25
λ = 1/2	81.88	38.21	11.07	81.75	29.78	1.84	84.73	29.73	4.04	73.21	38.71	11.02
λ = 1	91.75	35.84	11.44	97.02	31.06	5.29	98.67	14.71	1.22	78.90	33.06	6.22
λ = 2	97.94	38.06	7.74	81.68	36.51	12.71	91.34	28.53	4.13	98.37	23.61	1.97

Table 3. Average percentage of well-classified objects (Θ = 1.5)

	u = 0.5	u = 0.75	u = 0.9
Our model	97.39	71.38	33.69
D'Urso and Giordani	93.25	60.81	27.93
Hathaway et.al.	94.00	50.72	16.42
Yang et.al.	85.16	49.67	21.01

The mean square errors (MSE) between prototypes obtained by performing our clustering model and the ideal prototypes are shown in Tables 5 and 6. From these tables, it can be observed that MSE of the centers are more than MSE of left spreads and right spreads in case α, while MSE of spreads are higher than those of centers in case β. In both cases, small weighting factors ω_k (large values for $\frac{1}{w_k^q}$) are assigned to data points fitting well to one of the clusters whereas large ω_k (small values for $\frac{1}{w_k^q}$) are assigned to outliers. Thus, outliers can be easily identified by their large weighting factors.

CONCLUSION AND FUTURE WORK

This paper presented a fuzzy clustering model for fuzzy data based on a new distance. We have modified Keller's approach so that our model can be used in noisy environments. The weighting factors reduce the influence of outliers and enable us to identify them. Necessary conditions for the objective function to receive an optimum have been derived to calculate a partition of data. Also, in another approach, we transformed our distance to the Euclidean distance and reduced the problem of fuzzy clustering of fuzzy data to fuzzy clustering of crisp data. Finally, two simulation experiments were considered; one for comparing the performance of our model with those of other

Table 4. Average percentage of well-classified objects ($\Theta = 0.75$)

	$\alpha = 0.5$	$\alpha = 0.75$	$\alpha = 0.9$
Our model	91.12	37.70	11.09
D'Urso and Giordani	89.00	32.56	7.35
Hathaway et.al.	87.62	24.33	3.13
Yang et.al.	80.49	31.13	6.60

Table 5. Mean Square Error (MSE) for cluster prototypes – case (α) (with outliers)

	Centers (MSE)	Left Spreads (MSE)	Right Spreads (MSE)
$n = 100$	0.0035	0.0015	0.0016
$n = 200$	0.0025	0.0007	0.0007
$n = 300$	0.0022	0.0005	0.0005
$k = 2$	0.0025	0.0014	0.0013
$k = 8$	0.0024	0.0007	0.0007
$k = 16$	0.0034	0.0006	0.0006

Table 6. Mean Square Error (MSE) for cluster prototypes - case (β) (with outliers)

	Centers (MSE)	Left Spreads (MSE)	Right Spreads (MSE)
$n = 100$	0.0031	0.1365	0.1363
$n = 200$	0.0014	0.1447	0.1441
$n = 300$	0.0010	0.1481	0.1482
$k = 2$	0.0033	0.0140	0.0140
$k = 8$	0.0010	0.1232	0.1227
$k = 16$	0.0010	0.2937	0.2932

existing clustering models for fuzzy data and one for testing how well our model behaves in noisy environments.

Our model can be applied in settings where the presence of outliers can drastically affect the results. An example is the process control problem in which the presence of outliers usually represents that the process has been out of control.

Another problem that can be explored is to study in depth fuzzy clustering for interactive fuzzy data and determining the optimal weighting exponent (m).

REFERENCES

Berkhin, P. (2002). *Survey of clustering data mining techniques*. Retrieved from http://www.accrue.com/products/researchpapers.html

D'Urso, P., & Giordani, P. (2006). A weighted fuzzy c-means clustering model for fuzzy data. *Computational Statistics & Data Analysis*, *50*(6), 1496–1523. doi:10.1016/j.csda.2004.12.002

De Oliveira, J. V., & Pdrycz, W. (2007). *Advances in fuzzy clustering and its applications*. New York, NY: John Wiley & Sons. doi:10.1002/9780470061190

Gibbs, A. L., & Su, F. E. (2002). On choosing and bounding probability metrics. *International Statistical Review*, *70*(3), 419–435. doi:10.2307/1403865

Hathaway, R. J., Bezdek, J. C., & Pedrycz, W. (1996). A parametric model for fusing heterogeneous fuzzy data. *IEEE Transactions on Fuzzy Systems*, *4*(3), 1277–1282. doi:10.1109/91.531770

Hung, W. L., & Yang, M. S. (2005). Fuzzy clustering on LR-type fuzzy numbers with an application in Taiwanese tea evaluation. *Fuzzy Sets and Systems*, *150*(3), 561–577. doi:10.1016/j.fss.2004.04.007

Irpino, A., & Verde, R. (2008). Dynamic clustering for interval data using a Wasserstein-based distance. *Pattern Recognition Letters*, *29*(11), 1648–1658. doi:10.1016/j.patrec.2008.04.008

Keller, A. (2000). Fuzzy clustering with outliers. In *Proceedings of the 19th International Conference of the North American Fuzzy Information Processing Society* (pp. 143-147).

Näther, W. (2000). On random fuzzy variables of second order and their application to linear statistical inference with fuzzy data. *Metrika*, *51*, 201–221. doi:10.1007/s001840000047

Pedrycz, W., Bezdek, J. C., Hathaway, R. J., & Rogers, G. W. (1998). Two nonparametric models for fusing heterogeneous fuzzy data. *IEEE Transactions on Fuzzy Systems*, *6*(3), 411–425. doi:10.1109/91.705509

Tran, L., & Duckstein, L. (2002). Comparison of fuzzy numbers using a fuzzy distance measure. *Fuzzy Sets and Systems*, *130*, 331–341. doi:10.1016/S0165-0114(01)00195-6

Yang, M. S., Hwang, P. Y., & Chen, D. H. (2004). Fuzzy clustering algorithms for mixed feature variables. *Fuzzy Sets and Systems*, *141*(2), 301–317. doi:10.1016/S0165-0114(03)00072-1

Yang, M. S., & Ko, C. H. (1996). On a class of fuzzy c-numbers clustering procedures for fuzzy data. *Fuzzy Sets and Systems*, *84*(1), 4960. doi:10.1016/0165-0114(95)00308-8

Yang, M. S., & Liu, H. H. (1999). Fuzzy clustering procedures for conical fuzzy vector data. *Fuzzy Sets and Systems*, *106*(2), 189–200. doi:10.1016/S0165-0114(97)00277-7

Zimmermann, H. J. (2001). *Fuzzy set theory and its applications*. Amsterdam, The Netherlands: Kluwer Academic. doi:10.1007/978-94-010-0646-0

This work was previously published in the International Journal of Fuzzy System Applications, Volume 1, Issue 3, edited by Toly Chen, pp. 29-42, copyright 2010 by IGI Publishing (an imprint of IGI Global).

Chapter 4
A New Efficient and Effective Fuzzy Modeling Method for Binary Classification

T. Warren Liao
Louisiana State University, USA

ABSTRACT

This paper presents a new fuzzy modeling method that can be classified as a grid partitioning method, in which the domain space is partitioned by the fuzzy equalization method one dimension at a time, followed by the computation of rule weights according to the max-min composition. Five datasets were selected for testing. Among them, three datasets are high-dimensional; for these datasets only selected features are used to control the model size. An enumerative method is used to determine the best combination of fuzzy terms for each variable. The performance of each fuzzy model is evaluated in terms of average test error, average false positive, average false negative, training error, and CPU time taken to build model. The results indicate that this method is best, because it produces the lowest average test errors and take less time to build fuzzy models. The average test errors vary greatly with model sizes. Generally large models produce lower test errors than small models regardless of the fuzzy modeling method used. However, the relationship is not monotonic. Therefore, effort must be made to determine which model is the best for a given dataset and a chosen fuzzy modeling method.

INTRODUCTION

Binary classification refers to the task of classifying an unknown object into one of the two known classes. Many real world problems can be formulated as binary classification problems. The examples are numerous, which include determining whether a patient has certain disease or not, determining whether an artifact is defective or not, determining whether an email is a spam or not, and so on.

Before classification can be carried out, a classification system or a classifier must be built first. A classification system or classifier built for performing binary classification is called a binary classifier. Many machine learning methods

DOI: 10.4018/978-1-4666-1870-1.ch004

such as decision trees, neural networks, support vector machines, and fuzzy models are suitable for building binary classifiers. Of interest in this paper are fuzzy models.

A fuzzy model is comprised of a set of fuzzy if-then rules. Various forms of fuzzy rules are possible. Most well-known among them are Mamdani linguistic rules and TSK rules, as described below for an n-input and single-output system.

- **Mamdani linguistic rules** are of the form of R_i: If X_1 is A_{1i} and X_2 is A_{2i}, …, and X_n is A_{ni}, then Y is B_i. In each Mamdani rule, X_n denotes the nth input variable, A_{ni} the fuzzy set associated with A_n, and B_i the fuzzy set associated with output variable Y of rule R_i. B_i can be represented by its modal value or as a fuzzy singleton for fuzzy control and classification applications.
- **TSK rules** are of the form of R_i: If X_1 is A_{1i} and X_2 is A_{2i}, …, and X_n is A_{ni}, then $Y = b_{0i} + b_{1i}X_1 + b_{2i}X_2 + ... + b_{ni}X_n$. In each TSK rule, X_n denotes the nth input variable, A_{ni} the fuzzy set associated with A_n, and b_i is a parameter vector and b_{0i} is a scalar offset associated with rule R_i. If all parameters in the vector b_i are zeros, then TSK rules are identical to Mamdani rules with B_i being fuzzy singletons.

A fuzzy model built for classification consists of a set of fuzzy classification rules of the following form R_i: If X_1 is A_{1i} and X_2 is A_{2i}, …, and X_n is A_{ni}, then Y is B_i. In each classification rule, X_n denotes the nth input variable, A_{ni} the fuzzy set associated with A_n, and $B_i \in \{B_1, B_2, ..., B_m\}$ representing the class label of rule R_i. For binary classification problems, there are only two values, i.e., $B_i \in \{B_1, B_2\}$. Fuzzy classification rules are identical to Mamdani rules with B_i being fuzzy singletons or TSK rules with b_i being zeros. Each rule could be assigned a rule weight. If not explicitly assigned, each rule is considered having rule weight of one.

Fuzzy models can be built manually or interactively using a tool designed to do that, e.g. the FIS Editor in the Matlab fuzzy toolbox. These approaches, however, require *a priori* knowledge about the model. The alternative is to build fuzzy models from data for which numerous methods have been proposed. These data-driven methods can be grouped into the following categories (Liao, 2006):

- **Grid partitioning:** This method divides the domain space into overlapped rectangular parallelepiped grids by specifying the number and shape of membership functions along each dimension. Data is then used to choose rule consequent and to compute rule weight. The earliest and most famous paper in this category is the method proposed by Wang and Mendel (1992), called the WM method later.
- **Fuzzy clustering:** This method applies a fuzzy clustering algorithm to partition the domain space, either one dimension at a time or all dimensions together at one time. The former approach works like grid partitioning except that the data distribution is taken into account. The fuzzy c-means variant-based method is an example (Liao, 2004). The latter approach generates a set of rules with each corresponding to a cluster. The membership function along each dimension could be obtained by projection, but might not be interpretable. The *genfis2* function in the Matlab toolbox is an example, which implements the fuzzy subtractive clustering method.
- **Genetic-fuzzy modeling or using other metaheuristics:** This method involves the use of a genetic algorithm to help determine model structure and/or model parameters. Interested readers are referred to Cordón et al. (2004) for an overview of genetic fuzzy systems. The uses of other

metaheuristics for fuzzy modeling are relatively new to date.

- **Neurofuzzy modeling:** This method adapts a fuzzy model to the neural network structure and uses the capability of neural network to learn the model. The most famous method in this category is the adapted network-based fuzzy inference system (ANFIS) developed by Jang (1993).
- **Hybrid methods:** Methods in this category involve the use of more than one algorithm/method to learn the model.

Takagi and Sugeno (1985) published the first research work on fuzzy system identification, an interchangeable term for fuzzy modeling. Due to its significance, it is no surprise that their paper has the second highest citations in "fuzzy" in the Web of Science, only next to the classical paper of Zadeh. However, ANFIS might be the most popular one among all other fuzzy modeling methods because it is available through the Matlab fuzzy toolbox. But is ANFIS the best? To the best of our knowledge there has been no attempt in comparing ANFIS with other methods, not to mention comparing all of them, after 25 years research in this area. One possible reason is that not too many researchers are well verse in all methods. Another possible reason is that programs are not widely available and it requires high amount of effort for their implementation. Of course, effort continues on to keep developing more efficient and effective fuzzy modeling methods. Motivated by the need to answer the question whether ANFIS is the best, to evaluate the relative performances of different methods, and to keep searching for an efficient and effective method, the idea to carry out this study was born.

In essence, this paper presents a new fuzzy modeling method, compares it to six others, and reports the results obtained. The new method is fuzzy equalization based fuzzy modeling method and is compared with six other fuzzy modeling methods for the task of binary classification. It will be shown later that the new method is not only effective but also efficient than the other six methods. Details of the newly developed method and brief description of six other methods selected for comparison are presented. Five data sets chosen for testing are then described and test results are presented, followed by discussion. Finally the paper is concluded.

A NEW FUZZY MODELING METHOD AND SIX OTHERS FOR COMPARISON

In this section, a newly developed fuzzy modeling method is first presented. This method can be classified as a grid partitioning method. The domain space is partitioned one dimension at a time based on the idea of fuzzy equalization that was proposed by Pedrycz (2001). Let P(A) be the probability of a fuzzy event A. The idea of fuzzy equalization states that a family of fuzzy sets A = $\{A_1, A_2, \ldots, A_c\}$ should be generated in a way such that $P(A_1) = P(A_2) = \ldots = P(A_c) = 1/c$. It will be shown later that this idea works far better than the third method that performs grid partitioning in an ad hoc manner. Once the domain is partitioned, then fuzzy rules are built in the same way according to Liao (2004). Since this method has not been presented elsewhere, more details are given in the following.

Given a set of n-input and 1-output of 2-class training data, $\{X_1^t, \ldots, X_i^t, \ldots, X_n^t; y^t\}, t = 1, \ldots, T$. The following algorithm is followed, which is a modification of the original one developed by Pedrycz, to generate a family of triangular fuzzy sets with 0.5 overlapping between successive fuzzy sets one input variable at a time, as shown in Figure 1 (note that the first and last fuzzy sets are actually trapezoidal).

Step 0: Specify the number of fuzzy sets (membership functions) to be generated, *c*.

Step 1: Rank the values of X_i^t, $t = 1, \ldots, T$, in ascending order.

Step 2: Start with the lowest value, xmin = min(X_i^t) and proceed towards higher values to find the value of a_l so that $\sum_{x\min}^{a_1} \frac{1}{T} = \frac{1}{2c}$. Let j=1.

Step 3: Continue proceeding to higher values to find the value of a_{j+1} so that $\sum_{a_j}^{a_{j+1}} \frac{1}{2T} = \frac{1}{2c}$.

Step 4: Let j=j+1. If j >c stop; else repeat Step 3.

Once a set of membership functions for each input variable is obtained by the above algorithm, a set of fuzzy rules can then be constructed. For an *n*-input 1-output 2-class problem, the total number of possible rules is $2 * \prod_{i=1}^{n} c_i$, where c_i is the number of membership functions for input variable *i*. Let *B* takes 0 or 1 as the two values of *y*. Following Liao (2004), the weight of rule R_i, W_i, is learned using the same set of training data, according to either the max-min composition or the mean-min composition, as given below:

- **Max-min Composition:**

$$W_i = \underset{t}{Max}\, Min\{A_{1i}(x_1^t), \ldots, A_{ni}(x_n^t), B_i(y^t)\} \tag{1}$$

- **Mean-min Composition:**

$$W_i = \sum_{t=1}^{T} Min\{A_{1i}(x_1^t), \ldots, A_{ni}(x_n^t), B_i(y^t)\} \Big/ T \tag{2}$$

In the following, the six fuzzy modeling methods selected for comparison are briefly described. The first method is taken from Liao (2004), which applies a fuzzy c-means variant (FCMV) as proposed by Liao et al. (2003) to partition the domain space one dimension at a time and then to learn the rules and rule weights. In essence, the fuzzy c-means variant fixes the well-known counter-intuitive and non-convex membership functions generated by fuzzy c-means by two modifications: (i) by forcing the end terms to take extreme values as their centers, and (ii) by a post-processing operation that involves identifying the concave portion of a membership function and redistributing the associated membership values.

The second method is is similar to the first method, except that the alternated optimization scheme used in fuzzy c-means variant is replaced with an ant colony optimization algorithm developed by Socha and Dorigo (2008) for the real domain called ACO_R. In both the first and second methods, the generalized π-function (comprised of a generalized s- on the left side and a generalized z- on the right side) as proposed by Liao et al. (2003) is used. The generalized π-function, given

Figure 1. A family of 5 triangular fuzzy sets with 0.5 overlapping between successive fuzzy sets

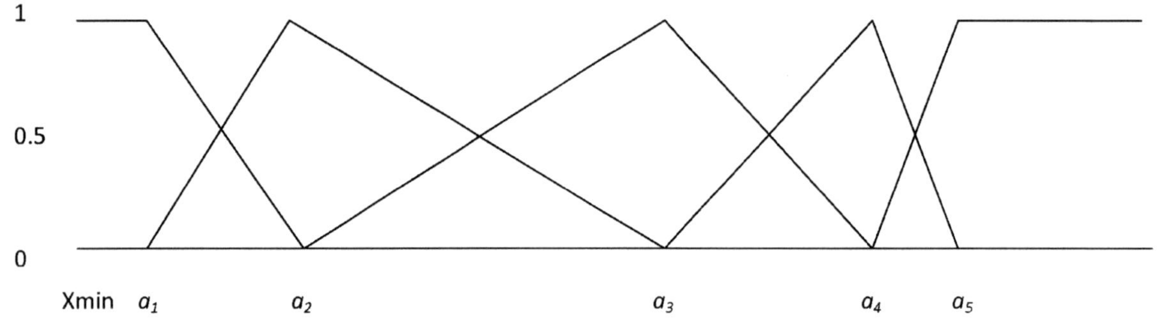

below for completeness, becomes a π-function (triangular) when the z-parameter is equal to 2 (1).

$$G\Pi MF = \begin{cases} 0 & x \le \alpha_L \\ \left(\frac{1}{2}\right)^{1-z}\left(\frac{x-\alpha_L}{\gamma-\alpha_L}\right)^z & \alpha_L < x \le \frac{\alpha_L+\gamma}{2} \\ 1-\left(\frac{1}{2}\right)^{1-z}\left(\frac{\gamma-x}{\gamma-\alpha_L}\right)^z & \frac{\alpha_L+\gamma}{2} < x \le \gamma \\ 1 & x = \gamma \\ 1-\left(\frac{1}{2}\right)^{1-z}\left(\frac{x-\gamma}{\alpha_R-\lambda}\right)^z & \gamma < x \le \frac{\gamma+\alpha_R}{2} \\ \left(\frac{1}{2}\right)^{1-z}\left(\frac{\alpha_R-x}{\alpha_R-\gamma}\right)^z & \frac{\alpha_R+\gamma}{2} < x \le \alpha_R \\ 0 & x > \alpha_R \end{cases} \quad (3)$$

In the next section, the best results of the first and second methods selected for comparison are shown in two ways: (i) using triangular function only (z=1), and (ii) using four different membership functions by setting the z-parameter to 0.5, 1, 2, and 5.

The third method selected for comparison is a modification of the algorithm originally proposed by Wang and Mendel (1992) for prediction and control. Following the original paper, triangular membership functions are used.

The fourth method employs the fuzzy subtractive clustering method to generate fuzzy rules by clustering of multi-dimensional data and then followed by projecting to each individual dimension (Chiu, 1994), which has been implemented as the *genfis2* function in the fuzzy toolbox. Gaussian membership functions were assumed in this method.

The fifth method employs the well-known ANFIS algorithm (Jang, 1993), implemented as the *anfis* function in the fuzzy toolbox. Before applying *anfis*, grid partitioning of the domain space is first carried out by the *genfis1* function that is also available in the fuzzy toolbox. Gaussian membership functions are chosen here for fair comparison with the fourth and six methods. Triangular membership functions were tried but failed.

The sixth method selected for comparison is a hybrid one, which involves using *gnefis2* to generate the initial set of fuzzy rules that are subsequently tuned with *anfis*. Gaussian membership functions are chosen here for fair comparison with the fourth and fifth methods.

DATASETS AND TEST RESULTS

Five binary classification datasets are selected for testing the performances of the seven fuzzy modeling methods. Two of them were obtained in previous research by the author. Three of them were taken from the UCI Machine Learning Archive. Table 1 summarizes the major characteristics of the five datasets. Note that to reduce the size of model the features selected in Ghazavi and Liao (2008) are used for three datasets with higher numbers of input variables. The training data and test data were generated from the original dataset based on random sampling with replacement. Each fuzzy model is tested with five sets of test data.

To test the performance of fuzzy models, a fuzzy reasoning method is needed. For fuzzy models built by the proposed, and the first and second methods selected for comparison, the fuzzy reasoning method proposed in the original papers to derive predicted y for a given $\mathbf{x}=\{x_1,\ldots,x_n\}$ is:

$$\hat{y} = \frac{\sum_{1i=1}^{c_1}\cdots\sum_{ni=1}^{c_n}\sum_{j=1}^{2}\min\{A_{1i}(x_1),\ldots,A_{ni}(x_n),w_i\}\cdot B_j}{\sum_{1i=1}^{c_1}\cdots\sum_{ni=1}^{c_n}\sum_{j=1}^{2}\min\{A_{1i}(x_1),\ldots,A_{ni}(x_n),w_j\}} \quad (4)$$

For fuzzy models built by the third fuzzy modeling method chosen for comparison, the fuzzy reasoning method used in the original paper that does not include rule weight and a modified

Table 1. Five datasets chosen for testing

Dataset	*n*-input variables	Selected features	Number of records	Number of training data	Number of test data	Source
Weld recognition	3	All	2493	600	300	Liao et al. (2000)
Welding flaw identification	25	{2, 6, 7}	399	300	100	Liao et al. (1999)
Haberman's survival	3	All	306	300	150	UCI
Pima diabetes	8	{2, 6, 8}	768	500	150	UCI
Wisconsin breast cancer	30	{21, 22, 28}	599	300	100	UCI

version that include rule weight are given in the following two equations, (5) and (6):

$$\hat{y} = \frac{\sum_{1i=1}^{c_1} \cdots \sum_{ni=1}^{c_n} \sum_{j=1}^{2} \prod \{A_{1i}(x_1), \ldots, A_{ni}(x_n), B_j\}}{\sum_{1i=1}^{c_1} \cdots \sum_{ni=1}^{c_n} \sum_{j=1}^{2} \prod \{A_{1i}(x_1), \ldots, A_{ni}(x_n)\}} \quad (5)$$

$$\hat{y} = \frac{\sum_{1i=1}^{c_1} \cdots \sum_{ni=1}^{c_n} \sum_{j=1}^{2} \prod \{A_{1i}(x_1), \ldots, A_{ni}(x_n), w_i, B_j\}}{\sum_{1i=1}^{c_1} \cdots \sum_{ni=1}^{c_n} \sum_{j=1}^{2} \prod \{A_{1i}(x_1), \ldots, A_{ni}(x_n), w_j\}} \quad (6)$$

For fuzzy models built by anfis related methods, the *evalfis* Matlab function that works identical to Equation (5) is used. In other words, the firing strengths (or matching degrees) of all premise terms are aggregated by the product operator. The implication operator that connects the premise part of the rule with the consequent part of the rule is also product. The defuzzification method is weighted average.

For each fuzzy model except that constructed by fuzzy subtractive clustering, the optimal number of membership functions for each input variable is an important parameter that needs to be determined, among others. In consideration of the size of training datasets, we elect to exhaustively test all possible 125 combinations of membership functions with each input variable taking 2, 3, 4, 5, or 6 membership functions of any chosen type. For fuzzy model constructed by fuzzy subtractive clustering, the model size is determined by the radius parameter. In this study, radius is varied from 0.1 to 0.9 at 0.05 increments. The smaller the radius, the larger the model size is. The performance of each model is measured primarily by the average test error rate based on five test results. Nevertheless, the training error, CPU time taken to built model, and average false positive and false negative rates of test results are also reported.

Weld Recognition

Table 2 summarizes the best results obtained by each method for the weld recognition dataset. The test results indicate that:

1. The best model with 1.53% average test error is built by anfis, which has the combination of {5,2,4}, indicating five Gaussian membership functions for the first variable, two for the second variable, and four for the third variable.

Table 2. Summary of best results for weld recognition data

Method	Rule Weight	Rank		Average Test Error Rate	Combination of Fuzzy Terms (radius for FS)	Shape of Membership Function
Fuzzy C-means Variant	Max-min	9	8	0.029333	{6,6,6}	Triangular
	Mean-min	8	7	0.028667	{6,3,6}	Triangular
ACO_R-Based Fuzzy Partitioning	Max-min	6	5	0.026	{4,4,6}	Triangular
	Mea-min	5	8	0.029333 0.024667	{4,2,6} {5,3,6}	Triangular Z=5
WM	No	2	2	0.022	{5,5,6}	Triangular
	Yes	3	3	0.022667	{4,4,6}	Triangular
Fuzzy Subtractive (FS)	No	11	11	0.031333	0.1	Gaussian
ANFIS	No	1	1	0.015333	{5,2,4}	Gaussian
FS-ANFIS	No	10	10	0.03	0.1	Gaussian
Fuzzy Equalization	Max-min	3	3	0.022667	{5,4,4}	Triangular
	Mean-min	7	6	0.026667	{5,6,6}	Triangular

2. The associated average false positive rate and average false negative rate are 1.56% and 1.58%, respectively.
3. It took 5.515 seconds of CPU time to build the best model with 1.83% training error, which is 0.33% higher than the lowest of 1.5% obtained for the {5, 5, 5} combination that yields 3.93% average test error. The correlation between average test errors and training errors for all 125 combinations is found to be 0.9556. However, a model with the lowest training error is not necessarily the one giving the lowest test error.
4. The standard deviation of average test errors for different combinations of membership functions used to build anfis models is 2.87%, with {2, 2, 2} being the worse combination at 15.47%.
5. Regardless the fuzzy modeling method a larger model generally has a lower average test error. However, the relationship is not monotonic. This makes it difficult to predict which model would turn out to be the best. Therefore, it is important to make an effort to determine the best model regardless which fuzzy modeling method is used.

Welding Flaw Identification

Table 3 summarizes the best results obtained by each method for the welding flaw identification dataset. The test results indicate that:

1. The best model with 13.2% average test error is built by anfis, which has the combination of {6, 6, 3}. Another best model with the same performance is built by fuzzy equalization with {6, 6, 5} combination and rule weights computed by the max-min composition rule.
2. The associated average false positive rate and average false negative rate are 11.32% and 15.72%, respectively, for the anfis model and 16.12% and 9.11%, respectively, for the fuzzy equalization-based model. Therefore, if false negative (missing welding flaws) is more serious than false positive (mistaking good welds as flawed), then the latter model is preferred and vice versa.
3. It took 15.39 seconds of CPU time to build the best anfis model with 8% training error, which is 0.67% higher than the lowest of 7.33% obtained for the {6, 3, 6} combina-

Table 3. Summary of best results for 3-feature welding flaw identification data

Method	Rule Weight	Rank		Average Test Error Rate	Combination of Fuzzy Terms (radius for FS)	Shape of Membership Function
Fuzzy C-means Variant	Max-min	9	9	0.184 0.178	{6,6,6} {6,6,6}	Triangular Z=2
	Mean-min	8	8	0.176 0.174	{5,5,2} {5,6,2},{6,6,2}	Triangular Z=0.5
ACO_R-Based Fuzzy Partitioning	Max-min	10	10	0.186 0.18	{6,2,4} {6,6,5}	Triangular Z=5
	Mean-min	3	3	0.15	{6,5,5}	Triangular
WM	No	6	6	0.172	{6,4,6}	Triangular
	Yes	6	6	0.172	{6,4,6}	Triangular
Fuzzy Subtractive (FS)	No	5	5	0.168	0.3	Gaussian
ANFIS	No	1	1	0.132	{6,6,3}	Gaussian
FS-ANFIS	No	4	4	0.158	0.25, 0.3	Gaussian
Fuzzy Equalization	Max-min	1	1	0.132	{6,6,5}	Triangular
	Mean-min	11	11	0.19	{6,3,2}	Triangular

tion that yields 15% average test error. The correlation between average test errors and training errors for all 125 combinations is found to be 0.1624, which indicates a very weak positive correlation. The reason is unclear why this correlation is particular low than others reported.

4. It took 0.093 seconds of CPU time to build the best fuzzy equalization-based fuzzy model with 7.33% training error, which is also the best among all combinations. Note that this CPU time is approximately two magnitudes lower than that taken by anfis. The correlation between average test errors and training errors for all 125 combinations is found to be 0.9267, which indicates a very strong positive correlation.
5. The standard deviation of average test errors for different combinations of membership functions used to build anfis models is 1.89%, with {5, 2, 3} being the worse combination at 23.2%. On the other hand, for fuzzy equalization-based fuzzy models the standard deviation is 4.43%, with {6, 2, 2} being the worse combination at 34%. Thus, fuzzy equalization-based modeling method has higher variance than anfis.
6. Regardless the fuzzy modeling method a larger model generally has a lower average test error. However, the relationship is not monotonic. This makes it difficult to predict which model would turn out to be the best. Therefore, it is important to make an effort to determine the best model regardless which fuzzy modeling method is used. This result is identical to the weld recognition dataset.

Haberman's Survival

Table 4 summarizes the best results obtained by each method for the Haberman's survival dataset. Since FCMV- and ACO_R-based methods are the best for this dataset in two different cases, the results are presented separately. In the case that using four z values on the generalized π functions for the FCMV- and ACO_R-based methods, the test results indicate that:

1. The best model with 14.8% average test error is built by ACO_R-based fuzzy partitioning method, which has the combination of {6, 6, 6} with each membership function being a generalized π-function with the z-parameter = 0.5 and rule weights computed by the max-min composition rule.
2. The associated average false positive rate and average false negative rate are 9.94% and 30.91%, respectively. Note that false negative (patients did within five years but mistaken as surviving) is approximately three times higher than false positive (patients survived more than five years but mistaken as not).
3. It took 0.406 seconds of CPU time to build the best model with 1.67% training error, which also to be the lowest among all combinations. The correlation between average test errors and train errors for all 500 (125x4) combinations is found to be 0.9765, which indicates a very strong positive correlation.
4. The standard deviation of average test errors for different combinations of membership functions used to build ACO_R-based fuzzy models is 11.82%, with {2, 2, 2} being the worse combination at 70.4% when z=0.5. The variance is thus very high.
5. Regardless the fuzzy modeling method a larger model generally has a lower average test error. However, the relationship is not monotonic. This makes it difficult to predict which model would turn out to be the best. Therefore, it is important to make an effort to determine the best model regardless which fuzzy modeling method is used. This result is identical to the previous two datasets presented.

In the case that using only one z value on the generalized π functions (i.e., z=1 equivalent to triangular membership functions) for the fuzzy c-means variant or ACO_R-based methods, the test results indicate that:

Table 4. Summary of best results for Haberman's survival data

Method	Rule Weight	Rank		Average Test Error Rate	Combination of Fuzzy Terms (radius for FS)	Shape of Membership Function
Fuzzy C-means Variant	Max-min	2	1	0.161333 0.154667	{6,4,6} {6,6,5}	Triangular Z=0.5
	Mean-min	6	8	0.190667 0.178667	{4,3,6} {5,6,5}	Triangular Z=0.5
ACO_R-Based Fuzzy Partitioning	Max-min	1	2	0.162667 0.148	{5,6,6} {6,6,6}	Triangular Z=0.5
	Mean-min	6	7	0.185333 0.178667	{5,5,4} {4,5,4}	Triangular Z=2
WM	No	10	10	0.198667	{3,6,5}	Triangular
	Yes	9	9	0.192	{3,5,5}	Triangular
Fuzzy Subtractive (FS)	No	11	11	0.205333	0.45	Gaussian
ANFIS	No	6	6	0.178667	{5,5,4}	Gaussian
FS-ANFIS	No	3	3	0.164	0.1	Gaussian
Fuzzy Equalization	Max-min	5	5	0.172	{6,6,3}	Triangular
	Mean-min	4	4	0.170667	{6,6,3}	Triangular

1. The best model with 16.13% average test error is built by FCMV-based fuzzy partitioning method, which has the combination of {6, 4, 6} with each membership function being a generalized π-function with the z-parameter = 1 (or triangular) and rule weights computed by the max-min composition rule.
2. The associated average false positive rate and average false negative rate are 10.26% and 30.2%, respectively. Note that false negative (patients did within five years but mistaken as surviving) is approximately three times higher than false positive (patients survived more than five years but mistaken as not).
3. It took 0.532 seconds of CPU time to build the best model with 1.67% training error, which also to be the lowest among all combinations. The correlation between average test errors and training errors for all 125 combinations is found to be 0.9812, which indicates a very strong positive correlation.
4. The standard deviation of average test errors for different combinations of membership functions used to build ACO_R-based fuzzy models is 11.91%, with {2, 2, 2} being the worse combination at 68.8% when z=1. The variance is thus very high.
5. Regardless the fuzzy modeling method a larger model generally has a lower average test error. However, the relationship is not monotonic. This makes it difficult to predict which model would turn out to be the best. Therefore, it is important to make an effort to determine the best model regardless which fuzzy modeling method is used. This result again is identical to the two datasets previously presented.

Table 5 summarizes the best results of both FCMV- and ACO_R-based fuzzy modeling method for five runs to capture the effect of initialization. The results indicate that:

Table 5. Initialization effect on FCM variant-based and ACO_R-based Fuzzy modeling methods for Haberman's survival data

Method	Rule Weight	Initialization	Average Test Error Rate with z=1 (four z values)	Combination of Fuzzy Terms	Shape of Membership Function
Fuzzy C-means Variant	Max-min	1	0.176(0.161333)	{5,4,6} ({6,6,5})	Tri (z=0.5)
		2	0.176 (0.16)	{5,4,6} ({6,6,6})	Tri (z=0.5)
		3	0.169333	{6,4,6}	Tri
		4	0.166667	{5,4,6}	Tri
		5	0.161333(**0.154667**)	{6,4,6} ({6,6,5})	Tri (z=0.5)
	Mean-min	1	0.193333(**0.178667**)	{5,4,4} ({5,6,5})	Tri (z=0.5)
		2	0.190667(0.181333)	{4,3,6} ({5,5,3})	Tri (z=0.5)
		3	0.193333(0.178667)	{5,4,4} ({5,6,5})	Tri (z=0.5)
		4	0.193333(0.178667)	{5,4,4} ({5,6,5})	Tri (z=0.5)
		5	0.193333(0.181333)	{5,4,4} ({5,5,3})	Tri (z=0.5)
ACO_R-Based Fuzzy Partitioning	Max-min	1	0.165333(0.156)	{6,4,6} ({6,5,6})	Tri (z=0.5)
		2	0.165333(0.154667)	{6,4,6} ({6,6,6})	Tri (z=0.5)
		3	0.162667(0.149333)	{6,6,6} ({5,6,6})	Tri (z=0.5)
		4	0.162667(**0.148**)	{5,6,6} ({6,6,6})	Tri (z=0.5)
		5	0.166667(0.158667)	{5,6,6} ({6,5,6})	Tri (z=0.5)
	Mean-min	1	0.188(**0.178667**)	{5,5,4} ({4,5,4})	Tri (z=2)
		2	0.186667(0.178667)	{5,5,4} ({6,5,5})	Tri (z=0.5)
		3	0.185333(0.181333)	{5,5,4} ({4,5,4})	Tri (z=2)
		4	0.189333(0.18)	{5,5,4} ({6,5,3})	Tri (z=0.5)
		5	0.188 (0.18)	{4,4,6} ({4,5,4})	Tri (z=2)

1. The best results often vary from run to run, in particular when the max-min composition rule is used to compute rule weights. This indicates that both FCMV- and ACO_R-based methods are not as stable as other fuzzy modeling methods.
2. Overall the max-min composition rule outperforms the mean-min composition rule for computing rule weights.
3. Tuning the z-parameter often leads to reduced error.
4. Better results than those reported in Table 4 might be possible if more runs are made.

Pima Diabetes

Table 6 summarizes the best results obtained by each method for the Pima diabetes dataset. Just like Haberman's survival dataset, two cases are distinguished. In the case that using four z values on the generalized π functions for the FCMV- and ACO_R-based methods, the test results indicate that:

1. The best model with 17.47% average test error is built by ACO_R-based fuzzy partitioning method, which has two equally best combinations obtained in separate runs: the combination of {5, 6, 6} with each membership function being a generalized π-function with the z-parameter = 5 and the combination of {6, 6, 5} with each membership function being a generalized π-function with the z-parameter=0.5. For both combinations, rule weights were computed by the max-min composition rule.
2. For the {5, 6, 6} combination, the associated average false positive rate and average false negative rate are 18.34% and 14.93%, respectively. Note that false negative (missed diagnosis of patients having diabetes) is slightly lower than false positive (patients not having diabetes mistaken as having). For the {6, 6, 5} combination, the associated average false positive rate and average false negative rate are 16.41% and 19.03%, respectively. Note that false negative (missed diagnosis of patients having diabetes) is

Table 6. Summary of best results for 3-feature Pima diabetes data

Method	Rule Weight	Rank		Average Test Error Rate	Combination of Fuzzy Terms (radius for FS)	Shape of Membership Function
Fuzzy C-means Variant	Max-min	3	3	0.185333 0.18	{6,5,6} {6,5,6}	Triangular Z=2
	Mean-min	10	11	0.226667 0.217333	{5,4,5} {5,5,4}	Triangular Z=0.5
ACO_R-Based Fuzzy Partitioning	Max-min	1	2	0.178667 0.174667	{6,6,6} {5,6,6}, {6,6,5}	Triangular Z=0.5, z=5
	Mean-min	6	6	0.210667	{5,5,5}	Triangular
WM	No	9	9	0.216	{6,4,6}	Triangular
	Yes	8	8	0.212	{4,5,4}	Triangular
Fuzzy Subtractive (FS)	No	5	5	0.209333	0.2	Gaussian
ANFIS	No	6	6	0.210667	{5,2,5}	Gaussian
FS-ANFIS	No	4	4	0.190667	0.2	Gaussian
Fuzzy Equalization	Max-min	2	1	0.176	{6,6,5}	Triangular
	Mean-min	11	10	0.225333	{4,5,6}	Triangular

slightly higher than false positive (patients not having diabetes mistaken as having), which is opposite to the other combination.
3. For the {5, 6, 6} combination, it took 0.578 seconds of CPU time to build the best model with 13.8% training error, which is 5.8% higher than the lowest of 8% obtained by the {6, 6, 6} combination with z=0.5 that yielded 19.87% average test error. The correlation between average test errors and training errors for all 500 (125x4) combinations is found to be 0.9610, which indicates a very strong positive correlation. For the {6, 6, 5} combination, it took 0.578 seconds of CPU time to build the best model with 9% training error, which happens also to be the lowest among all combinations. The correlation between average test errors and training errors for all 500 (125x4) combinations is found to be 0.9617, which indicates a very strong positive correlation.
4. For the {5, 6, 6} combination, the standard deviation of average test errors for different combinations of membership functions used to build ACO_R-based fuzzy models is 7.33%, with {2, 3, 2} being the worse combination at 59.47% when z=1. For the {6, 6, 5} combination, the standard deviation of average test errors for different combinations of membership functions used to build ACO_R-based fuzzy models is 7.29%, with {3, 2, 2} being the worse combination at 58.13% when z=1. The variance is thus very high.
5. Just like other datasets presented before this one, regardless the fuzzy modeling method a larger model generally has a lower average test error. However, the relationship is not monotonic. This makes it difficult to predict which model would turn out to be the best. Therefore, it is important to make an effort to determine the best model regardless which fuzzy modeling method is used.

In the case that using only one z value on the generalized π functions (i.e., z=1 equivalent to triangular membership functions) for the FCMV- and ACO_R-based methods, the test results indicate that:

1. The best model with 17.6% average test error is built by fuzzy equalization with [6, 6, 5} combination and rule weights computed by the max-min composition rule.
2. The associated average false positive rate and average false negative rate are 13.91% and 24.23%, respectively. This result indicates that false negative (missed diagnosis of patients having diabetes) is about 10% higher than false positive (patients not having diabetes mistaken as having).
3. (vii) It took 0.043 seconds of CPU time to build the best fuzzy equalization-based fuzzy model with 4% training error, which is also the best among all combinations. The correlation between average test errors and training errors for all 125 combinations is found to be 0.9739, which indicates a very strong positive correlation.
4. The standard deviation of average test errors for different combinations of membership functions used to build fuzzy equalization-based fuzzy models is 6%, with {4, 2, 2} being the worse combination at 46%.

Table 7 shows the best results of five runs for both the FCMV- and ACO_R-based fuzzy modeling methods. The same patterns observed in Table 5 are evident in Table 7 as well.

Wisconsin Breast Cancer

Table 8 summarizes the best results obtained by each method for the Wisconsin breast cancer dataset. The results indicate that:

1. The best model with 1.2% average test error is built by fuzzy equalization-based method

Table 7. Initialization effect on FCM variant-based and ACO_R-based Fuzzy modeling methods for Pima diabete data

Method	Rule Weight	Initialization	Average Test Error Rate	Combination of Fuzzy Terms	Shape of Membership Function
Fuzzy C-means Variant	Max-min	1	0.190667(0.1888)	{5,6,6} ({6,6,6})	Tri (z=0.5)
		2	0.192(**0.18**)	{6,6,6}({6,5,6})	Tri (z=2)
		3	0.192(0.186667)	{6,6,6,}({5,5,6})	Tri (z=5)
		4	0.185333	{6,5,6}	Triangular
		5	0.193333(0.190667)	{6,6,6}({6,6,5})	Tri (z=0.5)
	Mean-min	1	0.226667(**0.217333**)	{5,4,5} ({5,5,4})	Tri (z=0.5)
		2	0.226667(0.217333)	{5,4,5} ({5,5,4})	Tri (z=0.5)
		3	0.226667(0.217333)	{5,4,5} ({5,5,4})	Tri (z=0.5)
		4	0.226667(0.217333)	{5,4,5} ({5,5,4})	Tri (z=0.5)
		5	0.226667(0.217333)	{5,4,5} ({5,5,4})	Tri (z=0.5)
ACO_R-Based Fuzzy Partitioning	Max-min	1	0.185333(**0.174667**)	{6,6,6}({6,6,5})	Tri (z=0.5)
		2	0.178667	{6,6,6}	Triangular
		3	0.208 (0.196)	{5,5,5}({6,5,6})	Tri (z=0.5)
		4	0.2 (**0.174667**)	{5,6,6}({5,6,6})	Tri (z=5)
		5	0.197333(0.190667)	{6,6,6}({6,6,6})	Tri (z=2)
	Mean-min	1	0.22 (0.213333)	{5,5,5}({5,6,6})	Tri (z=2)
		2	**0.210667**	{5,5,5}	Triangular
		3	0.218667(0.210667)	{5,6,5}({6,5,4})	Tri (z=2)
		4	0.213333	{5,5,5}	Triangular
		5	0.213333(0.212)	{5,5,6}({6,6,6})	Tri (z=2)

with three different combinations, i.e., {4, 6, 6}, {5, 5, 3}, and {6, 6, 4}, and rule weights computed by the max-min composition rule.

2. For the {4, 6, 6} combination, the associated average false positive rate and average false negative rate are 0% and 2.95%, respectively. It took 0.079 seconds of CPU time to build this best fuzzy equalization-based fuzzy model with 0% training error, which is also one of the best among all combinations.
3. For the {5, 5, 3} combination, the associated average false positive rate and average false negative rate are 0.69% and 2%, respectively. It took 0.031 seconds of CPU time to build this best fuzzy equalization-based fuzzy model with 0.33% training error, which is 0.33% higher than the lowest of 0%.
4. For the {6, 6, 4} combination, the associated average false positive rate and average false negative rate are 0.67% and 1.96%, respectively. It took 0.062 seconds of CPU time to build this best fuzzy equalization-based fuzzy model with 0.67% training error, which is 0.67% higher than the lowest of 0%.
5. The correlation between average test errors and training errors for all 125 combinations is found to be 0.9254, which indicates a very strong positive correlation.
6. The standard deviation of average test errors for different combinations of membership functions used to build fuzzy equalization-based fuzzy models is 2.13%, with {2, 2, 2} being the worse combination at 15.2%.

DISCUSSION

The results presented clearly indicate that there is no one fuzzy model or one fuzzy modeling method that consistently dominates others. Three fuzzy modeling methods including the proposed fuzzy equalization-based method, anfis, and ACO_R-based method have been the winner in producing the lowest test error rate for more than one dataset.

Table 8. Summary of best results for 3-feature Wisconsin breast cancer data

Method	Rule Weight	Rank		Average Test Error Rate	Combination of Fuzzy Terms (radius for FS)	Shape of Membership Function
Fuzzy C-means Variant	Max-min	3	3	0.018 0.016	{6,5,5} {3,6,4},{6,6,5}	Triangular Z=0.5
	Mean-min	9	9	0.036 0.028	{6,5,5} {5,6,4}	Triangular Z=5
ACO_R-Based Fuzzy Partitioning	Max-min	2	2	0.014	{6,5,5}	Triangular
	Mean-min	7	8	0.032 0.024	{6,4,6} {5,5,5}	Triangular Z=5
WM	No	10	9	0.036	{6,5,6}	Triangular
	Yes	11	11	0.038	{6,4,5}	Triangular
Fuzzy Subtractive (FS)	No	4	4	0.022	0.25	Gaussian
ANFIS	No	4	4	0.022	{5,2,3}	Gaussian
FS-ANFIS	No	7	7	0.024	0.35, 0.4	Gaussian
Fuzzy Equalization	Max-min	1	1	0.012	{4,6, 6},{5,5,3}, {6, 6, 4}	Triangular
	Mean-min	4	4	0.022	{6,3,3}	Triangular

To rank these three and all other methods, their performances on all datasets must be considered together. Table 9 ranks all methods based on the sum of average test errors over all five datasets, for the case that only triangular functions are considered for the FCMV- and ACO_R-based fuzzy modeling methods. The results indicate that the proposed method is the best; anfis is the second best; and the hybrid of fuzzy subtractive clustering followed by anfis is the third best. If the average test-errors for each dataset are normalized to be between zero and one first before summing them, then some of the ranks are changed as noted in the parentheses. In particular, the first two ranks stay the same, but the ACO_R-based method with rule weights computed by the max-min composition rule becomes the third best.

Table 10 ranks all methods based on the sum of average test errors over all five datasets, for the case that the generalized π functions with four different z-parameter values are considered for the FCMV- and ACO_R-based fuzzy modeling methods. The results indicate that the proposed method is still the best; the ACO_R-based method with rule weights computed by the max-min composition rule becomes the second best; and anfis is the third best in this case. If the average test-errors for each dataset are normalized to be between zero and one first before summing them, then some of the ranks are changed as noted in the parentheses. Note that the first six ranks stay unchanged.

Based on Tables 9 and 10, it can be seen that using rule weights on the WM fuzzy modeling method according to Equation (6) performs slightly better than that without using rule weights at all according to Equation (5). Comparing the max-min compositional rule according to Equation (1) with the mean-min compositional rule according to Equation (2) used to compute rule weights, it is clear that the max-min rule outperforms the mean-min rule regardless whether the FCMV- or ACO_R-based or fuzzy equalization-based fuzzy modeling method is used.

Note that Equation (4) differs from Equations (5) & (6) primarily in the *t*-norm operator used

Table 9. Rank in sum of average test errors (using triangular on FCMV- and ACO_R-based methods)

	Rule weight	Weld recognition	Haberman	Pima diabetes	Breast Cancer	Weld flaw	Sum	Rank
fcmv	max-min	0.029333	0.161333	0.185333	0.018	0.184	0.577999	5
	mean-min	0.028667	0.190667	0.226667	0.036	0.176	0.658001	11
aco_r	max-min	0.026	0.162667	0.178667	0.014	0.186	0.567334	4(3)
	mean-min	0.029333	0.185333	0.210667	0.032	0.15	0.607333	6
wm	w/o	0.022	0.198667	0.216	0.036	0.172	0.644667	10(9)
	With	0.022667	0.192	0.212	0.038	0.172	0.636667	9(8)
fs	No	0.031333	0.205333	0.209333	0.022	0.168	0.635999	8(10)
anfis	No	0.015333	0.178667	0.210667	0.022	0.132	0.558667	2
fs-anfis	No	0.03	0.164	0.190667	0.024	0.158	0.566667	3(4)
Fuzzy equal.	max-min	0.022667	0.172	0.176	0.012	0.132	0.514667	1
	mean-min	0.026667	0.170667	0.225333	0.022	0.19	0.634667	7

to aggregate the matching degrees on all premise terms. To investigate whether the performance will improve or degrade, Equation (6) was used in place of Equation (4) on the best fuzzy modeling method identified above, i.e., the proposed method. To be consistent with the fuzzy reasoning method, the minimum operator used in Equations (1) & (2) in computing rule weights are also replaced with the product operator, which lead to two new composition rules.

- **Max-prod Composition:**

$$W_i = \underset{t}{Max} \prod \{A_{1i}(x_1^t), \ldots, A_{ni}(x_n^t), B_i(y^t)\} \quad (7)$$

- **Mean-prod Composition:**

$$W_i = \sum_{t=1}^{T} \prod \{A_{1i}(x_1^t), \ldots, A_{ni}(x_n^t), B_i(y^t)\} \Big/ T \quad (8)$$

Table 10. Rank in sum of average test errors (using generalized π functions with four z-parameter values on FCMV- and ACO_R-based methods)

	Rule weight	Weld recognition	Haberman survival	Pima diabetes	Breast Cancer	Weld flaw	Sum	Rank
fcmv	max-min	0.029333	0.159667	0.18	0.016	0.178	0.563	4
	mean-min	0.028667	0.178667	0.217333	0.028	0.174	0.626667	7(8)
aco_r	max-min	0.026	0.148	0.174667	0.014	0.18	0.542667	2
	mean-min	0.024667	0.178667	0.210667	0.024	0.15	0.588001	6
wm	w/o	0.022	0.198667	0.216	0.036	0.172	0.644667	11(10)
	With	0.022667	0.192	0.212	0.038	0.172	0.636667	10(9)
fs	No	0.031333	0.205333	0.209333	0.022	0.168	0.635999	9(11)
anfis	No	0.015333	0.178667	0.210667	0.022	0.132	0.558667	3
fs-anfis	No	0.03	0.164	0.190667	0.024	0.158	0.566667	5
Fuzzy equal.	max-min	0.022667	0.172	0.176	0.012	0.132	0.514667	1
	mean-min	0.026667	0.170667	0.225333	0.022	0.19	0.634667	8(7)

Table 11 summarizes the results, which indicate that the performance of using Equation (6) is worse than that of using Equation (4) overall, even though the opposite is true for the weld recognition dataset. Based on the sum of average test errors, the max-min composition remains to be the best, followed by the max-prod, the mean-prod, and lastly the mean-min rule.

There are other ways to compute rule weights. For comparison, two formulas used by Ishibuchi and Yamamoto (2005), called CF^{I} and CF^{II} as given in Equations (9) & (10) below were used together with the proposed fuzzy modeling method, in place of Equations (1) & (2).

$$CF_i^I = \sum_{t=1}^{T} B_i(y^t)\prod\{A_{1i}(x_1^t),\ldots,A_{ni}(x_n^t)\} \Big/ \sum_{t=1}^{T}\prod\{A_{1i}(x_1^t),\ldots,A_{ni}(x_n^t)\} \quad (9)$$

$$CF_i^I = \frac{\sum_{t=1}^{T} B_i(y^t)\prod\{A_{1i}(x_1^t),\ldots,A_{ni}(x_n^t)\}}{\sum_{t=1}^{T}\prod\{A_{1i}(x_1^t),\ldots,A_{ni}(x_n^t)\}} - \frac{\sum_{t=1}^{T} B_{i'}(y^t)\prod\{A_{1i}(x_1^t),\ldots,A_{ni}(x_n^t)\}}{\sum_{t=1}^{T}\prod\{A_{1i}(x_1^t),\ldots,A_{ni}(x_n^t)\}} \quad (10)$$

In Equation (10), both B_i and $B_{i'}$ are singletons and complementary to each other if taking 0 and 1 as their values. Table 12 summarizes the test results. Among all four rule weight computational methods the max-min compositional rule is the best, CF^{II} is the second best, followed by CF^{I}, and finally the mean-min rule. Our result that CF^{II} is better than CF^{I} is consistent with that reported by Ishibuchi and Yamamoto.

To give idea on how much CPU time each fuzzy modeling method takes to build fuzzy models, Table 13 lists some results obtained in the study, specifically the smallest model with {2, 2, 2} combination of fuzzy terms, the largest model with {6, 6, 6} combination of fuzzy terms, and the average of all model sizes varying from {2, 2, 2} to {6, 6, 6}. The results indicate that:

1. Overall, the anfis modeling method is most computationally intensive, followed by the hybrid of fuzzy subtractive clustering together with anfis. The most efficient group of modeling methods includes the WM method, the proposed fuzzy equalization-based method, and the fuzzy subtractive clustering-based method. The difference between the most efficient and the most inefficient model is 2 to 3 orders of magnitudes. The FCMV- and ACO_R-based methods lie in between, taking about 10 times higher computational time than the most efficient ones.

Table 11. Comparison of product and minimum operators

	max-min		max-prod		mean-min		mean-prod	
	Avg. err	comb	Avg. err	comb	Avg. err	comb	Avg. err	comb
Weld recog	0.022667	{3,5,5}	0.018	{5,5,5}	0.026667	{5,6,6}	0.018	{5,6,5}
Weld flaw	0.132	{6,6,5}	0.162	{2,4,5}	0.19	{6,3,2}	0.208	{2,5,2} {2,6,2}
Haberman	0.172	{6,6,3}	0.188	{5,6,3}	0.170667	{6,6,3}	0.174667	{4,4,2}
Pima diabetes	0.176	{6,6,5}	0.22	{5,6,5} {6,6,5}	0.225333	{4,5,6}	0.208	{6,6,5}
Wisconsin breast cancer	0.012	{4,6,6} {5,5,3} {6,6,4}	0.014	{6,3,6} {6,5,5}	0.022	{5,2,5}, {5,3,5} {6,3,3}, {6,4,4} {6,6,4}	0.018	{4,4,3} {4,5,3} {6,2,6} {6,3,5}
Total	0.514667		0.602		0.634667		0.626667	

Table 12. Comparison of different rule weight formulas used on the fuzzy equalization-based method

Dataset	Rule Weight	Average Test Error Rate	Combination of Fuzzy Terms
Weld recognition	Max-min	0.022667	{5,4,4}
	Mean-min	0.026667	{5,6,6}
	CF^{I}	0.017333	{3,5,5}
	CF^{II}	0.016670	{3,6,6}
Welding flaw identification	Max-min	0.132000	{6,6,5}
	Mean-min	0.190000	{6,3,2}
	CF^{I}	0.166000	{4,3,6}
	CF^{II}	0.162000	{6,3,5}
Haberman's survival	Max-min	0.172000	{6,6,3}
	Mean-min	0.170667	{6,6,3}
	CF^{I}	0.177333	{6,4,6}
	CF^{II}	0.173333	{4,2,3}
Pima diabetes	Max-min	0.176000	{6,6,5}
	Mean-min	0.225333	{4,5,6}
	CF^{I}	0.221333	{6,6,3}
	CF^{II}	0.212000	{6,6,5}
Wisconsin breast cancer	Max-min	0.012	{4,6, 6},{5,5,3},{6, 6, 4}
	Mean-min	0.022	{6,3,3}
	CF^{I}	0.018	{6,2,4}
	CF^{II}	0.018	{6,5,4}
Total	Max-min	0.5147	
	Mean-min	0.6347	
	CF^{I}	0.6000	
	CF^{II}	0.5820	

2. Due to the stochastic nature of the learning process, the time taken to learn a model is not necessarily monotonic with the model size. The smallest model size sometimes takes longer time than a larger model. On the same token, the largest model size sometimes takes less time than a smaller model.

CONCLUSION

This paper has presented a new fuzzy modeling method that is put together based on two existing studies. First, the domain space is partitioned one dimension at a time based on the idea of fuzzy equalization that was proposed by Pedrycz (2001). Once the domain is partitioned, then the rule weights of all possible rules were constructed according to Liao (2004). Based on the test results of five datasets, it can be concluded that the newly proposed method is not only the most effective in producing the lowest average test error rate but also one of the most efficient method in taking less time to build fuzzy models, in comparison with six other methods including the most popular method, i.e., anfis. Nevertheless, the proposed method is not the best for all five datasets. Both

Table 13. Average CPU times (in seconds) taken to build fuzzy models of selected and various sizes

Dataset Method	Model	Weld recognition	Weld flaw identification	Haberman's survival	Pima diabete	Wisconsin breast cancer
Fuzzy C-means Variant	{2,2,2} {6,6,6} All	0.063 0.859 0.482	0.156 0.953 0.335	0.140 0.703 0.512	0.031 0.844 0.610	0.078 0.485 0.420
ACO_R-Based Fuzzy Partitioning	{2,2,2} {6,6,6} All	0.250 0.735 0.329	0.125 0.39 0.179	0.156 0.406 0.183	0.203 0.625 0.276	0.157 0.406 0.184
WM	{2,2,2} {6,6,6} All	0.016 0.063 0.032	0.016 0.109 0.018	0.016 0.047 0.024	0.016 0.079 0.041	0.016 0.047 0.024
Fuzzy Subtractive (FS)	R=0.9 R=0.1 All	0.094 0.390 0.118	0.016 0.047 0.023	0.016 0.141 0.035	0.047 0.219 0.062	0.031 0.078 0.024
ANFIS	{2,2,2} {6,6,6} All	0.828 113.500 15.168	2.25 56.422 7.826	0.437 57.218 7.810	0.672 96.375 12.646	0.532 56.906 7.592
FS-ANFIS	R=0.9 R=0.1 All	0.703 7.297 1.220	0.516 8.047 1.459	0.313 163.360* 14.949	0.562 121.297* 9.263	0.343 44.453 3.552
Fuzzy Equalization	{2,2,2} {6,6,6} All	0.016 0.187 0.061	0.016 0.110 0.034	0.016 0.109 0.033	0.015 0.156 0.051	0.016 0.125 0.034

ACO_R-based method and anfis were the winners twice and might be useful in some cases.

Issues addressed in the Discussion section include the effect of rule weight as to the WM method, the effect of rule weight computational method as to the FCMV-, ACO_R-based, and proposed fuzzy equalization-based methods, the effect of *t*-norm and additional rule weight computational methods as to the proposed fuzzy equalization-based method. The CPU times taken by each method to build fuzzy models were also compared.

The only class of method not included in this study for comparison is the genetic-fuzzy method. The many varieties of genetic-fuzzy modeling methods, in our opinion, justify a stand-alone comparison among themselves. It is our hope that someone will take on this challenge in the near future.

REFERENCES

Chiu, S. (1994). Fuzzy model identification based on cluster estimation. *Journal of Intelligent and Fuzzy System*, *2*, 267–278.

Cordón, O., Gomide, F., Herrera, F., Hoffmann, F., & Magdalena, L. (2004). Ten years of genetic fuzzy systems: current framework and new trends. *Fuzzy Sets and Systems*, *141*, 5–31. doi:10.1016/S0165-0114(03)00111-8

Ghazavi, S. N., & Liao, T. W. (2008). Medical data mining by fuzzy modeling with selected features. *Artificial Intelligence in Medicine*, *43*, 195–206. doi:10.1016/j.artmed.2008.04.004

Ishibuchi, H., & Yamamoto, T. (2005). Rule weight specification in fuzzy rule-based classification systems. *IEEE Transactions on Fuzzy Systems*, *13*(4), 428–435. doi:10.1109/TFUZZ.2004.841738

Jang, J.-S. R. (1993). ANFIS: Adaptive network-based fuzzy inference system. *IEEE Transactions on Systems, Man, and Cybernetics*, *23*(3), 665–685. doi:10.1109/21.256541

Liao, T. W. (2004). Fuzzy reasoning based automatic inspection of radiographic welds: weld recognition. *Journal of Intelligent Manufacturing*, *15*, 69–85. doi:10.1023/B:JIMS.0000010076.56537.07

Liao, T. W. (2006). Mining Human Interpretable Knowledge Using Automatic Data-Driven Fuzzy Modeling Methods – A Review. In Triantaphyllou, E., & Felici, G. (Eds.), *Data Mining and Knowledge Discovery Approaches Based on Rule Induction Techniques* (pp. 495–550). New York: Springer. doi:10.1007/0-387-34296-6_15

Liao, T. W., Celmins, A. K., & Hammell, R. J. Jr. (2003). A fuzzy c-means variant for the generation of fuzzy term sets. *Fuzzy Sets and Systems*, *135*, 241–257. doi:10.1016/S0165-0114(02)00136-7

Liao, T. W., Li, D.-M., & Li, Y.-M. (1999). Detection of welding flaws from radiographic images using fuzzy clustering methods. *Fuzzy Sets and Systems*, *108*, 145–158. doi:10.1016/S0165-0114(97)00307-2

Liao, T. W., Li, D.-M., & Li, Y.-M. (2000). Extraction of welds from radiographic images using fuzzy classifiers. *Information Sciences*, *126*, 21–40. doi:10.1016/S0020-0255(00)00016-5

Pedrycz, W. (2001). Fuzzy equalization in the construction of fuzzy sets. *Fuzzy Sets and Systems*, *119*, 329–335. doi:10.1016/S0165-0114(99)00135-9

Socha, K., & Dorigo, M. (2008). Ant colony optimization for continuous domains. *European Journal of Operational Research*, *185*, 1155–1173. doi:10.1016/j.ejor.2006.06.046

Takagi, T., & Sugeno, M. (1985). Fuzzy identification of systems and its applications to modeling and control. *IEEE Transactions on Systems, Man, and Cybernetics*, *15*(1), 116–132.

Wang, L.-X., & Mendel, J. M. (1992). Generating fuzzy rules by learning from examples. *IEEE Transactions on Systems, Man, and Cybernetics*, *22*(6), 1414–1427. doi:10.1109/21.199466

This work was previously published in the International Journal of Fuzzy System Applications, Volume 1, Issue 1, edited by Toly Chen, pp. 17-35, copyright 2010 by IGI Publishing (an imprint of IGI Global).

Chapter 5
On Possibilistic and Probabilistic Information Fusion

Ronald R. Yager
Iona College, USA

ABSTRACT

This article discusses the basic features of information provided in terms of possibilistic uncertainty. It points out the entailment principle, a tool that allows one to infer less specific from a given piece of information. The problem of fusing multiple pieces of possibilistic information is and the basic features of probabilistic information are described. The authors detail a procedure for transforming information between possibilistic and probabilistic representations, and using this to form the basis for a technique for fusing multiple pieces of uncertain information, some of which is possibilistic and some probabilistic. A procedure is provided for addressing the problems that arise when the information to be fused has some conflicts.

INTRODUCTION

Information used in decision making generally comes from multiple sources. We are interested in the problem of multi-source information fusion in the case when the information provided has some uncertainty. Two important types of sources of information are electro-mechanical sensors and human observers/experts; this is particularly the case in security environments. We note that sensor-provided information generally has a probabilistic type of uncertainty. Human-observer-provided information, which is generally linguistic in nature, typically introduces a possibilistic type of uncertainty. Here we are faced with a problem in which we must fuse information with different modes of uncertainty. Here we shall discuss an approach to attaining capability based on the use of probability-possibility transformation. We first discuss the basic features of information provided in terms of possibilistic uncertainty. We point out the entailment principle, a tool that allows one to infer less specific from a given piece of information. We also discuss the problem of fusing multiple pieces of possibilistic information. We describe a procedure for transforming information between possibilistic and probabilistic representations. We use this to form the basis for a technique for fusing multiple pieces of uncertain information some of which is possibilistic and some probabilistic. We

DOI: 10.4018/978-1-4666-1870-1.ch005

provide a normalization procedure for addressing the problems that arise when the information to be fused has some conflict.

The most well-known uncertainty representation is the probabilistic model. Assume V is some variable taking its value in the space X. In probabilistic uncertainty we associate with each $x_i \in X$ a value $p_i \in [0, 1]$ indicating the probability that x_i is the value of V. We denote the collection of these as the probability distribution P. For any subset A of X the probability that V lies in A is $\Pr ob(A) = \sum_{i\ s.t.\ x_i \in A} p_i$. In the probabilistic framework the normalization property that Prob(X) = 1 requires that $\Sigma_i\, p_i = 1$. Probabilistic uncertainty is essentially based on ratio scale information. That is, if we know the ratio $\frac{p_j}{p_i}$ then we can completely determine the probability distribution. Within the framework of probability theory the Shannon entropy is used to quantify the amount of uncertainty associated with a probability distribution P

$$H(p) = -\sum_{i=1}^{n} p_i\ ln(p_i).$$

Possibilistic Uncertainty

Assume V is some variable taking its value in the space X. Formally under a possibilistic uncertainty model we associate with each $x_i \in X$ a value $\pi_i \in [0, 1]$ indicating the possibility that x_i is the value of V. We denote the collection of these as the possibilistic distribution $\prod$. For any subset A of X the possibility that V lies in A, $Poss[A] = \underset{x_i \in A}{Max}[\pi_i]$. In the possibilistic framework the normalization property that Poss[X] = 1 requires that $Max_i[\pi_i] = 1$. The implication here is that at least one of the x_i has possibility one. We also note from our definition of Poss[A] that if $B \subset C$ then $Poss[B] \leq Poss[C]$. Furthermore we have $Poss[\emptyset] = 0$.

Possibilistic uncertainty often arises from linguistically expressed information via a fuzzy set. Again assume V is some variable whose domain is the set X. A proposition or datum associated with this variable is a statement of the form V *is* ***Value***. Here ***Value*** is some constraint on the possible values of the variable (Zadeh, 2005). For example if V is John's age then ***Value*** could be "young" or "old", etc. Using the framework of computing with words introduced by Zadeh (1996, 1999) we can express the meaning of ***Value*** in terms of a fuzzy subset A of X. In the case of John's age we would define "young" as a fuzzy over ages. Using the fuzzy subset A we can induce a possibility distribution $\prod$ to express our knowledge of the value V by assigning $\pi_i = A(x_i)$. Thus the possibility that $V = x_i$ is the membership grade of x_i in A.

In passing we want to make one comment regarding the assignment statement V *is* ***Value***. In point of fact what we have denoted as the variable V can be more precisely viewed as consisting of an attribute and an object. Thus a variable is of the form Attribute(object). For example in the statement John *is* young, the more precise statement is age of John is young. Here we have a datum triple, attribute – object – value. Nevertheless in the following we shall continue to use the term variable.

Returning to our possibility distribution $\prod$ on V. We indicated that for any subset A we have $Poss[A] = \underset{x_i \in A}{Max}[\pi_i]$. We recall that for any subset A we can associate a function A: X → [0, 1] such that $A(x_i) = 1$ if $x_i \in A$ and $A(x_i) = 0$ if $x_i \notin A$. Using this we can express $Poss[A] = Max_i[A(x_i) \wedge \pi_i]$. This formulation allows us to calculate the possibility of fuzzy sets when $A(x_i) \in [0,1]$ instead of {0, 1}.

Within the framework of possibility theory the concept of specificity (Yager, 1998; Klir, 2006) plays a role analogous to the role of entropy in probability theory, it measures the uncertainty associated with a possibility distribution. Let ind(j) be the index of the value in X with the jth largest possibility. Using this we define the specificity of the distribution Π

$$Sp(\Pi) = \pi_{ind(1)} - \frac{1}{n-1}\sum_{j=2}^{n}\pi_{ind(j)}$$

It is the difference between the largest possibility and the average of the others. Since for a normalized possibility distribution $\pi_{ind(1)} = 1$ then $Sp(\Pi) = 1 - \frac{1}{n-1}\sum_{j=2}^{n}\pi_{ind(j)}$

The specificity indicates the amount of information we have about the variable. We note that if we know the exact value of the variable, let us say $V = x_1$ then $\Pi_{ind(1)} = 1$ and $\Pi_{ind(j)} = 0$ for all $j \neq 1$ and here we see $Sp(\Pi) = 1$. At the other extreme if $\pi_i = 1$ for all x_i then $Sp(\prod) = 0$. Here we have no information. We note this corresponds to the statement V *is* X, which is "I don't know."

Given two statements V *is* A and V *is* B with information about V then we can provide the combined information supplied by both of these by taking their conjunction. Here then

V *is* A **and** V *is* B ⇒ V *is* A ∩ B.

Using the standard interpretation of ∩ as the Min we get, V *is* D where $D(x_i) = Min(A(x_i), B(x_i))$. Here then possibility distribution is $\Pi(x_i) = A(x_i) \wedge B(x_i)$. If we denote Π_A as the possibility distribution induced by V *is* A and $\prod_B$ as the possibility distribution induced by V *is* B then $\Pi(x_i) = Min(\Pi_A(x_i), \Pi_B(x_i))$. It is the pointwise conjunction of the two possibility distributions. This can of course be extended to any number of statements. Thus

$$V\ is\ A_1\ and\ V\ is\ A_2\ and\ .....\ and\ V\ is\ A_q \Rightarrow V\ is\ D$$

where $D = A_1 \cap A_2 \cap ..... \cap A_q$. From this we get $\Pi_D(x_i) = Min_{j=1 to q}[\Pi_j(x_i)]$.

The fundamental inference mechanism available with possibilistic information makes use of the entailment principle. Consider the statement John's age is between 20 and 30. From this statement we can always infer that John's age is between 10 and 50. This is a manifestation of what is called the entailment principle (Zadeh, 1979; Yager, 1986; Yager & Kreinovich, 2007). Assume A and B are two normal fuzzy sets such that A ⊂ B. The entailment principle states that from the knowledge V *is* A we can always infer V *is* B. We recall A ⊂ B, if A(x) ≤ B(x) for all x. The entailment principle has an interesting manifestation in possibility theory. We first recall that the statement V *is* A induces a possibility distribution Π_A on X such that $\Pi_A(x_i) = A(x_i)$. Similarly V *is* B induces a possibility distribution Π_B. We observe here with A ⊂ B the relationship between these possibility distributions is that $\Pi_B(x_i) \geq \Pi_A(x_i)$. We see that the entailment principle has a direct version in the framework of possibility theory, called possibilistic entailment. Assume we have a possibility distribution Π_A on X. From this we can always infer a possibility distribution Π_B on X such that $\Pi_B(x) \geq \Pi_A(x)$ for all x. The use of this entailment doesn't come without a price. Consider the possibility distribution Π_A the amount of information contained in this distribution as calculated using the measure of specificity. If without loss of generality we assume $\Pi_A(x_1) = 1$ then

$$S_P(\Pi_A) = 1 - \frac{1}{n-1}\sum_{j=2}^{n}\Pi_A(x_j)).$$

It is one minus the average of all the possibilities except the first. Consider now Π_B. We see that since $\Pi_B(x_j) \geq \Pi_A(x_j)$ for all j then also $\Pi_B(x_1) = 1$. In this case we see

$$S_P(\Pi_B) = 1 - \frac{1}{n-1}\sum_{j=2}^{n}\Pi_B(x_j)).$$

Since

$\Pi_B(x_j) \geq \Pi_A(x_j)$ then

$$\sum_{j=2}^{n}\Pi_B(x_j) \geq \sum_{j=2}^{n}\Pi_A(x_j)$$

and hence $S_P(\Pi_B) \leq S_P(\Pi_A)$. We see that when applying the entailment principle we are essentially giving up some of the information we have. We are not adding information we are reducing the information we have. We also observe that the possibility distribution with ∏(x) = 1 for all x can always be inferred from any other distribution. It says I can also infer everything is possible.

When given a statement V *is* A one can qualify this statement by how certain we are of its truth we call this certainty qualification. Zadeh (1979) discussed this idea. Generally the less certain a statement the less information it provides. Consider a statement V *is* A **is** α certain where $\alpha \in [0, 1]$ indicates its certainty. Let us consider some properties we should require of the possibility distribution induced by the statement V *is* A **is** α certain. First we observe that if α = 1, then we should take the information on face value, this should have the same possibility distribution as the statement V *is* A. At the other extreme is the case where α = 0. This corresponds to the case where we have no information. This should induce a possibility distribution in which $\Pi(x_i) = 1$ for all x_i. We further note that as α decreases the amount of information we get should decrease. Yager (1985) suggested we can obtain a possibility distribution Π for V is A **is** α such that

$$\Pi(x) = S(A(x), \bar{\alpha})$$

where S is a t-conorm (Klement, Mesiar, & Paper, 2000) and $\bar{\alpha} = 1 - \alpha$. Let us look at this for some the notable situations. If α = 1 then $\bar{\alpha} = 0$ and $\Pi(x) = S(A(x), 0) = A(x)$. If α = 0 then $\bar{\alpha} = 1$ and $\Pi(x) = S(A(x), 1) = 1$ for all x. We also note that if A(x) = 1 then $S(A(x), \bar{\alpha}) = S(1, \bar{\alpha}) = 1$ for x. Consider now the case where $\alpha_1 > \alpha_2$. Here $\bar{\alpha}_1 < \bar{\alpha}_2$ and hence

$$\Pi_1(x) = S(A(x), \bar{\alpha}_1) \leq \Pi_2(x) = S(A(x), \bar{\alpha}_2).$$

Thus $\Pi_1(x_j) \leq \Pi_2(x_j)$ for all x_j. However we note if without loss of generality x_1 is such that $A(x_1) = 1$ then $\Pi_1(x_1) = \Pi_2(x_1) = 1$. From this we see

$$S_P(\Pi_1) = 1 - \frac{1}{n-1}\sum_{j=2}^{n}\Pi_1(x_j) \geq S_P(\Pi_2) = 1 - \frac{1}{n-1}\sum_{j=2}^{n}\Pi_2(x_j)$$

Thus we see that if $\alpha_1 > \alpha_2$ then $S_P(\Pi_1) \geq S_P(\Pi_2)$.

Actually we see that the interpretation V *is* A **is** α certain as inducing Π(x) = $S(A(x), \bar{\alpha})$ as an application of the entailment principle since $A(x) \leq S(A(x), \bar{\alpha})$.

While a large number of potential forms exist for the definition of the t-conorm S (Klement, Mesiar, & Paper, 2000) here we shall consider a class of the these called Sugeno (1974) - Weber (1983) t-conorms as it contains a number of important examples of t-conorms and is mathematically a well behaved family. The Sugeno-Weber t-conorms are defined such that

S(x, y) = Min[(x + y - λ x y), 1]

where λ ≤ 1. We note for λ = 1 this becomes S(x, y) = x + y – x y, the probabilistic sum. For λ = 0 this becomes S(x, y) = Min(x + y, 1), the bounded sum.

For the case of interest to us where y = 1- α and x = a then

$$S(a, \bar{\alpha}) = [(a + \bar{\alpha} - \lambda\, \bar{\alpha}\, a) \wedge 1]$$
$$= [((1-\alpha) + a(1 - \lambda(1-\alpha))) \wedge 1]$$

If $\lambda = 1$ then $S(a,\bar{\alpha}) = (1-\alpha) + \alpha\, a$, it is a weighted average of 1 and a. If $\lambda = 0$ then $S(a,\bar{\alpha}) = 1 \wedge ((1-\alpha)+a) = (\bar{\alpha}+a) \wedge 1$

POSSIBILITY-PROBABILITY TRANSFORMATIONS

With probability and possibility theory providing two main approaches to the representation of uncertainty considerable attempts have been made to provide methods for transforming a possibility distribution into a probability distribution and vice versa (Dubois & Prade, 1982, 1983; Delgado & Moral, 1987; Klir, 1989, 1990; Dubois, Prade, & Sandri, 1991; Geer & Klir, 1992; Klir & Parviz, 1992; Yager, 1993). Sambhoos, Guiffrida, and Llinas (2005) look at and compare a number of these transformations. For our purposes we shall use an approach to possibility-probability transformation that was initially suggested by Dubois and Prade (1982, 1983).

Assume P is a probability distribution on $X = \{x_1, \ldots, x_n\}$ where for convenience we indexed the elements in X so that $p_1 \geq p_2 \geq \ldots \geq p_n$. Here the elements have been indexed in descending order of probability. Dubois and Prade (1982, 1983) suggested we can translate this into a possibility distribution Π on X where π_j, the possibility of x_j, is defined such that

$$\pi_n = n\, p_n$$

$$\pi_j = j\,(p_j - p_{j-1}) + \pi_{j+1} \text{ for j = n –to 1} \qquad 1$$

If we let $p_{n+1} = 0$ we can more succinctly express the above as

$$\pi_j = \sum_{k=j}^{n} k(p_k - p_{k+1}) \text{ for all j } \mathbf{(I)}$$

From this we can also express the probability to possibility transformation as

$$\pi_j = j\; p_j + \sum_{k=j+1}^{n} p_k$$

Some properties of this transformation are the following

1. If $p_j = p_i$ then $\pi_j = \pi_i$
2. If $p_j > p_i$ then $\pi_j > \pi_i$
3. If $p_j = 0$ then $\pi_j = 0$
4. If $p_j = 1/n$ for all j then $\pi_j = 1$ for all j
5. $\pi_1 = 1$

We can get the possibility to probability transformation using the inverse of the above. Thus if we assume a possibility distribution with $\pi_1 \geq \pi_2 \geq \ldots\ldots \geq \pi_n$, here $\pi_1 = 1$, we can obtain an associated probability distribution P having

$$p_n = \frac{\pi_n}{n}$$

$$p_j = p_{j+1} + \frac{\pi_j - \pi_{j+1}}{j} \text{ for j = n -to 1} \qquad 1$$

If we denote $\pi_{n+1} = 0$ then we can more compactly express this transformation as

$$p_j = \sum_{k=j}^{n} \frac{\pi_k - \pi_{k+1}}{k} \text{ for all j = 1 to n } \mathbf{(II)}$$

An alternative form for the above is

$$p_j = \frac{\pi_j}{j} + \sum_{k=j+1}^{n} \left(\frac{\pi_k}{(k)(k-1)}\right) \text{ for all j = 1 to n}$$

We easily see $p_j \geq 0$ for all j. We can show the following properties

1. If $\pi_j = \pi_k$ then $p_j = p_k$
2. If $\pi_j > \pi_k$ then $p_j > p_k$

3. If $\pi_j = 0$ then $p_j = 0$
4. If $\pi_j = 1$ for all j then $p_j = 1/n$ for all j
5. $\Sigma_j p_j = 1$

Another method of possibility-probability transformation was suggested by Yager (1993). Here if we start with p_j then we obtain

$$\pi_j = \frac{p_j}{Max_i[p_i]}.$$

If we start with π_j then we obtain

$$p_j = \frac{\pi_j}{\sum_{i=1}^{n} \pi_i}.$$

We see this has similar properties as the preceding

1. If $p_j = p_i$ then $\pi_j = \pi_i$
2. If $p_j > p_i$ then $\pi_j > \pi_i$
3. If $p_j = 0$ then $\pi_j = 0$
4. If $p_j = 1/n$ for all j then $\pi_j = 1$ for all j
5. $Max_j[\pi_j] = 1$

Corresponding properties hold for the p_i determined from a possibility distribution.

ENTROPY CONSTRAINING PROBABILITY TRANSFORMATIONS

As we earlier indicated if Π is a possibility distribution then the entailment principle allows us to infer any other possibility distribution U such that $u_j \geq \pi_j$ for all j. Furthermore an important feature of this transformation is that Sp(Π) ≥ Sp(U), the information in U is not greater than that in Π. Thus the entailment principle allows us to obtain a new possibility distribution with less information than the original. It is a reflection of the disjunctive rule in logical reasoning where from the knowledge that *a* is true we can infer that *a or b* is true.

Yager and Kreinovich (2007) provided a method for going from one probability distribution to another probability distribution in a way that always has less information, larger entropy.

Assume we have a probability distribution P indexed so that $p_1 \geq p_2 \geq ..... \geq p_n$. We now proceed as follows:

1. We use the probability-possibility transfer **(I)** to obtain a possibility distribution Π on X such that $\pi_j = \sum_{k=j}^{n} k(p_k - p_{k+1})$
2. We apply the entailment principle on Π to obtain a new possibility distribution of U such that $u_j \geq \pi_j$ for all j.
3. Finally after reordering if necessary, we apply the possibility to probability transformation **(II)** to obtain a new probability distribution Q on X such that $q_i = \sum_{k=i}^{n} \frac{U_k - U_{k+i}}{k}$.

In Yager and Kreinovich (2007) it was shown that H(P) ≤ H(Q) thus Q always has less information than P. We can make some further observations about this transformation of P into Q. First we see there always exists the transformation of P to the distribution Q where $q_i = 1/n$ for i. This is obtained by simply making all the $u_i = 1$. A second property is that if x_i has the largest probability in P, then it will also have the largest probability in Q. We see this as follows, if p_1 is the largest probability then π_1 be equal to one. Since u_j is a possibility value then for all elements $u_j \leq 1$. Hence u_1 will still be one and therefore x_1 will be the largest possibilities in U and hence it will have one of the largest probabilities in Q.

ON THE RELATIONSHIP BETWEEN INPUT AND OUTPUT PROBABILITIES

As we have indicated we can start with a probability distribution P then using the transformation rule (I) to obtain a possibility distribution Π. We also indicated if we modify Π to obtain a new possibility distribution U such that $u_j \geq \pi_j$ for all j and then transform the possibility distribution U into a probability distribution Q then the entropies of P and Q are related so that H(Q) ≥ H(P), we have generated a probability with less information than the original. Here we want to look at the probability distributions resulting from a particular type of modification of the possibility distribution Π to U.

Let Π be a possibility distribution with components π_j and let U be a possibility distribution with components $u_j = S(\beta, \pi_j)$ where S is a t-conorm (Beliakov, Pradera, & Calvo 2007). Since $S(\beta, \pi_j) \geq \pi_j$ then this is an application of the entailment principle. Furthermore the probability distribution Q obtained from U will have more entropy. Here we shall consider using a special class of t-conorms called the Sugeno-Weber t-conorm (Sugeno, 1977; Weber, 1983) which as we earlier noted contain some important special members and is generally mathematically tractable. In this case

$$u_j = Min[(\pi_j + \beta - \lambda\beta\pi_j), 1] = (\pi_j + \beta - \lambda\beta\pi_j)\wedge 1$$

where $\lambda \leq 1$. We note as special cases of the family of t-conorms the following

$$\lambda = 1 \quad u_j = (\pi_j + \beta - \beta\pi_j) \wedge 1$$

$$\lambda = 0 \quad u_j = (\pi_j + \beta) \wedge 1$$

Our objective here is to start with the probability P use transformation rule I to get the possibility distribution Π then modify Π using the t-conorm S to obtain $u_j = (\pi_j + \beta - \lambda\beta\pi_j) \wedge 1$ and then apply the retransformation into a probability distribution to get q_j. We are interested in the relationship between p_j and q_j. Before proceeding we observe that

$$\pi_j + \beta - \lambda\beta\pi_j = \beta + (1 - \lambda\beta)\pi_j.$$

In the following we shall find it convenient to denote $(1 - \lambda\beta) = \delta$ and look at the form $u_j = ((\beta + \delta\pi_j) \wedge 1)$.

Here we start with the p_j which we will assume are indexed in decreasing order, $p_i \geq p_k$ if $i < k$. Furthermore for convenience we add the phantom probability $p_{n+1} = 0$. From the original p_j we obtain $\pi_j = jp_j + \sum_{k=j+1}^{n} p_k$ for j = 1 to n + 1. We note that $\pi_{n+1} = 0$. We also observe that the elements in Π are ordered so that $\pi_i \geq \pi_k$ if $i < k$. We now modify these possibility values using the Sugeno-Weber t–conorm to obtain $u_j = (\beta + \delta\pi_j) \wedge 1$.

We observe that the u_j are also ordered so that $u_i \geq u_k$ if $i < k$. We now let L be the largest index such that $u_L = 1$. In this case we have

$u_j = 1$ for j = 1 to L

$u_j = \beta + \delta\pi_j$ for $j \geq L + 1$

We also observe that since for $j \geq L + 1$ we have $\pi_j = jp_j + \sum_{k=j+1}^{n} p_k$ then for $j \geq L + 1$ we get

$$u_j = \beta + \delta(jp_j + \sum_{k=j+1}^{n} p_k)$$

Let us now look at the probabilities q_j that resulted from applying the possibility to probability transformation to the u_j

$$q_j = q_{j+1} + \frac{u_j - u_{j+1}}{j} \text{ for } j \leq n - 1$$

$$q_n = \frac{u_n}{n}$$

We first see that

$$q_n = \frac{\beta + \delta\pi_n}{n} = \frac{\beta + \delta(np_n + \sum_{k=n+1}^{n} p_k)}{n} = \frac{\beta}{n} + \delta p_n$$

$$q_{n-1} = q_n + \frac{u_{n-1} - u_n}{n-1} = \frac{\beta}{n} + \delta p_n + \frac{\delta(n-1)p_{n-1} + p_n - np_n}{n-1}$$

$$q_{n-1} = \frac{\beta}{n} + \delta p_n + \delta(p_{n-1} - p_n) = \frac{\beta}{n} + \delta p_{n-1}$$

If we conjecture $q_i = \frac{\beta}{n} + \delta p_i$ for $i > L+1$ then for $j > L$ we get

$$q_j = q_{j+1} + \frac{u_j - u_{j+1}}{j}$$

$$= \frac{\beta}{n} + \delta p_{j+1} + \frac{\delta(jp_j + \sum_{k=j+1}^{n} p_k - (j+1)(p_{j+1} + \sum_{k=j+2}^{n} p_k)}{j}$$

After some arithmetic manipulations we get

$$q_j = \frac{\beta}{n} + \delta p_{j+1} + \frac{\delta(jp_j - jp_{j+1})}{j} = \frac{\beta}{n} + \delta p_j$$

Thus by induction we have shown that $q_j = \frac{\beta}{n} + \delta p_j$ for $j \geq L+1$

We now consider $q_L = q_{L+1} + \frac{u_L - u_{L+1}}{L}$. Since we have just shown $q_{L+1} = \frac{\beta}{n} + \delta p_{L+1}$ and have previously obtained $u_L = 1$ and $u_{L+1} = \beta + \delta\pi_{L+1}$. Then here

$$q_L = \frac{\beta}{n} + \delta p_{L+1} + \frac{(1 - (\beta + \delta(L+1)(p_{L+1} + \sum_{k=L+2}^{n} p_k)))}{L}$$

$$q_L = \frac{1}{L} - (\frac{\beta}{L} - \frac{\beta}{n}) + \frac{\delta}{L}(L(p_{L+1} - (L+1)p_{L+1} - \sum_{k=L+2}^{n} p_k))$$

$$q_L = \frac{1}{L}(1 - \frac{\beta}{n}(n - L) - \delta \sum_{k=L+1}^{n} p_k)$$

Furthermore since for $j < L$ we have $q_j = q_{j+1} + \frac{u_j - u_{j+1}}{j}$ but since $u_j = u_{j+1}$ then $q_j = q_{j+1}$

Thus we see for $j < L$ that all $q_j = q_L = \frac{1}{L}(1 - \frac{\beta}{n}(n - L) - \delta \sum_{k=L+1}^{n} p_k$

Let us assure ourselves that we have a normalized probability distribution

$$\sum_{j=1}^{n} q_j = \sum_{j=1}^{L} q_L + \sum_{j=L+1}^{n} (\frac{\beta}{n} + \delta p_j)$$

$$\sum_{j=1}^{n} q_j = L\frac{1}{L}(1 - \frac{\alpha}{n}(n - L) - \delta \sum_{k=L+1}^{n} p_k + \frac{(n - L)\beta}{n} + \delta \sum_{j=L+1}^{n} p_j = 1$$

Thus here we have modified the probability for $j > L$ as $q_j = \frac{\beta}{n} + \delta p_j$ and divided the remaining probability equally between q_j for $j \leq L$. We further observe that since $\delta = 1 - \lambda\beta$ then for $j > L$ we have $q_j = \beta\frac{1}{n} + (1 - \lambda\beta)p_j$. We now see that these q_j's are a kind of "pseudo" weighted combination of $\frac{1}{n}$ and p_j. As a matter of fact if $\lambda = 1$ then we get exactly that

$q_j = \beta \frac{1}{n} + (1 - \beta)p_j$. In the case where $\lambda = 0$ we get $q_j = \beta \frac{1}{n} + p_j$. Here we add amount $\beta \frac{1}{n}$ to p_j. We further point out that when $\lambda = 1$, we have $u_j = 1 \wedge (\beta + \pi_j - \beta\pi_j)$. Since $\beta + \pi_j - \beta\pi_j \leq 1$ for all β and $\pi_j \in [0, 1]$ then $u_j = \beta + \pi_j - \beta\pi_j \leq 1$ and hence

$$q_j = \beta \frac{1}{n} + (1 - \beta)p_j \text{ for all j.}$$

We here want to make some further observation regarding the effects of λ. Consider we have some normal possibility distribution Π, with components π_j. Here we assume again the indexing is such that $\pi_1 = 1$ and $\pi_i \geq \pi_j$ for i < j. In this case consider the modification

$$u_j = 1 \wedge (\pi_j + \beta - \lambda\beta\pi_j) = 1 \wedge (\beta + (1 - \lambda\beta)\pi_j)$$

where $\lambda \leq 1$. We see that for fixed β, u_j gets larger as λ gets smaller. Since the specificity or information in U is expressed as $S_P(U) = 1 - \sum_{j=2}^{n} u_j$ we see that the smaller λ is the less information we have in U.

If P is the probability distribution which we transform into a new probability distribution Q the procedure just discussed is

$P \Rightarrow \Pi \Rightarrow$ modify the $\pi_j \Rightarrow u_j \Rightarrow q_j$

We recall

$$q_j = \frac{1}{n}\beta + (1 - \lambda\beta)p_j \text{ for } j > L$$

$$q_j = \frac{1}{L}(1 - \sum_{j=L+1}^{n} p_j) \text{ for } j \leq L$$

where L is the first place where $u_j = 1 \wedge (\beta + (1 - \lambda\beta)\pi_j) \geq 1$. We note that since the π_j are such that $\pi_j \geq \pi_k$ for j < k then $u_j \geq u_k$ for j < k. We see that as λ decreases the value of L increases. This means that more of the q_j are equal. The essential effect here of decreasing λ decreases is to bring an increase in the entropy.

MULTI-SOURCE FUSION

We now turn to the issue of the fusion of multiple sources of information, some of which are expressed in terms of possibility distributions and others are expressed in terms of probability distributions. Here we shall investigate one approach to this fusion problem.

Assume we have r_1 pieces of information expressed in terms of a possibility distribution. We denote these as, V *is* Π_i for i = 1 to r_1 Here for each possibility distribution Π_i we denote $\Pi_i(x_j)$ as the possibility associated with the value x_j. In addition we have r_2 pieces of data expressed in terms of probability distributions. We denote this V *is* P_i for $i = r_1 + 1$ to r where $r = r_1 + r_2$. Here for each probability distribution we denote $P_i(x_j)$ as the probability associated with x_j. We note here for each P_i we have $\sum_{j=1}^{n} P_i(x_j) = 1$ and for each Π_i we have $Max_j[\Pi_i(x_j)] = 1$. Each of the pieces of information is normalized.

We now suggest an approach to the fusion of these multiple pieces of information.

1. Convert each probabilistically expressed information into a possibilistic type of uncertainty using the transformation rule I. We shall denote these converted distributions as V *is* Π_i for i = r_{i+1} to r.

Using the available "and", conjunction, operation in possibility theory obtain Π the conjunction of all r possibility distributions. In particular we get

$$\Pi(x_j) = Min_i[\Pi_i(x_j)]$$

Once having the possibility distribution Π we can provide it to the user or we can transform it into a fused probabilistic distribution using transformation (II).

Example: *Assume V is a variable that takes value in the space X= {x_1, x_2, x_3, x_4, x_5}. Assume we have two pieces of information expressing our knowledge about the value of V. One of these uses a possibilistic representation and the other uses a probabilistic representation. In the case of the possibilistic information we have*

$\Pi_1(x_1) = 1$, $\Pi_1(x_2) = 0.2$, $\Pi_1(x_3) = 1$, $\Pi_1(x_4) = 0.1$ and $\Pi_1(x_5) = 0.8$

In the case of the probabilistic representation we have:

$P(x_1) = 0.4$, $P(x_2) = 0.3$, $P(x_3) = 0.15$, $P(x_4) = 0.1$ and $P(x_5) = 0.05$

Using transformation rule I we get the associated possibility distribution Π_2 such that

$\Pi_2(x_1) = 1$, $\Pi_2(x_2) = 0.9$, $\Pi_2(x_3) = 0.6$, $\Pi_2(x_4) = 0.45$ and $\Pi_2(x_5) = 0.25$

Fusing these two pieces of information we get a possibility distribution Π such that $\Pi(x_j) = Min[\Pi_1(x_j), \Pi_2(x_j)]$, In this case our fused possibility distribution is

$\Pi(x_1) = 1$, $\Pi(x_2) = 0.2$, $\Pi(x_3) = 0.6$, $\Pi(x_4) = 0.1$ and $\Pi(x_5) = 0.25$

If we like we can convert this into a probability distribution P* in which

$P^*(x_1) = 0.637$, $P^*(x_2) = 0.045$, $P^*(x_3) = 0.237$, $P^*(x_4) = 0.02$ and $P^*(x_5) = 0.061$

In implementing this fusion process one problem that can arise is that the resulting possibility distribution Π is not normalized. This would occur in the above example if $\Pi_1(x_1) \neq 1$. In this case we have $Max_j[\Pi(x_j)] < 1$. One approach to address this problem is to uniformly reduce the confidence we assign to the distributions being fused so as to obtain a resulting normalized fused distribution.

Thus for each piece of data that is being fused we reduce its confidence to α. Here as we discussed earlier the effect of this is that for each of the r constituent possibility distributions Π_i we formulate a new possibility distribution U_i such that

$$U_i(x_j) = S_\lambda(1-\alpha,\ \Pi_i(x_j)]$$

where α is the confidence we shall assign to each of the possibility distributions and S_λ is a Sugeno-Weber type t-conorm with parameter $\lambda \leq -1$. Thus if we denote $\beta = 1 - \alpha$, as the amount we reduce the certainty, we obtain $U_i(x_j) = Min(1, (\Pi_i(x_j) + \beta - \lambda\beta\Pi_i(x_j)))$. Thus we have relaxed the constraints imposed by each of the pieces of data being fused. Initially we shall assume the value for λ has been selected, subsequently we shall discuss the method for selecting the value of λ.

Using this transformation we obtain a new fused possibility distribution U such that

$$U(x_j) = Min_i[S_\lambda(\beta,\ \Pi_i(x_j))] = Min_i[1 \wedge (\Pi_i(x_j) + \beta - \lambda\beta\Pi_i(x_j))]$$

As noted our objective in performing this confidence qualification is to obtain a normalized fused possibility distribution. That is we need at least one x_j such that U(x$_j$) = 1. Let us look at the properties of the $U(x_j)$. Because of the monotonicity of the t-conorm operator we see that $U_i(x_j) \geq U_i(x_k)$ if $\Pi_i(x_j) > \Pi_i(x_k)$. From this it follows that

$$U(x_j) = Min_i[U_i(x_j)] = S_\lambda((\beta, Min_i[\Pi_i(x_j))$$

Thus $U(x_j)$ is determined by the minimal value of $\Pi_i(x_j)$ for all i. The least reduction in confidence, the largest α, is obtained by finding the $Max_j[U(x_j)]$ and then selecting α = 1 – λ, so that $Max_j[U(x_j)]$ equals one. In particular the $Max_j[U(x_j)]$ are obtained by first finding ordering $Max_j[Min_i[\Pi_i(x_j)]]$.

Thus we must find the value x_j that has the largest intersection of $\Pi_i(x_j)$. Let us denote $\Pi^* = Max_j[Min_i[\Pi_i(x_j)]]$, it is the maximum possibility value of the conjunction of all the distribution Π_i. Once we have this we require that

$$S_\lambda(1-\alpha, \Pi^*) = S_\lambda(\beta, \Pi^*) = 1$$

In particular we require that $\Pi^* + \beta - \lambda\beta\Pi^* = 1$. Here then $(1-\lambda\Pi^*)\beta = 1-\Pi^*$ and hence

$$\beta = \frac{1 - \Pi^*}{1 - \lambda\Pi^*}$$

With β = 1 - α then our required confidence must satisfy $1-\alpha = \frac{1 - \Pi^*}{1 - \lambda\Pi^*}$ and hence

$$\alpha = 1 - \frac{1 - \Pi^*}{(1 - \lambda\Pi^*)} = \frac{(1 - \lambda)\Pi^*}{1 - \lambda\Pi^*} = \frac{\Pi^* - \lambda\Pi^*}{1 - \lambda\Pi^*} \in [0,1]$$

We observe that $\frac{\partial\alpha}{\partial\Pi^*} = \frac{1 - \lambda}{(1 - \lambda\Pi^*)^2}$ and $\frac{\partial\alpha}{\partial\Pi^*} \geq 0$ since $\lambda \leq 1$. Thus the bigger Π^* is the less we have to reduce the confidence.

Note: *We observe that for the case where $\lambda = 1$ then we let $\beta + (1 - \beta)\Pi^* = 1$. We see that if $\Pi^* \neq 1$ then the only way to get this equal to one is for $\beta = 1$, that is $\alpha = 0$.*

What we have just shown is that for any value of λ, the appropriate value for the maximal certainty, α, we can assign to the data and still obtain a normalized fused possibility distribution is $\alpha = \frac{\Pi^*(1-\lambda)}{1-\lambda\Pi^*}$. The next question is what value should we select for λ. We note Π^* is independent of λ. A reasonable criterion for selecting the value of λ is to obtain a fused possibility distribution with the most information. That is we want to select λ to maximize the specificity of U. Let us now look at the specificity of U, Sp(U).

For simplicity in the following discussion we shall assume that Π^* occurs as x_1. In this case we see

$$Sp(U) = 1 - \frac{1}{n-1}\sum_{j=2}^{n} U(x_j)$$

The largest value of Sp(U) is obtained by making each of $U(x_j)$ as small as possible.

With $T_1(x_j) = Min_i[\Pi_i(x_j)]$ we have

$$U(x_j) = S_\lambda(\beta, \Pi(x_j)) = (\beta + \Pi(x_j) - \lambda\beta\Pi(x_j)) \wedge 1$$

for $\lambda \leq 1$. Here $\beta = \frac{1 - \Pi^*}{1 - \Pi^*\lambda}$. Note that $0 < \beta \leq 1$ since $0 < 1 - \Pi^* \leq 1 - \lambda\Pi^*$. Now to find λ to minimize

$$U(x_j) = \Pi(x_j) + \frac{1 - \Pi^*}{1 - \lambda\Pi^*} - \lambda\Pi(x_j)\frac{(1 - \Pi^*)}{1 - \lambda\Pi^*}) = \frac{\Pi(x_j)(1 - \lambda) + (1 - \Pi^*)}{1 - \lambda\Pi^*}$$

We take the derivative of $U(x_j)$ with respect to λ

$$\frac{dU(x_j)}{d\lambda} = \frac{(\Pi^* - \Pi(x_j))(1 - \Pi^*)}{(1 - \Pi^*\lambda)^2}$$

Since

$$\Pi^* > \Pi(x_j) \text{ and } 1 > \Pi(x_j) \geq \Pi(x_j)\lambda$$

then $\frac{dU(x_j)}{d\lambda} > 0$ thus $U(x_j)$ is an increasing function of λ. In this case the minimum value of $U(x_j)$ is obtained by making λ as small as possible, $\lambda \to -\infty$. We see using L'Hopitals rule that

$$\underset{\lambda \to -\infty}{Lim}\, U(x_j) = \underset{\lambda \to -\infty}{Lim} \frac{\Pi(x_j)(1-\lambda)+(1-\Pi^*)}{1-\lambda\Pi^*} = \frac{\Pi(x_j)}{\Pi^*}$$

Thus the greatest lower bound of $U(x_j)$ is

$$\frac{\Pi(x_j)}{\Pi^*} = \frac{\Pi(x_j)}{Max_i(\Pi(x_j))}$$

Thus here we see the maximal entropy normalization is to let $\lambda \to \infty$. In this case we get as our normalized possibility distribution U such that $U(x_j) = \frac{\Pi(x_j)}{Max_i(\Pi(x_j))}$.

CONCLUSION

We discussed the basic features of information provided in terms of possibilistic uncertainty. We pointed out the entailment principle, a tool that allows one to infer less specific from a given piece of information. We also discussed the problem of fusing multiple pieces of possibilistic information. The basic features of probabilistic information were described. We described a procedure for transforming information between possibilistic and probabilistic representations. We used this to form the basis for a procedure for fusing multiple pieces of uncertain information some of which is possibilistic and some probabilistic. We provided a procedure for addressing the problems that arise when the information to be fused has some conflicts.

REFERENCES

Beliakov, G., Pradera, A., & Calvo, T. (2007). *Aggregation functions: A guide for practitioners*. Heidelberg, Germany: Springer-Verlag.

Delgado, M., & Moral, S. (1987). On the concept of possibility-probability consistency. *Fuzzy Sets and Systems*, *21*, 311–318. doi:10.1016/0165-0114(87)90132-1

Dubois, D., & Prade, H. (1982). On several representations of an uncertain body of evidence. In Gupta, M. M., & Sanchez, E. (Eds.), *Fuzzy information and decision processes* (pp. 167–181). Amsterdam, The Netherlands: North-Holland.

Dubois, D., & Prade, H. (1983). Unfair coins and necessary measures: A possible interpretation of histograms. *Fuzzy Sets and Systems*, *10*, 15–20. doi:10.1016/S0165-0114(83)80099-2

Dubois, D., Prade, H., & Sandri, S. (1991). On possibility/probability transformations. In *Proceedings of the Fourth IFSA Conference* (pp. 50-53).

Geer, J. F., & Klir, G. J. (1992). A mathematical analysis of information-preserving transformations between probabilistic and possibilistic formulations of uncertainty. *International Journal of General Systems*, *20*, 143–176. doi:10.1080/03081079208945024

Klement, E. P., Mesiar, R., & Pap, E. (2000). *Triangular norms*. Dordrecht, The Netherlands: Kluwer Academic.

Klir, G. J. (1989). Probability-possibility conversion. In *Proceedings of the Third IFSA Congress* (pp. 408-411).

Klir, G. J. (1990). A principle of uncertainty and information invariance. *International Journal of General Systems*, *17*, 249–275. doi:10.1080/03081079008935110

Klir, G. J. (2006). *Uncertainty and information*. New York, NY: John Wiley & Sons.

Klir, G. J., & Parviz, B. (1992). Probability-possibility transformations: A comparison. *International Journal of General Systems*, *21*, 291–310. doi:10.1080/03081079208945083

Sambhoos, K., Guiffrida, A. L., & Llinas, J. (2005). *Research on possibility-probability transformations in support of innovative approaches to fusion 2*. Buffalo, NY: University of Buffalo.

Sugeno, M. (1974). *Theory of fuzzy integrals and its application*. Unpublished doctoral dissertation, Tokyo Institute of Technology, Tokyo, Japan.

Sugeno, M. (1977). Fuzzy measures and fuzzy integrals: A survey. In Gupta, M. M., Saridis, G. N., & Gaines, B. R. (Eds.), *Fuzzy automata and decision process* (pp. 89–102). Amsterdam, The Netherlands: North-Holland.

Weber, S. (1983). A general concept of fuzzy connectives, negations and implications based on t-norms. *Fuzzy Sets and Systems*, *11*, 115–134. doi:10.1016/S0165-0114(83)80073-6

Yager, R. R. (1985). On the question of credibility of evidence for expert systems. In *Proceedings of the Conference on the Analysis of Decision Problems within an Uncertain and Imprecise Environment* (pp. 1-15).

Yager, R. R. (1986). The entailment principle for Dempster-Shafer granules. *International Journal of Intelligent Systems*, *1*, 247–262. doi:10.1002/int.4550010403

Yager, R. R. (1993). Element selection from a fuzzy subset using the fuzzy integral. *IEEE Transactions on Systems, Man, and Cybernetics*, *23*, 467–477. doi:10.1109/21.229459

Yager, R. R. (1998). On measures of specificity. In Kaynak, O., Zadeh, L. A., Turksen, B., & Rudas, I. J. (Eds.), *Computational intelligence: Soft computing and fuzzy-neuro integration with applications* (pp. 94–113). Berlin, Germany: Springer-Verlag. doi:10.1007/978-3-642-58930-0_6

Yager, R. R., & Kreinovich, V. (2007). Entropy conserving transforms and the entailment principle. *Fuzzy Sets and Systems*, *158*, 1397–1405. doi:10.1016/j.fss.2007.01.019

Zadeh, L. A. (1979). A theory of approximate reasoning. In Hayes, J., Michie, D., & Mikulich, L. I. (Eds.), *Machine intelligence 9* (pp. 149–194). New York, NY: Halstead Press.

Zadeh, L. A. (1996). Fuzzy logic = computing with words. *IEEE Transactions on Fuzzy Systems*, *4*, 103–111. doi:10.1109/91.493904

Zadeh, L. A. (1999). Outline of a computational theory of perceptions based on computing with words. In Sinha, N. K., & Gupta, M. M. (Eds.), *Soft computing and intelligent systems* (pp. 3–22). Boston, MA: Academic Press.

Zadeh, L. A. (2005). Toward a generalized theory of uncertainty (GTU)-An outline. *Information Sciences*, *172*, 1–40. doi:10.1016/j.ins.2005.01.017

This work was previously published in the International Journal of Fuzzy System Applications, Volume 1, Issue 3, edited by Toly Chen, pp. 1-14, copyright 2010 by IGI Publishing (an imprint of IGI Global).

Chapter 6
Applying a Fuzzy and Neural Approach for Forecasting the Foreign Exchange Rate

Toly Chen
Feng Chia University, Taiwan

ABSTRACT

Accurately forecasting the foreign exchange rate is important for export-oriented enterprises. For this purpose, a fuzzy and neural approach is applied in this study. In the fuzzy and neural approach, multiple experts construct fuzzy linear regression (FLR) equations from various viewpoints to forecast the foreign exchange rate. Each FLR equation can be converted into two equivalent nonlinear programming problems to be solved. To aggregate these fuzzy foreign exchange rate forecasts, a two-step aggregation mechanism is applied. At the first step, fuzzy intersection is applied to aggregate the fuzzy forecasts into a polygon-shaped fuzzy number to improve the precision. A back propagation network is then constructed to defuzzify the polygon-shaped fuzzy number and generate a representative/crisp value to enhance accuracy. To evaluate the effectiveness of the fuzzy and neural approach, a practical case of forecasting the foreign exchange rate in Taiwan is used. According to the experimental results, the fuzzy and neural approach improved both the precision and accuracy of the foreign exchange rate forecasting by 79% and 81%, respectively.

INTRODUCTION

The foreign-exchange rate between two currencies specifies how much one currency is worth in terms of the other. There are three types of exchange transactions: spot transaction, forward transaction, and swap transaction. Accurately forecasting the foreign exchange rate is very important for export-oriented enterprises in Taiwan. Unfavorable foreign exchange rates also result in the increase of raw material costs and the decrease of gross margin for these enterprises.

The fluctuation in the foreign exchange rate can be treated as a type of time series. Theoretically there are many approaches, e.g. moving average (MA), weighted moving average (WMA), exponential smoothing (ES), linear regression (LR), artificial neural network (ANN), auto-regressive

DOI: 10.4018/978-1-4666-1870-1.ch006

integrated moving average (ARIMA), and others that can be applied to forecast the foreign exchange rate. Recently, Tseng et al. (2001) proposed fuzzy ARIMA for this purpose. In this study, Chen and Lin's hybrid fuzzy linear regression (FLR) and back propagation network (BPN) approach (Chen & Lin, 2008) is applied in this study. The FLR-BPN approach is a general approach like Mamdani's or Takagi-Sugeno's fuzzy inference systems, and can be applied to forecast any phenomena in various fields of research or applications, e.g. semiconductor yield forecasting (Chen & Lin, 2008), job cycle time estimation (Chen, 2009), the book-to-bill ratio forecasting (Chen & Wang, 2010), etc.

Theoretically, FLR problems can be classified into four categories, according to whether the inputs and outputs are fuzzy or not (D'Urso, 2003). Traditionally, there are two ways of solving an FLR problem: linear programming (LP) methods and fuzzy least-squares methods. The first one attempts to minimize the average fuzziness (Tanaka & Watada, 1988) or to maximize the average membership (Peters, 1994) given that the fuzzy forecast contains the actual value to a certain degree. Redden and Woodall (1994) compared various FLR methods and discussed the differences between fuzzy and traditional regression approaches. Some researchers (e.g. Chang & Lee, 1994; Donoso et al., 2006) pointed out the weakness of Tanaka's approach, which did not consider the optimization of the central tendency. In addition, Tanaka's approach usually derived a high number of crisp estimates. Bardossy (1990) used different fuzziness measures. The second way tries to minimize the sum of the square of residuals, which is quadratic in nature (Diamond, 1988; Tanaka & Lee, 1998). Recently, Donoso et al. (2006) constructed a quadratic non-possibilistic (QNP) model in which the quadratic error for both the central tendency and each one of the spreads is minimized.

On the other hand, a BPN (or a feed forward neural network) is a kind of artificial neural network with three types of layers – the input layer, the hidden layer, and the output layer. A BPN is frequently used to imitate the relationship among the inputs and outputs/responses of a complex system (Hsiao et al., 2008; Chen, 2009). Each layer in a BPN contains some neurons. Each neuron receives a signal from the neurons in the previous layer. These signals are multiplied by their weights. The weighted inputs are summed up, and then the result is scaled into a fixed interval before it is outputted to the neurons of the next layer. Finally, the network output can be compared with the actual response for which the deviation is calculated. Such a deviation is then passed backward throughout the network to adjust the weights/parameters. That's why it is called a back propagation network.

BPNs have been universally applied to forecast various phenomena in many research fields. For example, Piramuthu (1991) evaluated the financial credit risk with a BPN. Chang et al. (2005) and Chang and Hsieh (2003) both forecasted the cycle time of every job in a semiconductor manufacturing factory with the BPN approach. Li et al. (2006) constructed a BPN for network flow forecasting and diagnosis. Al-Deek (2007) combined a BPN and time series to forecast the inbound and outbound movements of heavy trucks at a seaport. The advantages of a BPN include the tolerance of noise, the speed of the application, and the capability of simulating complex systems (Piramuthu, 1991). FLRs and BPNs are linear and nonlinear in nature, respectively, and therefore their combination has the potential for dealing with data with hybrid or unclear patterns.

In the fuzzy and neural approach, multiple experts (or decision makers) construct their own FLR equations from various viewpoints to forecast the foreign exchange rate. Each FLR equation is then converted into two equivalent nonlinear programming (NP) problems to be solved. The future foreign exchange rate forecasted with an FLP equation is a fuzzy value. The fuzzy foreign exchange rate forecasts by different experts might not be equal and therefore need to be aggregated. Therefore, a two-step aggregation mechanism is applied. At the first step, fuzzy intersection (FI) is applied to aggregate the fuzzy forecasts by different experts into a polygon-shaped fuzzy number, in order to improve the precision. Afterwards, considering the special shape of the polygon-shaped fuzzy number, a BPN is constructed to defuzzify the polygon-shaped fuzzy number and to generate a representative/crisp value, in order to enhance the accuracy.

The rest of this paper is organized as follows. The fuzzy and neural approach that is composed of three steps is introduced. A practical example is used to demonstrate the application of the fuzzy and neural approach. The performance of the fuzzy and neural approach is evaluated and compared with those of some existing approaches. Based on the analysis results, some points are made. Finally, the concluding remarks and some directions for future research are given.

THE FLR-BPN APPROACH

The FLR-BPN approach has three steps which will be described in the following sections:

1. Constructing multiple FLR equations to generate fuzzy foreign exchange rate forecasts. Each FLR equation is converted into two equivalent NP models.
2. Applying FI to aggregate fuzzy foreign exchange rate forecasts into a polygon-shaped fuzzy number.
3. Defuzzify the aggregation result with a BPN.

Step 1: Constructing Multiple FLR Equations

In the FLR-BPN approach multiple experts construct their own FLR equations to forecast the foreign exchange rate of a future day.

$$\tilde{y}_n = \tilde{w}_0(+)\sum_{k=1}^{K}\tilde{w}_k \cdot y_{n-k}, \quad (1)$$

$$y_n = D(\tilde{y}_n), \quad (2)$$

where $\tilde{y}_n$ is the fuzzy foreign exchange rate forecast of day n; $\tilde{w}_k$ are constants or coefficients, $k = 0 \sim K$; (+) denotes fuzzy addition; $D()$ is the defuzzification function. Assuming all variables are given in triangular fuzzy numbers (TFNs):

$$\tilde{y}_n = (y_{n1}, y_{n2}, y_{n3}), \quad (3)$$

$$\tilde{w}_k = (w_{k1}, w_{k2}, w_{k3}). \quad (4)$$

In Chen and Lin's approach, some experts are asked to submit their opinions about the fuzzy forecasts. Afterwards, equation (1) can be fitted by solving the following two NP models:

(Model NP I)

$$\text{Min } Z_1 = \sum_{n=1}^{N} d_n^{o_l} \quad (5)$$

subject to

$$d_n = w_{03} + \sum_{k=1}^{K} w_{k3} y_{n-k} - w_{01} - \sum_{k=1}^{K} w_{k1} y_{n-k}, \quad (6)$$

$$y_n \geq w_{01} + \sum_{k=1}^{K} w_{k1} y_{n-k} + s_l(w_{02} + \sum_{k=1}^{K} w_{k2} y_{n-k} - w_{01} - \sum_{k=1}^{K} w_{k1} y_{n-k}), \quad (7)$$

$$y_n \leq w_{03} + \sum_{k=1}^{K} w_{k3} y_{n-k} + s_l(w_{02} + \sum_{k=1}^{K} w_{k2} y_{n-k} - w_{03} - \sum_{k=1}^{K} w_{k3} y_{n-k}), \quad (8)$$

$n = 1 \sim N$,

$$w_{k1} \leq w_{k2} \leq w_{k3}, \quad (9)$$

$k = 0 \sim K$.

(Model NP II)

$$\text{Max } Z_2 = \bar{s} \tag{10}$$

subject to

$$Z_1 \leq (T - K)d_l^{o_l}, \tag{11}$$

$$y_n \geq w_{01} + \sum_{k=1}^{K} w_{k1} y_{n-k} + s_n(w_{02} + \sum_{k=1}^{K} w_{k2} y_{n-k} - w_{01} - \sum_{k=1}^{K} w_{k1} y_{n-k}), \tag{12}$$

$$y_n \leq w_{03} + \sum_{k=1}^{K} w_{k3} y_{n-k} + s_n(w_{02} + \sum_{k=1}^{K} w_{k2} y_{n-k} - w_{03} - \sum_{k=1}^{K} w_{k3} y_{n-k}), \tag{13}$$

$$\bar{s} = \sqrt[m_l]{\frac{\sum_{n=1}^{N} s_n^{m_l}}{N}}, \tag{14}$$

$$0 \leq s_n \leq 1, \tag{15}$$

$i = 1 \sim N$,

$$w_{k1} \leq w_{k2} \leq w_{k3}, \tag{16}$$

$k = 0 \sim K$.

where y_n is the actual foreign exchange rate of day n; N is the number of days; K indicates the number of moving periods; o_l reflects the sensitivity of expert l to the uncertainty of the fuzzy foreign exchange rate forecast; o_l ranges from 0 (not sensitive) to ∞ (extremely sensitive); s_l indicates the satisfaction level required by expert l; $0 \leq s_l \leq 1$; d_l is the desired range of every fuzzy foreign exchange rate forecast by expert l; $0 \leq d_l$; m_l represents the relative importance of the outliers in fitting the FLR equation to expert l; $m_l \in Z^+$. When $m_l = 1$, the relative importance of the outliers is the highest and is equal to that of the non-outliers; $l = 1 \sim L$ (the number of experts). o_l should be set within [0 1] if the variation in the variable is less than 1, otherwise, it should be greater than 1. In this way, any difference from the optimal objective function value will be magnified, which improves the sensitivity.

NP model I is based on Tanaka and Watada's philosophy (Tanaka & Watada, 1988). The objective function is to minimize the high-order sum of the ranges of fuzzy foreign exchange rate forecasts. Constraints (7) and (8) request that the membership of the actual value in the fuzzy foreign exchange rate forecast has to be greater than or equal to s. Constraint (9) shows the sequence of the three corners of a TFN. This NP model has one objective, $N + 3k + 3$ variables, and $3N + 2k + 2$ constraints.

On the other hand, NP model II is based on Peters's philosophy (Peters, 1994). In Constraint (14), the generalized average formula is applied to calculate the average satisfaction level, so as to lessen the relative importance of the outliers. This NP model has one objective, $2N + 3k + 3$ variables, and $5N + 2k + 4$ constraints. Theoretically, these nonlinear objective functions and constraints can be easily converted into quadratic ones that are more tractable. For example,

$\text{Min } Z_1 = \sum_{n=1}^{N} d_n^4$ can be replaced with

$$\text{Min } Z_1 = \sum_{n=1}^{N} g_n^2, \tag{17}$$

$$g_n = d_n^2. \tag{18}$$

In either model, every fuzzy foreign exchange rate forecast contains the actual value. As a result, the intersection of the two fuzzy foreign exchange rate forecasts generated by the two models also contains the actual value. Besides, the intersection has a range narrower than those of the two regions. Therefore, the forecasting precision measured in terms of the average range is indeed improved after intersection. If there are L experts, then after incorporating experts' opinions into the two NP models there will be at most $2L$ NP problems to be solved. After solving each NP problem, the optimal solution is used to construct a corresponding FLR equation. Eventually, there will be at most $2L$ FLR equations, each of which generates a fuzzy foreign exchange rate forecast. A mechanism is therefore required to aggregate these fuzzy foreign exchange rate forecasts. In the fuzzy and neural approach, a two-step aggregation mechanism is applied.

Step 2: Applying FI to Aggregate the Fuzzy Foreign Exchange Rate Forecasts

The aggregation mechanism is composed of two steps. In the first step, FI is applied to aggregate the fuzzy foreign exchange rate forecasts into a polygon-shaped fuzzy number, in order to improve the precision of the foreign exchange rate forecasting. FI aggregates n fuzzy foreign exchange rate forecasts in the following manner:

$$\mu_{I(\tilde{y}_i(1),\ \tilde{y}_i(2),\ \ldots,\ \tilde{y}_i(L))}(x) = \min(\mu_{\tilde{y}_i(1)}(x),\ \mu_{\tilde{y}_i(2)}(x),\ \ldots,\ \mu_{\tilde{y}_i(L)}(x)), \quad (19)$$

where $I(\tilde{y}_i(1),\ \tilde{y}_i(2),\ \ldots,\ \tilde{y}_i(L))$ indicates the result of obtaining the FI of the fuzzy foreign exchange rate forecasts by L experts. If these fuzzy foreign exchange rate forecasts are approximated with TFNs, then the FI is a polygon-shaped fuzzy number.

The output of this step is a polygon-shaped fuzzy number that specifies the narrowest range of the fuzzy foreign exchange rate forecast. However, in practical applications a crisp foreign exchange rate forecast is usually required. Therefore, a crisp fuzzy foreign exchange rate forecast has to be generated from the polygon-shaped fuzzy number. For this purpose, many defuzzification methods are applicable (Liu, 2007). After obtaining the defuzzified value, it is compared with the actual foreign exchange rate to evaluate the accuracy. However, among the existing defuzzification methods, no one method is superior to all other methods in every case. Besides, the most suitable defuzzification method for a fuzzy variable is often chosen from the existing methods, which cannot guarantee the optimality of the chosen method. In addition, the shape of the polygon-shaped fuzzy number is special. These phenomena provided the motive for proposing a tailored defuzzification method. To this purpose, a BPN is applied in the fuzzy and neural approach, because theoretically a well-trained BPN (without being stuck in a local minima) with a good selected topology can successfully map any complex distribution.

Step 3: Constructing a BPN to Defuzzify the Aggregation Result

The configuration of the BPN used in the fuzzy and neural approach is established as follows:

1. **Inputs:** $2m$ parameters corresponding to the m corners of the polygon-shaped fuzzy number and the membership function values of these corners. All input parameters have to be normalized using the partial normalization approach before they are fed into the network.
2. **Single hidden layer:** Generally speaking, one or two hidden layers are more beneficial for the convergence property of the BPN.
3. Number of neurons in the hidden layer is chosen from 1~$4m$ according to a preliminary analysis, considering both effectiveness (forecasting accuracy) and efficiency (execution time).
4. **Output:** the crisp foreign exchange rate forecast.
5. **Network learning rule:** Delta rule.
6. **Transformation function:** Sigmoid function, $f(x) = \frac{1}{1+e^{-x}}$. (20)

7. Learning rate (η): 0.01~1.0.
8. Batch learning.
9. **Number of epochs per replication:** 60000.
10. **Number of initial conditions/replications:** 100. Among the results, the best one is chosen for the subsequent analyses.

The procedure for determining the parameter values refers to Chen and Lin (2008).

A PRACTICAL EXAMPLE

To demonstrate the application of the fuzzy and neural approach, the problem of foreign exchange rate forecasting in Tseng et al. (2001) is used. The original data are shown in Figure 1, which is a time-series problem. After applying MA to the collected data, the number of moving period was determined to be eight days (Figure 2). The original values were normalized into [0.1, 0.9] before applying the fuzzy and neural approach.

Three experts were asked to submit their opinions about the fuzzy foreign exchange rate forecasts, which were summarized in Table 1. As a result, there were six NP problems to be solved. The results are shown in Figures 3 through 5. From the optimization result of each NP problem, a corresponding FLR equation was constructed. All the six FLR equations were applied to forecast the foreign exchange rate. Subsequently, the FI of the forecasting results was obtained (see Figure 6), which was a polygon-shaped fuzzy number for each period.

The data of the corners of all polygon-shaped fuzzy numbers, i.e. the aggregated fuzzy exchange rate forecasts (summarized in Table 2), were used to train and/or test the BPN defuzzifier. There were at most 6 corners, and therefore the number of inputs to the BPN defuzzifier was set to be 12. The optimal number of hidden-layer nodes were chosen from 1~24. The 4-fold cross-validation was adopted. Finally, the BPN was applied to defuzzify a polygon-shaped fuzzy number input to the network to generate the representative value, i.e. the crisp foreign exchange rate forecast. The results are shown in Figure 7.

In the experiment, the experts' opinions were stored into the expert opinions database constructed using Microsoft Excel 2003. The Optimization Toolbox of MATLAB 2006a was applied to solve the NP problems. To exchange and synchronize the data between the expert opinions database and the optimizer, the Excel Link add-in

Figure 1. The exchange rate forecasting problem

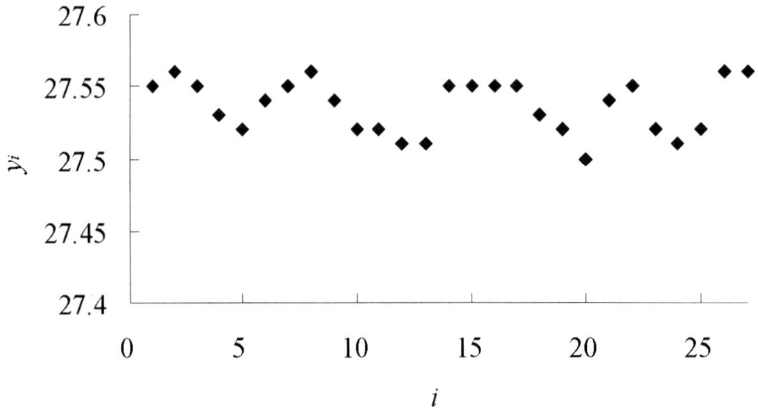

Figure 2. Determining the moving period

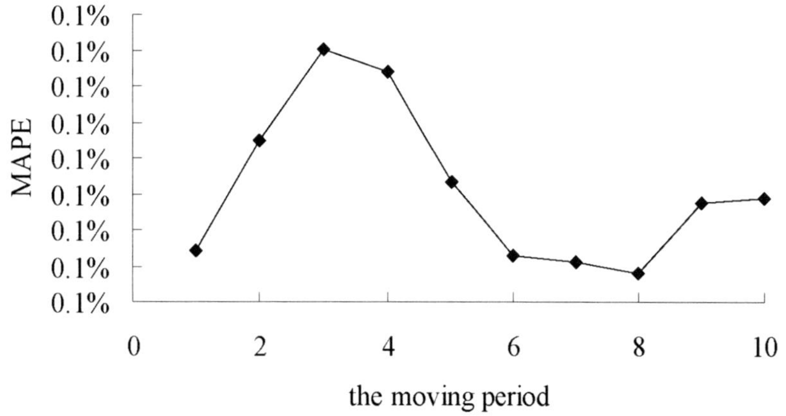

of MATLAB was used, which communicates between the Excel workspace and the MATLAB workspace and positions Excel as a front end to MATLAB. Finally, the Neural Network Toolbox of MATLAB 2006a was applied to implement the BPN defuzzification approach. The forecasting accuracy was measured with root mean squared error (RMSE), mean absolute error (MAE), and mean absolute percentage error (MAPE). On the other hand, the forecasting precision of a non-biased crisp approach can be measures with 6σ (or 6 RMSE), because theoretically the probability that such an interval contains the actual value is about 99.7%, under the assumption that residuals follow a normal distribution. Comparatively, the precision of a fuzzy approach can be measured with the average spread (or range) of fuzzy foreign exchange rate forecasts if all of them contain the actual value.

Table 1. The opinions of three experts

l	o_l	d_l	s_l	m_l
1	2	1	0.1	2
2	1	0.8	0.2	1
3	3	0.9	0.2	1

To compare with some existing approaches, MA, ES, BPN, ARIMA, and FARIMA (Tseng et al., 2001) were also applied to the collected data. In MA, various numbers of moving periods (from 10 to 1 step -1) were tried. Among them, the best one was chosen for the subsequent analyses. The number of inputs in BPN was determined in a similar way. In ES, the value of the smoothing constant changed from 0 to 1 with an interval of 0.1, and then the value contributing to the best performance was adopted. In the BPN approach, there was one hidden layer with nodes that were twice as many as the number of inputs. The number of epochs was set to be 60000 epochs. In addition, the initial values of all parameters in the BPN were randomly generated, 100 times. Among them, the best setting was kept for the subsequent analyses. A 4-fold evaluation process was applied to cross-validate the data. In ARIMA, there were three stages: model identification, model estimation, and model checking. The minimum information criterion (MINIC) method (Hannan & Rissanen, 1982) was used to identify the order in the ARIMA process. Besides, the augmented dickey fuller (ADF) unit root tests (Dickey & Fuller, 1981) were used to test the stationarity and seasonal stationarity in the foreign exchange

Figure 3. The optimization result (expert #1) (UL: upper limit; LL: lower limit)

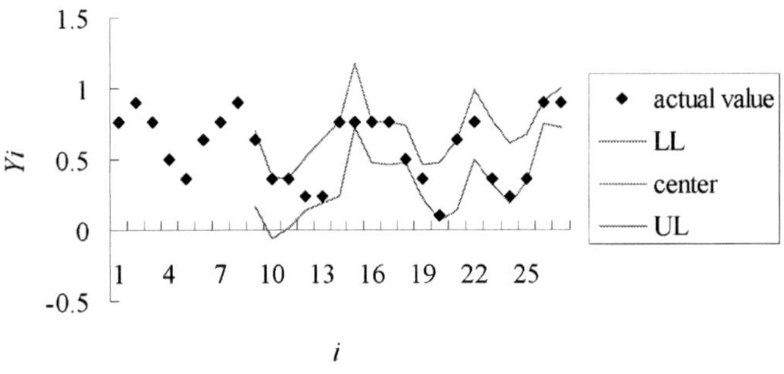

(NP I)

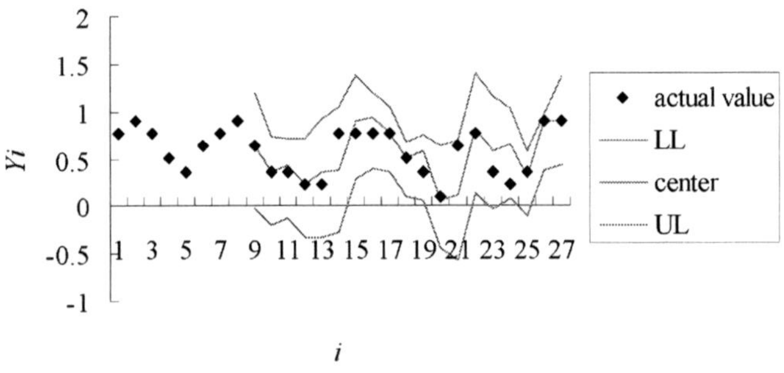

(NP II)

rate data. The performances achieved by applying these approaches are compared in Table 3. In FARIMA, a fuzzy foreign exchange rate forecast was defuzzified with the center-of-gravity (COG) formula:

MA was adopted as the comparison basis, and the percentage of improvement on the performance measure by applying another approach is enclosed in parentheses following the performance measure. According to experimental results, the following points can be made

1. The accuracy of foreign exchange rate forecasting, measured in terms of MAPE, of the fuzzy and neural approach, was significantly better than those of the other approaches by achieving a 81% reduction in MAPE over the comparison basis – MA. The advantages over ES and ARIMA were 62% and 29%, respectively. The performance of the fuzzy and neural approach with respect to MAE or RMSE was also significantly better than those of the other approaches.
2. ES outperformed MA because there were sudden changes in direction in the collected data.
3. The performance of a nonlinear approach like BPN was not satisfactory because the pattern of the collected data was not nonlinear but considerably stable.
4. Compared with the fuzzy approach, FARIMA, the accuracy of the fuzzy and

Figure 4. The optimization result (expert #2)

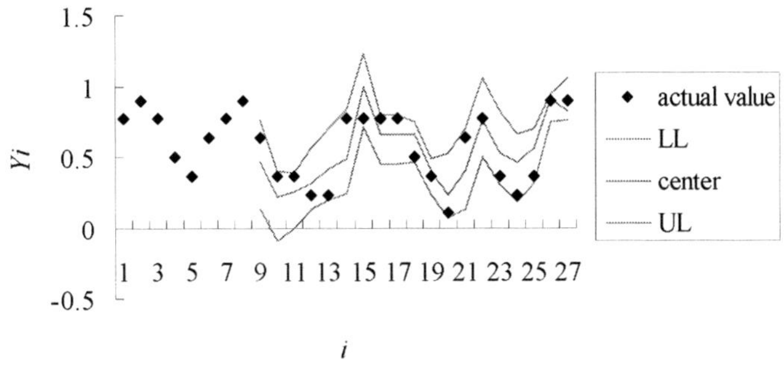

(NP I)

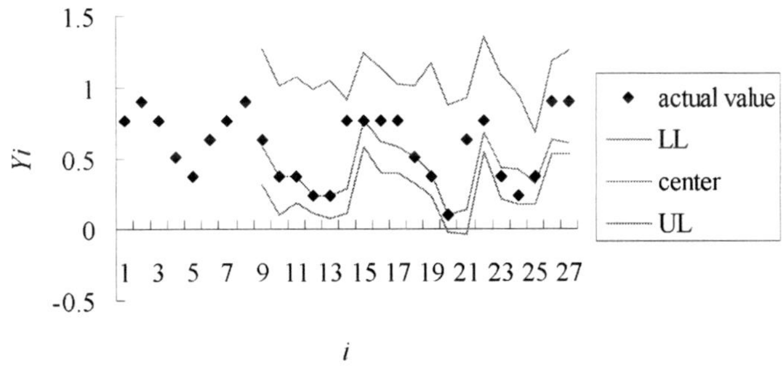

(NP II)

neural approach was also better with respect to each of the three performance measures. Nevertheless, FARIMA and ARIMA still achieved satisfactory performances in this experiment.

5. On the other hand, the precision of the fuzzy and neural approach was significantly better than those of the other approaches. The advantage over the baseline approach (MA) was up to 79%. In other words, with the fuzzy and neural approach, it was possible to give a very small range for the future foreign exchange rate.
6. The advantage of the fuzzy and neural approach came from two sources. First, it was shown to be a successful way of hybridizing linear and nonlinear approaches (FLR and BPN). In addition, incorporating multiple experts' opinions and then aggregating these opinions in an effective way also contributed to the good performance of the fuzzy and neural approach.

CONCLUSION AND DIRECTIONS FOR FUTURE RESEARCH

To improve the precision and accuracy of the foreign exchange rate forecasting, a fuzzy and neural approach is applied in this study. In the fuzzy and

Figure 5. The optimization result (expert #3)

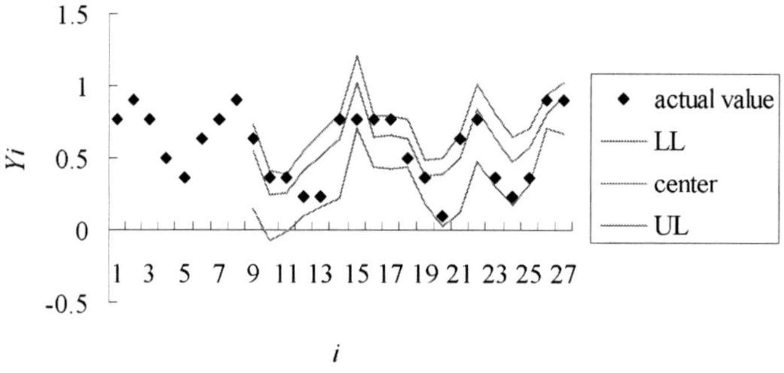

(NP I)

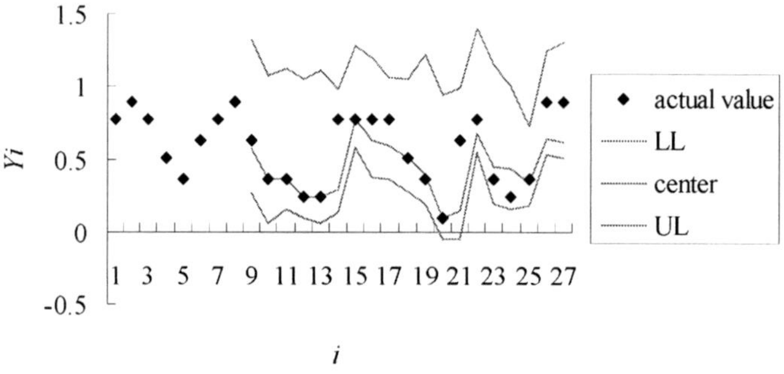

(NP II)

Figure 6. The results of FI

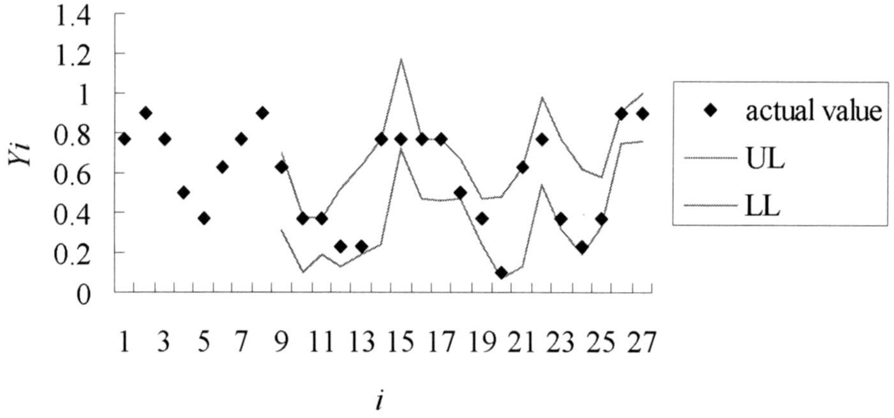

Table 2. The corners of the polygon-shaped fuzzy forecasts

Period #	(corner, membership)
9	(0.31, 0.00), (0.47, 0.57), (0.55, 0.71), (0.70, 0.18), (0.71, 0.00)
10	(0.10, 0.00), (0.28, 0.66), (0.38, 0.13), (0.39, 0.00)
11	(0.18, 0.00), (0.30, 0.63), (0.36, 0.25), (0.37, 0.00)
12	(0.13, 0.00), (0.39, 0.67), (0.51, 0.20), (0.53, 0.00)
13	(0.19, 0.00), (0.50, 0.67), (0.51, 0.65), (0.64, 0.20), (0.65, 0.00)
14	(0.23, 0.00), (0.55, 0.57), (0.76, 0.22), (0.77, 0.00)
15	(0.72, 0.00), (0.98, 0.55), (1.17, 0.14), (1.18, 0.00)
16	(0.47, 0.00), (0.56, 0.29), (0.71, 0.57), (0.77, 0.20), (0.78, 0.00)
17	(0.45, 0.00), (0.69, 0.75), (0.76, 0.25), (0.77, 0.00)
18	(0.47, 0.00), (0.59, 0.44), (0.68, 0.00)
19	(0.24, 0.00), (0.37, 0.58), (0.41, 0.66), (0.47, 0.20), (0.48, 0.00)
20	(0.07, 0.00), (0.07, 0.04), (0.31, 0.58), (0.40, 0.41), (0.48, 0.16), (0.49, 0.00)
21	(0.13, 0.00), (0.40, 0.52), (0.63, 0.11), (0.64, 0.00)
22	(0.54, 0.00), (0.56, 0.13), (0.85, 0.70), (0.96, 0.32), (0.98, 0.17), (0.99, 0.00)
23	(0.33, 0.00), (0.63, 0.65), (0.77, 0.16), (0.78, 0.00)
24	(0.19, 0.00), (0.50, 0.72), (0.52, 0.74), (0.62, 0.18), (0.63, 0.00)
25	(0.33, 0.00), (0.47, 0.40), (0.59, 0.00)
26	(0.75, 0.00), (0.85, 0.60), (0.87, 0.57), (0.91, 0.23), (0.92, 0.00)
27	(0.76, 0.00), (0.76, 0.12), (0.89, 0.56), (1.00, 0.23), (1.01, 0.00)

neural approach, multiple experts construct their own FLR equations from various viewpoints to forecast the foreign exchange rate. Each FLR equation can be fitted with two equivalent NP problems to be solved. To aggregate the fuzzy foreign exchange rate forecasts by different experts, a two-step aggregation mechanism is applied. At the first step, FI is applied to aggregate the fuzzy foreign exchange rate forecasts into a polygon-shaped fuzzy number, in order to improve the precision of the foreign exchange rate forecasting. The polygon-shaped fuzzy number contains the actual value. After that, considering the special shape of the polygon-shaped fuzzy number, a BPN is constructed to defuzzify the polygon-shaped

Table 3. The forecasting performances of different methods

	MA	ES	BPN	ARIMA	FARIMA	The proposed methodology
RMSE	0.021	0.02(-2%)	0.030(+45%)	0.012(-41%)	0.006(-70%)	0.005(-74%)
MAE	0.019	0.015(-19%)	0.025(+36%)	0.009(-52%)	0.004(-77%)	0.004(-81%)
MAPE	0.068%	0.055%(-19%)	0.092%(+36%)	0.033%(-52%)	0.016%(-77%)	0.013%(-81%)
Precision (average range)	0.124	0.122(-2%)	0.180(+45%)	0.073(-41%)	0.046(-63%)	0.026(-79%)

$$D(\tilde{y}_i) = \frac{y_{i,1} + y_{i,2} + y_{i,3}}{3}. \qquad (21)$$

fuzzy number and to generate a representative/crisp value to enhance the accuracy.

A practical case containing the historical data of the foreign exchange rate from NTD to USD was used to demonstrate the application of the fuzzy and neural approach. According to experimental results, we found that:

1. The forecasting accuracy (measured in terms of MAE, MAPE, or RMSE) of the fuzzy and neural approach was significantly better than those of some existing crisp approaches (MA, ES, BPN, and ARIMA).
2. The fuzzy and neural approach also surpassed the fuzzy approach, FARIMA, in forecasting the foreign exchange rate.
3. At the same time, the proposed methodology outperformed the existing approaches in forecasting precision measured with the average range.
4. The aggregation mechanism in the fuzzy and neural approach effectively narrowed the region of searching the actual value, which not only elevated the precision, but also increased the probability of finding out the actual value.

More sophisticated fuzzy and neural approaches can be developed for the foreign exchange rate forecasting in future studies.

REFERENCES

Al-Deek, H. M. (2007). Use of vessel freight data to forecast heavy truck movements at seaports. *Transportation Research Record: Journal of the Transportation Research Board, 1804*, 217–224. doi:10.3141/1804-29

Bárdossy, A. (1990). Note on fuzzy regression. *Fuzzy Sets and Systems, 65*, 65–75. doi:10.1016/0165-0114(90)90064-D

Chang, P.-C., & Hsieh, J.-C. (2003). A neural networks approach for due-date assignment in a wafer fabrication factory. *International Journal of Industrial Engineering, 10*(1), 55–61.

Chang, P.-C., Hsieh, J.-C., & Liao, T. W. (2005). Evolving fuzzy rules for due-date assignment problem in semiconductor manufacturing factory. *Journal of Intelligent Manufacturing, 16*, 549–557. doi:10.1007/s10845-005-1663-4

Chang, P. T., & Lee, E. S. (1994). Fuzzy linear regression with spreads unrestricted in sign. *Computers & Mathematics with Applications (Oxford, England), 28*, 61–70. doi:10.1016/0898-1221(94)00127-8

Chen, T. (2008). A SOM-FBPN-ensemble approach with error feedback to adjust classification for wafer-lot completion time prediction. *International Journal of Advanced Manufacturing Technology, 37*(7-8), 782–792. doi:10.1007/s00170-007-1007-y

Chen, T. (2009). A fuzzy-neural knowledge-based system for job completion time prediction and internal due date assignment in a wafer fabrication plant. *International Journal of Systems Science, 40*(8), 889–902. doi:10.1080/00207720902974553

Chen, T., & Lin, Y. C. (2008). A fuzzy-neural system incorporating unequally important expert opinions for semiconductor yield forecasting. *International Journal of Uncertainty, Fuzziness, and Knowledge-based Systems, 16*(1), 35–58. doi:10.1142/S0218488508005030

Chen, T., & Wang, Y. C. (2010). A hybrid fuzzy and neural approach for forecasting the book-to-bill ratio in the semiconductor manufacturing industry. *International Journal of Advanced Manufacturing Technology*.

Chen, T., Wang, Y. C., & Tsai, H. R. (2009a). Lot cycle time prediction in a ramping-up semiconductor manufacturing factory with a SOM-FBPN-ensemble approach with multiple buckets and partial normalization. *International Journal of Advanced Manufacturing Technology*, *42*(11-12), 1206–1216. doi:10.1007/s00170-008-1665-4

Chen, T., Wang, Y. C., & Wu, H. C. (2009b). A fuzzy-neural approach for remaining cycle time estimation in a semiconductor manufacturing factory – a simulation study. *International Journal of Innovative Computing. Information and Control*, *5*(8), 2125–2139.

Chen, T., Wu, H.-C., & Wang, Y.-C. (2009c). Fuzzy-neural approaches with example post-classification for estimating job cycle time in a wafer fab. *Applied Soft Computing*, 1225–1231. doi:10.1016/j.asoc.2009.03.006

Cheng, I. S., Tsujimura, Y., Gen, M., & Tozawa, T. (1995). An efficient approach for large scale project planning based on fuzzy Delphi method. *Fuzzy Sets and Systems*, *76*, 277–288. doi:10.1016/0165-0114(94)00385-4

D'Urso, P. (2003). Linear regression analysis for fuzzy/crisp input and fuzzy/crisp output data. *Computational Statistics & Data Analysis*, *42*, 47–72. doi:10.1016/S0167-9473(02)00117-2

Diamond, P. (1988). Fuzzy least squares. *Information Sciences*, *46*, 141–157. doi:10.1016/0020-0255(88)90047-3

Dickey, D. A., & Fuller, W. A. (1981). Likelihood ratio statistics for autoregressive time series with a unit. *Econometrica*, *49*(4), 1057–1072. doi:10.2307/1912517

Donoso, S., Marin, N., & Vila, M. A. (2006). Quadratic programming models for fuzzy regression. In *Proceedings of the International Conference on Mathematical and Statistical Modeling in Honor of Enrique Castillo*.

Eraslan, E. (2009). The estimation of product standard time by artificial neural networks in the molding industry. *Mathematical Problems in Engineering*.

Hsiao, F. H., Xu, S. D., Lin, C. Y., & Tsai, Z. R. (2008). Robustness design of fuzzy control for nonlinear multiple time-delay large-scale systems via neural-network-based approach. *IEEE Transactions on Systems, Man, and Cybernetics. Part B, Cybernetics*, *38*(1), 244–251. doi:10.1109/TSMCB.2006.890304

Kaufmann, A., & Gupta, M. M. (1998). *Fuzzy Mathematical Models in Engineering and Management Science*. Amsterdam, The Netherlands: North-Holland.

Li, Q., Xu, M., Zhang, H., & Liu, F. (2006). Neural network based flow forecast and diagnosis. In *Computational Intelligence and Security* (LNCS 3802, pp. 542-547).

Liu, X. (2007). Parameterized defuzzification with maximum entropy weighting function - another view of the weighting function expectation method. *Mathematical and Computer Modelling*, *45*, 177–188. doi:10.1016/j.mcm.2006.04.014

Peters, G. (1994). Fuzzy linear regression with fuzzy intervals. *Fuzzy Sets and Systems*, *63*, 45–55. doi:10.1016/0165-0114(94)90144-9

Piramuthu, S. (1991). Theory and methodology – financial credit-risk evaluation with neural and neuralfuzzy systems. *European Journal of Operational Research*, *112*, 310–321. doi:10.1016/S0377-2217(97)00398-6

Redden, D. T., & Woodall, W. H. (1994). Properties of certain fuzzy linear regression methods. *Fuzzy Sets and Systems*, *64*, 361–375. doi:10.1016/0165-0114(94)90159-7

Tanaka, H., & Lee, H. (1998). Interval regression analysis by quadratic programming approach. *IEEE Transactions on Fuzzy Systems*, *6*(4), 473–481. doi:10.1109/91.728436

Tanaka, H., Uejima, S., & Asai, K. (1982). Fuzzy limear regression model. *IEEE Transactions on Systems, Man, and Cybernetics*, *12*, 903–907. doi:10.1109/TSMC.1982.4308925

Tanaka, H., & Watada, J. (1988). Possibilistic linear systems and their application to the linear regression model. *Fuzzy Sets and Systems*, *272*, 275–289. doi:10.1016/0165-0114(88)90054-1

Tseng, F. M., Tzeng, G. H., Yu, H. C., & Yuan, B. J. C. (2001). Fuzzy ARIMA model for forecasting the foreign exchange market. *Fuzzy Sets and Systems*, *118*, 9–19. doi:10.1016/S0165-0114(98)00286-3

This work was previously published in the International Journal of Fuzzy System Applications, Volume 1, Issue 1, edited by Toly Chen, pp. 36-48, copyright 2010 by IGI Publishing (an imprint of IGI Global).

Chapter 7
A Novel Fuzzy Associative Memory Architecture for Stock Market Prediction and Trading

Chai Quek
Nanyang Technological University, Singapore

Zaiyi Guo
Nanyang Technological University, Singapore

Douglas L. Maskell
Nanyang Technological University, Singapore

ABSTRACT

In this paper, a novel stock trading framework based on a neuro-fuzzy associative memory (FAM) architecture is proposed. The architecture incorporates the approximate analogical reasoning schema (AARS) to resolve the problem of discontinuous (staircase) response and inefficient memory utilization with uniform quantization in the associative memory structure. The resultant structure is conceptually clearer and more computationally efficient than the Compositional Rule Inference (CRI) and Truth Value Restriction (TVR) fuzzy inference schemes. The local generalization characteristic of the associative memory structure is preserved by the FAM-AARS architecture. The prediction and trading framework exploits the price percentage oscillator (PPO) for input preprocessing and trading decision making. Numerical experiments conducted on real-life stock data confirm the validity of the design and the performance of the proposed architecture.

INTRODUCTION

Financial engineering is a rapidly expanding research area. Trading systems based on computational intelligence techniques for financial asset management, notably in the areas of equities trading and risk management for derivatives like options and swaps has received considerable interest from both researchers and financial traders. Neural networks (NN) have been used extensively for market forecasting (White, 1988; Chiang et al., 1996). More recently, time-delay, recurrent and probabilistic NNs have been used to analyze the predictive capability of the networks using "live

DOI: 10.4018/978-1-4666-1870-1.ch007

data" for several highly volatile and consumer stocks (Saad et al., 1998; Moody & Saffell, 2001; Chen et al., 2003).

Trading systems using a chaos-based modeling procedure to construct alternative price prediction models based on technical, adaptive, and statistical models have been proposed (Wilson, 1994). An intelligent stock trading decision support system has been developed (Chou et al., 1996), that can forecast buying and selling signals for the prediction of short-term and long-term trends using rule-based NNs. NNs and genetic algorithms were used in Baba et al. (2000) to construct an intelligent decision support system (DSS) to analyze the Tokyo Stock Exchange Stock Price Index (TOPIX). The DSS was able to project the high and low TOPIX values four weeks into the future and suggested buy/sell decisions based on the average value. Similar techniques were adapted to perform "bull flag" pattern recognition and to learn the trading rules from price and volume of the New York Stock Exchange Composite Index (NYSE-CI) (Leigh et al., 2002). Predictor attributes, including financial statement variables and macroeconomic variables, were used with a back-propagation NN to integrate fundamental and technical analysis for financial performance prediction (Lam, 2004).

Fuzzy logic and neural networks are both motivated by human learning and interpretation. Fuzzy Logic gives a framework for approximate reasoning and allows qualitative knowledge about the problem to be translated into an executable set of rules, but by itself cannot automatically construct or acquire the rules used to make those decisions. NNs are known for their computational power, fault tolerance and learning capabilities, but are inadequate for explaining how they arrive at their decisions. Fuzzy neural networks (FNNs) are hybrid systems that attempt to combine the advantages of both techniques. Numerous approaches to integrate fuzzy systems and NNs have been proposed. An extensive bibliography on FNNs can be found in (Buckley & Hayashi, 1995). In recent years, there has been an increasing emphasis on neuro-fuzzy systems in financial engineering (Trippi & Turban, 1993) and stock price forecasting (Thammano, 1999; Wang, 2003; Ang & Quek, 2006).

This paper examines the application of a neuro-fuzzy associative memory (FAM) architecture which incorporates the approximate analogical reasoning schema (AARS). This novel fuzzy neural network architecture, referred to as FAM-AARS, is applied to the stock prediction and trading problem. The rest of the paper is organized as follows. The following section presents the details of the proposed FAM-AARS architecture and introduces the direct-output FAM-AARS. The prediction and trading framework which exploits the price percentage oscillator (PPO) for input preprocessing and trading decision making is presented. The performance of the direct-output FAM-AARS network is evaluated based on experiments conducted on real-life stock data.

THE FAM-AARS ARCHITECTURE

Associative memory (AM) is a class of neural networks which forms a content-addressable structure which is able to map a set of input patterns to a set of output patterns and is characterized by its excellent storage, recall and error correction properties. The cerebellar model articulation controller (CMAC) (Albus, 1975) is a sub-class of AM inspired by the neuro-physiological theory of the cerebellum (Albus, 1972, 1975). The CMAC neural network is an alternative to the well-known backpropagation-trained multi-layer NN which has disadvantages of: requiring many iterations to converge; requiring a large amount of computation so that the algorithm runs slowly unless implemented in expensive custom hardware; having an error surface which can have relative minima; not allowing successful incremental learning if one has finite time to train, etc (Glanz et al., 1991; Miller et al., 1990). The CMAC structure is attractive because of its localized generalization, rapid algorithmic computation based on least mean square (LMS) training, incremental training, functional representation, output superposition and fast practical hardware realization (Miller et al., 1990). CMAC was originally developed for control applications, however, its usage is not restricted to control. As a neural network structure, its application area can be as wide as that of the conventional feed-forward multiplayer neural networks.

CMAC is characterized by a piecewise-constant response because the network response is computed by the summation of activated cells, where each cell contributes evenly. Close inputs may be mapped to identical sets of cells, generating identical outputs. The rectangular shape of the CMAC receptive field functions produces a discontinuous (staircase) response, which can produce control fluctuations in the output. A number of techniques have been developed (Stephen et al., 1992; Ang & Quek, 2000) to improve the accuracy of the function approximation capability. In addition to the undesirable output characteristics, it has also been observed that uniform quantization of the input space can result in inefficient data storage due to uneven variations of the target function in the problem space (Moody, 1989; Kim & Lin, 1992). Ker et al. (1997) proposed a new model called fuzzy CMAC (FCMAC) to fully exploit the advantages of fuzzy set theory and the local generalization feature of the CMAC model. This FCMAC model has a number of advantages because it results in a hybrid system where the fuzzy logic handles the vagueness and uncertainty and the neural network provides computational power, fault tolerance and learning capabilities. Such hybrid systems are able to model highly complex, dynamic and non-linear problem domains using a linguistic model where traditional mathematical formulations from the first principle are too enigmatic or simply unrealizable.

One of the major problems associated with existing fuzzy systems is that the inference models used are conceptually unclear. For instance, the Mamdani inference model has been widely accepted in control applications. It employs the min operator as the interpretation operator for ease of computation (Mendel, 1995). However, this interpretation does not agree with the material implication in the classical Modus Ponens schema (Dubois & Prade, 1996). The approximate analogical reasoning schema (AARS) (Turksen & Zhong, 1990) is an appealing fuzzy inference scheme for its intuitive simplicity and computational efficiency compared to other well-known fuzzy inference schemes such as the compositional rule of inference (CRI) (Zadeh, 1975) and the truth-value restriction (TVR) (Mantaras, 1990). Though CRI is dominant in the field of approximate reasoning, two issues were mentioned in (Turksen & Zhong, 1990). One was the difficulty in choosing the implication function I_M, and the other was the composition operation. They argued that the matrix operations were conceptually complicated and unclear. One example is that CRI has the undesirable outcome $A \circ (A \rightarrow B) \neq B$. This has been readily verified using numeric examples. TVR is similar to CRI in that it uses a fuzzy implication function to link up antecedent and consequence, which is essentially based on the matrix operation whose effect is conceptually unclear (Zhou & Quek, 1996, 1999). An AARS-based generalized modus ponens (GMP) proposed in (Turksen & Zhong, 1990) takes the form described by Equation (1):

Rule: *IF x* is *A*, *THEN y* is *B*, threshold= τ (1)
Fact: *x* is *A*
Conclusion:

$$y \text{ is } \begin{cases} MF(B, SM(A, A')) & \text{if } SM(A, A;) >= \tau \\ no \ \text{conclusion} & \text{otherwise} \end{cases}$$

There are three key items for this scheme: a similarity measure (*SM*) for antecedent matching; a threshold value (τ) for the choice of rule firing; and a modification function (*MF*) to deduce the consequence set.

THE FAM-AARS INFERENCE SCHEME

1. Similarity Measure (*SM*)

This term evaluates the similarity between a rule's antecedent *A* and the observation *A'*, (Turksen & Zhong, 1990), the *SM* between fuzzy sets is derived from the distance measure (*DM*), given by Equations (2-3):

$$SM = (1 + DM)^{-1} \rightarrow SM \in [0,1] \quad (2)$$

$$DM(A, A') = 1 - \sup_{x \in X} \mu_{A \cap A'}(x) \quad (3)$$

The definition in Equation (2) essentially translates any positive *DM* into the range [0, 1], but if Eq (2) is used together with the disconsistency measure defined in Equation (3) (Turksen & Zhong, 1990), it will create a problem that *SM* is biased into the range [0.5, 1]. This is undesirable in the context of FAM-AARS. The contributions of consequences from different rules depend on *SMs* with these rules and such bias lessens the difference among these *SMs*. A way to normalize the *SM* as proposed in (Quek & Zhou, 2006) is to modify Equation (2) as shown in Equation (4):

$$SM = 1 - DM \quad (4)$$

This preserves the effect that the larger the *DM* value, the smaller the value of *SM*. When Equation (4) is used with Equation (3), *SM* becomes Equation (5):

$$SM(A, A') = \sup_{x \in X} \mu_{A \cap A'}(X) \quad (5)$$

This definition of *DM* has the advantage over other definitions (Turksen & Zhong, 1990), such as the Hausdórff Measure as it takes complete information from the two fuzzy sets into consideration rather than only considering the centroids. AARS uses a single value of *SM*, unlike in TVR where the matching between *A* and *A'* is represented as a function. This is considered a drawback in terms of accuracy, but it reduces computational complexity and makes the whole process simpler and easier to integrate into a neural-network-like structure (Ang & Quek, 2006; Zhou & Quek, 1999, 1996; Quek & Zhou, 2006; Tung et al., 2004; Quek & Singh, 2005).

2. AARS Threshold (τ)

The threshold (τ) is the value below that the rule is considered irrelevant to the facts and should be filtered for consequence deduction. This parameter is currently determined through trial and error.

3. Modification Function (*MF*)

A modification function is a function of the rule's consequence B and the similarity measure SM, i.e., $\widetilde{B} = MF(B, SM)$. There are two types of modification function (Turksen & Zhong, 1990); namely: the expansion and the reduction forms. The inferred fuzzy set using the expansion form corresponds to the linguistic hedge "more or less". The reduction form has the opposite effect to the expansion form, and it is in some sense an imitation of CRI (Zhou & Quek, 1996). It has the effect of the linguistic hedge "very". Generally speaking, the smaller the value of *SM*, the greater the difference between *B* and $\widetilde{B}$ (large expansion or large reduction), and for the extreme case when SM=1, $\widetilde{B}$ matches *B* exactly. This is one of the advantages of AARS over CRI, as CRI does not guarantee that exactly matched inputs generate exactly matched outputs. One of the problems with the *MF* defined in (Turksen & Zhong, 1990) is that it may result in a non-normalized fuzzy set. To achieve the effect of expansion or reduction, rather than scaling the membership value, the width of the fuzzy set can be scaled instead (Guo et al., 2006), as shown in Figure 1.

Let A denote the fuzzy set to be modified, then the property of the MFs can be described by Equations (6-7) (Guo et al., 2006):

- (expansion form):
- $\forall \alpha \in [0,1] \quad \frac{c(A) - \min({}^{\alpha}\tilde{A})}{c(A) - \min({}^{\alpha}A)} = \frac{\max({}^{\alpha}\tilde{A}) - c(A)}{\max({}^{\alpha}A) - c(A)} = \frac{1/SM}{1}$

  (6)
- MF^{r} (reduction form):
- $\forall \alpha \in [0,1] \quad \frac{c(A) - \min({}^{\alpha}\tilde{A})}{c(A) - \min({}^{\alpha}A)} = \frac{\max({}^{\alpha}\tilde{A}) - c(A)}{\max({}^{\alpha}A) - c(A)} = \frac{SM}{1}$

  (7)

where ${}^{\alpha}A$ is the $\alpha - cut$ of fuzzy set A, and c(A) is the centroid of A, with $\tilde{A}$ and A sharing the same centroid.

The modified set $\tilde{A}$ generated from A by MFs is defined (Guo et al., 2006) as shown in Equations (8-9):

MF^{X} (expansion form):

$$\mu_{\tilde{A}}(x) = \mu_{A}(c(A) + SM \cdot (x - c(A))) \quad (8)$$

MF^{r} (reduction form):

$$\mu_{\tilde{A}}(x) = \mu_{A}\left(c(A) + \frac{x - c(A)}{SM}\right) \quad (9)$$

where $c(A)$ is the centroid of fuzzy set A.

THE FAM-AARS ARCHITECTURE

The FAM-AARS is a FAM architecture that incorporates AARS, and belongs to the category of fuzzy neural networks which are functionally equivalent to a fuzzy inference system and attempts to computationally mimic the associative semantic memory of the human neocortical structure (Sim et al., 2006; Nhut et al., 2006). FAM-AARS consists of two main components: a knowledge base realized by the FCMAC network structure and an AARS inference scheme as the decision-making unit.

THE FAM-AARS NETWORK STRUCTURE

The network structure serves as the knowledge base (semantic memory) where information such as fuzzy rules are stored. An FAM has two options for storing the weight in a memory cell; one is to store real numbers as in the original FAM model, while the other is to store fuzzy sets so that the entire network is converted into a fuzzy inference system and defuzzification is required at the end. Both models are restricted to multiple-input-single-output (MISO) networks, as a multiple-input-multiple-output (MIMO) network

Figure 1. The FAM-AARS modification functions

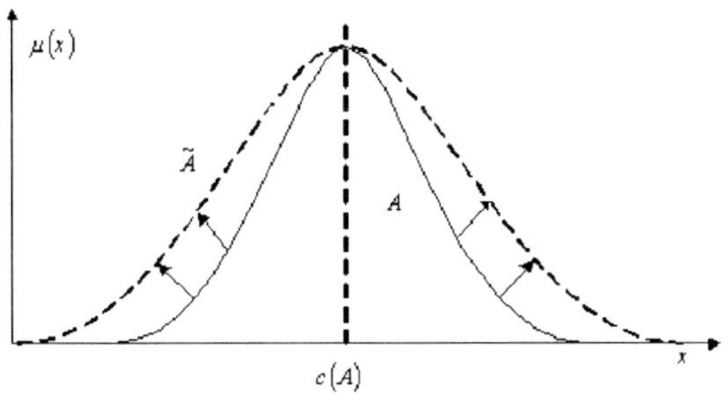

(a) MF^{x}

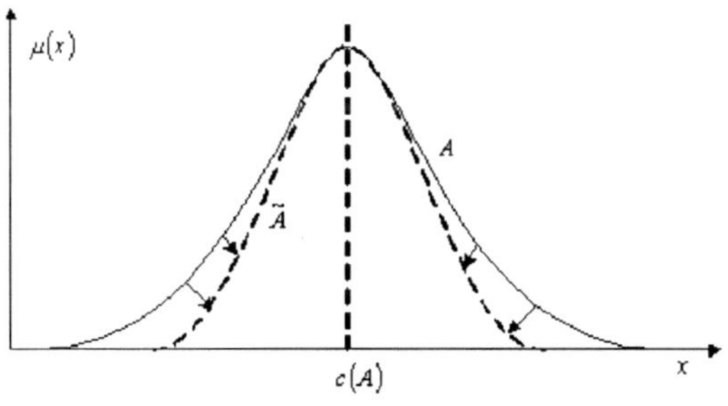

(b) MF^{r}

can always be represented as an aggregation of multiple MISOs. In this paper, we only discuss the direct-output FAM-AARS (which stores real numbers) because of its superior performance in time series prediction (Guo et al., 2006). A detailed description of the fuzzy-output FAM-AARS can be found in (Guo et al., 2006).

The input space is pre-clustered using discrete incremental clustering (DIC) (Tung & Quek, 2002). This achieves a non-linear quantization of the input space by building clusters, or fuzzy sets based on the input data distribution. Hence the quantization ratio is proportional to the number of fuzzy sets that cover the region, which provides the linguistic coverage of the input space of the fuzzy rule system and thus resolves the memory wastage issue present in the uniform quantization method.

The network is constructed in a form of a multi-dimensional lookup table based on the centroids of these clusters. A memory cell content is denoted by $Z_{[j_1 j_2 \cdots j_i \cdots j_I]^T}$, where $[j_1 j_2 \cdots j_i \cdots j_I]^T$ is the index pattern, or the addressing code of this particular cell. Typically one cell stores one output for a MISO structure, but this can be easily generalized to an output vector for MIMO. Figure 2 illustrates the idea for a simple direct-output FAM-AARS network with 2 input dimensions and one output dimension (MISO). With 4 clusters in first dimension and 5 clusters for the second, a total number of 20 cells is created; each is indexed with a unique index pattern. The arrow at each cell denotes that the storage is a real number, hence a direct output.

THE FAM-AARS INFERENCE RULES

Given an input tuple $X = [x_1, x_2, \ldots, x_i, \ldots, x_I]^T$, the inference rule steps to derive the output are shown in the flow diagram of Figure 3. Additional details related with the FAM-AARS inference rules, including that for the fuzzy-output FAM-AARS, can be found in (Guo et al., 2006). For the direct-output FAM-AARS, where only crisp values or singleton fuzzy sets are stored, the *MF* does not have any impact on the consequence, as the set widths are zero. Therefore for the direct-output FAM-AARS, step 5 in Figure 3 is skipped, resulting in a single output o as defined in Equation (10):

$$o = \frac{\sum_{n=1}^{N} Z_{d_n} \cdot ASM_n \cdot \delta_n}{\sum_{n=1}^{N} ASM_n \cdot \delta_n} \tag{10}$$

where δ_n and ASM_n are the nth firing decision and the nth aggregate similarity measure for the nth selected rule, both derived in step 4 of Figure 3, d_n is the index of $_{\mathbf{Z}}$derived in step 2 and Z_{d_n} returns the crisp value stored in the cell.

Figure 4 illustrates the inference process for a simple direct-output FAM-AARS with 2 input dimensions, 4 clusters for the 1st dimension and 5 clusters for the 2nd dimension. For a particular input tuple, the fuzzification is performed (step 1). In this instant, a singleton fuzzifier is illustrated. Based on the kernel of the fuzzified inputs, a set of rules is selected and their cell locations are identified (step 2). For this example, the cells are $\mathbf{Z}_{[2\,2]T}$, $\mathbf{Z}_{[2\,3]T}$, $\mathbf{Z}_{[3\,2]T}$ and $\mathbf{Z}_{[3\,3]T}$. Similarity measures (SM) are computed for each input dimension (step 3) and they are combined to produce the aggregate similarity measures (ASM) with selected rules (step 4). Because the direct-output FAM-AARS requires no consequence derivation, it jumps to step 6 to derive the final output as a combination of the consequences of the selected rules.

THE FAM-AARS LEARNING RULES

By the nature of the direct-output structure, the learning rule takes the form similar to a pure neural network. In the context of FAM, learning

Figure 2. DoFAM-AARS network structure with 2 input dimensions

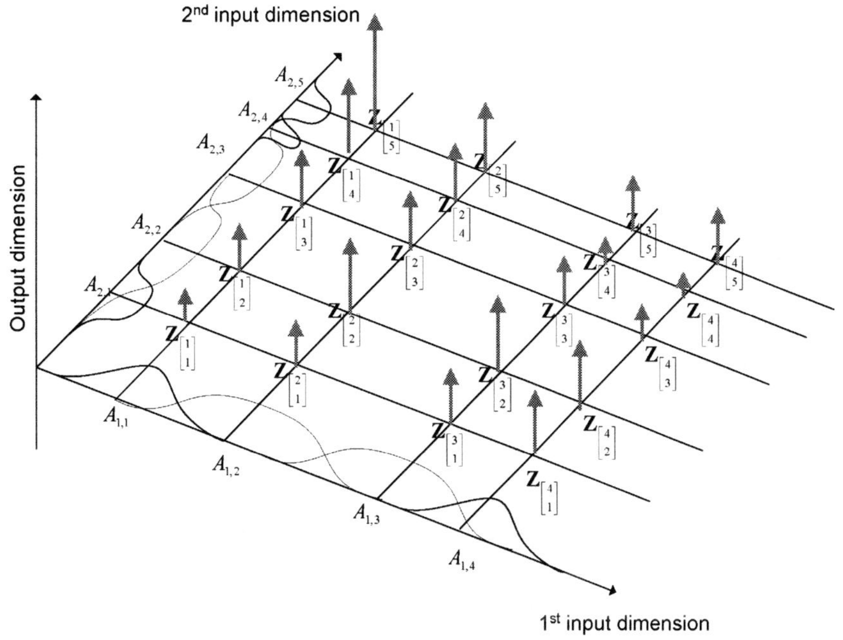

and inference are localized, hence batch-learning (Jang, 1996) is inappropriate because it is illogical for a cell to learn based on the irrelevant errors produced by faraway cells. Thus only pattern learning (Jang, 1996), where weights are updated after each training instance, is implemented.

The supervised learning procedure updates the stored output value to reduce the error ε between the desired output o^d and the deduced output o. The cell update is proportional to the effect during the inference phase, as described by Equations (11-14):

$$\varepsilon = o^d - o; \forall \hat{n} \in N \tag{11}$$

$$Z_{d_{\hat{n}}}^{(T+1)} = Z_{d_{\hat{n}}}^{(T)} + \frac{ASM_{\hat{n}} \cdot \delta_{\hat{n}}}{\sum_{n=1}^{N} ASM_n \cdot \delta_n} \cdot (\lambda \cdot (1-\xi) \cdot \varepsilon + \xi \cdot \Delta Z_{d_{\hat{n}}}^{(T)}) \tag{12}$$

$$\Delta Z_{d_{\hat{n}}}^{(T)} = \begin{cases} Z_{d_{\hat{n}}}^{(T)} - Z_{d_{\hat{n}}}(T-1) & \text{if } \delta_{\hat{n}} = 1 \\ \Delta Z_{d_{\hat{n}}}(T-1) & \text{otherwise} \end{cases} \tag{13}$$

$$ASM_n = \frac{1}{I}\sum_{i=1}^{I} SM_{i,q_{n,i}} \tag{14}$$

where N is the number of rules; I is the number of input dimensions; $d_{\hat{n}}$ is the index to the memory location currently being updated; $Z_{d_{\hat{n}}}^{(T)}$ is the content in the cell indexed by $d_{\hat{n}}$ after the T th update to the cell; δ_n is the firing decision for the nth selected rule; λ is the learning constant and ξ is the momentum constant. The introduction of the momentum term can be used to increase the learning speed if learning constant needs to be set to a very small value. The momentum constant can be set to 0 if this term is not required.

FAM-AARS PERFORMANCE EVALUATION

In this section, we evaluate the performance of our proposed FAM-AARS architecture by comparing against Turksen's version of point-valued AAR (PVAAR) and interval-valued AAR (IVAAR) (Na-

Figure 3. The FAM-AARS inference process flow diagram

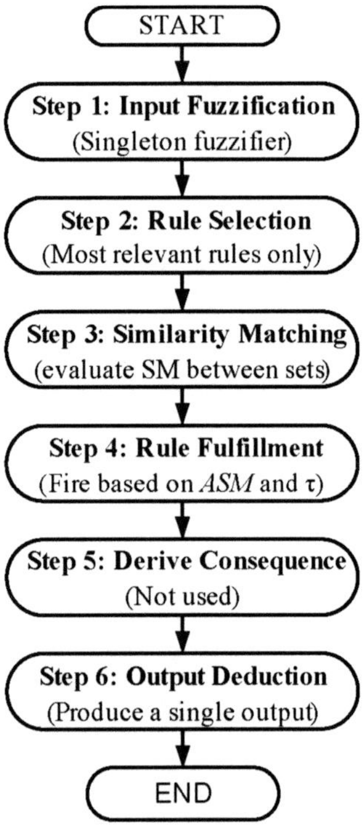

kanishi et al., 1993). We use one of the classical benchmark problems that have been developed for evaluating neural networks, namely the Nakanishi data set (Nakanishi et al., 1993). This data set consists of 3 individual experiments: a non-linear system, the human operation of a chemical plant, and the daily price of a stock in a stock market. In our experiments, each data set is organized into a training set and a testing set of the same size. Two commonly used performance metrics, the root-mean-squared error (RMSE) and the Pearson product moment correlation coefficient (r) are used for performance evaluation. The results are shown in Table 1.

It can be seen from Table 1 that FAM-AARS performed significantly better, in terms of both RSME and r, for the stock market example, with all networks reporting a small number of "invalid cases". For the other examples, it was observed that IVAAR and PVAAR appeared to perform slightly better. However, both networks have discarded a lot of "invalid cases" where there is no strong rule to fire due to the high threshold setting, which is the obvious reason for the im-

Figure 4. Inference procedure of DoFAM-AARS (2 input dimensions)

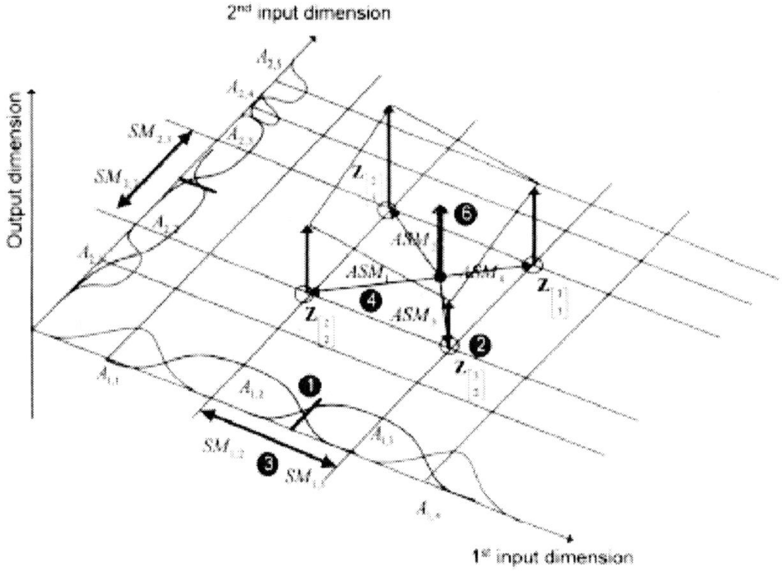

Table 1. Comparisons between FAM-AARS, PVAAR and IVAARI for the Nakanishi data set

Architecture	No. of Rules	Recall		Generalization		
		RMSE	*r*	Invalid	*RMSE*	*r*
A non-linear system						
FAM-AARS	**115**	**0.4792**	**0.8914**	**2**	**0.5398**	**0.8599**
PVAAR	-	-	-	8	0.4318	0.9202
IVAAR	-	-	-	9	0.5718	0.8724
Human operation of a chemical plant						
FAM-AARS	**212**	**267.77**	**0.9928**	**1**	**412.49**	**0.9851**
PVAAR	-	-	-	8	584.27	0.9921
IVAAR	-	-	-	2	463.38	0.9941
Daily price of a stock in a stock market						
FAM-AARS	**201**	**2.4822**	**0.9887**	**4**	**4.9519**	**0.9078**
PVAAR	-	-	-	2	5.9827	0.8876
IVAAR	-	-	-	4	6.2089	0.8681

provement. The excellent performance of FAM-AARS for the stock market example led us to further explore its performance in this area.

STOCK MARKET PREDICTION AND TRADING

1. Time Series Data

Stock prices and their derivatives such as moving averages, price momentum and oscillators are typical time series. A time series is a chronological sequence of observations on a particular variable. Time series are often examined in the hope of discovering a historical pattern that can be exploited in the preparation of a forecast, and generally consist of 4 components (Bowerman & O'Connell, 1993).

- **Trend:** The upward or downward movement that characterizes a time series over a period of time.
- **Cycle:** The recurring up and down movements around trend levels.
- **Seasonal variations:** Periodic patterns in a time series.
- **Irregular fluctuations:** Erratic movements in a time series that follow no recognizable or regular patterns.

The main approach in financial trading is to identify trends and maintain an investment position (long, short or hold) until evidence indicates that the trend has reversed. It is evident that trends in stock prices can be very volatile, almost haphazard at times. To identify the trends in a financial market, investors usually resort to two types of market analysis; namely: fundamental and technical. Fundamental analysis studies the overall economy, financial conditions and other data to determine a company's underlying value and potential for future growth. Technical analysis, on the other hand, does not attempt to measure a security's intrinsic value. This approach evaluates securities and predicts future movements by analyzing past statistics generated by market activity. The primary difference between these two types is that the fundamental analysis focuses on the studies of the causes of market movements, while the technical focuses on their effects. The fundamental analysis (Abarbanell & Bushee, 1997) generally involves the study of the economic forces of supply and demand and often results in long term recommendations. However, three main difficulties are found in fundamental analysis. Firstly, it is difficult to trace the large amount of relevant historical facts and data to make a complete analysis. Secondly, the analysis procedures vary from one case to another. Thirdly, information must be consistently obtained and the qualitative facts must be correctly verified by the experts. The technical analysis (Pring, 2002), on the other hand, assumes that shifts in supply and demand forces will be exhibited in market movements and therefore, the market movements can be used as a close proxy to determine changes in the ebb and flow of supply and demand. Compared against fundamental analysis, technical analysis can be conveniently conducted for any stock because it only analyzes the historical quantitative data that are relatively easy to obtain, such as the price and volume. Therefore, our proposed financial trading system generates the trading signals according to the results of the technical analysis.

Among the various technical indicators, the moving average (MA) is an important one that is used to reveal the trend line by smoothing the cyclical fluctuations in the price line and to identify trend reversals through crossovers. There are many variations of MAs employed in technical analysis. The three most common ones are: simple, weighted and exponential. Compared with the other two MAs, the exponential MA (EMA) assigns larger weights to more recent periods, which is considered more responsive to the evolving dynamics in the stock. An n-day EMA at time t represented by $\text{EMA}_n(t)$ can be derived using Equation (15).

$$EMA_n(t) = EMA_n(t-1) + \frac{2}{n+1}[y(t) - EMA_n(t-1)] \quad (15)$$

where $y(t)$ is the current stock price.

Used for signaling the trend reversal, price oscillator is an indicator based on the absolute or percentage difference between a short period MA and a long period MA. Using the percentage difference between a S-day EMA and a L-day EMA, the PPO is computed using Equation (16), where $S<L$. The main advantage of PPO is that its values are comparable across time periods regardless of change in price levels in the long run.

$$PPO_{S,L}(t) = \frac{EMA_S(t) - EMA_L(t)}{EMA_L(t)} \quad (16)$$

The advantage of using the MA and MA derived indicators, such as PPO, to generate trend reversal signal is that they are easy to use and understand and they guarantee that the system will catch every major price move. Therefore, the PPO is used to generate trading signal in our proposed trading model.

2. FNN Data Selection

The data selection (input/output definition) of the FNN must consider the following:

- **Data availability:** The availability of data suggests that a technical analysis approach be considered.
- **FNN input clustering:** The range of inputs should be relatively stable across time due to the restrictions imposed by the input clustering techniques of the FNN. This suggests that inputs like raw prices are inappropriate as the price can increase or decrease to a level that has never been seen in the training samples.
- **Input/output dependencies:** As noisy data does not provide meaningful learning patterns, and instead just confuses the network, there should be reasonable dependencies between inputs and outputs.

The *PPO* trading model is chosen as it can be easily constructed from historical data and has a stable value range across time. Because *PPO* suffers less from noise, it is expected to exhibit more regular and predictable patterns than price series or *MACD*. The main problem with *PPO* is the delay in trading signal generation, which is offset by the prediction capability of the FNN for early trading signal detection. Thus, the problem is formulated as follows:

- The trading model is based on *PPO*, i.e., go long when $PPO>0$ and go short when $PPO<0$.
- The FNN is to predict *PPO* in the expectation that the trading signal can be detected earlier, and thus improve the performance of the pure *PPO* model.
- The inputs to the FNN are the past n *PPO* values: $PPO(t_0)$, $PPO(t_0-1)$, ..., $PPO(t_0-n+1)$.
- The outputs of the FNN are the next m *PPO* values: $PPO(t_0+1)$, $PPO(t_0+2)$, ..., $PPO(t_0+m)$. The predicted *PPO* value is used in place of the current *PPO* value.

3. Performance Evaluation

RMSE and r are commonly used for the performance evaluation of FNNs. However, in the context of stock forecasting and trading they can be inadequate. A high r or a small *RMSE* between actual and predicted values cannot show if the trader can make any actual improvement in profits. From a trader's viewpoint, a small prediction value error with an incorrect trend direction is worse than a large prediction error but with a correct trend direction.

Moody et al (1998) proposed a method to evaluate trading performance in terms of multiplicative profits. The long, short and neutral positions are represented by 1, -1 and 0 respectively. The initial wealth $W(0)$ is set to 1, and wealth at time t is defined in Equations (17) – (18):

$$W(t) = W(t-1)(1 + r(t)F(t))(1 - \delta\left|F(t) - F(t-1)\right|) \tag{17}$$

$$r(t) = \frac{P(t)}{P(t-1)} - 1 \tag{18}$$

where $P(t)$ is the price at time t; $r(t)$ is the price interest (or the profit by only considering price change); $W(t)$ is the wealth at time t ($W(0)=1$); $F(t)$ is the position held at time t (1 = long, -1 = short, 0 = neutral) and δ is the transaction cost expressed as a percentage of the total transaction. The assumptions are that only one stock is being traded with the entire wealth being reinvested into the market.

This method is superior to RMSE or r because:

- Transaction cost is considered. This is closer to the real-life situation where overly frequent trade actions may be unprofitable because of the existence of transaction cost.
- Short position is considered. Though highly risky selling short is part of real-life trading. A bad trading decision that causes the trader to enter into a wrong trade position will get penalized, which is reflected as profit reduction.
- Clean and comparable measurement. Because the wealth is expressed as a ratio of initial wealth, it can be compared across time and stocks.

4. Experiments on Real-Life Data

We use 6 past *PPO* values as inputs and predict the next 4 *PPO* values. The periods of the MA window are selected empirically depending on the noisiness of the price data. The data is split into a training set and a testing set. The trading position is determined from the sign of the *PPO* values and the transaction cost is fixed at $\delta = 0.002$. The final wealth is computed as:

W^0: The final wealth with the pure *PPO* trading system, i.e., without prediction.
W^i: The final wealth with trading based on the *i*-day prediction of *PPO*.

The direct-output FAM-AARS is used because of its superior performance in time-series prediction (Guo et al., 2006), with the input fuzzy sets being evenly partitioned.

IBM (International Business Machines Corp) NYSE: The experimental data is constructed from 4548 price values for IBM obtained from Yahoo Finance during the period from 1st June 1979 to 28th May 1997 and consists of 4490 data points in total. The first 3000 data points are used for training with the remaining data used for testing. Figure 5(a) shows the result with MA period setting: Long = 50, Short = 30, AARS threshold = 0.5, and trading band = 0.01. Trading with a 4 day ahead prediction can accumulate wealth of 2.4235, while trading without prediction accumulates only 1.7586. It is shown that the more advanced the prediction, the earlier the trader can switch to the correct position, although there is a risk of inaccuracy when the prediction horizon increases. Figure 5(b) shows the trading decision and wealth accumulation without prediction and with 4 day ahead prediction. The inaccurate predictions of *PPO* at peak values is inconsequential as trading decisions rely more on the values that are close to the zero line. It is confirmed that the extra profit earned with prediction is due to the early action in trading, e.g., at $t = 250$, 1050, 1150, etc.

NKE (Nike Inc) NYSE: The experimental data is constructed from 3011 price values for NKE obtained from Yahoo Finance for the period 19th July 1987 to 19th July 1999 and consists of 2953 data points in total. The first 2000 data points are used for training with the remaining data used for testing. Figure 6(a) shows the result with MA period setting: Long = 50, Short = 30, AARS threshold = 0.5, and trading band = 0.01. NKE turns out to be non-profitable probably due to the change in the underlying dynamics of the stock market in the later part of the data set. Nevertheless, trading without prediction loses 35% while trading with 4 day ahead prediction is able to achieve a small profit (2%). A noise trading signal (Figure 6(b) at $t = 450$) represents the main reason that prediction is worse. Despite the noise, the prediction version catches up and outperforms the W^0 system because of the early trading decisions (e.g., $t = 720$, 780, 800, 820) later in the data set.

STAN (Standard Chartered PLC) London: The data is constructed from 3288 price values for STAN.L obtained from Yahoo for the period 6th January 1986 to 23rd July 1999 and consists of 3230 data points in total. The first 3000 data points are used for training. Figure 7(a) shows the result with MA period setting: Long = 50, Short = 30, AARS threshold = 0.5, and trading band = 0.01. Figure 7(b) shows the trading decision and wealth accumulation without prediction and with 4 day ahead prediction. As in the NKE data, the noise signal at t=440 causes a slight degradation in the performance. However, there is still a significant benefit from the prediction and early trading decisions at t=770, 870, 950, 1050.

STI (Straits Time Index) Singapore: The experimental data is constructed from 3878 price values for ^STI obtained from Yahoo Finance for the period 28th December 1987 to 25th July 2003 and consists of 3820 data points in total. The first 2500 data points are used for training with the remaining data used for testing. Figure 8(a) shows the result with MA period setting: Long = 50, Short = 25, AARS threshold = 0.5, and trading band = 0.01. The drop in wealth for 4 day ahead prediction suggests that the noise due to the increase of the prediction horizon and the corresponding decrease in prediction accuracy actually starts affecting the performance. Figure 8(b) shows the trading decision and wealth accumulation without prediction and with 3 day ahead prediction.

HSI (Hang Seng Index) Hong Kong: The experimental data is constructed from 4097 price

Figure 5. (a) Trading result (multiplicative wealth) on IBM, (b) Trading decisions and wealth on IBM

W^0	W^1	W^2	W^3	W^4
1.7586	1.7429	1.8415	2.1224	**2.4235**

(a)

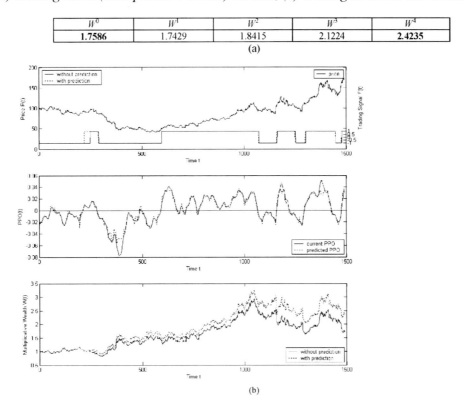

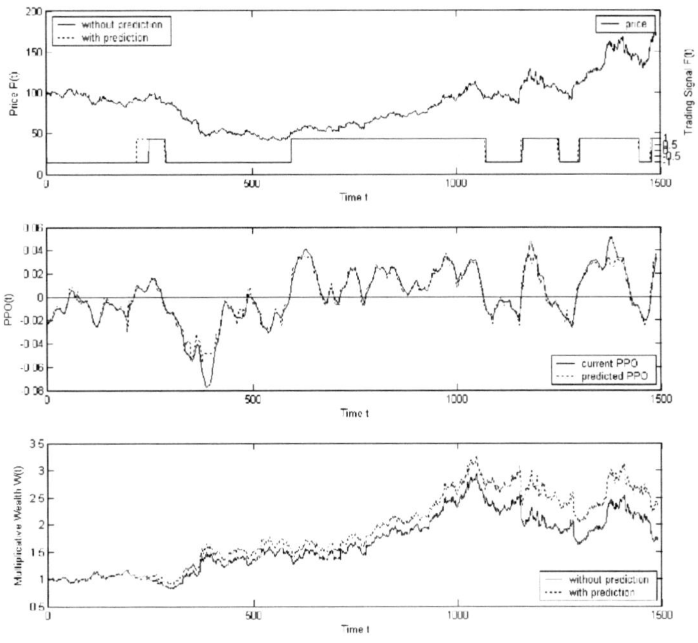

(b)

values for ^HSI obtained from Yahoo Finance for the period 31st December 1986 to 25th July 2003 and consists of 4039 data points in total. The first 2500 data points are used for training with the remaining data used for testing. Figure 9(a) shows the result with MA setting: Long = 50, Short = 25, AARS threshold = 0.5, and trading band = 0.01. Figure 9(b) is the trading decision and wealth accumulation without prediction and with 4 day ahead prediction. Wealth is negative; however, early trading action can cut the losses. Fine-tuning the MA periods does improve the results, showing that optimal settings vary with stock.

CONCLUSION

FAM-AARS, a novel neuron-fuzzy associative memory for a stock trading framework has been proposed and implemented. A redefinition of the modification functions and the similarity measure in AARS is performed to adapt AARS to FAM memory structure. The AARS threshold aims to reduce the effect from irrelevant rules, resulting in an improved throughput. The prediction and trading framework exploits the PPO for input preprocessing and decision making in trading. Numerical experiments conducted on real-life stock data confirm the validity of the design and the performance of the FAM-AARS system.

Further improvements such as automatic tuning of the AARS threshold value using techniques such as GA or objective function formulation and employing moving windows for training so that the network is more adaptive over time are being considered for future work. The relationship between the threshold value and the performance is being investigated. The threshold can be viewed as a way to increase the localization level for the network. Whether the system performance improves or degrades depends on if the original network is sufficiently or overly localized.

Figure 6. (a) Trading result (multiplicative wealth) on NKE, (b) Trading decisions and wealth on NKE

W^0	W^1	W^2	W^3	W^4
0.6556	0.57202	0.63618	0.84815	**1.0219**

(a)

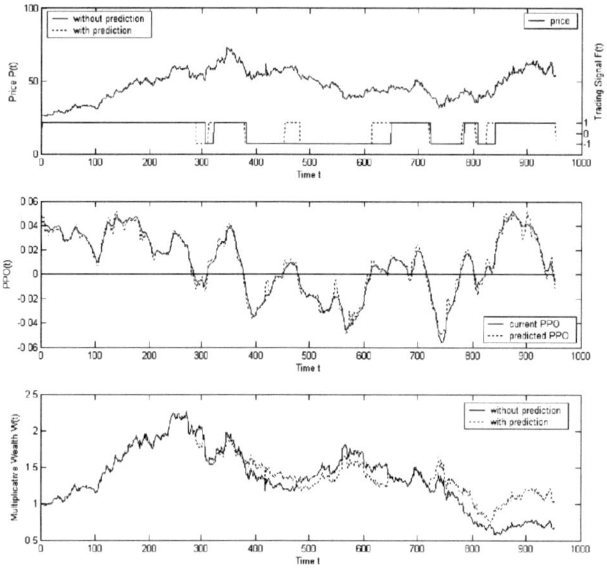

(b)

Figure 7. (a) Trading result (multiplicative wealth) on STAN, (b) Trading decisions and wealth on STAN

W^0	W^1	W^2	W^3	W^4
2.4761	2.5491	3.5822	3.1462	**3.8118**

(a)

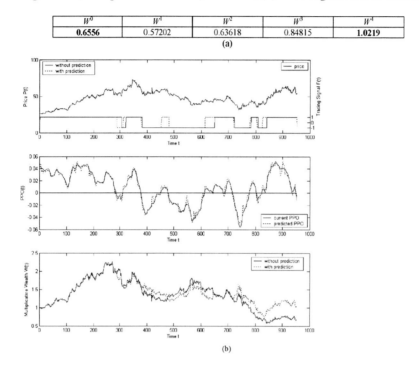

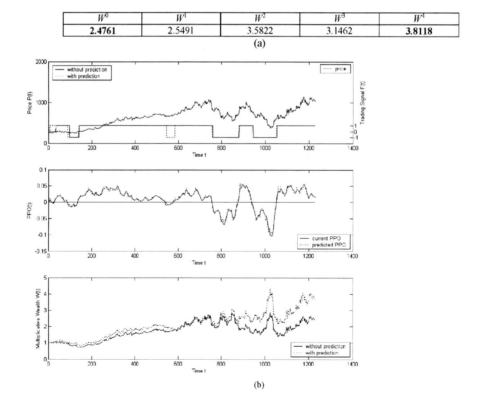

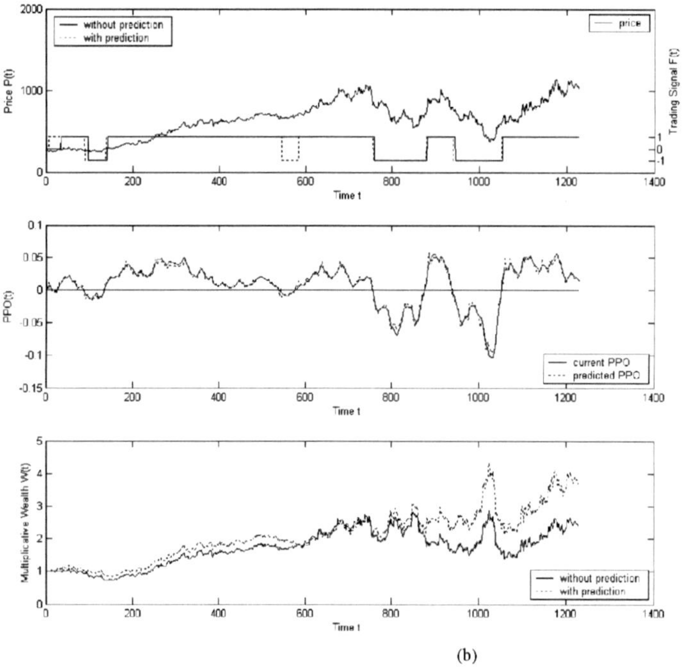

(b)

Figure 8. (a) Trading result (multiplicative wealth) on STI, (b) Trading decisions and wealth on STI

W^0	W^1	W^2	W^3	W^4
2.5932	3.1471	3.2256	**3.2599**	2.7293

(a)

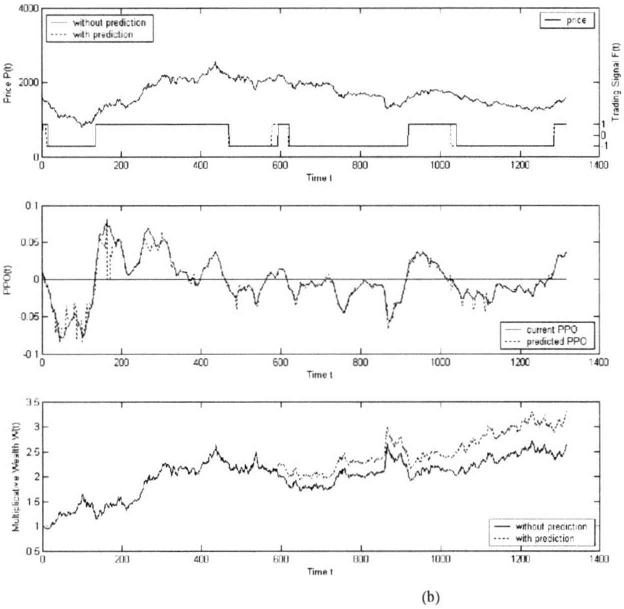

(b)

Figure 9. (a) Trading result (multiplicative wealth) on HSI, (b) Trading decisions and wealth on HSI

W^0	W^1	W^2	W^3	W^4
0.40613	0.5373	0.71318	0.80966	**0.82594**

(a)

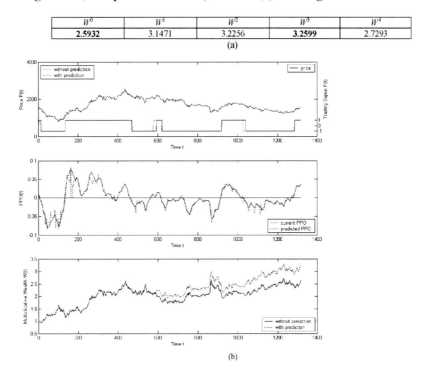

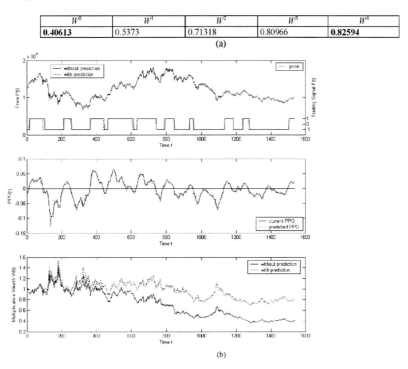

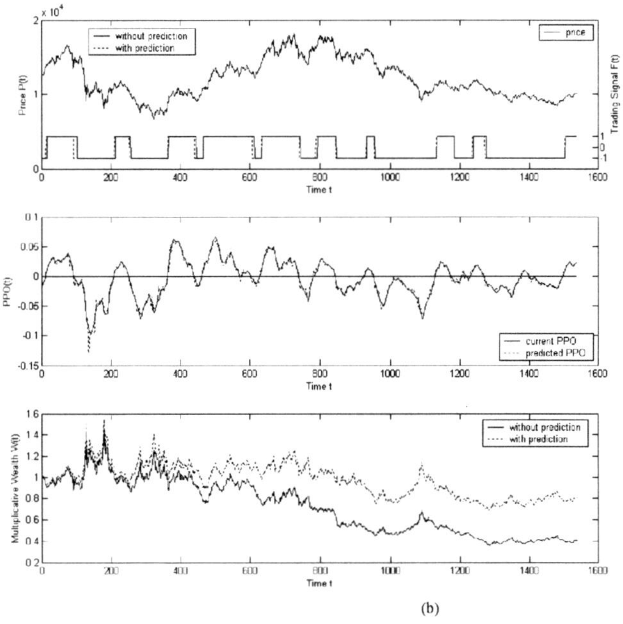

(b)

REFERENCES

Abarbanell, J. S., & Bushee, B. J. (1997). Fundamental analysis, future earnings, and stock prices. *Journal of Accounting Research, 35*, 1–24. doi:10.2307/2491464

Albus, J. S. (1972). *Theoretical and Experimental Aspects of a Cerebellar Model*. Unpublished PhD dissertation, University of Maryland.

Albus, J. S. (1975). A new approach to manipulator control: the cerebellar model articulation controller (CMAC). *Journal of Dynamics Systems, Measurement, and Control. Transactions of ASME, 97*(3), 220–227.

Ang, K. K., & Quek, C. (2000). Improved MCMAC with momentum, neighbourhood, and averaged trapezoidal output. *IEEE Transactions on Systems. Man and Cybernetics: Part B, 30*(3), 491–500. doi:10.1109/3477.846237

Ang, K. K., & Quek, C. (2006). Stock trading using RSPOP: a novel rough set neuro-fuzzy approach. *IEEE Transactions on Neural Networks, 17*(5), 1301–1315. doi:10.1109/TNN.2006.875996

Baba, N., Inoue, N., & Asakawa, H. (2000). *Utilization of neural networks & gas for constructing reliable decision support systems to deal stocks*. Osaka, Japan: Osaka-Kyoiku University.

Bowerman, B. L., & O'Connell, R. T. (1993). *Forecasting and Time Series: An Applied Approach*. Pacific Grove, CA: Duxbury Press.

Buckley, J. J., & Hayashi, Y. (1995). Neural nets for fuzzy systems. *Fuzzy Sets and Systems, 71*, 265–276. doi:10.1016/0165-0114(94)00282-C

Chen, A.-S., Leung, M. T., & Daouk, H. (2003). Application of neural networks to an emerging financial market: forecasting and trading the Taiwan Stock Index. *Computers & Operations Research, 30*(6), 901–923. doi:10.1016/S0305-0548(02)00037-0

Chiang, W.-C., Urban, L., & Baldridge, G. W. (1996). A neural network approach to mutual fund asset value forecasting. *Omega. International Journal of Management Science, 24*(2), 205–215.

Chou, T. S.-C., Yang, C.-C., Chen, C.-H., & Lai, F. (1996). A rule-based neural stock trading decision support system. In *Proceedings of the IEEE/IAFE Conference on Computational Intelligence for Financial Engineering*, New York (pp.148-154).

Dubois, D., & Prade, H. (1996). What are fuzzy rules and how to use them. *Fuzzy Sets and Systems, 84*, 169–185. doi:10.1016/0165-0114(96)00066-8

Glanz, F. H., Miller, W. T., & Kraft, L. G. (1991). An overview of the CMAC neural network. In *Proceedings of the IEEE Conference on Neural Networks for Ocean Engineering* (pp. 301-308).

Guo, Z., Quek, C., & Maskell, D. L. (2006). FCMAC-AARS: a novel FNN architecture for stock market prediction and trading. In *Proceedings of the International Conference on Evolutionary Computation*, Vancouver, BC, Canada.

Jang, J.-S. R. (1996). Input selection for ANFIS learning. In *Proceedings of the IEEE International Conference on Fuzzy Systems* (pp. 1493-1499).

Ker, J., Hsu, C., Kuo, Y., & Liu, B. (1997). A fuzzy CMAC model for color reproduction. *Fuzzy Sets and Systems, 91*(1), 53–68. doi:10.1016/S0165-0114(96)00083-8

Kim, H., & Lin, C. S. (1992). Use of adaptive resolution for better CMAC learning. In *Proceedings of the International Joint Conference on Neural Networks* (pp. 517-522).

Lam, M. (2004). Neural network techniques for financial performance prediction: integrating fundamental and technical analysis. *Decision Support Systems, 37*(4), 567–581. doi:10.1016/S0167-9236(03)00088-5

Leigh, W., Purvis, R., & Ragusa, J. M. (2002). Forecasting the NYSE composite index with technical analysis, pattern recognizer, neural network, and genetic algorithm: a case study in romantic decision support. *Decision Support Systems*, *32*(4), 361–377. doi:10.1016/S0167-9236(01)00121-X

Mantaras, R. L. (1990). *Approximate Reasoning Models*. Chichester, UK: Ellis Horwood Limited.

Mendel, J. M. (1995). Fuzzy logic systems for engineering: A tutorial. *Proceedings of the IEEE*, *83*(3), 345–377. doi:10.1109/5.364485

Miller, W. T., Glanz, F. H., & Kraft, L. G. (1990). CMAC: an associative neural network alternative to backpropagation. *Proceedings of the IEEE*, *78*, 1561–1567. doi:10.1109/5.58338

Moody, J. (1989). Fast learning in multi-resolution hierarchies. *Advances in Neural Information Processing Systems*, *1*, 29–38.

Moody, J., & Saffell, M. (2001). Learning to trade via direct reinforcement. *IEEE Transactions on Neural Networks*, *12*(4), 875–889. doi:10.1109/72.935097

Moody, J., Wu, L., Liao, Y., & Saffell, M. (1998). Performance functions and reinforcement learning for trading systems and portfolios. *Journal of Forecasting*, *17*(5-6), 441–470. doi:10.1002/(SICI)1099-131X(1998090)17:5/6<441::AID-FOR707>3.0.CO;2-#

Nakanishi, H., Turksen, I. B., & Sugeno, M. (1993). A review and comparison of six reasoning methods. *Fuzzy Sets and Systems*, *57*, 257–294. doi:10.1016/0165-0114(93)90024-C

Nhut, M., Shi, D., & Quek, C. (2006). FCMAC-BYY: fuzzy CMAC using Bayesian Ying-Yang learning. *IEEE Transactions on Systems, Man and Cybernetics. Part B*, *36*(5), 1180–1190.

Pring, M. J. (2002). *Technical Analysis Explained: The Successful Investor's Guide to Spotting Investment Trends and Turning Points*. New York: McGraw-Hill.

Quek, C., & Singh, A. (2005). POP-Yager: a novel self-organising fuzzy neural network based on the Yager inference. *Expert Systems with Applications*, *29*(1), 229–242. doi:10.1016/j.eswa.2005.03.001

Quek, C., & Zhou, R. W. (2006). Structure and learning algorithms of a nonsingleton input fuzzy neural network based on the approximate analogical reasoning schema. *Fuzzy Sets and Systems*, *157*(13), 1814–1831. doi:10.1016/j.fss.2005.12.010

Saad, E. W., Prokhorov, D. V., & Wunsch, D. C. II. (1998). Comparative study of stock trend prediction using time delay, recurrent and probabilistic neural networks. *IEEE Transactions on Neural Networks*, *9*(6), 1456–1470. doi:10.1109/72.728395

Sim, J., Tung, W. L., & Quek, C. (2006). FC-MAC-Yager: A novel Yager inference scheme based fuzzy CMAC. *IEEE Transactions on Neural Networks*, *17*(6), 1394–1410. doi:10.1109/TNN.2006.880362

Stephen, H. L., David, A. H., & Jack, J. G. (1992). Theory and development of higher-order CMAC neural networks. *IEEE Control Systems*, *12*(2), 23–30. doi:10.1109/37.126849

Thammano, A. (1999). Neuro-fuzzy model for stock market prediction. In *Proceedings of the Artificial Neural Networks in Engineering Conference (ANNIE'99)* (pp. 587-591).

Trippi, R. R., & Turban, E. (1993). *Neural Networks in Finance and Investing: Using Artificial Intelligence to Improve Real-World Performance*. Chicago: Probus Publishing Company.

Tung, W. L., & Quek, C. (2002). GenSoFNN: a generic self-organizing fuzzy neural network. *IEEE Transactions on Neural Networks*, *13*(5), 1075–1086. doi:10.1109/TNN.2002.1031940

Tung, W. L., Quek, C., & Cheng, P. (2004). GenSo-EWS: a novel neural-fuzzy based early warning system for predicting bank failures. *Neural Networks*, *17*(4), 567–587. doi:10.1016/j.neunet.2003.11.006

Turksen, I. B., & Zhong, Z. (1990). An approximate analogical reasoning schema based on similarity measures and interval-valued fuzzy sets. *Fuzzy Sets and Systems*, *34*, 223–346. doi:10.1016/0165-0114(90)90218-U

Wang, Y. (2003). Mining stock price using fuzzy rough set system. *Expert Systems with Applications*, *24*, 13–23. doi:10.1016/S0957-4174(02)00079-9

White, H. (1988). Economic prediction using neural networks: the case of IBM daily stock returns. In *Proceedings of the Second Annual IEEE Conference on Neural Networks* (pp. 451-458).

Wilson, C. L. (1994). Self-organizing neural network system for trading common stocks. In *Proceedings of the IEEE International Conference on Neural Networks* (pp. 3651-3654).

Zadeh, L. A. (1975). Calculus of fuzzy restrictions. In Zadeh, L. A.,(Eds.), *A.Fuzzy Sets and Their Applications to Cognitive and Decision Processes* (pp. 1–39). New York: Academic Press.

Zhou, R. W., & Quek, C. (1996). POPFNN: A pseudo outer-product based fuzzy neural network. *Neural Networks*, *9*(9), 1569–1581. doi:10.1016/S0893-6080(96)00027-5

Zhou, R. W., & Quek, C. (1999). POPFNN-AAR(S): A pseudo outer-product based fuzzy neural network. *IEEE Transactions on Systems, Man, and Cybernetics*, *29*(6), 859–870. doi:10.1109/3477.809038

This work was previously published in the International Journal of Fuzzy System Applications, Volume 1, Issue 1, edited by Toly Chen, pp. 61-78, copyright 2010 by IGI Publishing (an imprint of IGI Global).

Chapter 8
Hand Gesture Recognition Using Multivariate Fuzzy Decision Tree and User Adaptation

Moon-Jin Jeon
Korea Aerospace Research Institute, Korea

Sang Wan Lee
Massachusetts Institute of Technology, USA

Zeungnam Bien
Ulsan National Institute of Science and Technology, Korea

ABSTRACT

As an emerging human-computer interaction (HCI) technology, recognition of human hand gesture is considered a very powerful means for human intention reading. To construct a system with a reliable and robust hand gesture recognition algorithm, it is necessary to resolve several major difficulties of hand gesture recognition, such as inter-person variation, intra-person variation, and false positive error caused by meaningless hand gestures. This paper proposes a learning algorithm and also a classification technique, based on multivariate fuzzy decision tree (MFDT). Efficient control of a fuzzified decision boundary in the MFDT leads to reduction of intra-person variation, while proper selection of a user dependent (UD) recognition model contributes to minimization of inter-person variation. The proposed method is tested first by using two benchmark data sets in UCI Machine Learning Repository and then by a hand gesture data set obtained from 10 people for 15 days. The experimental results show a discernibly enhanced classification performance as well as user adaptation capability of the proposed algorithm.

INTRODUCTION

Human computer interaction (HCI) technology has been widely used in various assistive systems for the disabled and the elderly. One of the recent highlighted topics is "understanding a user's intention" from natural human signals such as voice or gesture. Those signals, if successfully recognized, can provide a comfortable and convenient means for the user to interact with an engineering system.

DOI: 10.4018/978-1-4666-1870-1.ch008

For example, a vision-based hand gesture recognition technique can be used to control a multitude of home appliances. Do et al. (2005) developed the Soft Remote Control System which enables the disabled user to control various home appliances using a set of simple hand gestures. Positions of a face and one hand are calculated using images obtained by stereo cameras. A concatenation of those positions constitutes a 3D trajectory of hands, from which the system recognizes those user's commands.

Critical factors that affect the performance of such systems are known to be false positive errors and inter-person variation / intra-person variation. False positive errors are caused by hand gestures that are meaningless but similar to some hand command gestures. To cope with this problem of false positive error, Yang (2007) proposed a gesture spotting method using the fuzzy garbage model in which an input gesture is classified either as a command gesture or a garbage gesture. The experimental results of the study shows that the command gestures such as "up" or "left" are effectively distinguished from the garbage gestures such as "eating" or "reading". In this paper, we deal with the latter problem of inter-person variation and intra-person variation.

When multiple users access to the system, the user independent (UI) recognition algorithms cannot compete with the user dependent (UD) recognition algorithms in the recognition rate. Furthermore, even for the same user, some characteristics of hand motion vary over time, which results in degradation of performance. The inter-person variation problem can be tackled by properly invoking some of the UD model techniques, model selection methods, or user adaptation strategies. The intra-person variation problem can be tackled by using fuzzy logic owing to its robustness property against uncertainty and ambiguity of human motion.

In particular, fuzzy decision tree learning has been widely used in classification problems due to its two merits: (1) interpretability of the decision tree and (2) capability of fuzzy logic in handling uncertainty and ambiguity (Janikow, 1998). Though the fuzzy decision tree is known to show a good performance in learning and classification tasks, however, it can be vulnerable to an overfitting situation that degrades prediction and adaptation performance. The higher the degree of overlap among membership functions is, the bigger the structure of the fuzzy decision tree becomes.

In this paper, we propose a multivariate fuzzy decision tree (MFDT) structure which effectively prunes the decision tree so as to enhance the classification and adaptation performance. The fuzzy decision tree model can be simplified by using a multivariate concept. Specifically, several recognition models are first built, and the best model that fits a new user is selected by using the maximum likelihood model comparison. Subsequently a user adaptation algorithm is presented, based on the gradient descent method.

To demonstrate the performance of the proposed algorithm, we use IRIS and WINE data set in UCI Machine Learning Repository (Merz et al., 1996) and a hand gesture data set which is collected from 10 people for 15 days. The experimental results show that the classification and user adaptation performances of the proposed method are better than those of other well-known fuzzy decision tree techniques.

VISION-BASED HAND GESTURE RECOGNITION FOR THE SOFT REMOTE CONTROL SYSTEM

Vision-based hand gesture recognition for the soft remote control system (Do et al., 2005) is carried out in the following steps:

1. Face and hand "region of interests" (ROI) are extracted from camera images.
2. A trajectory of the hand position relative to the face position is calculated.

3. A start position and an end position of the trajectory are segmented.
4. Segmented trajectories are classified.

A set of hidden Markov models (HMM) can be used to recognize the hand gestures that contain temporal and spatial information (Rabiner, 1989). In the learning process, the parameters of the HMMs are optimized to fit the training sequences that correspond to the hand gestures. In the recognition process, the best matching HMM model for the observed sequence is chosen (Yamato et al., 1992). Without any painstaking feature selection process, vector quantized (VQ) label and discrete HMM can be used for hand gesture recognition. It can be readily applied to complex gesture recognition problems. In spite of these merits, the HMM-based method has been known to require excessive amount of training data (Bourlard, 1990).

Hong et al. (2000) used finite state machine (FSM) for hand gesture recognition. Gestures are modeled as sequences of states in spatial-temporal space. The trajectories of a hand gesture are a set of points represented by a set of Gaussian spatial regions. In learning hand gestures, the number of states and their spatial parameters are calculated. The temporal information from the segmented hand gesture is then added to the states. The resulting state sequence can be regarded as a FSM recognizer. When a new sample is presented, each gesture recognizer decides whether to remain in the current state or to go to the next state. The sample is classified as the gesture whose FSM recognizer reaches to a final state. What distinguishes FSM from HMM is that FSM aborts a corresponding recognition process when a next point of the trajectory of the sample gesture is located far from the cluster of the FSM model. Hence, FSM is often believed to be simpler and faster than HMM.

Yang (2007) used fuzzy logic for hand gesture recognition due to its capability of easily dealing with ambiguity and uncertainty of human signals such as hand gestures. Using fuzzy sets, the weakness of the system caused by a fixed crisp decision boundary can be overcome. Also multiple membership functions can be utilized to effectively classify multiple classes (Su, 2000).

There are two major stumbling blocks that can affect the performance of a hand gesture recognition system. The first is the variability of the characteristics of the hand gestures among different individuals. It is called inter-person variation (Yang et al., 1996). Figure 1 illustrates the trajectories of the hand gestures (including up, left, and clockwise circle) of three subjects. We can easily find that the angle and length of line-type hand gestures and the radius of the clockwise circle gesture of these subjects are different. Figure 2 represents the inter-person variation in a feature space. It can be seen from the Figure 2 that distributions of the feature vectors of user 1 and user 2 are very different.

Secondly, even for the same person, hand gesture characteristics can vary from time to time, which is called "Intra-person variation". Figure 3 illustrates the characteristics of hand gesture which were recorded from one person for 15 days. It can be seen from Figure 3 that these features vary over time. This phenomenon is known to cause misclassification (Su, 2000).

One possible method for reducing the phenomena of inter-person variation and intra-person variation is a technique of loosening the decision boundaries of a hand gesture while not damaging the decision boundaries of other hand gestures (Jung, 2006). However, the false positive error caused by misclassification of unintended motions might be also increased. Figure 4 shows that classification rate is inversely proportional to false positive rejection rate.

Model construction methods can be stratified according to how training data are organized, such as user independent (UI) model construction and user dependent (UD) model construction.

The UI model is trained with the data of multiple users. On the contrary, each UD model is

Figure 1. Hand gesture trajectories of three subjects

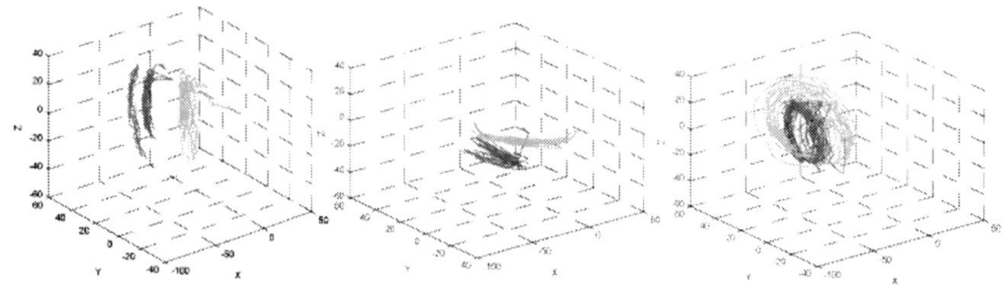

trained with the data of the corresponding user; the number of users is thus equal to the number of the UD models. Though it is easy to collect training data for the UI model construction, the risk of having lower classification rate is higher than that of the UD model. Conversely, though it is easy to achieve high classification rate with UD models, it is often difficult to gather training data for a new user. If we use a UI model to recognize a new user's hand gestures, the performance could be hampered by inter-person variation. To cope with this problem, various methods of user adaptation and personalized recognition have been studied. Jung (2007) suggested an adaptation method for a UI model and successfully applied to Korean sign language recognition problem. For the facial expression recognition problem, Kim (2004) suggested an example of a user dependent model and a model selection method.

In the next section, we propose a hand gesture recognition system which is capable of handling multiple users and the case of adding a new user. We then construct a personalized hand gesture recognition system, which includes the proposed MFDT learning and classification method. The UD model for each user is trained using the MFDT learning method. A maximum likelihood-based model comparison method is used to select a model that is fitted for a new user's patterns. Adaptation of this model for a new user is conducted by a gradient descent-based adaptation method.

MULTIVARIATE FUZZY DECISION TREE

A fuzzy decision tree provides a powerful means to overcome a major limitation of the decision tree, which stems from a crisp decision boundary. The key characteristic of fuzzy decision tree learning is that a 'single' sample can be designated as a reference point in 'multiple' nodes. Hence, the size and complexity of the trained fuzzy decision tree can be increased excessively, which brings about a poor generalization performance. This weakness of a fuzzy decision tree can be resolved by introducing an attribute vector, as opposed to using a separate attribute that branches each node of a fuzzy decision tree.

In a learning process, each node branches out to its child nodes by the points which maximize information gain of the node. In Figure 5, for example, if each node of the decision tree is branched by only one attribute, the decision tree may need six nodes for a complete separation of the data points. A multivariate decision tree suggested by Yildiz et al. (2001), however, needs just one split. We assert that the same principle can be effectively applied to the fuzzy decision tree. The resulting algorithm we propose in this paper is called a MFDT.

Figure 2. Inter-person variation in feature space

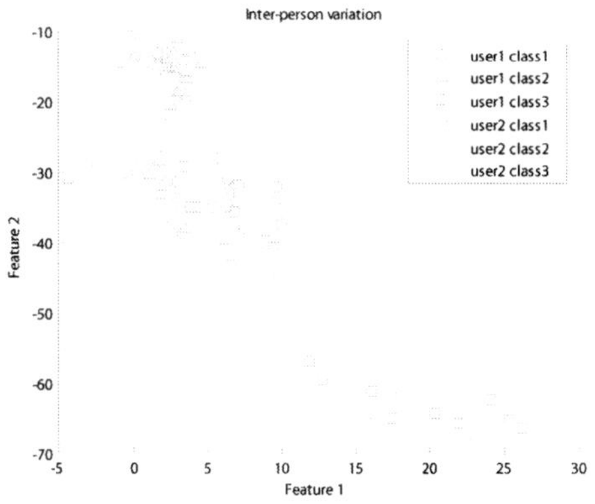

Figure 3. Variation of hand gesture characteristic for 15 days

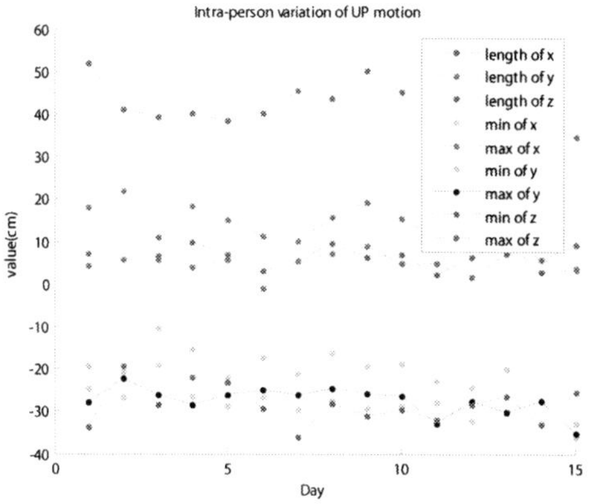

MFDT Learning

While FDT uses a single attribute to split each node, MFDT uses an attribute vector (i.e., multiple attributes) which is obtained by using linear discriminant analysis (LDA). LDA is a dimension reduction method that can be applied to classification problems (Alpaydin, 2004). It finds a projection vector **w** such that separability of the projected data is maximized as follows:

$$\text{Maximize } J(\mathbf{w}) = \frac{\mathbf{w}^T S_B \mathbf{w}}{\mathbf{w}^T S_W \mathbf{w}}, \quad (1)$$

$$\text{where, } S_B = \sum_{i=1}^{K} (\mathbf{m}_i - \mathbf{m})(\mathbf{m}_i - \mathbf{m})^T, \quad (2)$$

$$S_W = \sum_{i=1}^{K} S_i, \quad (3)$$

$$S_i = \sum_{\mathbf{x} \in \text{ class } i} (\mathbf{x} - \mathbf{m}_i)(\mathbf{x} - \mathbf{m}_i)^T, \quad (4)$$

$$\mathbf{m} = \frac{1}{K} \sum_{i=1}^{K} \mathbf{m}_i. \quad (5)$$

Here **x** denotes a sample point, K is the total number of classed, and $\mathbf{m}_i$ is the mean of samples of the i_{th} class.

The largest eigenvector of $S_W^{-1} S_B$ is the solution **w** that maximizes $J(\mathbf{w})$ (Alpaydin, 2004). This vector is used as an attribute vector onto which the given data are projected. Figure 6 represents LDA and projection results of Iris data from UCI Repository (Merz et al., 1996).

The whole process of MFDT learning is as follows:

Step 1: Generate a root node. After generating a root node, assign all training data to the root node.

Step 2: Build a node such that the information gain is maximized.

1. Obtain **w** using LDA, where **w** = the maximum eigenvector of $S_W^{-1} S_B$. (6)
2. Calculate the attribute value using **w**: $z = \mathbf{w}^T \mathbf{x}$. (7)
3. Calculate the entropy of the current node:

$$Entropy(S) = -\sum_i P_i^S \log_2 P_i^S, \ (8)$$

$$\text{where } P_i^S = \frac{N_S^i}{N_S}, \ N_S = \sum_i N_S^i \quad (9)$$

Figure 4. Relationship between classification rate and false positive rejection rate in hand gesture recognition task

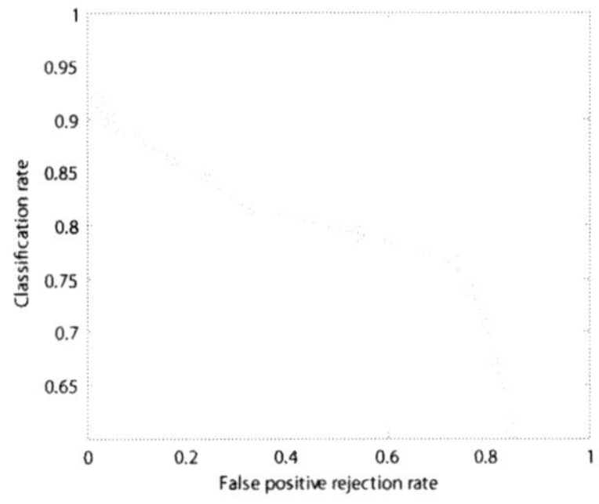

Figure 5. Example of Univariate split and multivariate split for Iris data (Merz et al., 1996)

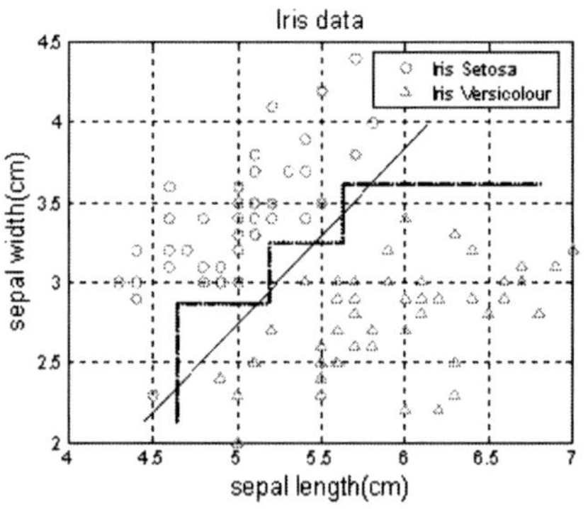

with N_S^i = number of data of the i_{th} class and S=a set of the attribute values z assigned in the current node.

4. Find a membership function which maximizes the information gain when the set of the attribute value z in the current node is split by using the membership function.

If the attribute values are arranged in an ascending order, there should be some points by which the class changes when crossing them. If we set N membership functions, we can get all combinations of N-1 such points. Then those combinations are used to make the membership functions. The middle points of the triangular membership functions are calculated as follows:

$$m^{v,c} = \frac{1}{n_v}\sum_{j=1}^{n_v} z_j^v \ , \ v = 1, ..., N \qquad (10)$$

where n_v denotes the number of the attribute value z in the v_{th} area; z_j^v denotes the j_{th} value of the set of z when z in the v_{th} area are arranged in ascending order; $m^{v,c}$ denotes the middle point of the v_{th} membership function.

We simply define the left and right points of each triangular membership function with their middle points as:

$$m^{v,l} = m^{v,c} - 0.5(1+\gamma)(m^{v,c} - m^{v-1,c}),$$
$$m^{v,r} = m^{v,c} + 0.5(1+\gamma)(m^{v+1,c} - m^{v,c}), \qquad (11)$$

where γ is the degree of overlapping in a membership function (Figure 7).

Using the above parameters of the membership function, we can calculate the fuzzy membership values for the attribute values in the current node.

If we define $\mu_{S_{v|\mathbf{w}}}(x)$ as the v_{th} membership function of current node, the information gain of the current node is calculated as follows:

$$Gain(S, \mathbf{w}) = Entropy(S) - \sum_v \frac{N_{S_{v|\mathbf{w}}}}{N_S} Entropy(S_{v|\mathbf{w}}), \qquad (12)$$

where $Entropy(S_{v|\mathbf{w}}) = -\sum_i P_i^{S_{v|\mathbf{w}}} \log_2 P_i^{S_{v|\mathbf{w}}}$, (13)

$$P_i^{S_{v|\mathbf{w}}} = \frac{C^i_{S_{v|\mathbf{w}}}}{C_{S_{v|\mathbf{w}}}}, C_{S_{v|\mathbf{w}}} = \sum_i C^i_{S_{v|\mathbf{w}}}, \qquad (14)$$

$$C^{i}_{S_{v|\mathbf{w}}} = \sum_{\substack{class\ of\ z=i \\ z\in Supp\left(S_{v|\mathbf{w}}\right)}} \mu_{S_{v|\mathbf{w}}}\left(z\right), \tag{15}$$

$$S_{v|\mathbf{w}} = \{(z, \mu_{S_{v|\mathbf{w}}}(z)) \mid \mu_{S_{v|\mathbf{w}}}(z) : \text{the membership value of the } v_{th} \text{ membership function.}\} \tag{16}$$

5. Calculate the attribute and the membership function in case that the current node is branched by using a single attribute (univariate case); the corresponding attribute vector **w** is as follows:

$$\mathbf{w} = [1\ 0\ .\ .\ 0]^T, [0\ 1\ .\ .\ 0]^T, ..., [0\ 0\ .\ .\ 1]^T \tag{17}$$

6. Select a univariate or multivariate node which has bigger information gain. The branch and child nodes are generated by using the membership function and the **w** that is selected.

Step 3: If the termination conditions are satisfied, make the current node a leaf node which refers to a class label of the majority of training data in the leaf node. If the termination conditions are not satisfied, go back to Step 2 for each child node.

The termination conditions are as follows:

1. The class labels of all data in a current node are the same.
2. The depth of the current node is larger than a predefined maximum value. The depth of a node is the number of nodes from the root node to the current node.

Figure 8 represents a membership function generated by the procedure. Figure 9 represents a MFDT trained by using Iris data set. The trained MFDT achieved 98% classification accuracy.

MFDT Classification

The classification procedure is as follows.

Step 1: Calculate the T-norm of the membership functions and attribute vectors of the nodes from the root node to each leaf node.

1. Projection of input data into the attribute vector in each node $z_i = \mathbf{w}_i^T \mathbf{x}$ (18)
2. The T-norm from the root node to the n_{th} leaf node is defined as follows: $T_n = \prod_{i=\text{root node}}^{n_{th}\ \text{leaf node}} \mu^{i}_{S_{v|\mathbf{w}}}\left(z_i\right)$ (19)

The T-norm value is calculated by this method.

Step 2: Calculate the average T-norm of the leaf nodes that belong to the same class. Then classify the input data as the class with the maximum average T-norm value.

Figure 6. LDA results of Iris data

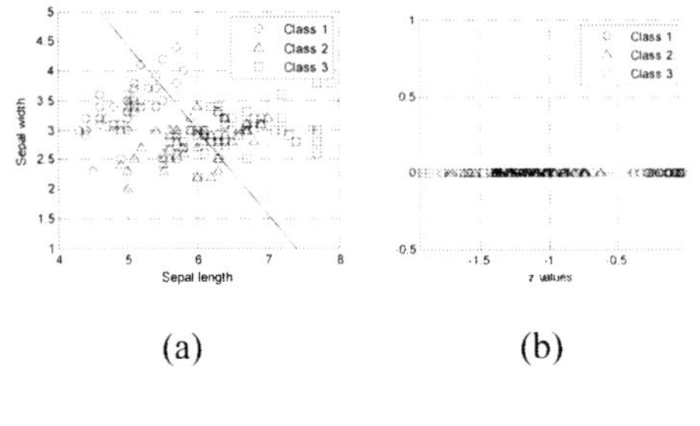

(a) (b)

Figure 7. Degree of overlap γ

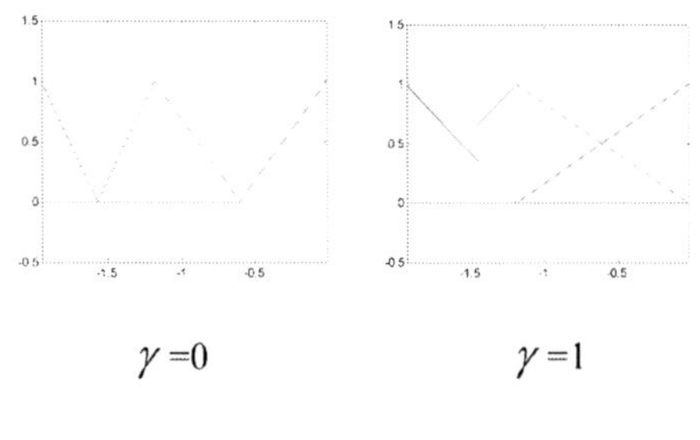

γ =0 γ =1

Figure 8. Automatically generated membership function

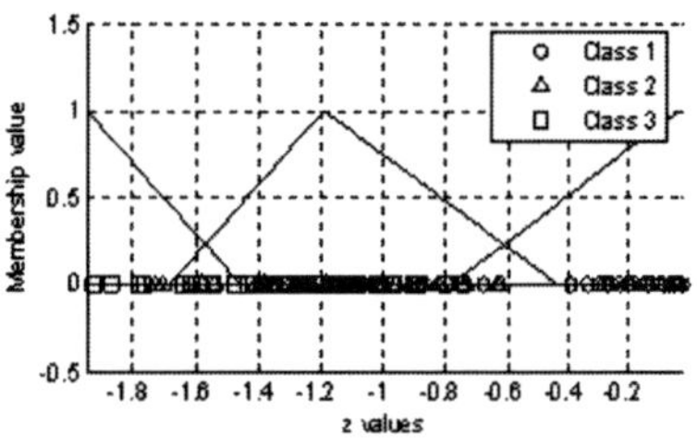

A_i : The average value of T-norms of leaf nodes which have the i_{th} class label.

$$class\, C = \arg\max_i A_i \tag{20}$$

MODEL SELECTION AND USER ADAPTATION

Model Selection

A multitude of recognition models that are generated by using multiple users' data sets are kept in a model pool and are used to recognize the input hand gestures.

When a new user starts to use the system, the most appropriate model is selected by using a model selection method (Figure 10). We measure how well the recognition model fits a hand gesture using the maximum likelihood model comparison (Duda et al., 2001).

$$P(m_i \mid D) = \frac{P(D \mid m_i)\, p(m_i)}{p(D)} \propto P(D \mid m_i)\, p(m_i) \tag{21}$$

$$\text{Selected Model } \hat{m} = \arg\max_{m_i} P(D \mid m_i) \tag{22}$$

Figure 9. Trained MFDT

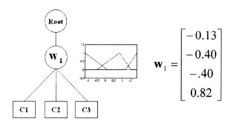

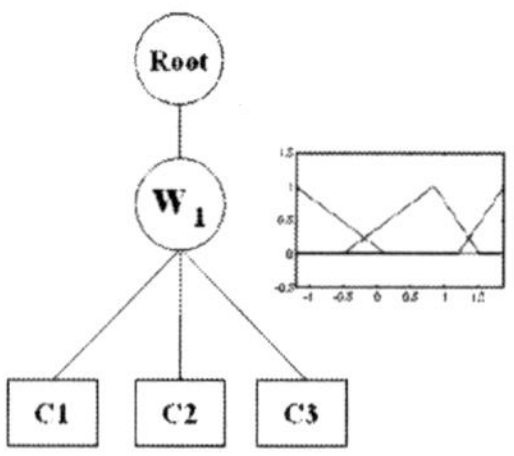

$$\mathbf{W}_1 = \begin{bmatrix} -0.13 \\ -0.40 \\ -.40 \\ 0.82 \end{bmatrix}$$

D : data, m_i : i_{th} model

The model pool is built by using hand gestures of multiple users; A next step is to adapt to new users' patterns or to a change in existing users' patterns.

User Adaptation

The MFDT model selected in the model selection phase is used to recognize the new user's hand gestures. It also can adapt to the patterns of the new user's hand gestures; for instance, it could be adapted by a gradient descent based adaptation method. The MFDT adaptation is a kind of incremental adaptation methods (Fu et al., 2000).

An MFDT model has information about the membership function (refer to Figure 11). The average T-norm value of the leaf nodes which have the same class label as that of input data has to be bigger than any other average T-norm values of the leaf nodes which have other classes. We can select an error function which becomes small if the T-norm of the leaf nodes which have the same class label as that of input data is closed to 1. Then the error function can be adapted to minimize the error.

The adaptation process for the whole leaf nodes is as follows.

1. The input data is projected onto attribute vector of each node which is on the route from the root node to each leaf node. The projected value is given by $z_n = \mathbf{w}_n^T \mathbf{x}$, (23) where $\mathbf{w}_n$ is the attribute vector of the n_{th} node from the root node, and $T = \prod_{n=1}^{N} \mu_n^i(z_n)$, (24) where N is the depth of the parent node.

Update the parameters of the membership function of each node by using a gradient descent method. While the center value of a membership function is calculated by using training data, the left and right values of the membership function are calculated based on the relationship of the center values of successive membership functions:

$$m_n^{i,l} = m_n^{i,c} - 0.5(1+\gamma)\left(m_n^{i,c} - m_n^{i-1,c}\right), \tag{25}$$

$$m_n^{i,r} = m_n^{i,c} + 0.5(1+\gamma)\left(m_n^{i+1,c} - m_n^{i,c}\right), \tag{26}$$

where n: node number; i: membership function number; γ: the degree of overlapping in a membership function (Figure 7).

The membership value is calculated as follows:

$$\mu_n^i(z_n) = \begin{cases} \dfrac{z_n - (1-\gamma)m_n^{i,c} - \gamma m_n^{i-1,c}}{\gamma\left(m_n^{i,c} - m_n^{i-1,c}\right)}, & z_n < m_n^{i,c} \\ \dfrac{z_n - (1-\gamma)m_n^{i,c} - \gamma m_n^{i+1,c}}{\gamma\left(m_n^{i,c} - m_n^{i+1,c}\right)}, & z_n > m_n^{i,c} \end{cases} \tag{27}$$

$\mu_n^i(z_n)$: membership value of input data

An error function whose minimization increases T-norm is defined as follows:

$$E = \frac{1}{2}(T-1)^2 = \frac{1}{2}\left(\prod_{n=1}^{N} \mu_n^i(z_n) - 1\right)^2. \tag{28}$$

And its adaptation rule is given by a gradient descent method:

Figure 10. Model Selection process in MFDT-based hand gesture recognition system

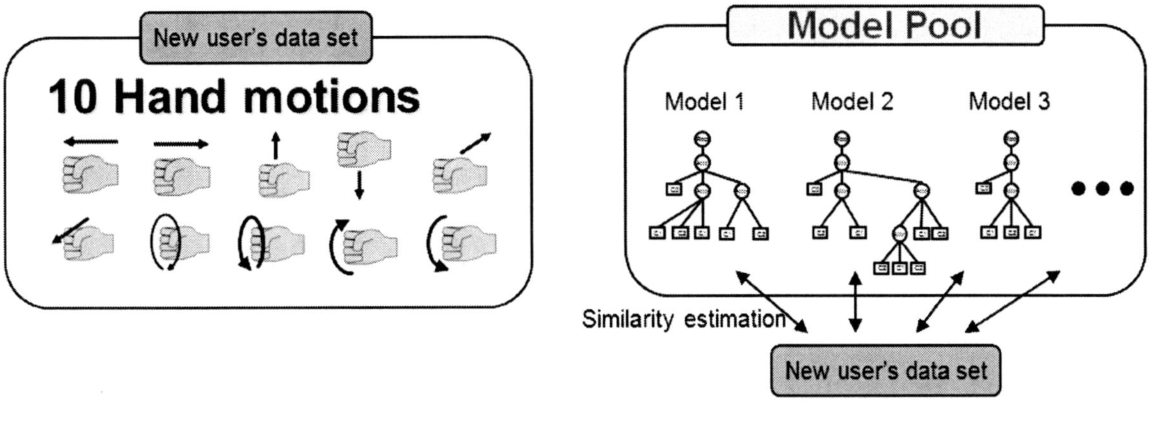

$$\frac{\partial E}{\partial m_n^{i,c}} = \begin{cases} \left(\prod_{n=1}^{N} \mu_n^i(z_n) - 1\right) \cdot \left(\prod_{\substack{k=1 \\ k \neq n}}^{N} \mu_k^i(z_n)\right) \cdot \frac{m_n^{i-1,c} - z_n}{\gamma\left(m_n^{i,c} - m_n^{i-1,c}\right)^2}, z_n < m_n^{i,c} \\ \left(\prod_{n=1}^{N} \mu_n^i(z_n) - 1\right) \cdot \left(\prod_{\substack{k=1 \\ k \neq n}}^{N} \mu_k^i(z_n)\right) \cdot \frac{m_n^{i+1,c} - z_n}{\gamma\left(m_n^{i,c} - m_n^{i+1,c}\right)^2}, z_n > m_n^{i,c} \end{cases}, \quad (29)$$

$$m_n^{i,c} \leftarrow m_n^{i,c} - \eta \frac{\partial E}{\partial m_n^{i,c}}, \quad (30)$$

$$m_n^{i,l} \leftarrow m_n^{i,c} - 0.5(1+\gamma)\left(m_n^{i,c} - m_n^{i-1,c}\right), \quad (31)$$

$$m_n^{i,r} \leftarrow m_n^{i,c} + 0.5(1+\gamma)\left(m_n^{i+1,c} - m_n^{i,c}\right), \quad (32)$$

$$m_n^{i+1,l} \leftarrow m_n^{i+1,c} - 0.5(1+\gamma)\left(m_n^{i+1,c} - m_n^{i,c}\right), \quad (33)$$

$$m_n^{i-1,r} \leftarrow m_n^{i-1,c} + 0.5(1+\gamma)\left(m_n^{i,c} - m_n^{i-1,c}\right), \quad (34)$$

where η denotes the adaptation size for an adaptation step.

Table 1. Benchmark data sets

Data set	Classes	Instances	Features
Iris	3	150	4
Wine	3	178	13

3. Repeat 2) until the T-norm becomes smaller than the predefined value.

EXPERIMENTAL RESULTS

Classification Test using Benchmark Data

We tested the MFDT learning and classification method using Iris and Wine data set of UCI Machine Learning Repository (Merz et al., 1996). Table 1 provides the specification of the data sets.

We used a 5x2 fold cross validation method to test the proposed method. Test results for Iris data and Wine data are shown in Tables 2 and 3. Note that MFDT achieved higher classification rate with fewer nodes than FDT. It can be seen from Table 4 that MFDT outperforms several conventional decision tree algorithms with a crisp decision boundary such as C4.5, and C5.0. γ is set to be 0.5. Table 5 illustrates how the value of γ influences the number of nodes and the classification rate.

Classification Test using Hand Gesture Data

The hand gesture data were collected using 3 stereo camera units equipped on the ceiling of an intelligent residential space (Bien et al., 2002). The images were saved at 10 frames per second. The start and end points of hand gestures were manually segmented. We collected 10 kinds of hand gestures from 10 people for 15 days. Table 6 shows the 10 kinds of hand gestures and the 16 kinds of features used to build MFDT models.

The unintended or meaningless motion should be taken into consideration in a learning process of a hand gesture recognition model because a careful consideration of these factors can reduce false positive error. To obtain the data of unin-

Figure 11. Parameters of membership function in a node of MFDT

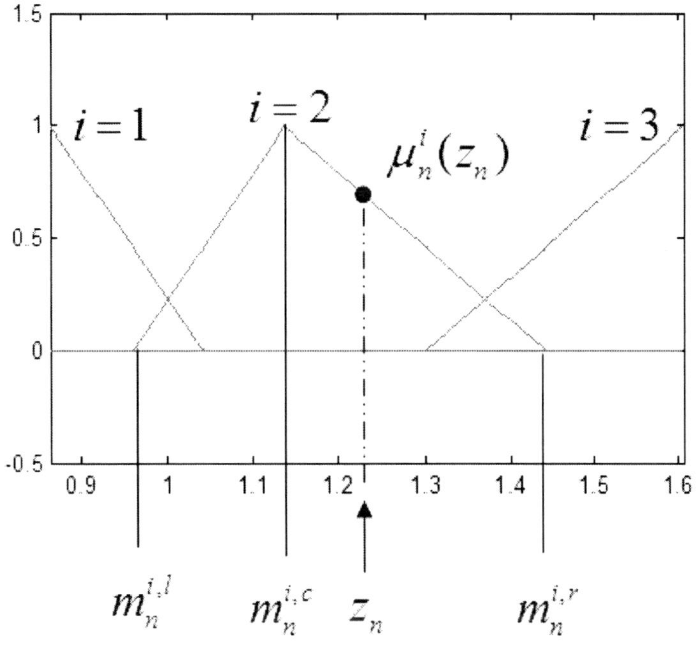

tended motions, we have collected 'garbage' data (Table 7) and have assigned them as the 11th class data.

The classification rates and the false positive rejection rates using the UD model of FDT and MFDT are compared in Table 8. Each user's recognition model is trained by using five sets of hand gesture data and two sets of garbage data.

Nine UD models except the selected user's model and a UI model are used for model selection. We have used one data set of a new user for model selection. The selected model then under-

Table 2. Classification rate of Iris data

Trial	FDT		MFDT	
	Number of nodes	CR (%)	Number of nodes	CR (%)
1	3	92.0	1	96.0
2	16	94.7	10	98.7
3	4	90.7	1	94.7
4	16	93.3	10	98.7
5	4	93.3	3	93.3
6	9	94.7	6	97.3
7	8	96.0	7	96.0
8	7	86.7	3	96.0
9	4	90.7	3	93.3
10	9	96.0	9	100.0
Average	8	92.8	5.3	96.4

Table 3. Classification rate of Wine data

Trial	FDT		MFDT	
	Number of nodes	CR (%)	Number of nodes	CR (%)
1	9	84.1	6	84.1
2	13	85.2	5	92.0
3	10	92.0	5	93.2
4	10	87.5	6	87.5
5	8	94.3	5	89.8
6	13	87.5	7	89.8
7	9	87.5	5	93.2
8	12	81.8	8	89.8
9	12	93.2	6	93.2
10	12	95.5	7	93.2
Average	10.8	88.9	6	90.6

Table 4. Comparison of classification rate of decision trees

Set	**C4.5**	**C5.0**	**FDT**	**MFDT**
Iris	92.9	92.9	92.8	96.4
Wine	86.6	89.2	88.9	90.6

Table 5. Influence of γ for classification of benchmark data

γ	**Iris data**		**Wine data**	
	Number of nodes	CR (%)	Number of nodes	CR (%)
0	2	94.4	2.5	91.6
0.1	2	94.8	3.3	92.6
0.2	2	96.5	3.9	90.9
0.3	3	97.1	4.2	91.0
0.4	4.3	95.9	4.8	89.7
0.5	5.3	96.4	6	90.6
0.6	5.9	96.1	6.4	89.7
0.7	7	96.0	7.6	88.8
0.8	9.2	94.7	9.4	90.9
0.9	11.9	95.5	11	92.7
1	17.9	94.9	15.1	91.9

Table 6. Classes and features of hand gesture data

10 Hand motion classes	**16 Attributes**
1. Up 2. Down 3. Left 4. Right 5. Forward 6. Backward 7. Clockwise circle 8. Counter clockwise circle 9. Clockwise half circle 10. Counter clockwise half circle	Length on x, y, z axis Minimum value on x, y, z axis Maximum value on x, y, z axis Time index of minimum value on x, y, z axis Time index of maximum value on x, y, z axis Eccentricity

Figure 12. Average classification rate during user adaptation

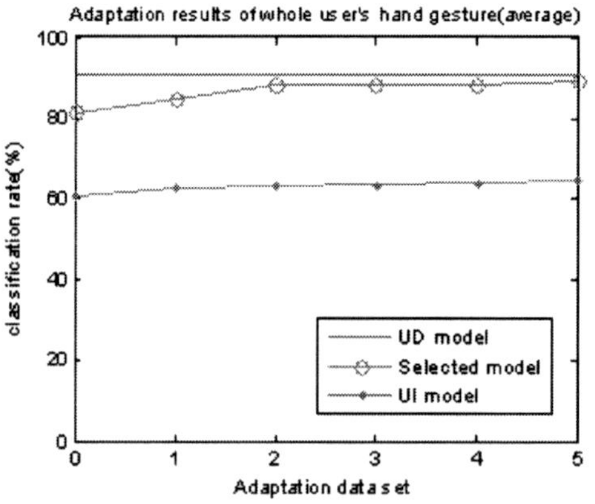

Table 7. Unintended garbage motions

Drinking water Reading a newspaper Fold a blanket	Stretching Raising one arm

goes adaptation. Figure 12 shows user adaptation performances. As the adaptation proceeds, the average classification rate increases.

Table 9 shows an increase in recognition rates of 10 user's hand gestures when using the proposed model selection and user adaptation method. After user adaptation, the recognition rates of the selected model are increased and become close to that of the UD model; however, the recognition rate of UI model does not change much because the adaptation efficiency is decreased due to a huge size of the UI model.

CONCLUDING REMARKS

In this paper, we have proposed the MFDT learning and classification algorithm for robust hand gesture recognition. We have shown that the proposed MFDT method has a better generalization performance than the univariate fuzzy deci-

Table 8. Hand gesture recognition rate

User	FDT			MFDT		
	Number of nodes	RR (%)	FPRR (%)	Number of nodes	RR (%)	FPRR (%)
User 1	28	90.8	26.9	21	88.3	26.7
User 2	24	77.7	63.1	18	90.8	35.0
User 3	16	86.9	58.5	12	98.3	41.7
User 4	17	66.9	73.1	11	90.8	58.3
User 5	18	76.5	61.5	17	93.3	58.3
User 6	23	89.2	51.5	15	86.7	50.0
User 7	16	74.6	62.3	18	86.7	31.7
User 8	22	85.0	53.8	17	85.0	26.7
User 9	21	92.3	75.4	17	89.2	21.7
User 10	20	90.8	53.1	12	96.7	71.7
Average	20.5	83.1	57.9	15.8	90.6	42.2

sion tree. The simulation results of classification tests using benchmark data set have shown that MFDT has a better classification performance than a typical general decision tree and fuzzy decision tree. In classification, model selection, and user adaptation tests using hand gesture data, the selected UD models show the best recognition performance.

The adaptation method proposed in this paper adjusts the parameter values of the membership function of each node. For more reliable adaptation of the MFDT, attribute vectors also should be able to adapt to the patterns of the hand gestures.

The proposed MFDT can be applied for various recognition problems. The robust classification and adaptation capability of MFDT can be applied to human signal recognition systems such

Table 9. Recognition rate using model selection and user adaptation

User	UI model	UI model + Adaptation	Selected model	Selected model + Adaptation
User 1	42.0	55.0	71.7	83.3
User 2	61.0	63.0	80.8	90.8
User 3	69.0	72.5	92.5	95.0
User 4	73.0	68.5	81.7	88.3
User 5	52.5	66.5	67.5	86.7
User 6	72.5	72.5	81.7	90.8
User 7	63.5	59.0	84.2	85.8
User 8	54.0	66.5	75.8	84.2
User 9	52.5	54.0	93.3	94.2
User 10	66.0	71.0	85.0	94.2
Average	60.6	64.9	81.4	89.3

as a hand posture recognition system or a gait recognition system in the future.

REFERENCES

Alpaydin, E. (2004). *Introduction to machine learning*. Cambridge, MA: MIT Press.

Bien, Z., Park, K.-H., Bang, W.-C., & Stefanov, D. H. (2002). LARES: An intelligent sweet home for assisting the elderly and the handicapped. In *Proceedings of the 1st Cambridge Workshop on Universal Access and Assistive Technology* (pp.43-46).

Bourlard, H., & Wellekens, C. J. (1990). Links between Markov models and multilayer perceptrons. *IEEE Transactions on Pattern Analysis and Machine Intelligence*, *12*(12), 1167–1178. doi:10.1109/34.62605

Do, J.-H., Jang, H.-Y., Jung, S.-H., Jung, J.-W., & Bien, Z. (2005). Soft remote control system in the intelligent sweet home. In *Proceedings of the IEEE/RSJ International Conference on Intelligent Robots and Systems* (pp. 3984-3989).

Duda, R. O., Hart, P. E., & Stork, D. G. (2001). *Pattern classification* (2nd ed.). Hoboken, NJ: John Wiley & Sons.

Fu, H.-C., Chang, H.-Y., Xu, Y. Y., & Pao, H.-T. (2000). User adaptive handwriting recognition by self-growing probabilistic decision-based neural networks. *IEEE Transactions on Neural Networks*, *11*(6), 1373–1384. doi:10.1109/72.883451

Hong, P., Turk, M., & Huang, T. S. (2000). Gesture modeling and recognition using finite state machines. In *Proceedings of the 4th IEEE International Conference on Automatic Face and Gesture Recognition* (pp. 410-415).

Janikow, C. Z. (1998). Fuzzy decision trees: Issues and methods. *IEEE Transactions on Systems, Man, and Cybernetics*, *28*(1), 1–14. doi:10.1109/3477.658573

Jung, S.-H. (2007). *Incremental user adaptation in Korean sign language recognition using motion similarity and prediction from adaptation history.* Unpublished master's thesis, Korea Advanced Institute of Science and Technology, Daejeon, Korea.

Jung, S.-H., Do, J.-H., & Bien, Z. (2006). Adaptive hand motion recognition in soft remote control system. In *Proceedings of the 7th International Workshop on Human-friendly Welfare Robotic Systems.*

Kim, D.-J. (2004). *Image-based personalized facial expression recognition system using fuzzy neural networks.* Unpublished doctoral dissertation, Korea Advanced Institute of Science and Technology, Daejeon, Korea.

Merz, C. J., & Murphy, P. M. (1996). *UCI repository for machine learning data-bases.* Retrieved from http://archive.ics.uci.edu/ml/

Nam, Y., & Wohn, K. (1996). Recognition of. space-time hand-gesture using hidden Markov model. In *Proceedings of the ACM Symposium on Virtual Reality Software and Technology* (pp. 51-58).

Rabiner, L. R. (1989). A tutorial on hidden Markov models and selected applications in speech recognition. *Proceedings of the IEEE*, *77*(2), 257–285. doi:10.1109/5.18626

Su, M. C. (2000). A fuzzy rule-based approach to spatio-temporal hand gesture recognition. *IEEE Transactions on Systems, Man, and Cybernetics Part C*, *30*(2), 276–281.

Yamato, J., Ohya, J., & Ishii, K. (1992). Recognizing human action in time-sequential images using hidden Markov model. In *Proceedings of the IEEE Conference on Computer Vision and Pattern Recognition* (pp. 375-385).

Yang, S.-E. (2007). *Gesture spotting using fuzzy garbage model and user adaptation.* Unpublished master's thesis, Korea Advanced Institute of Science and Technology, Daejeon, Korea.

Yildiz, O. T., & Alpaydin, E. (2001). Omnivariate decision trees. *IEEE Transactions on Neural Networks, 12*(6), 1539–1546. doi:10.1109/72.963795

This work was previously published in the International Journal of Fuzzy System Applications, Volume 1, Issue 3, edited by Toly Chen, pp. 15-31, copyright 2010 by IGI Publishing (an imprint of IGI Global).

Chapter 9
Gesture Spotting Using Fuzzy Garbage Model and User Adaptation

Seung-Eun Yang
Korea Aerospace Research Institute, Korea

Kwang-Hyun Park
Kwangwoon University, Korea

Zeungnam Bien
Ulsan National Institute of Science and Technology, Korea

ABSTRACT

Thanks to the rapid advancement of human-computer interaction technologies it is becoming easier for the elderly and/or people with disabilities to operate various electrical systems. Operation of home appliances by using a set of predefined hand gestures is an example. However, hand gesture recognition may fail when the predefined command gestures are similar to some ordinary but meaningless behaviors of the user. This paper uses a gesture spotting method to recognize a designated gesture from other similar gestures. A fuzzy garbage model is proposed to provide a variable reference value to determine whether the user's gesture is the command gesture or not. Further, the authors propose two-stage user adaptation to enhance recognition performance: that is, off-line (batch) adaptation for inter-person variation and on-line (incremental) adaptation for intra-person variation. For implementation of the two-stage adaptation method, a genetic algorithm (GA) and the steepest descent method are adopted for each stage. Experimental results were obtained for 5 different users with left and up command gestures.

INTRODUCTION

The shortage of caregivers for the elderly and/or people with disabilities has become a critical social issue in many countries (Kim, 2003; Stefanov, 2004). To cope with the trend, many studies have been conducted on user-friendly human-computer interaction (HCI) techniques by which the elderly and people with disabilities can operate various electrical systems more easily. A soft remote control system developed at HWRS-ERC in KAIST is an application example of enabling the users to operate home appliances by hand gestures (Do, 2002, 2005). The system

DOI: 10.4018/978-1-4666-1870-1.ch009

utilizes ten different predefined hand gestures as a given set of commands as shown in Figure 1.

In spite of these attempts to create a competitive and useful application of hand gestures, the recognition of command gestures is faced with various practical problems: the 3 imperative issues are listed.

- A set of features and decision rules to distinguish command gestures should be carefully obtained through observation of the user's behaviors. To cope with this requirement, Han (2006) proposed a feature selection method based on a separability index matrix (SIM) and Jeon (2008) constructed a multivariate fuzzy decision tree by using linear discriminant analysis (LDA) and information gain.
- The system needs to be adapted to a specific user to enhance the performance since each person behaves in different way (inter-person variation). In addition, the characteristic of a single user may also change in a different environment. Therefore, an additional adaptation method is required for intra-person variation which occurs for a single user. This adaptation should be executed in real time to modify the temporal difference from a single user.
- In actual operation of the system, it is often observed that some of the user's ordinary but meaningless behaviors are mistakenly recognized as command gestures. In fact, those command gestures are very likely to be basic natural motions and thus similar motions can appear in the user's ordinary behaviors. More complicated command gestures may be adopted as an alternative to maximize the different aspects between the command gestures and ordinary motions. Such an approach, however, is prone to decrease the usability of the system and can cause a cognitive burden for the user to memorize the larger set of command gestures.

Given these problems and issues, we introduce a gesture spotting method using the fuzzy garbage model to recognize the designated gesture from the user's ordinary behaviors, while the existing approach focused on the construction of decision rules with user adaption to achieve high classification performance (Jeon, 2008).

Human gestures observed in ordinary behaviors are likely to be unstructured and occur unconsciously. Moreover, the characteristics of human behavior are different from person to person. It is thus rather difficult to express human gestures by a

Figure 1. Ten different hand gestures

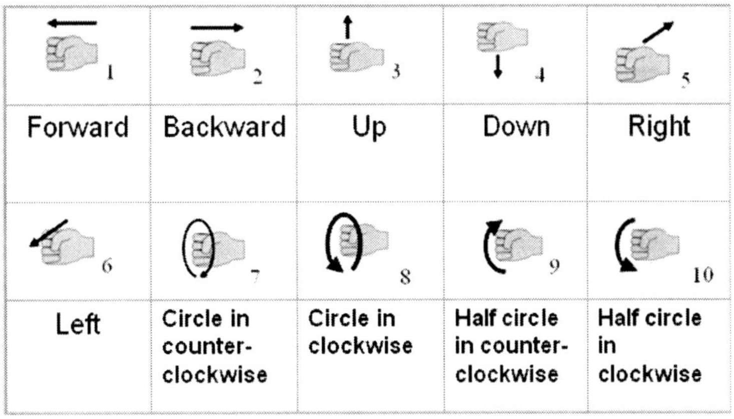

stochastic or statistical model. As noted in Passino and Yurkovich (1998), the sensitivity to variation of parameters or environment can be reduced by using the fuzzy logic which uses fuzzified values instead of crisp values. In addition, much of the acquired knowledge through observation is easily applied to the system. The usage of linguistic values, in fact, makes the fuzzy logic superior to other methods, since the ambiguous characteristic of unconscious human behaviors can be expressed and manipulated by linguistic variables.

To recognize a specific gesture, our proposed method utilizes two fuzzy models: one for a command gesture and the other for garbage gestures. The motion data of the user are applied to both models, and then each model gives a score which shows the degree of belongingness to each model. By comparing the scores from these models, the system decides whether the input data represent a command gesture or not.

To achieve these two goals, a two-stage adaptation method is proposed. At the first stage, a genetic algorithm (GA) is adopted to overcome the inter-person variation of human behaviors, due to its excellent searching capability for multiple parameters by exploration and exploitation (Burns, 2001). The problem of slow searching in GA is not a critical issue for this stage since the adaptation is executed offline before the user operates the system. At the second stage, the parameters are modified online while a steepest descent method is used to minimize an objective function (Nomura, 1992).

The objective and assumptions for this study are described in the next section, and the details of the proposed method are followed. A feature selection method using SIM (Han, 2006) is discussed and the concept of a fuzzy garbage model is described. Then, two-stage user adaptation by adopting GA and the steepest descent is discussed. Finally, some experimental results are shown and the concluding remarks are stated.

OBJECTIVE AND ASSUMPTIONS

The objective of this study is to build a recognition system capable of discriminating a set of predefined hand command gestures from ordinary command-like meaningless gestures in daily life. As in the previous soft remote control system (Do, 2002, 2005), we shall specifically consider four

Figure 2. Command gestures to be discriminated from ordinary gestures

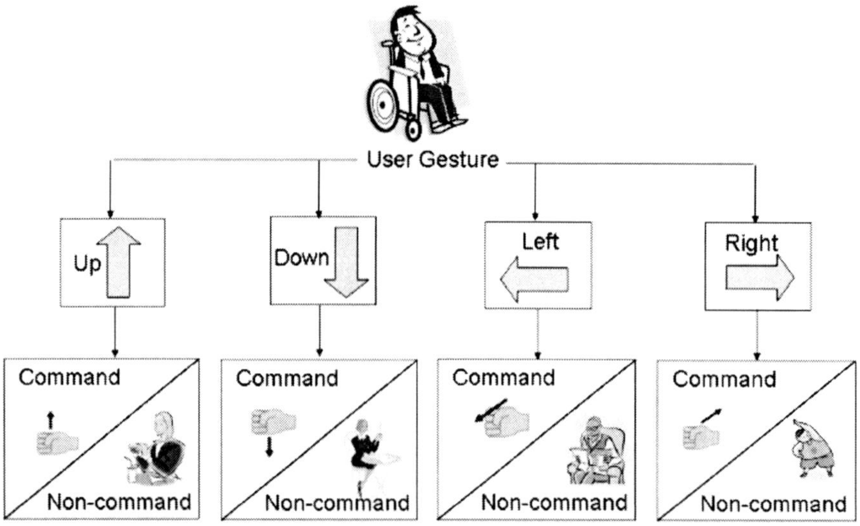

kinds of primitive command gestures of up, down, left and right as shown in Figure 2.

In our system, the user's gesture is recognized by using information of the user's face and hand trajectories observed by two USB cameras. It is assumed that only one user uses the system at a time and that the face and hands of the user are correctly detected.

GESTURE SPOTTING METHOD

The proposed gesture spotting method utilizes a fuzzy garbage model and two-stage user adaptation. Figure 3 shows the overall procedure of the method.

Feature Selection

Feature selection is an indispensable procedure to recognize command gestures. Eleven features are considered as candidates through observation of the user's behavior as follows:

- Distance change between a face and a hand while the hand is moving
- Distance between a face and a hand when a user finishes a gesture
- Elapsed time at a standstill of a hand when a user finishes a gesture
- Length of hand movement
- Elapsed time for hand movement
- Speed of hand movement
- Eccentricity of hand trajectory
- Peak frequency of hand movement
- Median speed of hand movement
- Standard deviation of hand movement speed
- Mean speed of hand movement

It is not easy to implement a recognition system with well-defined fuzzy rules by using all features. Further, too many features tend to increase the complexity and execution time. Therefore, it is desired to select an appropriate number of features which are essential to spot meaningful gestures. In this study, the method of SIM-based feature (SIMF) (Han, 2006) is adopted to select

Figure 3. Overall procedure of gesture spotting

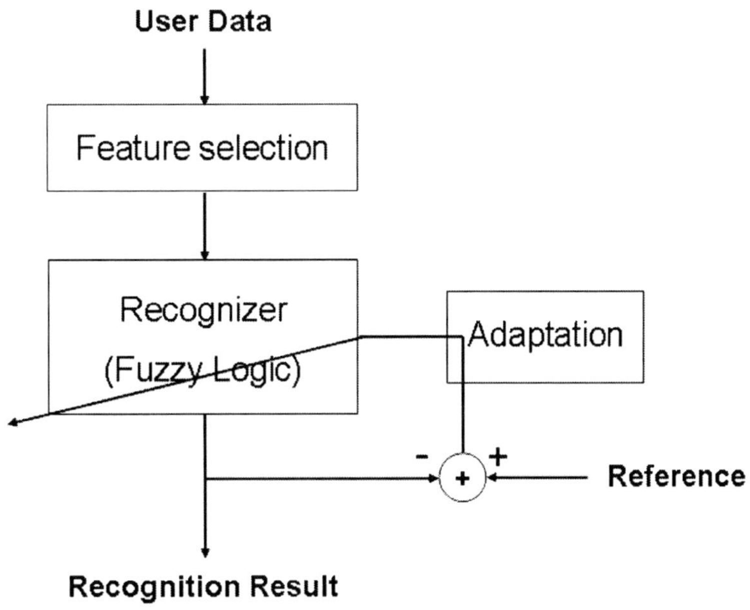

appropriate features. SIMF is a feature selection algorithm based on the criterion function of classifiability which provides relevant features with very low computational cost and in reference to the forward search technique. Classifiability is defined by SIM described in Table 1 (Han, 2006).

In Table 1, C is the total number of classes and N is the total number of features. w_i and w_j denote *i*-th and *j*-th class, respectively, and $J(\cdot)$ is a criterion function of distance-measure type such as Bhattacharyya distance where x_k denotes the *k*-th feature. SVM_{avg} is defined as $SVM_{avg} = \sum_k SVM_k \,/\, N$ and $SVM_{threshold}$ is a $C \times C$ matrix filled with a threshold value as its components. $\geq$, $\varnothing$, $\vee$ and $\otimes$ are component-wise operator for $\geq$, $\vee$ (logical OR), /(division) and $\times$ (multiplication), respectively.

An example of feature selection using SIMF is described in Table 2, and three classes and three features are given in Figure 4. Figure 4 shows the separability value matrix (SVM) for each feature. Among three features, feature 3 has the highest classifiability. Therefore, feature 3 is selected as the most important feature. Then, the relevance of remaining features with respect to the selected feature is updated, and the second important feature is selected as described in Table 3. When two features have the same classifiability, the separability is utilized to select an appropriate feature.

At the next step, an appropriate number of features should be decided. Fuzzy rules are increased by the factor of 4^n as the number n of features increases by one when 4 membership values are used. Therefore, the overall complexity increases abruptly as the number of features increases. To find an appropriate number of features, Bhattacharyya distance is measured by increasing the number of features. In Table 4, the increase in distance for more than two features is not significant. From this observation, the number of features is decided as two for gesture spotting.

Thus, the following two features are selected to discriminate "up" and "eating" command gestures. Figure 5 shows the distribution of the selected feature data for "up" command.

- **A:** Distance change between a face and a hand while the hand is moving

Table 1. Definitions for each criterion (Han, 2006)

Criterion	Definition
Separability Value Matrix	$SVM_k \leftarrow \left[J(w_i, w_j; \{x_k\})\right]_{C\times C}$ $, 1 \leq i, j \leq C, \quad 1 \leq k \leq N$
Separability Index Matrix	$SIM_k \leftarrow \left(SVM_k \geq SVM_{avg}\right) \vee \left(SVM_k \geq SVM_{threshold}\right)$
Weight Matrix	$WM_k \leftarrow SIM_k \varnothing \left(\sum_{i=1}^{N} SIM_k\right)$
Classifiability	$C(x_k) = \sum_i \sum_j \left(WM_k \otimes SIM_k\right)$
Separability	$S(x_k) = \sum_i \sum_j \left(WM_k \otimes SVM_k\right)$

Table 2. Feature selection using SIMF

$$SVM_1 = \begin{bmatrix} 0 & 9.9 & 18.1 \\ & 0 & 1.1 \\ & & 0 \end{bmatrix}, SVM_2 = \begin{bmatrix} 0 & 4.7 & 10.5 \\ & 0 & 1.2 \\ & & 0 \end{bmatrix}, SVM_3 = \begin{bmatrix} 0 & 0.1 & 4.6 \\ & 0 & 3.2 \\ & & 0 \end{bmatrix}$$

$$SVM_{avg} = \begin{bmatrix} 0 & 4.9 & 11.1 \\ & 0 & 1.8 \\ & & 0 \end{bmatrix}, SVM_{threshold} = \begin{bmatrix} 3 & 3 & 3 \\ & 3 & 3 \\ & & 3 \end{bmatrix}$$

$$SIM_1 = \begin{bmatrix} 0 & 1 & 1 \\ & 0 & 0 \\ & & 0 \end{bmatrix}, SIM_2 = \begin{bmatrix} 0 & 1 & 1 \\ & 0 & 0 \\ & & 0 \end{bmatrix}, SIM_3 = \begin{bmatrix} 0 & 0 & 1 \\ & 0 & 1 \\ & & 0 \end{bmatrix}$$

$$WM_1 = \begin{bmatrix} 0 & 0.5 & 0.33 \\ & 0 & 0 \\ & & 0 \end{bmatrix}, WM_2 = \begin{bmatrix} 0 & 0.5 & 0.33 \\ & 0 & 0 \\ & & 0 \end{bmatrix}, WM_3 = \begin{bmatrix} 0 & 0 & 0.33 \\ & 0 & 1 \\ & & 0 \end{bmatrix}$$

$$C(x_1) = 0.83, \quad C(x_2) = 0.83, \quad C(x_3) = 1.33$$

- **B:** Distance between a face and a hand when a user finishes a gesture

Four membership values (ZO: Zero, PS: Positive Small, PM: Positive Medium, PB: Positive Big) are used as shown in Figure 6, and sixteen fuzzy rules are generated through observation of command gestures.

- If A is ZO and B is ZO then Out is ZO
- If A is ZO and B is PS then Out is ZO
- If A is ZO and B is PM then Out is ZO
- If A is ZO and B is PB then Out is PS
- If A is PS and B is ZO then Out is PS
- If A is PS and B is PS then Out is ZO
- If A is PS and B is PM then Out is ZO
- If A is PS and B is PB then Out is PB

Figure 4. Each component of SVM for all features

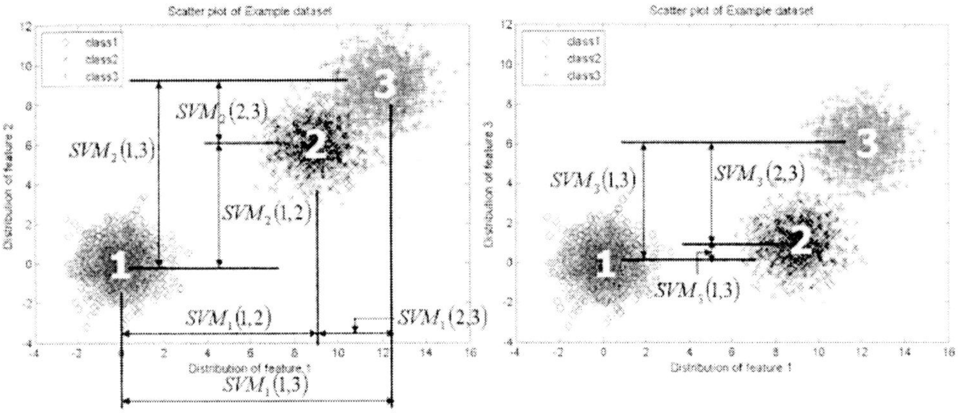

Table 3. Feature selection using SIMF after updating relevance for remaining features

$$SIM_1 = \begin{bmatrix} 0 & 1 & 1 \\ & 0 & 0 \\ & & 0 \end{bmatrix} \wedge \left\{ \sim \begin{bmatrix} 0 & 0 & 1 \\ & 0 & 1 \\ & & 0 \end{bmatrix} \right\} = \begin{bmatrix} 0 & 1 & 0 \\ & 0 & 0 \\ & & 0 \end{bmatrix}$$

$$SIM_2 = \begin{bmatrix} 0 & 1 & 1 \\ & 0 & 0 \\ & & 0 \end{bmatrix} \wedge \left\{ \sim \begin{bmatrix} 0 & 0 & 1 \\ & 0 & 1 \\ & & 0 \end{bmatrix} \right\} = \begin{bmatrix} 0 & 1 & 0 \\ & 0 & 0 \\ & & 0 \end{bmatrix}$$

$$WM_1 = \begin{bmatrix} 0 & 0.5 & 0 \\ & 0 & 0 \\ & & 0 \end{bmatrix}, WM_2 = \begin{bmatrix} 0 & 0.5 & 0 \\ & 0 & 0 \\ & & 0 \end{bmatrix}$$

$$C(x_1) = 0.5, \quad C(x_2) = 0.5$$
$$S(x_1) = 4.95, \quad S(x_2) = 2.35$$

Table 4. Bhattacharyya distance between two different gestures

# of features	1	2	3	4	5
Up/ eating	0.00594	0.00784	0.00896	0.01063	0.01221
Up/garbage	0.44218	0.50563	0.52321	0.55970	0.56262

Figure 5. Distribution of selected feature data for "up" command and non-command gesture

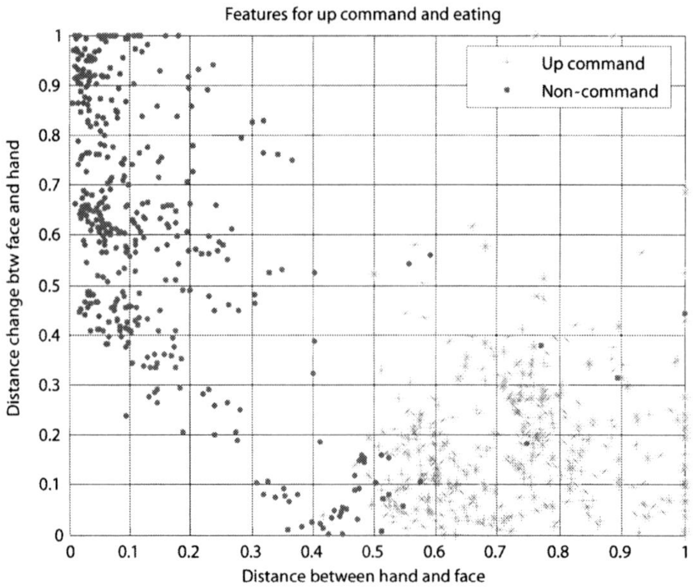

- If A is PM and B is ZO then Out is PB
- If A is PM and B is PS then Out is PM
- If A is PM and B is PM then Out is PB
- If A is PM and B is PB then Out is PS
- If A is PB and B is ZO then Out is PB
- If A is PB and B is PS then Out is PB
- If A is PB and B is PM then Out is PM
- If A is PB and B is PB then Out is PS

The defuzzified values using the center of sum method for "up" command are shown in Figure 7 in which the membership functions in Figure 6 and the sixteen fuzzy rules above are utilized. Note that the command gestures can be recognized with the defuzzified values different from each data. When only one threshold value is used for recognition, however, incorrect recognition may occur since the defuzzified value of user 3 for non-command data is bigger than the defuzzified value of user 2 for "up" command. Even though the threshold value can be changed, false positive (FP: recognition of a non-command gesture as a command gesture) or false negative (FN: recognition of a command gesture as a non-command gesture) error is inevitable in this case. To resolve this difficulty due to a single threshold value, a fuzzy garbage model is proposed.

Fuzzy Garbage Model

The gesture spotting procedure using a fuzzy garbage model is described in Figure 8. Two different fuzzy models exist for the command gesture and the garbage gesture. Each model is defined by their specific rules and membership functions. When a feature vector from the input data is applied to the system, the system returns the scores from each model. By comparing the scores, the system decides if it is a command gesture or a command-like garbage gesture that has actually occurred. In the example shown in Figure 9, the score from a fuzzy garbage model is lower than the score from a command gesture model. In this case, the score from a garbage model performs a role as a threshold value in recognizing the user's gesture, since the result is decided by comparing the scores from various models.

To utilize a fuzzy garbage model, the fuzzy rules for garbage gestures are defined as follows. The same features selected by SIMF for "up" command are used for garbage gestures.

- If A is ZO and B is ZO then Out is ZO
- If A is ZO and B is PS then Out is PM
- If A is ZO and B is PM then Out is PB

Figure 6. Fuzzy membership functions for the system

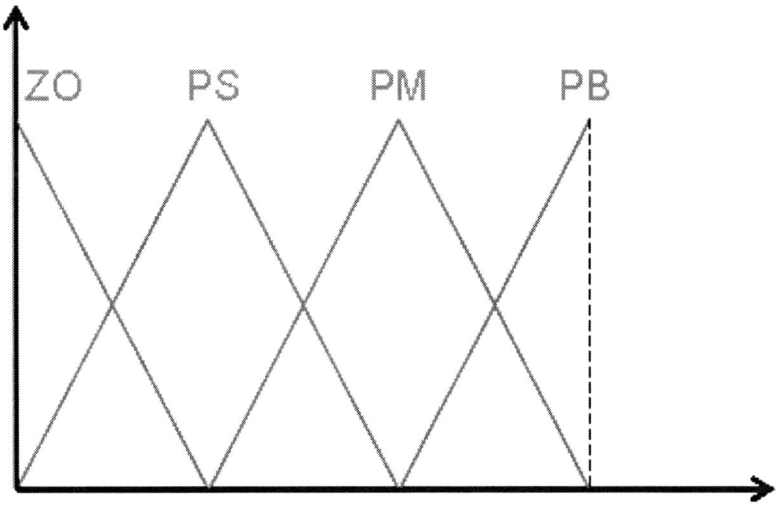

Figure 7. Defuzzified values for five different users

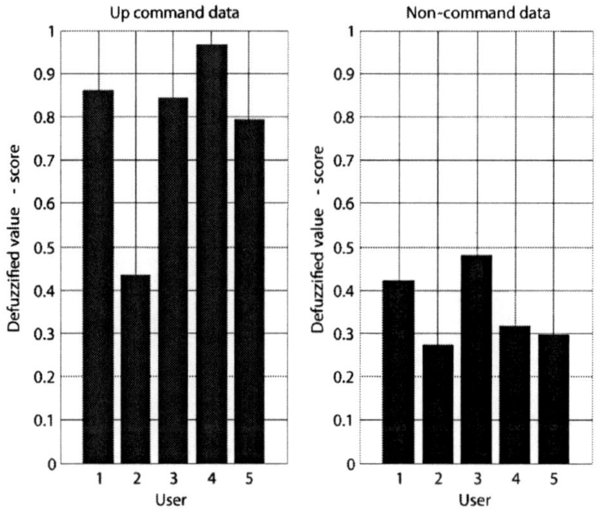

- If A is ZO and B is PB then Out is PB
- If A is PS and B is ZO then Out is PM
- If A is PS and B is PS then Out is PM
- If A is PS and B is PM then Out is PM
- If A is PS and B is PB then Out is PM
- If A is PM and B is ZO then Out is PS
- If A is PM and B is PS then Out is PS
- If A is PM and B is PM then Out is PS
- If A is PM and B is PB then Out is ZO
- If A is PB and B is ZO then Out is ZO
- If A is PB and B is PS then Out is PS
- If A is PB and B is PM then Out is ZO

Figure 8. Gesture spotting using fuzzy garbage model

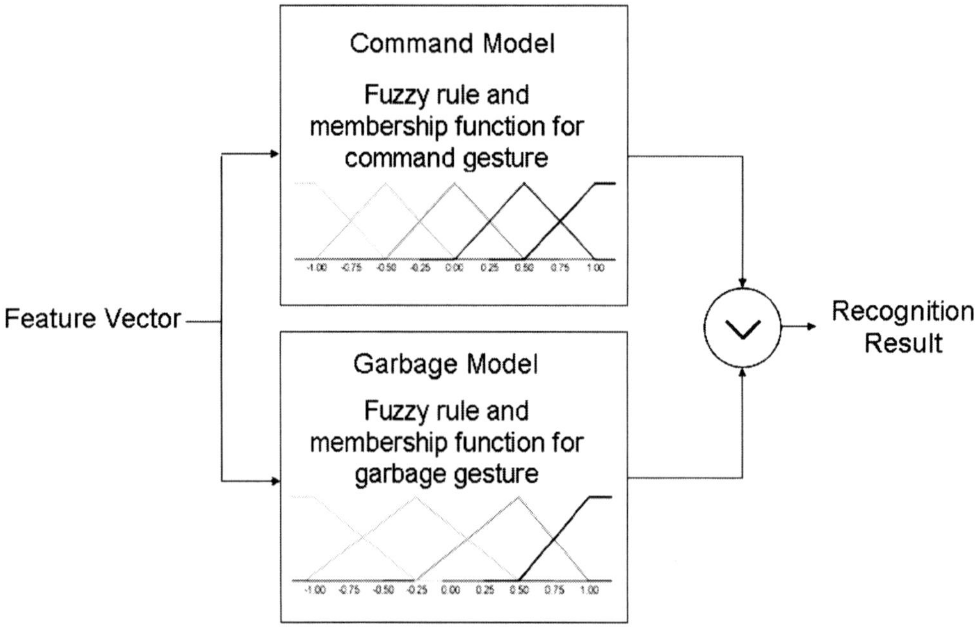

Figure 9. Variable threshold by using a fuzzy garbage model

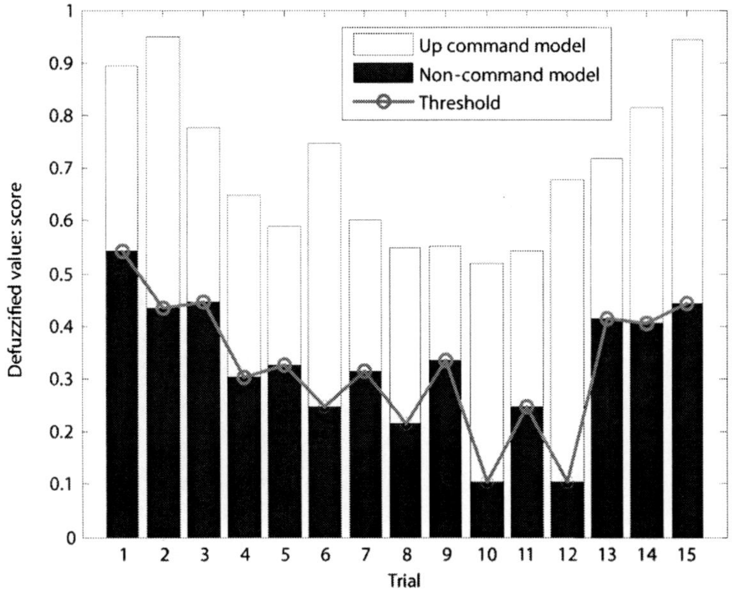

- If A is PB and B is PB then Out is ZO

The characteristic of human behavior is uncertain and time-varying, and thus an optimization process is required to enhance the recognition performance. Among the optimization methods, genetic algorithm (GA) is adopted due to the capability of effective searching and robustness (Teng et al., 2003).

A set of four membership functions are used for two inputs and one output of the system. For ZO and PB, the width of the right side and the width of the left side of membership functions are optimized respectively. For PS and PM, there are three parameters to be optimized: the center position, and the widths of the left and right sides of the membership function. Parameters for optimization are listed in Table 5.

The optimization is executed for a command gesture model and a garbage model at the same time. Therefore, all the parameters are optimized at once. The purpose of parameter optimization

Table 5. Parameters for the membership functions

Linguistic Value	Feature 1	Feature 2	Output
ZO	w_{r_1}	w_{r_2}	w_r
PS	c_1, w_{l_1}, w_{r_1}	c_2, w_{l_2}, w_{r_2}	c, w_l, w_r
PM	c_1, w_{l_1}, w_{r_1}	c_2, w_{l_2}, w_{r_2}	c, w_l, w_r
PB	w_{l_1}	w_{l_2}	w_l

is to increase the separability between a command gesture model and a garbage model. Thus, the fitness function is set as Equation (1). The optimization is executed in the way that the output score of a command model is increased for command data and decreased for garbage data. Also in parallel, the output score of a garbage model for garbage data is increased and decreased for command data. Big penalty is given to the fitness when there occurs a range covered by no membership function.

$$\frac{\sum_{i=1}^{N_A}\{X_A(i)-Y_A(i)\}+\sum_{i=1}^{N_B}\{X_B(i)-Y_B(i)\}}{N_A+N_B} \quad (1)$$

N_A: Total number of command data
N_B: Total number of garbage data
X_A: Output score of a command model for command data
Y_A: Output score of a garbage model for command data
X_B: Output score of a garbage model for garbage data
Y_B: Output score of a command model for gargabe data

Data for command gestures and garbage gestures are applied to optimize two models. At the initial stage of the optimization, no specific user is considered. Therefore, a user independent (UI) model is developed throughout this optimization. In this study, 70 training samples are gathered from 5 different users. The optimization condition and the optimized membership functions are shown in Table 6 and Figure 10 respectively. The width of PM in feature 1 from non-command data is very small and it means PM is rarely used for the system. However, the linguistic value corresponding to the membership function with a very small width is not removed since various characteristics of the user's hand gestures may lead different optimization results.

Table 6. Conditions for optimization using GA

Conditions	Defined Value
Number of population for each parameter	50
Assigned number of bits for each parameter	20
Rate of mutation	0.02
Rate of crossover	0.25
Number of generation	10000

The recognition error using the optimized membership functions is shown in Figure 11. The left part of Figure 11 shows the result for the case of using only one threshold value, and the right part for the case of using a fuzzy garbage model. As shown in Figure 11, the recognition performance is enhanced when a fuzzy garbage model is utilized. However, user 5 shows a remarkably high error rate compared with the other users due to the fact that each user has different characteristic in their behavior. Therefore, the utilization of the same system to various users may degrade the recognition rate. For this problem, user adaptation of the system is required to develop an appropriate system for different users. In this study, two-stage user adaptation is adopted.

Two-Stage User Adaptation

As mentioned in the previous section, the problem of inter-person variation may deteriorate the recognition performance. To resolve this deficiency, batch adaptation is applied offline. Batch adaptation converts a user-independent (UI) model to a user-dependant (UD) model. Recall that the fuzzy garbage model is an UI model. An UD system usually outperforms a UI system, as long as a sufficient amount of training data is available to obtain UD models (Fu, 2000).

For inter-person variation, many cases should be considered to define an optimal condition for a specific user. This process should be done before the user uses the system. Therefore, execution time for adaptation is not a critical problem. GA is appropriate for this problem because of its efficient searching capability.

Figure 10. Optimized membership functions using GA

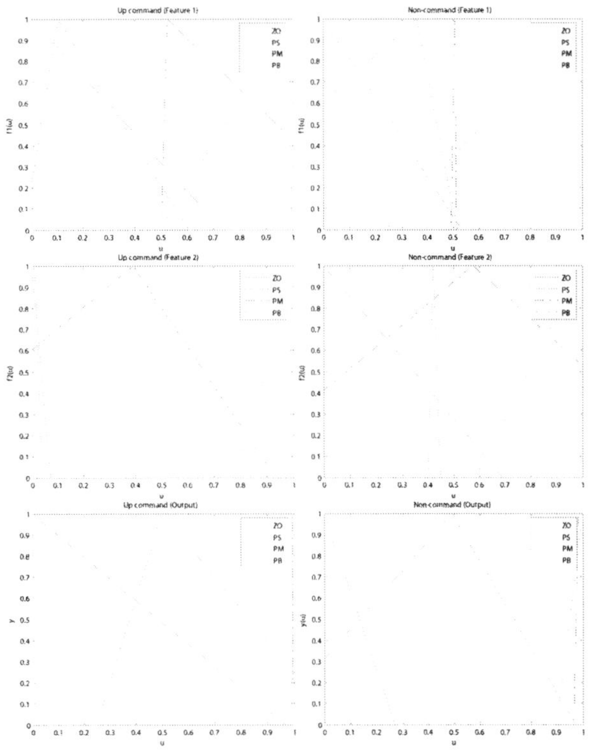

The condition for adaptation is the same as described in the previous section. Data for a single user are applied since it is a conversion process from an UI model to an UD model. The initial population for each parameter is adopted from the optimized value of the UI model.

In addition, we observe that the characteristic of a gesture made by a single person is often dependent on the interacting environment or on

Figure 11. Recognition error for a command gesture

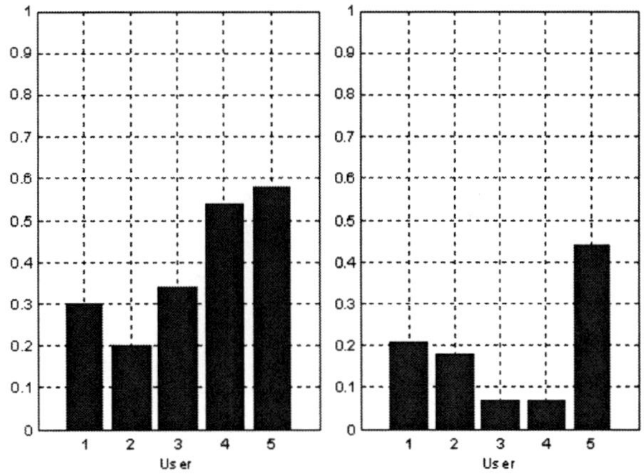

the user's emotion. For example, when the user is angry, his/her hand movement may appear faster than ordinary. This intra-person variation requires another way of user adaptation. Incremental adaptation is applied for this, in which user-specific training data are acquired over a long period of time, and the UD model is adapted incrementally as a new training data set is acquired (Fu, 2000).

The incremental adaptation process should be done in real time since the parameter modification process should not affect the overall performance of the system. Because of the time constraint, a steepest descent method is applied to the adaptation. As one of the simplest descent methods, the steepest descent method finds the fastest direction to minimize a given function quickly (Marks, 1992; Arabshahi, 1997; Habbi & Zelmat, 2003).

All the parameters described in Table 5 may not be adapted simultaneously since the membership functions fired by the input data only are adapted. A fuzzy command gesture model and a fuzzy garbage model are adapted separately by one step.

The error function is defined as Equation (2) where the defuzzified value is defined as *D*. The parameter update rule is described as Equation (3). The defuzzified value *D* is calculated with the fuzzy parameters such as the center point, the left and right widths of the membership function.

$$J = \frac{1}{2}(1 - D)^2 \tag{2}$$

$$\alpha_{new} = \alpha_{old} - \eta \frac{\partial J}{\partial \alpha_{old}} \tag{3}$$

The output membership function is considered first, and each parameter of the output membership function is described in Figure 12 and Equations (4)-(6). The derivative form of the error function *J* is found to adjust the parameters. The updating rule for each output parameter is described in Equation (7). To update each parameter, the partial differential equations in (8)-(14) are required.

$$L(u) = \frac{u}{w_l} + \frac{1}{w_l}(w_l - c) = \frac{u}{c - u_0} - \frac{u_0}{c - u_0} \tag{4}$$

$$R(u) = -\frac{1}{w_r}(u - c) + 1 = \frac{u}{c - u_3} - \frac{u_3}{c - u_3} \tag{5}$$

$$\begin{aligned} u_0 &= c - w_l, \quad u_1 = w_l \cdot y + c - w_l, \\ u_2 &= c + w_r - w_r \cdot y, \quad u_3 = c + w_r \end{aligned} \tag{6}$$

$$\alpha_{new} = \alpha_{old} - \eta(1 - D)\frac{\partial D}{\partial \alpha_{old}} \tag{7}$$

$$\frac{\partial D}{\partial \alpha_{old}} = \frac{\left(\sum_{i=1}^{N} \frac{\partial}{\partial \alpha_{old}} \int_0^1 u \cdot f_i(u)du\right) \times \left(\sum_{i=1}^{N} \int_0^1 f_i(u)du\right)}{\left(\sum_{i=1}^{N} \int_0^1 f_i(u)du\right)^2} - \frac{\left(\sum_{i=1}^{N} \int_0^1 u \cdot f_i(u)du\right) \times \left(\sum_{i=1}^{N} \frac{\partial}{\partial \alpha_{old}} \int_0^1 f_i(u)du\right)}{\left(\sum_{i=1}^{N} \int_0^1 f_i(u)du\right)^2}$$

$$S_u = \int_0^1 u \cdot f_i(u)du, \quad S = \int_0^1 f_i(u)du \tag{8}$$

$$\frac{\partial S_u}{\partial c} = \frac{(u_0 - u_1)^2}{2w_l} + y(u_2 - u_1) + \frac{(u_2 - u_3)^2}{2w_r} \tag{9}$$

$$\frac{\partial S_u}{\partial w_l} = \frac{2u_1^2(y-1) - u_0^2 + u_1^2 - 2u_0u_1(y-1)}{2w_l} + \frac{2u_1^3 + u_0^3 - 3u_0u_1^2}{6w_l^2} - y(y-1)u_1 \tag{10}$$

Figure 12. Output membership function of the model

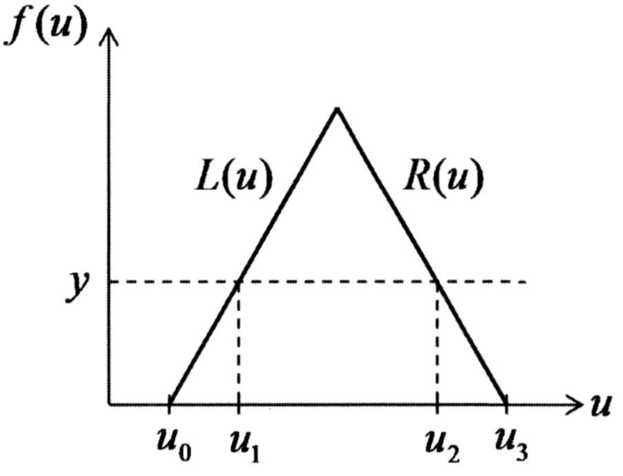

$$\frac{\partial S_u}{\partial w_r} = \frac{2u_2^2(1-y) - u_2^2 + u_3^2 - 2u_2u_3(1-y)}{2w_r} - \frac{2u_2^3 + u_3^3 - 3u_2u_3^2}{6w_r^2} + y(1-y)u_2 \quad (11)$$

$$\frac{\partial S}{\partial c} = 0 \quad (12)$$

$$\frac{\partial S}{\partial w_l} = \frac{(u_1 - u_0)(2yw_l - u_1 + u_0)}{2w_l^2} + y(1-y) \quad (13)$$

$$\frac{\partial S}{\partial w_r} = -\frac{(u_2 - u_3)(2yw_r + u_2 - u_3)}{2w_r^2} + y(1-y) \quad (14)$$

To adjust the parameters of the input membership function, two cases should be considered for the left and right sides of the membership func-

Figure 13. Input membership function of the model

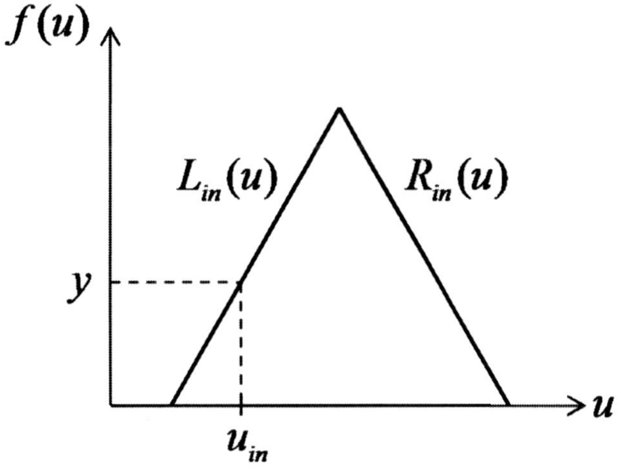

tion. Since the triangular membership functions are used, only left or right side of the membership function is fired at a time depending on the feature value. The input membership function is shown in Figure 13 for the case the left side of membership function is fired. In this case, Equation (15) is utilized for parameter adaptation. On the other hand, when the right side is fired, Equation (16) is used. The partial differential equations for input membership function are described in Table 7 (Case 1: the left side of the membership function is fired, Case 2: the right side of the membership function is fired).

$$L_{in}(u) = \frac{u}{w_{in,l}} + \frac{1}{w_{in,l}}(w_{in,l} - c_{in}) \qquad (15)$$

$$R_{in}(u) = -\frac{1}{w_{in,r}}(u - c_{in}) + 1 \qquad (16)$$

The overall procedure of two-stage adaptation is shown in Figure 14 and Figure 15. The first stage is batch adaptation using GA for inter-person variation, and the second stage is incremental adaptation using the steepest descent method for intra-person variation.

EXPERIMENTAL RESULTS

For the experiment, we have collected data from 5 different users for 5 days in consideration of the fact that one day is too short to reveal the time-varying characteristics of the user's behavior. From each person, 75 data are collected for each gesture. 25 data are used for training and 50 data are used for testing.

Four experiments are conducted. In the first experiment, the previously mentioned soft remote control system is used. The second experiment is conducted using the fuzzy logic with one threshold value. The third experiment is for the case of using a fuzzy garbage model, and finally user adaptation is applied with a fuzzy garbage model for the fourth experiment. In each case,

Table 7. Partial differential equations for input membership functions

Case 1	Case 2
$\frac{\partial S}{\partial c_{in}} = -\frac{1}{w_{in,l}}(w_r + w_l)(1-y)$ $\frac{\partial S}{\partial w_{in,l}} = \frac{c_{in} - u_{in}}{w_{in,l}^2}(w_r + w_l)(1-y)$ $\frac{\partial S}{\partial w_{in,r}} = 0$ $\frac{\partial S_u}{\partial c_{in}} = -\frac{1}{2w_{in,l}}(u_2^2 - u_1^2)$ $\frac{\partial S_u}{\partial c_{in}} = \frac{c_{in} - u_{in}}{2w_{in,l}^2}(u_2^2 - u_1^2) \quad \frac{\partial S_u}{\partial w_{in,r}} = 0$	$\frac{\partial S}{\partial c_{in}} = \frac{1}{w_{in,r}}(w_r + w_l)(1-y)$ $\frac{\partial S}{\partial w_{in,l}} = 0$ $\frac{\partial S}{\partial w_{in,r}} = -\frac{c_{in} - u_{in}}{w_{in,r}^2}(w_r + w_l)(1-y)$ $\frac{\partial S_u}{\partial c_{in}} = \frac{1}{2w_{in,r}}(u_2^2 - u_1^2)$ $\frac{\partial S_u}{\partial c_{in}} = 0$ $\frac{\partial S_u}{\partial w_{in,r}} = -\frac{c_{in} - u_{in}}{2w_{in,r}^2}(u_2^2 - u_1^2)$

Figure 14. The first stage of user adaptation

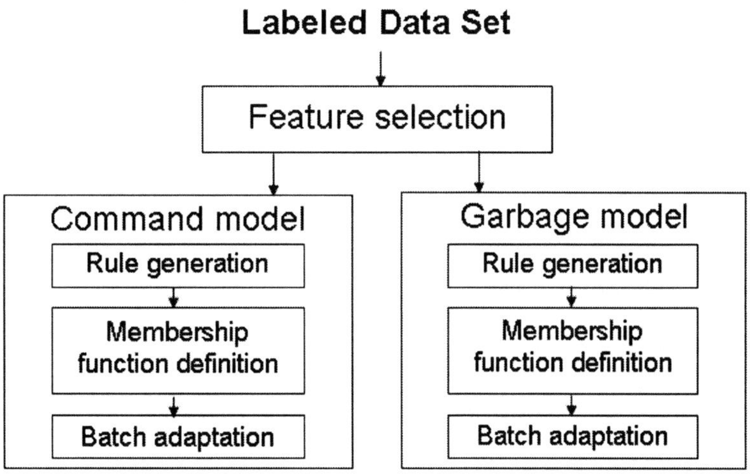

the selected features and experimental results are presented in Table 8, Table 9, Table 10, and Table 11 (RR: Recognition Rate, FN: False Negative, FP: False Positive).

The false positive error is high when the previous system is applied for the testing data. This means the previous system is not capable of distinguishing the garbage gestures from the command gestures. The recognition rate increases when the result is compared from the left column to the right column. This test confirms the superiority of the proposed system.

The recognition capability decreases when the command gestures are mixed with many other similar gestures. In this case, it is more difficult to find appropriate features which discriminate command and non-command gestures as shown in Figure 16. It is thus desired to develop a method which complements the recognition system.

Figure 15. The second stage of user adaptation

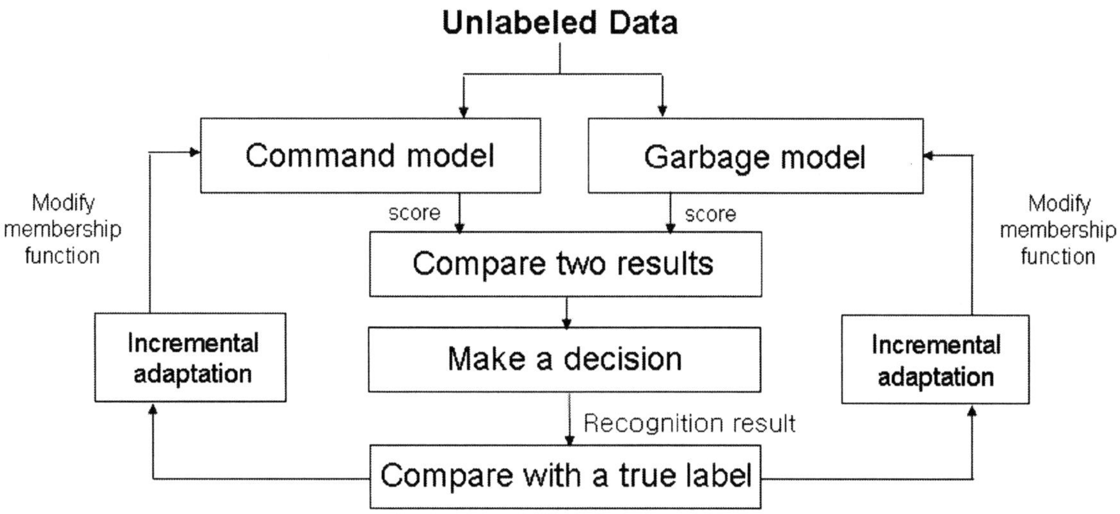

Table 8. Experimental results for discrimination of up command and eating

Person	Previous Soft Remote Control System			Fuzzy Logic by Threshold			Garbage Model (UI)			Garbage Model + Adaptation(UD)		
	RR	FN	FP	RR	FN	FP	RR	FN	FP	RR	FN	FP
1	82	2	34	91	10	8	98	4	0	100	0	0
2	48	4	100	74	32	20	74	50	2	100	0	0
3	77	0	46	72	36	20	93	8	6	96	0	8
4	53	0	94	92	6	10	100	0	0	100	0	0
5	72	0	56	41	44	74	80	38	2	99	0	2

Table 9. Experimental results for discrimination of up command and garbage gestures

Person	Previous System			Fuzzy Logic by Threshold			Garbage Model (UI)			Garbage Model + Adaptation(UD)		
	RR	FN	FP	RR	FN	FP	RR	FN	FP	RR	FN	FP
1	57	2	84	61	16	62	62	64	12	85	16	14
2	50	4	96	64	28	44	62	58	18	88	6	18
3	56	0	88	57	24	62	53	74	20	99	2	0
4	61	0	78	55	34	56	87	4	22	98	2	2
5	59	0	82	62	6	70	82	6	30	94	4	8

Table 10. Experimental results for discrimination of left command and reading

Person	Previous System			Fuzzy Logic by Threshold			Garbage Model (UI)			Garbage Model + Adaptation(UD)		
	RR	FN	FP	RR	FN	FP	RR	FN	FP	RR	FN	FP
1	59	0	82	49	26	76	95	8	2	97	6	0
2	52	8	88	87	22	4	85	26	4	98	2	2
3	50	2	98	75	36	14	94	10	2	99	2	0
4	49	4	98	78	20	24	97	4	2	100	0	0
5	50	0	100	83	4	30	95	0	10	100	0	0

CONCLUDING REMARKS

In this paper, we handled the difficulty of recognizing a meaningful gesture from command-like garbage gestures. As a solution, a fuzzy garbage model was proposed for gesture spotting and we have shown that the proposed system successfully determines if the input data are command gestures or not.

A feature selection technique called SIMF (Han, 2006) is adopted by which important features are selected to discriminate command gestures and garbage gestures. By observing the characteristic of the user's gestures, a set of fuzzy rules are also generated for each case, and the membership functions are optimized by GA. The proposed system is further improved by two-stage user adaptation: the first stage is batch adaptation using GA for

Table 11. Experimental results for discrimination of left command and garbage gestures

Person	Previous System			Fuzzy Logic by Threshold			Garbage Model (UI)			Garbage Model + Adaptation(UD)		
	RR	FN	FP	RR	FN	FP	RR	FN	FP	RR	FN	FP
1	53	0	94	61	48	30	78	8	36	88	2	22
2	48	8	96	56	42	46	58	2	82	96	2	6
3	51	2	96	82	16	20	78	0	44	95	4	6
4	56	4	84	45	36	74	45	82	28	98	4	0
5	54	0	92	45	78	32	90	0	20	97	0	6

Figure 16. Feature space when up command gesture is mixed with eating action and many other similar gestures

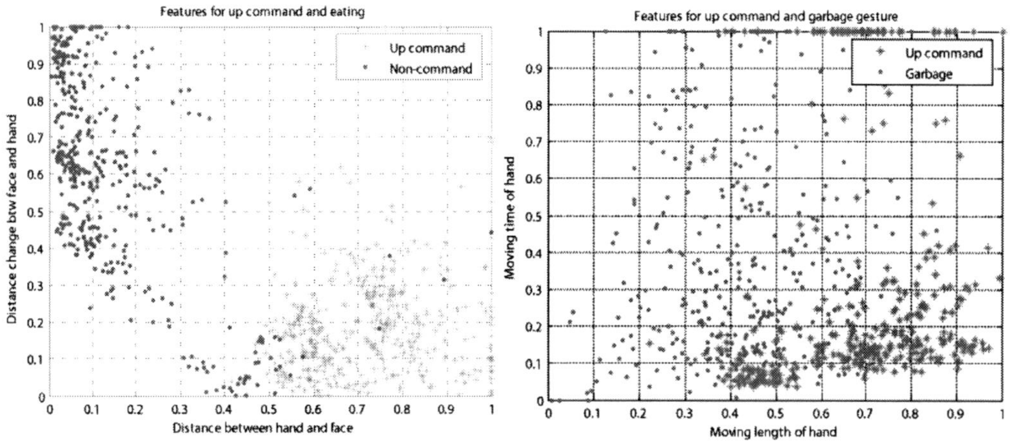

inter-person variation and the second stage is incremental adaptation using the steepest descent method for intra-person variation. By utilizing the two-stage user adaptation, the system is optimized for the target user.

The notion of a fuzzy garbage model for discriminating meaningful gestures from other similar gestures can be utilized for many other applications such as voice recognition, sign language recognition, and activity discovery. Also by analyzing the experimental results, we have noted that, if gesture spotting is divided into two phases, that is, recognition of defined primitive motion and spotting command gestures, the spotting performance is expected to be further improved.

REFERENCES

Arabshahi, P., Marks, R. J., Oh, S., Caudell, T. P., Choi, J. J., & Song, B. G. (1997). Pointer adaptation and pruning of min-max fuzzy inference and estimation. *IEEE Transactions on Circuits and Systems-II: Analog and Digital Signal Processing*, *44*(9), 696–709. doi:10.1109/82.624992

Burns, R. S. (2001). *Advanced control engineering*. Oxford, UK: Butterworth-Heinemann.

Do, J.-H., Jang, H., Jung, S. H., Jung, J. W., & Bien, Z. (2002). Soft remote control system using hand pointing gesture. *International Journal of Human-friendly Welfare Robotic Systems*, *3*(1), 27–30.

Do, J.-H., Kim, J.-B., Park, K.-H., Bang, W.-C., & Bien, Z. Z. (2005). Soft remote control system in the intelligent sweet home. In *Proceedings of the IEEE/RSJ International Conference on Intelligent Robots and Systems*, Edmonton, AB, Canada (pp. 3984-3989).

Fu, H.-C., Chang, H.-Y., Xu, Y. Y., & Pao, H.-T. (2000). User adaptive handwriting recognition by self-growing probabilistic decision-based neural networks. *IEEE Transactions on Neural Networks*, *11*(6), 1373–1384. doi:10.1109/72.883451

Habbi, A., & Zelmat, M. (2003). An improved self-tuning mechanism of fuzzy control by gradient descent method. In *Proceedings of the 17th European Simulation Multiconference*, Nottingham, UK.

Han, J.-S. (2006). *New feature subset selection method and its application for EMG recognition*. Unpublished doctoral dissertation, Korean Advanced Institute of Science and Technology, Daejeon, Korea.

Jeon, M.-J. (2008). *Hand gesture recognition using multivariate fuzzy decision tree and user adaptation*. Unpublished master's thesis, Korean Advanced Institute of Science and Technology, Daejeon, Korea.

Kim, B. S. (2003). Rehabilitation and assistive devices. In *Proceedings of the 8th International Conference on Rehabilitation Robotics*, Daejeon, Korea.

Marks, R. J., II, Oh, S., Arabshahi, P., Caudell, T. P., Choi, J. J., & Song, B. G. (1992). Steepest descent adaptation of min-max fuzzy if-then rules. In *Proceedings of the IEEE/INNS International Joint Conference on Neural Networks*, Beijing, China.

Nomura, H., Hayashi, I., & Wakami, N. (1992). A learning method of fuzzy inference rules by descent method. In *Proceedings of the IEEE International Conference on Fuzzy Systems*, San Diego, CA (pp. 203-210).

Passino, K. M., & Yurkovich, S. (1998). *Fuzzy control*. Reading, MA: Addison-Wesley.

Stefanov, D. H., Bien, Z. Z., & Bang, W.-C. (2004). The smart house for older persons and persons with physical disabilities: Structure, technology arrangements, and perspectives. *IEEE Transactions on Neural Systems and Rehabilitation Engineering*, *12*(2), 228–250. doi:10.1109/TNSRE.2004.828423

Teng, T. K., Shieh, J. S., & Chen, C. S. (2003). Genetic algorithms applied in online autotuning PID parameters of a liquid-level control system. *Transactions of the Institute of Measurement and Control*, *25*(5), 433–450. doi:10.1191/0142331203tm0098oa

This work was previously published in the International Journal of Fuzzy System Applications, Volume 1, Issue 3, edited by Toly Chen, pp. 47-65, copyright 2010 by IGI Publishing (an imprint of IGI Global).

Chapter 10
Indirect Adaptive Fuzzy Control for a Class of Uncertain Nonlinear Systems with Unknown Control Direction

Salim Labiod
University of Jijel, Algeria

Hamid Boubertakh
University of Jijel, Algeria

Thierry Marie Guerra
Université de Valenciennes et du Hainaut-Cambrésis, France

ABSTRACT

In this paper, the authors propose two indirect adaptive fuzzy control schemes for a class of uncertain continuous-time single-input single-output (SISO) nonlinear dynamic systems with known and unknown control direction. Within these schemes, fuzzy systems are used to approximate unknown nonlinear functions and the Nussbaum gain technique is used to deal with the unknown control direction. This paper first presents a singularity-free indirect adaptive control algorithm for nonlinear systems with known control direction, and then this control algorithm is generalized for the case of unknown control direction. The proposed adaptive controllers are free from singularity, allow initialization to zero of all adjustable parameters of the used fuzzy systems, and guarantee asymptotic convergence of the tracking error to zero. Simulations performed on a nonlinear system are given to show the feasibility of the proposed adaptive control schemes.

INTRODUCTION

During the last two decades, there has been significant progress in the area of adaptive control design for nonlinear systems (Slotine & Li, 1991; Krstic et al., 1995; Spooner et al., 2002). Most of the developed adaptive control approaches assume that an accurate model of the system is available and the unknown parameters appear linearly with respect to known nonlinear functions. However, this assumption is not sufficient for many practical situations, because it is difficult to precisely describe a nonlinear system by known nonlinear

DOI: 10.4018/978-1-4666-1870-1.ch010

functions and, therefore, the problem of controlling uncertain nonlinear systems remains a challenging task.

As a model-free design method, fuzzy control has found extensive applications for complex and ill-defined plants (Wang, 1994; Passino & Yurkovich, 1998). In most of these applications, the rule base of the fuzzy controller is constructed from expert knowledge. However, it is sometimes difficult to build the rule base of some plants, or the need may arise to tune the controller parameters if the plant dynamics change. In the hope to overcome this problem, based on the universal approximation theorem and on-line learning ability of fuzzy systems, several stable adaptive fuzzy control schemes have been developed to incorporate the expert knowledge systematically (Wang, 1994). The stability study in such schemes is performed by using the Lyapunov design approach. Conceptually, there are two distinct approaches that have been formulated in the design of a fuzzy adaptive control system: direct and indirect schemes. In the direct scheme, the fuzzy system is used to approximate an unknown ideal controller (Wang, 1994; Spooner & Passino, 1996; Ordonez & Passino, 1999; Chang, 2000; Essounbouli & Hamzaoui, 2006; Labiod & Guerra, 2007). On the other hand, the indirect scheme uses fuzzy systems to estimate the plant dynamics and then synthesizes a control law based on these estimates (Wang, 1994; Ordonez & Passino, 1999; Chang, 2000; Tong et al., 2000; Labiod et al., 2005; Essounbouli & Hamzaoui, 2006; Labiod & Boucherit, 2006). It is worth mentioning that the indirect adaptive approach has the drawback of the controller singularity problem (Labiod et al., 2005; Labiod & Boucherit, 2006).

However, in the aforementioned papers, the control direction is assumed known a priori, i.e., the sign of the control gain is assumed known for the designer. Without this assumption, adaptive controllers design becomes much more difficult, because in this case, one cannot decide the direction along which the control operates and/or the direction of the search of controller parameters. For instance, in indirect fuzzy adaptive control, the control direction is used in the modification of adaptive laws in order to avoid the controller singularity problem and/or in the design of robust control terms to accommodate for approximation errors. In the adaptive control literature, the unknown control direction problem was addressed by (1) using the Nussbaum-type function in the control design for linear and nonlinear systems (Nussbaum, 1983; Ye & Jiang, 1998; Ge et al., 2004; Liu & Huang, 2006; Zhang & Ge, 2007; Liu & Huang, 2008; Boulkroune et al., 2010; Chen & Zhang, 2010) (2) directly estimating unknown parameters involved in the control directions (Kaloust & Qu, 1995; Kaloust & Qu, 1997) (3) the correction vector method for first order nonlinear systems (Brogliato & Lozano, 1994). (4) a switching scheme based on a monitoring function for variable structure model reference adaptive control for linear plants (Lin et al., 2003; Dong et al., 2007) (5) an hysteresis-type function for indirect adaptive fuzzy control (Park et al., 2006; Labiod & Guerra, 2010).

In this paper we present two indirect adaptive fuzzy control schemes for a class of uncertain nonlinear systems where fuzzy systems are used to estimate on-line the unknown nonlinear functions of the system. Based on these adaptive fuzzy approximations, we first present a singularity-free indirect adaptive control algorithm for uncertain nonlinear systems with known control direction. Then, by using the Nussbaum gain technique, we generalize the first algorithm for the case of unknown control direction.

The remainder of this paper is organized as follows. A class of SISO nonlinear systems and control objective are first described, followed by a brief description of the used fuzzy systems. Then, the proposed indirect adaptive control schemes are presented with their adaptive laws and stability analysis. The effectiveness of the proposed adaptive controllers is demonstrated via a simulation example. Finally, some conclusions are drawn.

PROBLEM FORMULATION

Consider the class of single-input single-output (SISO) nonlinear systems that can be described by the following differential equations

$$\begin{cases} \dot{x}_i = x_{i+1}, i = 1, \dots, n-1 \\ \dot{x}_n = f(\mathbf{x}) + g(\mathbf{x}) u \\ y = x_1 \end{cases} \quad (1)$$

or, equivalently

$$y^{(n)} = f(\mathbf{x}) + g(\mathbf{x}) u \quad (2)$$

where $\mathbf{x} = [x_1, \dots, x_n]^T \in \mathbb{R}^n$, is the state vector of the system; $u \in \mathbb{R}$ is the scalar control input; $y \in \mathbb{R}$ is the scalar system output; $f(\mathbf{x})$ and $g(\mathbf{x})$ are unknown smooth nonlinear functions.

In respect of the dynamic system (1), the following assumptions will be made:

Assumption 1: The order n of the system is known.
Assumption 2: The state vector $\mathbf{x}$ is available for measurement.
Assumption 3: The control gain $g(\mathbf{x})$ and its sign are unknown with $0 < \underline{g} \leq |g(\mathbf{x})| \leq \overline{g}$, where $\underline{g}$ and $\overline{g}$ are positive constants.

The objective is to design an adaptive fuzzy controller for system (1) such that the system output $y(t)$ follows a desired trajectory $y_d(t)$ while all signals in the closed-loop system remain bounded.

Regarding the development of the control law, the following assumption should also be made:

Assumption 4: The desired trajectory $y_d(t)$ and its time derivatives $y_d^{(i)}(t)$, $i = 1, \dots, n$, are smooth and bounded.

Now, let us define the tracking error vector as

$$\mathbf{e} = \left[e, \dot{e}, \dots, e^{(n-1)}\right]^T \in \mathbb{R}^n \quad (3)$$

where

$$e = y_d - y \quad (4)$$

and let s be a sliding surface defined as

$$s = \left(\frac{d}{dt} + \lambda\right)^{n-1} e, \quad \lambda > 0 \quad (5)$$

From (5), $s(t) = 0$ represents a linear differential equation whose solution implies that the tracking error $e(t)$ converges to zero with a time constant $(n-1)/\lambda$. In addition, the time-derivatives of $e(t)$ up to $n-1$ also converge to zero (Slotine & Li, 1991). Thus, the control objective becomes the design of a controller to keep $s(t)$ at zero.

The time derivative of s is given by

$$\dot{s} = v - f(\mathbf{x}) - g(\mathbf{x}) u \quad (6)$$

where

$$v = y_d^{(n)} + \kappa_{n-1} e^{(n-1)} + \dots + \kappa_1 \dot{e} \quad (7)$$

with
$\kappa_j = (n-1)!/((n-j)!(j-1)!)\lambda^{n-j}$, $j = 1, \dots, n-1$.

Now, if the nonlinear functions $f(\mathbf{x})$ and $g(\mathbf{x})$ are known, the following control law can be derived to achieve the tracking control objective

$$u = \frac{1}{g(\mathbf{x})}\left(v - f(\mathbf{x}) + ks\right) \quad (8)$$

where k is a positive design parameter.

As a result, the closed-loop error dynamic becomes

$$\dot{s} = -ks \tag{9}$$

From which one can conclude that $s(t) \to 0$ as $t \to \infty$ and, therefore, $e(t)$ and all its time derivatives up to $n-1$ converge to zero (Slotine & Li, 1991).

However, in this paper, the nonlinear functions $f(\mathbf{x})$ and $g(\mathbf{x})$ are unknown, so control law (8) is not implementable. In this case, the purpose is to use fuzzy systems to approximate the unknown functions and, based on these fuzzy approximations, we develop a new robust adaptive controller to meet control objectives.

BRIEF DESCRIPTION OF THE USED FUZZY SYSTEMS

The used fuzzy logic system is a zero-order Takagi-Sugeno fuzzy system that performs a mapping from an input vector $\mathbf{z} = [z_1, \ldots, z_m]^T \in \Omega_{\mathbf{z}} \subset \mathbb{R}^m$ to a scalar output variable $y_f \in \mathbb{R}$, where $\Omega_{\mathbf{z}} = \Omega_{z_1} \times \cdots \times \Omega_{z_m}$ and $\Omega_{z_i} \subset \mathbb{R}$. If we define M_i fuzzy sets F_i^j, $j = 1, \ldots, M_i$, for each input z_i, then the fuzzy system will be characterized by a set of if-then rules of the form (Passino & Yurkovich, 1998; Wang, 1994)

$$\mathrm{R}^k : \text{If } z_1 \text{ is } G_1^k \text{ and} \ldots \text{and } z_m \text{ is } G_m^k \text{ Then } y_f \text{ is } c^k \quad (k = 1, \ldots, N) \tag{10}$$

where $G_i^k \in \{F_i^1, \ldots, F_i^{M_i}\}$, $i = 1, \ldots, n$; c^k is the crisp output of the *k*-th rule, and N is the total number of rules.

By using the singleton fuzzifier, product inference engine, and center-average defuzzifier, the final output of the fuzzy system is given as follows (Wang, 1994)

$$y_f(\mathbf{z}) = \frac{\sum_{k=1}^{N} \mu_k(\mathbf{z}) c^k}{\sum_{k=1}^{N} \mu_k(\mathbf{z})} \tag{11}$$

where

$$\mu_k(\mathbf{z}) = \prod_{i=1}^{m} \mu_{G_i^k}(z_i), \text{ with}$$

$$\mu_{G_i^k} \in \left\{\mu_{F_i^1}, \ldots, \mu_{F_i^{M_i}}\right\}$$

with $\mu_{F_i^j}(x_i)$ is the membership function of the fuzzy set F_i^j.

By introducing the concept of fuzzy basis functions (Wang, 1994), the output given by (11) can be rewritten in the following compact form

$$y_f(\mathbf{z}) = \mathbf{w}^T(\mathbf{z})\theta \tag{12}$$

where $\theta = [c^1, \ldots, c^N]^T$ is a vector grouping all consequent parameters, and

$$\mathbf{w}(\mathbf{z}) = [w_1(\mathbf{z}), \ldots, w_N(\mathbf{z})]^T$$

is a set of fuzzy basis functions defined as

$$w_k(\mathbf{z}) = \frac{\mu_k(\mathbf{z})}{\sum_{j=1}^{N} \mu_j(\mathbf{z})}, \quad k = 1, \ \ldots, N \tag{13}$$

The fuzzy system (12) is assumed to be well-defined such that $\sum_{j=1}^{N} \mu_j(\mathbf{z}) \neq 0$ for all $\mathbf{z} \in \Omega_{\mathbf{z}}$.

The fuzzy system (12) is a universal approximator of continuous functions over a compact set if its parameters are suitably selected (Wang, 1994). In particular, it has been proved that given a smooth function $f(\mathbf{z})$ defined from $\Omega_{\mathbf{z}} \subset \mathbb{R}^n$ to $\mathbb{R}$ and $\varepsilon_f > 0$, there exists a fuzzy system in the form of (12) such that

$\sup_{\mathbf{z}\in\Omega_{\mathbf{z}}} \left|f(\mathbf{z}) - \mathbf{w}^T(\mathbf{z})\theta\right| \leq \varepsilon_f$.

In this paper, it is assumed that the structure of the fuzzy system and the fuzzy basis function parameters are properly specified in advance by the designer. This means that the designer decision is needed to determine the structure of the fuzzy system (that is, determine relevant inputs, number of membership functions for each input, membership function parameters, number of rules), and the consequent parameters should be calculated by learning algorithms.

ADAPTIVE FUZZY CONTROLLER DESIGN

In this section, we propose to use fuzzy systems to approximate the unknown functions $f(\mathbf{x})$ and $g(\mathbf{x})$, and to use these approximations for developing a well-defined adaptive controller with its adaptive laws in order to meet control objectives and guarantee boundedness of all internal signals of the closed-loop system.

First, assume that the nonlinear functions $f(\mathbf{x})$ and $g(\mathbf{x})$ can be approximated, over a compact set $\Omega_{\mathbf{x}}$, by fuzzy systems of the form of (12) as follows

$$f(\mathbf{x}) = \mathbf{w}_f^T(\mathbf{x})\theta_f^* + \varepsilon_f(\mathbf{x}) \tag{14}$$

$$g(\mathbf{x}) = \mathbf{w}_g^T(\mathbf{x})\theta_g^* + \varepsilon_g(\mathbf{x}) \tag{15}$$

where $\varepsilon_f(\mathbf{x})$ and $\varepsilon_g(\mathbf{x})$ are fuzzy approximation errors, θ_f^* and θ_g^* are optimal parameter vectors that minimize functions $\left|\varepsilon_f(\mathbf{x})\right|$ and $\left|\varepsilon_g(\mathbf{x})\right|$, respectively, and $\mathbf{w}_f(\mathbf{x})$ and $\mathbf{w}_g(\mathbf{x})$ are fuzzy basis function vectors assumed suitably specified by the designer.

In this paper, we assume that the used fuzzy systems do not violate the universal approximation property on the operating compact set $\Omega_{\mathbf{x}}$, which is assumed large enough so that the state vector $\mathbf{x}$ remains inside it under closed-loop control.

Assumption 5: The fuzzy approximation errors $\varepsilon_f(\mathbf{x})$ and $\varepsilon_g(\mathbf{x})$ are bounded for all $\mathbf{x}\in\Omega_{\mathbf{x}}$ as $\left|\varepsilon_f(\mathbf{x})\right| \leq \bar{\varepsilon}_f$ and $\left|\varepsilon_g(\mathbf{x})\right| \leq \bar{\varepsilon}_g$, where $\bar{\varepsilon}_f$ and $\bar{\varepsilon}_g$ are unknown positive constants.

Since the optimal parameter vectors θ_f^* and θ_g^* are unknown, so they should be estimated by suitable adaptation laws. Let θ_f and θ_g be the estimates of the ideal vectors θ_f^* and θ_g^*, respectively. Now, define the adaptive fuzzy approximations $\hat{f}(\mathbf{x})$ and $\hat{g}(\mathbf{x})$ of the actual functions $f(\mathbf{x})$ and $g(\mathbf{x})$, respectively, as follows

$$\hat{f}(\mathbf{x}) = \mathbf{w}_f^T(\mathbf{x})\theta_f \tag{16}$$

$$\hat{g}(\mathbf{x}) = \mathbf{w}_g^T(\mathbf{x})\theta_g \tag{17}$$

From Equations (14)-(17), one obtains

$$f(\mathbf{x}) - \hat{f}(\mathbf{x}) = \mathbf{w}_f^T(\mathbf{x})\tilde{\theta}_f + \varepsilon_f(\mathbf{x}) \tag{18}$$

$$g(\mathbf{x}) - \hat{g}(\mathbf{x}) = \mathbf{w}_g^T(\mathbf{x})\tilde{\theta}_g + \varepsilon_g(\mathbf{x}) \tag{19}$$

where $\tilde{\theta}_f = \theta_f^* - \theta_f$ and $\tilde{\theta}_g = \theta_g^* - \theta_g$ are the parameter estimation error vectors.

Using the certainty equivalence principle (Wang, 1994), the following fuzzy adaptive control law can be derived

$$u = \frac{1}{\hat{g}(\mathbf{x})}\left(v - \hat{f}(\mathbf{x}) + ks\right) \tag{20}$$

This control term results from (8) by replacing actual functions $f(\mathbf{x})$ and $g(\mathbf{x})$ by their respective fuzzy approximations $\hat{f}(\mathbf{x})$ and $\hat{g}(\mathbf{x})$.

It should be noted that the certainty equivalent controller (20) cannot guarantee the stability of the closed-loop system due to the unavoidable reconstruction errors and to the fact that this control law is not well-defined when $\hat{g}(\mathbf{x})$ approaches zero. Therefore, the problem we are investigating becomes that of finding a well-defined adaptive fuzzy controller with its parameter adaptation laws to meet control objective in spite of the presence of approximation errors and the unknown control direction.

In the following subsections, this problem will be solved by considering two adaptive fuzzy controllers. The first one will be developed for the known control direction case and the second one for the unknown control direction case. The first controller will serve as a basis for the second one.

Controller Design with Known Control Direction

The following theorem gives the solution of the adaptive fuzzy tracking control problem for the SISO uncertain nonlinear system (1) for the case where the control direction ($\operatorname{sgn}(g(\mathbf{x}))$ is known.

Theorem 1: *Consider the uncertain nonlinear system (1) with* $\operatorname{sgn}(g(\mathbf{x}))$ *assumed known and suppose that Assumptions 1-5 are satisfied. Then the feedback control law is given by*

$$u = u_{ad} + u_r \tag{21}$$

where

$$u_{ad} = \frac{|\hat{g}(\mathbf{x})|^{\alpha} \operatorname{sgn}(\hat{g}(\mathbf{x}))}{\varepsilon_0 + |\hat{g}(\mathbf{x})|^{1+\alpha}}\left(v - \hat{f}(\mathbf{x}) + \operatorname{sgn}(g(\mathbf{x}))k_0\hat{g}(\mathbf{x})s + ks\right) \tag{22}$$

$$u_r = \frac{\varphi s}{|s| + \delta^2 \exp(-\varphi)} \operatorname{sgn}(g(\mathbf{x})) \tag{23}$$

$$\varphi = \hat{\varepsilon}\left(1 + |u_0| + \left|u_{ad} - \operatorname{sgn}(g(\mathbf{x}))k_0 s\right|\right) \tag{24}$$

$$u_0 = \frac{\varepsilon_0}{\varepsilon_0 + |\hat{g}(\mathbf{x})|^{1+\alpha}}\left(v - \hat{f}(\mathbf{x}) + \operatorname{sgn}(g(\mathbf{x}))k_0\hat{g}(\mathbf{x})s + ks\right) \tag{25}$$

with the following parameter adaptation laws

$$\dot{\theta}_f = -\gamma_f \mathbf{w}_f(\mathbf{x}) s \tag{26}$$

$$\dot{\theta}_g = -\gamma_g \mathbf{w}_g(\mathbf{x}) s\left(u_{ad} - \operatorname{sgn}(g(\mathbf{x}))k_0 s\right) \tag{27}$$

$$\dot{\hat{\varepsilon}} = \gamma_0\left(1 + |u_0| + \left|u_{ad} - k_0 \operatorname{sgn}(g(\mathbf{x}))s\right|\right)|s| \tag{28}$$

$$\dot{\delta} = -\gamma_0 \delta \tag{29}$$

$$\operatorname{sgn}(x) = 1 \text{ if } x > 0, 0 \text{ if } x = 0, -1 \text{ if } x < 0$$

where ε_0, α, k, k_0, γ_f, γ_g, and γ_0 are positive design parameters; $\delta(0) > 0$, and $\hat{\varepsilon}$ the estimate of $\varepsilon^* = (1/\underline{g})\max(1, \bar{\varepsilon}_f, \bar{\varepsilon}_g)$, guarantees the boundedness of all signals in the closed loop system and the asymptotic convergence of the tracking error.

Proof: *Using (21) and substituting* $\hat{f}(\mathbf{x}) + \hat{g}(\mathbf{x})u_{ad}$ *to the right-hand side of (6), one obtains*

$$\dot{s} = v - \hat{f}(\mathbf{x}) - \hat{g}(\mathbf{x})u_{ad} - \left(f(\mathbf{x}) - \hat{f}(\mathbf{x})\right) - \left(g(\mathbf{x}) - \hat{g}(\mathbf{x})\right)u_{ad} - g(\mathbf{x})u_r \tag{30}$$

Using the control law (22), (30) can be simplified to

$$\dot{s} = -ks + u_0 - k_0\hat{g}(\mathbf{x})\operatorname{sgn}(g(\mathbf{x}))s$$
$$-(f(\mathbf{x}) - \hat{f}(\mathbf{x})) - (g(\mathbf{x}) - \hat{g}(\mathbf{x}))u_{ad} - g(\mathbf{x})u_r \tag{31}$$

Substituting $k_0|g(\mathbf{x})|s$ to the right-hand side of (32) and using (18) and (19), (31) becomes

$$\dot{s} = -ks - k_0|g(\mathbf{x})|s - \mathbf{w}_f^T(\mathbf{x})\tilde{\theta}_f$$
$$-\mathbf{w}_g^T(\mathbf{x})\tilde{\theta}_g(u_{ad} - k_0\operatorname{sgn}(g(\mathbf{x}))s) - g(\mathbf{x})u_r - \varepsilon(\mathbf{x}) \tag{32}$$

where

$$\varepsilon(\mathbf{x}) = -u_0 + \varepsilon_f(\mathbf{x}) + \varepsilon_g(\mathbf{x})(u - k_0\operatorname{sgn}(g(\mathbf{x}))s) \tag{33}$$

Consider now the following Lyapunov function candidate

$$V = \frac{1}{2}s^2 + \frac{1}{2\gamma_f}\tilde{\theta}_f^T\tilde{\theta}_f + \frac{1}{2\gamma_g}\tilde{\theta}_g^T\tilde{\theta}_g + \frac{\underline{g}}{2\gamma_0}\tilde{\varepsilon}^2 + \frac{\underline{g}}{2\gamma_0}\delta^2 \tag{34}$$

with $\tilde{\varepsilon} = \varepsilon^* - \hat{\varepsilon}$.

The derivative of V along (32) will be

$$\dot{V} = -k\,s^2 - k_0|g(\mathbf{x})|s^2 + W_1 + W_2 \tag{35}$$

where

$$W_1 = -s\mathbf{w}_f^T(\mathbf{x})\tilde{\theta}_f - s\mathbf{w}_g^T(\mathbf{x})\tilde{\theta}_g(u_{ad} - k_0\operatorname{sgn}(g(\mathbf{x}))s)$$
$$-\frac{1}{\gamma_f}\tilde{\theta}_f^T\dot{\theta}_f - \frac{1}{\gamma_g}\tilde{\theta}_g^T\dot{\theta}_g \tag{36}$$

$$W_2 = -sg(\mathbf{x})u_r - s\varepsilon(\mathbf{x}) - \frac{\underline{g}}{\gamma_0}\tilde{\varepsilon}\dot{\hat{\varepsilon}} + \frac{\underline{g}}{\gamma_0}\delta\dot{\delta} \tag{37}$$

Using the adaptation laws (26) and (27), (36) becomes

$$W_1 = 0 \tag{38}$$

From assumptions 3 and 5 and the fact that $su_r \geq 0$, (37) can be upper bounded as

$$W_2 \leq -s\underline{g}u_r + \underline{g}\varepsilon^*(1 + |u_0| + |u_{ad} - k_0\operatorname{sgn}(g(\mathbf{x}))s|)|s|$$
$$-\frac{\underline{g}}{\gamma_0}(\varepsilon^* - \hat{\varepsilon})\dot{\hat{\varepsilon}} + \frac{\underline{g}}{\gamma_0}\delta\dot{\delta} \tag{39}$$

Equations (23), (24), (28) and (29), can be used to further upper bound (39) as

$$W_2 \leq 0 \tag{40}$$

Using (38) and (40), (35) can be upper bounded as

$$\dot{V} \leq -k\,s^2 - k_0|g(\mathbf{x})|s^2 \tag{41}$$

From which one can conclude the boundedness of $V(t)$, $s(t)$, $\tilde{\theta}_f(t)$, $\tilde{\theta}_g(t)$, $\hat{\varepsilon}(t)$ and $\delta(t)$, and the asymptotic convergence of $s(t)$ and $e(t)$ (Labiod & Boucherit, 2006).

Remark 1: *In order to obtain a singularity-free adaptive controller, we have used in the adaptive control component (22) the function* $|\hat{g}(\mathbf{x})|^{\alpha}\operatorname{sgn}(\hat{g}(\mathbf{x}))\big/\left(\varepsilon_0 + |\hat{g}(\mathbf{x})|^{1+\alpha}\right)$ *instead of the function* $1/\hat{g}(\mathbf{x})$. *The term* $\operatorname{sgn}(g(\mathbf{x}))k_0\hat{g}(\mathbf{x})s$ *is introduced in (22) to allow initialization to zero of the parameters* θ_g *(Labiod & Boucherit, 2006). The robust control component (23) is introduced to compensate for approximation errors and improve the closed-loop system robustness.*

Controller Design with Unknown Control Direction

In the preceding subsection, the control direction (sign of $g(\mathbf{x})$) is required to be known for the adaptive fuzzy controller design. Obviously, the algorithm cannot work if $\operatorname{sgn}\left(g\left(\mathbf{x}\right)\right)$ is unknown. In this case, the Nussbaum function technique will be used to tackle this problem. The Nussbaum function technique was originally proposed in (Nussbaum, 1983) and has been effectively used in controller design in solving the difficulty of unknown control directions (Ioannou & Sun, 1996; Ge et al., 2004). A function is called a Nussbaum-type function if it has the following properties (Nussbaum, 1983; Liu & Huang, 2006):

$$\lim_{\nu\to+\infty} \sup\left(\frac{1}{\nu}\int_0^{\nu} N\left(\tau\right)d\tau\right) = +\infty \tag{42}$$

$$\lim_{\nu\to+\infty} \inf\left(\frac{1}{\nu}\int_0^{\nu} N\left(\tau\right)d\tau\right) = -\infty \tag{43}$$

For example, the continuous functions $e^{\tau^2}\cos\left(\tau\right)$ and $\tau^2\cos\left(\tau\right)$ are Nussbaum-type functions. In this paper, the even Nussbaum function $\tau^2\cos\left(\tau\right)$ is used.

Lemma 1: (Liu & Huang, 2008), Le $V\left(\cdot\right)$ and $\tau\left(\cdot\right)$ be smooth functions defined on $\left[0,t_f\right)$ with $V\left(t\right)\geq 0$, $\forall t\in\left[0,t_f\right)$; $N\left(\cdot\right)$ be an even smooth Nussbaum-type function. If the following inequality holds for $\forall t\in\left[0,t_f\right)$:

$$V\left(t\right)\leq c_0+\int_0^t\left(g\left(\zeta\right)N\left(\tau\left(\zeta\right)\right)+c_1\right)\dot{\tau}\left(\zeta\right)d\zeta,$$

where $g\left(t\right)$ is a piecewise continuous time function which takes values in the unknown closed interval $I=\left[\underline{g},\overline{g}\right]$ with $0\notin I$; c_1 is any positive number and c_0 represents some suitable constant, then $V\left(\cdot\right)$, $\tau\left(\cdot\right)$ and

$$\int_0^t\left(g\left(\zeta\right)N\left(\tau\left(\zeta\right)\right)+c_1\right)\dot{\tau}\left(\zeta\right)d\zeta$$

are bounded on $\left[0,t_f\right)$.

The following theorem gives the solution of the adaptive fuzzy tracking control problem for the SISO uncertain nonlinear system (1) for the case where the control direction ($\operatorname{sgn}\left(g\left(\mathbf{x}\right)\right)$ is unknown.

Theorem 2: Consider the uncertain nonlinear system (1) with $\operatorname{sgn}\left(g\left(\mathbf{x}\right)\right)$ assumed unknown and suppose that Assumptions 1-5 are satisfied. Then the feedback control law and parameter adaptive laws are given by

$$u = u_{ad} + u_r \tag{44}$$

$$u_{ad} = \frac{\left|\hat{g}\left(\mathbf{x}\right)\right|^{\alpha}\operatorname{sgn}\left(\hat{g}\left(\mathbf{x}\right)\right)}{\varepsilon_0+\left|\hat{g}\left(\mathbf{x}\right)\right|^{1+\alpha}}\left(\upsilon-\hat{f}\left(\mathbf{x}\right)+k_0N\left(\tau\right)\hat{g}\left(\mathbf{x}\right)s+ks\right) \tag{45}$$

$$u_r = N\left(\tau\right)\overline{u}_r \tag{46}$$

$$\overline{u}_r = \frac{\varphi s}{\left|s\right|+\delta^2\exp\left(-\varphi\right)} \tag{47}$$

$$\varphi = \hat{\varepsilon}\left(1+\left|u_0\right|+\left|u_{ad}-k_0N\left(\tau\right)s\right|\right) \tag{48}$$

$$u_0 = \frac{\varepsilon_0}{\varepsilon_0+\left|\hat{g}\left(\mathbf{x}\right)\right|^{1+\alpha}}\left(\upsilon-\hat{f}\left(\mathbf{x}\right)+k_0N\left(\tau\right)\hat{g}\left(\mathbf{x}\right)s+ks\right) \tag{49}$$

$$\dot{\theta}_f = -\gamma_f\mathbf{w}_f\left(\mathbf{x}\right)s \tag{50}$$

$$\dot{\theta}_g = -\gamma_g\mathbf{w}_g\left(\mathbf{x}\right)s\left(u_{ad}-k_0N\left(\tau\right)s\right) \tag{51}$$

$$\dot{\hat{\varepsilon}} = \gamma_0\left(1+\left|u_0\right|+\left|u_{ad}-k_0N\left(\tau\right)s\right|\right)\left|s\right| \tag{52}$$

$$\dot{\delta} = -\gamma_0 \delta \tag{53}$$

$$N(\tau) = \tau^2 \cos(\tau) \tag{54}$$

$$\dot{\tau} = k_0 s^2 + s\bar{u}_r \tag{55}$$

where ε_0, α, k, k_0, γ_f, γ_g, γ_0 are positive design parameters, $\delta(0) > 0$ and $\hat{\varepsilon}$ is the estimate of $\varepsilon^* = \max(1, \bar{\varepsilon}_f, \bar{\varepsilon}_g)$, guarantee the boundedness of all signals in the closed loop system and the asymptotic convergence of the tracking error.

Proof: Using (44) and substituting $\hat{f}(\mathbf{x}) + \hat{g}(\mathbf{x}) u_{ad}$ to the right-hand side of (6), one obtains

$$\dot{s} = \upsilon - \hat{f}(\mathbf{x}) - \hat{g}(\mathbf{x}) u_{ad} - \left(f(\mathbf{x}) - \hat{f}(\mathbf{x})\right) - \left(g(\mathbf{x}) - \hat{g}(\mathbf{x})\right) u_{ad} - g(\mathbf{x}) u_r \tag{56}$$

Using the control law (45), (56) becomes

$$\dot{s} = -ks + u_0 - k_0 \hat{g}(\mathbf{x}) N(\tau) s - \left(f(\mathbf{x}) - \hat{f}(\mathbf{x})\right) - \left(g(\mathbf{x}) - \hat{g}(\mathbf{x})\right) u_{ad} - g(\mathbf{x}) u_r \tag{57}$$

Substituting $k_0 g(\mathbf{x}) N(\tau) s$ to the right-hand side of (57) and using (18) and (19), (57) becomes

$$\dot{s} = -ks - k_0 g(\mathbf{x}) N(\tau) s - \mathbf{w}_f^T(\mathbf{x}) \tilde{\theta}_f - \mathbf{w}_g^T(\mathbf{x}) \tilde{\theta}_g \left(u_{ad} - k_0 N(\tau) s\right) - g(\mathbf{x}) u_r - \varepsilon(\mathbf{x}) \tag{58}$$

where

$$\varepsilon(\mathbf{x}) = -u_0 + \varepsilon_f(\mathbf{x}) + \varepsilon_g(\mathbf{x}) \left(u - k_0 N(\tau) s\right) \tag{59}$$

Consider now the following Lyapunov function candidate

$$V = \frac{1}{2} s^2 + \frac{1}{2\gamma_f} \tilde{\theta}_f^T \tilde{\theta}_f + \frac{1}{2\gamma_g} \tilde{\theta}_g^T \tilde{\theta}_g + \frac{1}{2\gamma_0} \tilde{\varepsilon}^2 + \frac{1}{2\gamma_0} \delta^2 \tag{60}$$

with $\tilde{\varepsilon} = \varepsilon^* - \hat{\varepsilon}$.

Calculating the derivative of V along (58) yields

$$\dot{V} = -k s^2 - k_0 g(\mathbf{x}) N(\tau) s^2 + W_1 + W_2 \tag{61}$$

where

$$W_1 = -s\mathbf{w}_f^T(\mathbf{x}) \tilde{\theta}_f - s\mathbf{w}_g^T(\mathbf{x}) \tilde{\theta}_g \left(u_{ad} - k_0 N(\tau) s\right) - \frac{1}{\gamma_f} \tilde{\theta}_f^T \dot{\theta}_f - \frac{1}{\gamma_g} \tilde{\theta}_g^T \dot{\theta}_g \tag{62}$$

$$W_2 = -sg(\mathbf{x}) u_r - s\varepsilon(\mathbf{x}) - \frac{1}{\gamma_0} \tilde{\varepsilon} \dot{\hat{\varepsilon}} + \frac{1}{\gamma_0} \delta \dot{\delta} \tag{63}$$

Using the adaptation laws (50) and (51), we obtain

$$W_1 = 0 \tag{64}$$

From assumptions 2 and 5, (63) can be upper bounded as

$$W_2 \le -sg(\mathbf{x}) u_r + \varepsilon^* \left(1 + |u_0| + |u_{ad} - k_0 N(\tau) s|\right)|s| - \frac{1}{\gamma_0} \left(\varepsilon^* - \hat{\varepsilon}\right) \dot{\hat{\varepsilon}} + \frac{1}{\gamma_0} \delta \dot{\delta} \tag{65}$$

Substituting $s\bar{u}_r$ to the right-hand side of (65), and using (47), (48), (52) and (53), (65) can be reduced to

$$W_2 \le -sg(\mathbf{x})u_r + s\bar{u}_r \tag{66}$$

Substituting (64) and (66) in (61) results in

$$\dot{V} \le -k s^2 - k_0 g(\mathbf{x}) N(\tau) s^2 - sg(\mathbf{x})u_r + s\bar{u}_r \tag{67}$$

Moreover, (67) can be rewritten as

$$\dot{V} \le -(k-k_0)s^2 - g(\mathbf{x})N(\tau)(k_0 s^2 + su_r) + (k_0 s^2 + s\bar{u}_r) \tag{68}$$

Then, it follows from (55) that

$$\dot{V} \le -(k-k_0)s^2 - g(\mathbf{x})N(\tau)\dot{\tau} + \dot{\tau} \tag{69}$$

With the condition $k > k_0$, we can obtain the following inequality

$$V(t) \le V(0) + \int_0^t (-g(\nu)N(\nu)+1)\dot{\tau}(\nu)d\nu \tag{70}$$

Using Lemma 1, we can conclude from (70) the boundedness of $V(t)$, $N(t)$ and $\int_0^t (-g(\nu)N(\nu)+1)\dot{\tau}(\nu)d\nu$ for $t \in [0, t_f)$. According to Liu & Huang (2006) and Liu & Huang (2008), since no finite-time escape phenomenon may happen, then $t_f = \infty$. Therefore, $s(t)$, $\tilde{\theta}_f(t)$, $\tilde{\theta}_g(t)$, $\mathbf{x}(t)$, $u(t)$, $\hat{\varepsilon}(t)$ and $\delta(t)$ are bounded. As an intermediate result, $s(t)$ is square integrable and $\dot{s}(t)$ is bounded. Moreover, by invoking Barbalat's lemma, we can conclude the asymptotic convergence of $s(t)$ and $e(t)$.

SIMULATION RESULTS

In this section, to demonstrate the effectiveness of the proposed adaptive fuzzy control algorithms, we consider the tracking control problem of the second-order system

$$\begin{cases} \dot{x}_1 = x_2 \\ \dot{x}_2 = f(\mathbf{x}) + g(\mathbf{x})u \end{cases} \tag{71}$$

where $f(\mathbf{x}) = 1.5(1-x_1^2)x_2 - x_1$ and $g(\mathbf{x}) = 1 + x_1^2 + x_2^2 + x_1x_2$.

The control objective is to force the system output $y(t) = x_1(t)$ to track the desired trajectory $y_d(t) = \sin(t)$.

Within this simulation, two fuzzy systems in the form of (12) are used to approximate the unknown smooth functions $f(\mathbf{x})$ and $g(\mathbf{x})$. The input variables of the used fuzzy systems are x_1 and x_2. For each input variable we have defined five Gaussian membership functions with centers -1, -0.5, 0, 0.5 and 1, and a variance equal to 4.

The system initial conditions are $\mathbf{x}(0) = [0.5, 0]^T$, and the initial values of the adjustable parameters $\theta_f(0)$ and $\theta_g(0)$ are set equal to zero. The design parameters that intervening in the control law are selected as: $\lambda = 2$, $k = 10$, $k_0 = 5$, $\alpha = 0.8$, $\varepsilon_0 = 0.01$, $\sigma = 5$, $\gamma_f = 1$, $\gamma_g = 1$, $\gamma_0 = 0.001$, $\delta(0) = 1$ and $\tau(0) = 0$.

For the first controller where the sign of $g(\mathbf{x})$ is assumed known, simulation results are shown in Figures 1 and 2. It can be seen that actual trajectories converge to the desired ones.

For the second controller where the sign of $g(\mathbf{x})$ is assumed unknown, simulation results for the case of a positive control direction are shown in Figures 3 through 5, and for the case of a negative control direction

$$(g(\mathbf{x}) = -(1 + x_1^2 + x_2^2 + x_1x_2))$$

are shown in Figures 6 through 8. The results show the convergence of the state variables to their desired trajectories, that after a short transient period the Nussbaum function $N(\tau)$ takes the

Figure 1. System's response (known positive control direction): actual (solid lines); desired (dotted lines)

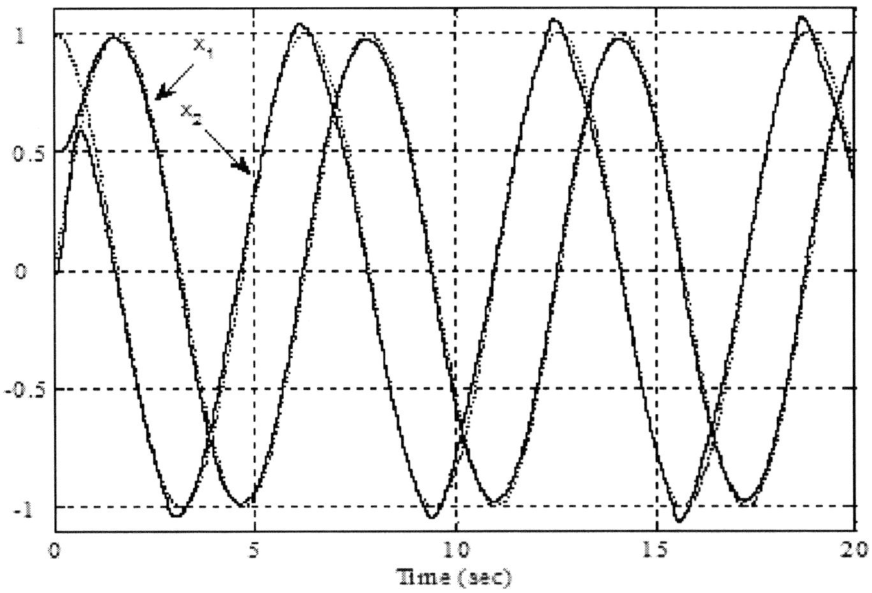

Figure 2. Control input signal (known positive control direction)

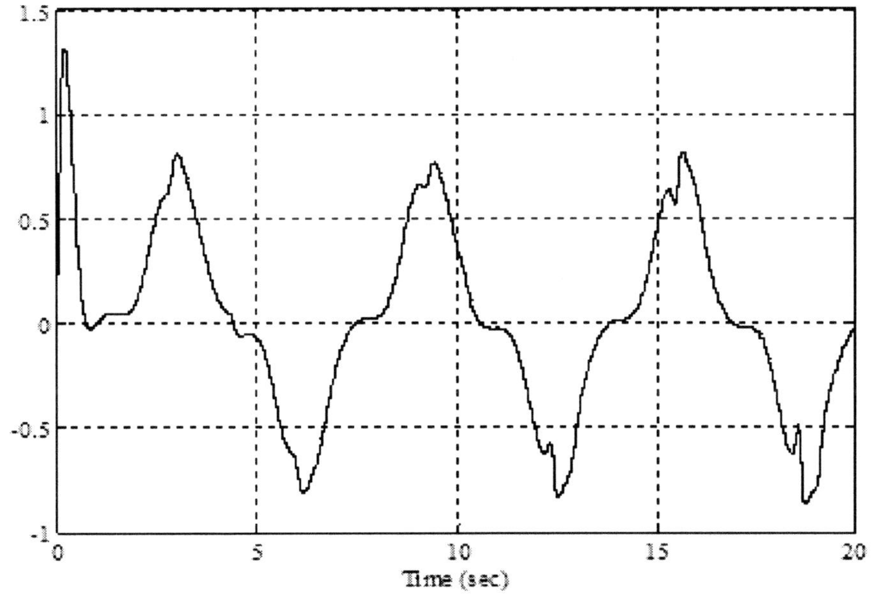

Figure 3. System's response (unknown positive control direction): actual (solid lines); desired (dotted lines)

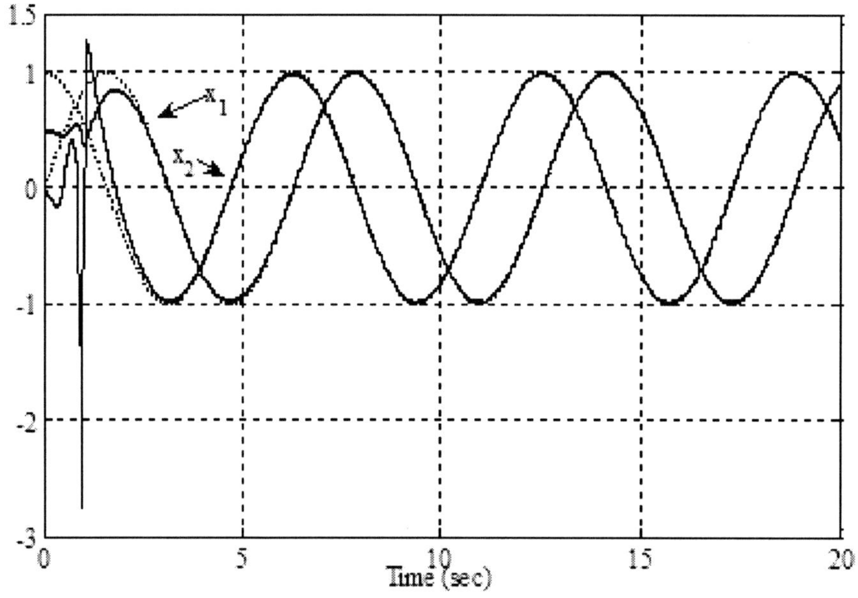

Figure 4. Control input signal (unknown positive control direction)

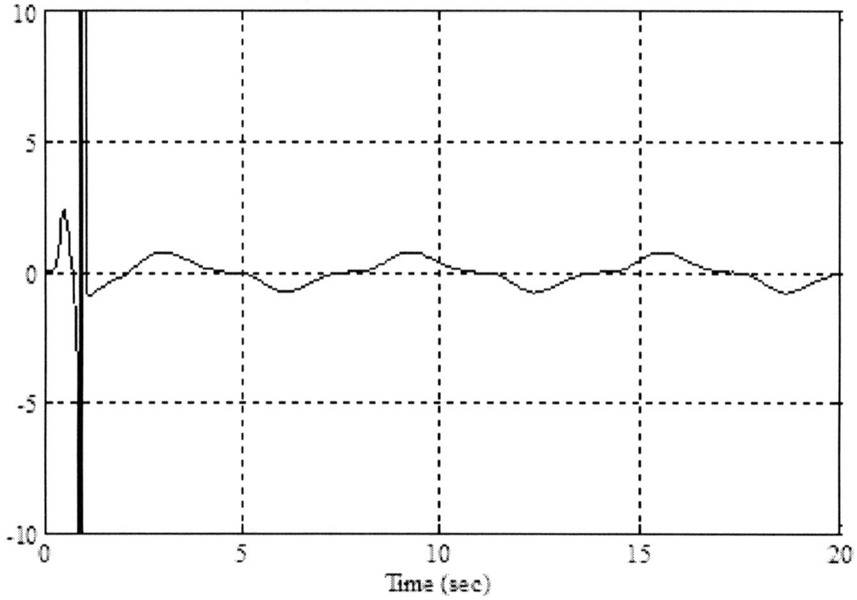

Figure 5. Nussbaum function $N(\tau)$ *(unknown positive control direction)*

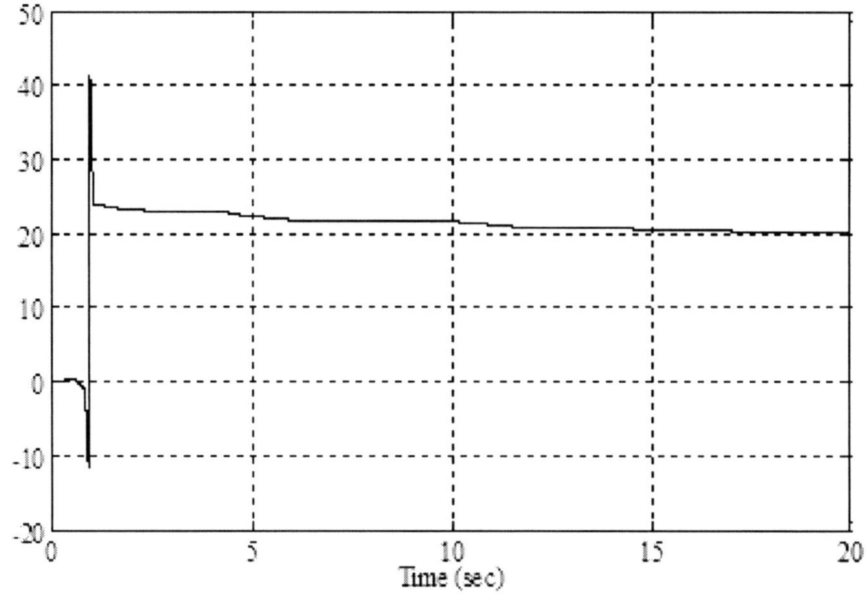

Figure 6. System's response (unknown negative control direction): actual (solid lines); desired (dotted lines)

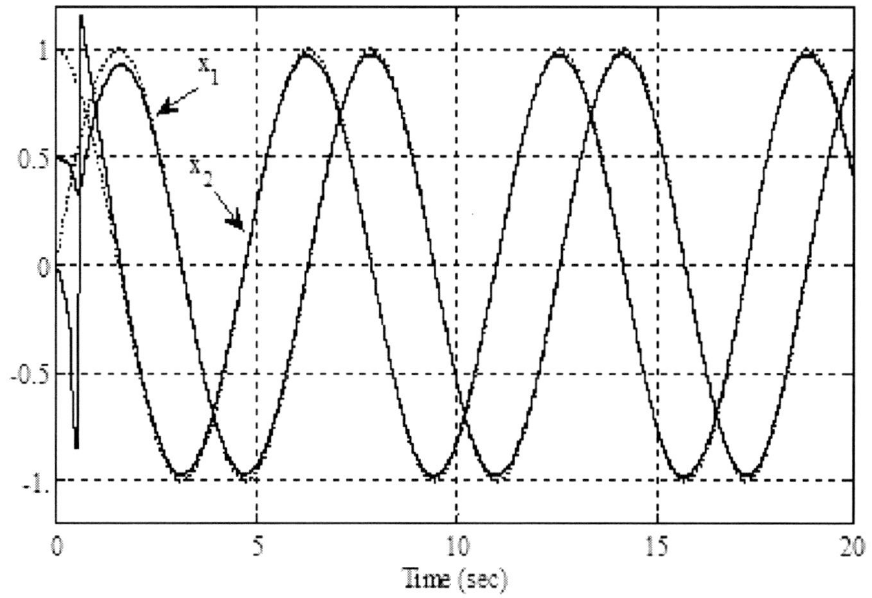

Figure 7. Control input signal (unknown negative control direction)

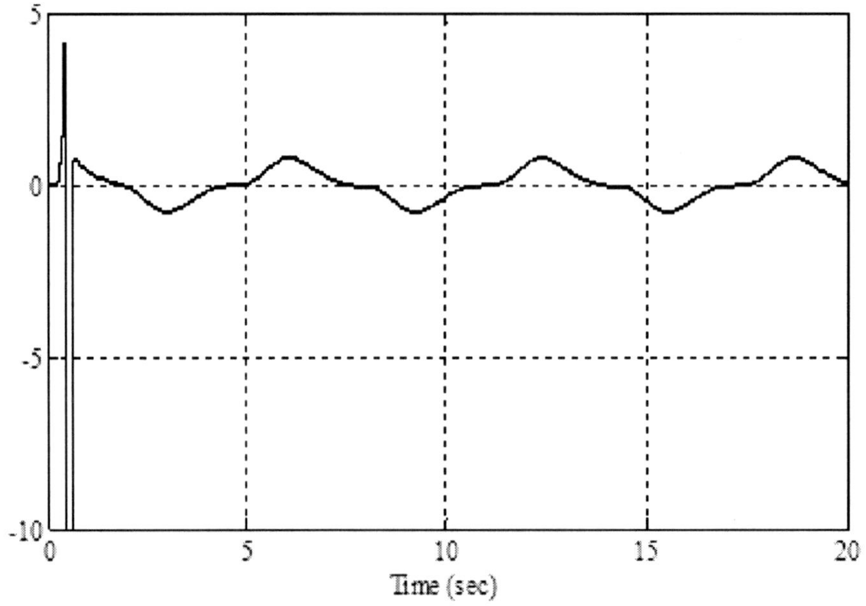

Figure 8. Nussbaum function $N(\tau)$ *(unknown negative control direction)*

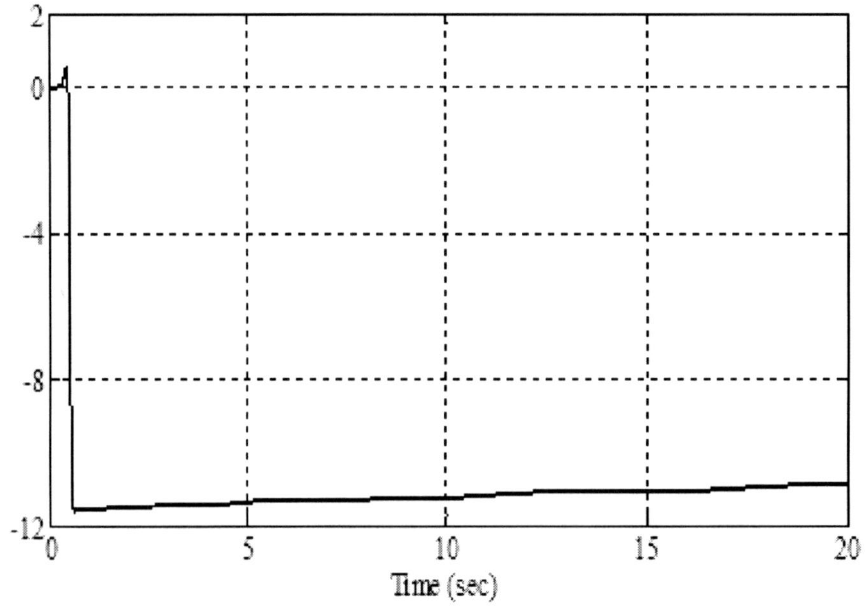

same sign as $g(\mathbf{x})$, and that the control input has peaking phenomenon during this transient period.

CONCLUSION

In this paper, firstly it is presented a stable indirect adaptive fuzzy control scheme for a class of SISO uncertain nonlinear systems with known control direction. The adaptive scheme consists of an adaptive fuzzy controller with its adaptive laws and a robust control term used to compensate for approximation errors. The adaptive scheme is free from singularity. This is achieved through the modification of the inverse of the fuzzy approximation of the control gain. Secondly, by using the Nussbaum-type function, the first adaptive controller is generalized to the case of SISO nonlinear systems with unknown control direction. The proposed adaptive schemes allow initialization to zero of all adjustable parameters of the fuzzy systems, guarantee the boundedness of all signals in the closed-loop system, and ensure the convergence of the tracking error to zero. Simulation results performed on a simple nonlinear system demonstrate the feasibility of the proposed adaptive control schemes. Future work will focus on the extension of the proposed approach to the case of MIMO nonlinear systems.

REFERENCES

Boulkroune, A., Tadjine, M., M'Saad, M., & Farza, M. (2010). Fuzzy adaptive controller for MIMO nonlinear systems with known and unknown control direction. *Fuzzy Sets and Systems*, *161*, 797–820. doi:10.1016/j.fss.2009.04.011

Brogliato, B., & Lozano, R. (1994). Adaptive control of first-order nonlinear systems with reduced knowledge of the plant parameters. *IEEE Transactions on Automatic Control*, *39*, 1764–1768. doi:10.1109/9.310070

Chang, Y. C. (2001). Adaptive fuzzy-based tracking control for nonlinear SISO systems via VSS and H^{∞} approaches. *IEEE Transactions on Fuzzy Systems*, *9*, 278–292. doi:10.1109/91.919249

Chen, W., & Zhang, Z. (2010). Globally stable adaptive backstepping fuzzy control for output-feedback systems with unknown high-frequency gain sign. *Fuzzy Sets and Systems*, *161*, 821–836. doi:10.1016/j.fss.2009.10.026

Dong, W. H., Sun, X. X., & Lin, Y. (2007). Variable structure model reference adaptive control with unknown high frequency gain sign. *Acta Automatica Sinica*, *33*(4), 404–408.

Essounbouli, E., & Hamzaoui, A. (2006). Direct and indirect robust adaptive fuzzy controllers for a class of nonlinear systems. *International Journal of Control, Automation, and Systems*, *4*(2), 146–154.

Ge, S. S., Hong, F., & Lee, T. H. (2004). Adaptive neural control of nonlinear time-delay systems with unknown virtual control coefficients. *IEEE Transactions on Systems, Man, and Cybernetics – Part B*, *34*(1), 499-516.

Ioannou, P. A., & Sun, J. (1996). *Robust adaptive control*. Upper Saddle River, NJ: Prentice Hall.

Kaloust, J., & Qu, Z. (1995). Continuous robust control design for nonlinear uncertain systems without a priori knowledge of control direction. *IEEE Transactions on Automatic Control*, *40*, 276–282. doi:10.1109/9.341792

Kaloust, J., & Qu, Z. (1997). Robust control design for nonlinear uncertain systems with an unknown time-varying control direction. *IEEE Transactions on Automatic Control*, *42*(3), 393–399. doi:10.1109/9.557583

Krstic, M., Kanellapoulos, I., & Kokotovic, P. (1995). *Nonlinear and adaptive control design*. New York, NY: Wiley Interscience.

Labiod, S., & Boucherit, M. S. (2006). Indirect fuzzy adaptive control of a class of SISO nonlinear systems. *Arabian Journal for Science and Engineering, 31*(1B), 61–74.

Labiod, S., Boucherit, M. S., & Guerra, T. M. (2005). Adaptive fuzzy control of a class of MIMO nonlinear systems. *Fuzzy Sets and Systems, 151*(1), 59–77. doi:10.1016/j.fss.2004.10.009

Labiod, S., & Guerra, T. M. (2007). Direct adaptive fuzzy control for a class of MIMO nonlinear systems. *International Journal of Systems Science, 38*(8), 665–675. doi:10.1080/00207720701500583

Labiod, S., & Guerra, T. M. (2010). Indirect adaptive fuzzy control for a class of nonaffine nonlinear systems with unknown control directions. *International Journal of Control, Automation, and Systems, 8*(4), 903–907. doi:10.1007/s12555-010-0425-z

Lin, Y., Hsu, L., Costa, R. R., & Lizarralde, F. (2003). Variable structure model reference adaptive control for systems with unknown high frequency gain. In *Proceedings of the 42nd IEEE Conference on Decision and Control* (pp. 3525-3530).

Liu, L., & Huang, J. (2006). Global robust stabilization of cascade-connected systems with dynamic uncertainties without knowing the control direction. *IEEE Transactions on Automatic Control, 51*, 1693–1699. doi:10.1109/TAC.2006.883023

Liu, L., & Huang, J. (2008). Global robust output regulation of lower triangular systems with unknown control direction. *Automatica, 44*, 1278–1284. doi:10.1016/j.automatica.2007.09.014

Nussbaum, R. D. (1983). Some remarks on the conjecture in parameter adaptive control. *Systems & Control Letters, 3*, 243–246. doi:10.1016/0167-6911(83)90021-X

Ordonez, R., & Passino, K. M. (1999). Stable multi-input multi-output adaptive fuzzy/neural control. *IEEE Transactions on Fuzzy Systems, 7*(3), 3453–3453. doi:10.1109/91.771089

Park, J. H., Kim, S. H., & Moon, C. J. (2006). Adaptive fuzzy controller for the nonlinear system with unknown sign of the input gain. *International Journal of Control, Automation, and Systems, 4*(2), 178–186.

Passino, K. M., & Yurkovich, S. (1998). *Fuzzy control*. Reading, MA: Addison-Wesley.

Slotine, J. E., & Li, W. (1991). *Applied nonlinear control*. Upper Saddle River, NJ: Prentice Hall.

Spooner, J. T., Maggiore, M., Ordonez, R., & Passino, K. M. (2002). *Stable adaptive control and estimation for nonlinear systems*. Chichester, UK: John Wiley & Sons. doi:10.1002/0471221139

Spooner, J. T., & Passino, K. M. (1996). Stable adaptive control using fuzzy systems and neural networks. *IEEE Transactions on Fuzzy Systems, 4*, 339–359. doi:10.1109/91.531775

Tong, S. C., Tang, J., & Wang, T. (2000). Fuzzy adaptive control of multivariable nonlinear systems. *Fuzzy Sets and Systems, 111*, 153–167. doi:10.1016/S0165-0114(98)00058-X

Wang, L. X. (1994). *Adaptive fuzzy systems and control: Design and stability analysis*. Upper Saddle River, NJ: Prentice Hall.

Ye, X., & Jiang, J. (1998). Adaptive nonlinear design without a priori knowledge of control directions. *IEEE Transactions on Automatic Control, 43*, 1617–1621. doi:10.1109/9.728882

Zhang, T. P., & Ge, S. S. (2007). Adaptive neural control of MIMO nonlinear state time-varying delay systems with unknown dead-zones and gain signs. *Automatica, 43*, 1021–1033. doi:10.1016/j.automatica.2006.12.014

This work was previously published in the International Journal of Fuzzy System Applications, Volume 1, Issue 4, edited by Toly Chen, pp. 1-17, copyright 2010 by IGI Publishing (an imprint of IGI Global).

Chapter 11
Approximated Simplest Fuzzy Logic Controlled Shunt Active Power Filter for Current Harmonic Mitigation

Rambir Singh
Motilal Nehru National Institute of Technology, India

Asheesh K. Singh
Motilal Nehru National Institute of Technology, India

Rakesh K. Arya
M. P. Council of Science and Technology, India

ABSTRACT

This paper examines the size reduction of the fuzzy rule base without compromising the control characteristics of a fuzzy logic controller (FLC). A 49-rule FLC is approximated by a 4-rule simplest FLC using compensating factors. This approximated 4-rule FLC is implemented to control the shunt active power filter (APF), which is used for harmonic mitigation in source current. The proposed control methodology is less complex and computationally efficient due to significant reduction in the size of rule base. As a result, computational time and memory requirement are also reduced significantly. The control performance and harmonic compensation capability of proposed approximated 4-rule FLC based shunt APF is compared with the conventional PI controller and 49-rule FLC under randomly varying nonlinear loads. The simulation results presented under transient and steady state conditions show that dynamic performance of approximated simplest FLC is better than conventional PI controller and comparable with 49-rule FLC, while maintaining harmonic compensation within limits. Due to its effectiveness and reduced complexity, the proposed approximation methodology emerges out to be a suitable alternative for large rule FLC.

DOI: 10.4018/978-1-4666-1870-1.ch011

INTRODUCTION

In the recent decades, the applications of semiconductor devices have increased manifold due their numerous advantages such as better controllability, higher efficiency, fast switching and better current handling capacity. But at the same time their inherent nonlinear characteristics introduce many undesirable features in the system, such as harmonics, poor voltage regulation, poor power factor, low system efficiency, and interference in nearby communication system etc. All these undesirable features affect the power quality. Shunt passive filters were traditionally used to provide current harmonic compensations. But these passive filters suffer from certain demerits such as fixed compensation, large size, detuning problems due to aging effect and resonance as pointed out by Singh, Al-Haddad, and Chandra (1999) in their review paper on active filters for power quality improvement. Shunt active power filters (APF) have emerged as an undisputed alternative over their passive counterparts for current harmonics mitigation and reactive power compensation.

The voltage source inverter (VSI) based shunt APF is preferred for harmonic and reactive power compensation as discussed by Akagi, Kanajawa, and Nabae (1984), Peng, Akagi, and Nabae (1990), Grady, Samotyj, and Noyola (1990), Singh, Chandra, and Al-Haddad (1999), and Singh, Al-Haddad, et al. (1999). A control scheme is proposed by Dixon, Contrado, and Morán (1999) which require sensing of line currents only for generation of reference current. The scheme is simple and easy to implement. Recently most of the reported work uses this scheme, which includes the papers of Jain, Agrawal, and Gupta (2002) on control of shunt APF using fuzzy logic controller (FLC) and Mishra and Bhende (2007) on bacterial foraging based optimized based control of shunt APF.

Artificial intelligence (AI) is getting popularity among control engineers due to its ability to handle complex problems at randomly varying operating conditions. Bose (1994) in his invited paper has explored the possibilities of expert system, fuzzy logic and neural network applications in power electronics and motion control. This work has provided a new space of opportunities for control engineers. Anis Ibrahim and Morcos (2002) focus on the effectiveness of AI and advanced mathematical tools for power quality applications. A fuzzy logic based shunt APF shows better dynamic response and higher control precision as compared with the PI controller as concluded by Dixon et al. (1999), Jain et al. (2002), An, Zhikang, Wenji, Ruixiang, and Chunming (2009), and Karuppanan and Mahapatra (2011). All these papers have proposed 49-rule FLC for the control of shunt APF. The 49-rule FLC has the drawback of a large number of fuzzy sets and control rules. Due to this the complexity of the controller increases, as large computational time and large memory is required to execute the desired control action. This drawback is overcome in this paper by approximating a 49-rule FLC with a simplest 4-rule FLC.

In the recent past some studies regarding reduction of rule base size have been reported. Bezine, Derbel, and Alimi (2002) explained some issues on design and rule base size reduction for the fuzzy control of robot manipulators. Ciliz (2005) explained some concepts regarding resizing of rule base by removing inconsistent and redundant rules for the application of vacuum cleaner. These two studies were application specific. Hampel and Chaker (1998) provided some conclusions for minimization of number of variable parameters for optimization of fuzzy controller. Moser and Navrata (2002) proposed a fuzzy controller with conditionally firing rules, in this work number of rules were not minimized only the conditions under which they will fire were reduced. Zeng and Singh (1994, 1995) proposed a mathematical description of approximation theory of fuzzy systems for single input single output (SISO) and multi input multi output (MIMO) cases. The main focus of these papers was the approximation capabilities of the fuzzy systems for approximating a mathematical polynomial rather than on rule reduction.

In this work an approximation methodology is proposed with reduced rule base size, which is implemented on a shunt APF. The dc link voltage of shunt APF is compared with a reference voltage. The error and change of error are used as inputs to FLC. The output of FLC is the incremental change in peak value of reference source current, which will be further utilized to generate reference source current. In current control loop this current will be compared with actual source current to generate the gate pulses for shunt APF to regulate the dc link voltage and provide efficient harmonic compensation. The approximation technique is independent of process to be controlled because the methodology is based on comparing the output of large rule FLC with reduced rule FLC at central values of membership functions and then designing the compensating factors to minimize the deviation between the responses of two controllers. The proposed FLC based control of shunt APF results in better dynamic performance than conventional PI controller and harmonic compensation is comparable with 49-rule FLC during transient state and even better during steady state operation of shunt APF.

The rest of the paper is organized in the following sequence. The next section covers the introduction of 49-rule FLC and simplest 4-rule FLC. The proposed approximation technique and derivation of compensating factors are discussed in a subsequent section. The basic compensation principle and control scheme of shunt APF with proposed 4-rule approximated FLC follows next. The simulation results and comparison of performances of proposed scheme with conventional PI and 49-rule FLC are discussed before making the concluding remarks. Finally concluding remarks are made and scope of future research is presented.

FUZZY LOGIC CONTROLLER (FLC)

Fuzzy control provides the expertise of human experience into machine. The FLC basically comprise of the following modules: fuzzification module, rule base, inference mechanism and defuzzification module as shown in Figure 1. In the fuzzification module, the input signal to the controller which is in the form of crisp value is converted into the form suitable to inference mechanism. Based on this fuzzified input the inference mechanism activates and applies rules from the rule base which contains rules in the form of if-then statements. The inference mechanism then passes on this expert's linguistic description to defuzzification module, which converts the linguistic information into crisp value as an output of FLC.

49-Rule FLC

In 49-rule FLC seven (say N) triangular membership functions are used for each input i.e. error (e) and change in error (ce) of dc link voltage in the universe of discourse (UOD) of (–L, L). The output variable i.e. incremental change in peak value of reference source current (δI_{max}) is also taken care of by seven triangular membership functions in the UOD of (–H, H). Out of these seven triangular membership functions, three membership functions (say J) represents each positive and negative UOD, while one membership function takes care of zero linguistic variables. The UOD for both input and output variables have been taken from –1 to 1. The defuzzification is performed using the centre-of-gravity method. The membership functions for each input variables (e and ce) and output variable δI_{max} are shown in Figure 2(a) and (b) respectively. Seven linguistic variables used are negative large (NL), negative medium (NM), negative small (NS), zero (ZE), positive small (PS), positive medium (PM) and positive large (PL). The control rules used for 49-rule FLC are shown in Table 1.

Table 1. Rule base for 49 rules FLC

e→ ce↓	NL	NM	NS	ZE	PS	PM	PL
NL	NL	NL	NL	NL	NM	NS	ZE
NM	NL	NL	NL	NM	NS	ZE	PS
NS	NL	NL	NM	NS	ZE	PS	PM
ZE	NL	NM	NS	ZE	PS	PM	PL
PS	NM	NS	ZE	PS	PM	PL	PL
PM	NS	ZE	PS	PM	PL	PL	PL
PL	ZE	PS	PM	PL	PL	PL	PL

Let the central values of membership function for output variable is denoted by γ_i, where i ranges from –J to J, such that γ_{-J} = -H, γ_0 = 0, γ_J = H. The output membership functions with their central values are shown in Figure 2(b). For an equidistant space W between the central values of two adjacent linguistic membership functions, W can be given by:

$$W = \frac{H}{J} \tag{1}$$

where H is the span of positive or negative UOD and J represents number of membership functions used to cover the corresponding UOD.

The membership grades for output variable δI_{max} (t) can be defined in terms of corresponding central values γ_i, for i = 0,1,2,..,J (i.e. in the positive UOD) as shown in (2) to (4).

For i = J

$$\mu_i(\delta I_{max}(t)) = \begin{cases} 1, & \text{for } \delta I_{max}(t) \geq \gamma_J \\ \frac{\delta I_{max}(t) - \gamma_{J-1}}{W}, & \text{for } \gamma_{J-1} \leq \delta I_{max}(t) < \gamma_J \\ 0, & \text{otherwise} \end{cases} \tag{2}$$

For i = J-1, J-2, …1

$$\mu_i(\delta I_{max}(t)) = \begin{cases} 1, & \text{for } \delta I_{max}(t) = \gamma_i \\ \frac{\gamma_{i+1} - \delta I_{max}(t)}{W}, & \text{for } \gamma_i < \delta I_{max}(t) \leq \gamma_{i+1} \\ \frac{\delta I_{max}(t) - \gamma_{i-1}}{W}, & \text{for } \gamma_{i-1} \leq \delta I_{max}(t) < \gamma_i \\ 0, & \text{otherwise} \end{cases} \tag{3}$$

For i = 0

$$\mu_i(\delta I_{max}(t)) = \begin{cases} 1, & \text{for } \delta I_{max}(t) = \gamma_0 \\ \frac{\gamma_1 - \delta I_{max}(t)}{W}, & \text{for } \gamma_0 < \delta I_{max}(t) \leq \gamma_1 \\ \frac{\delta I_{max}(t) - \gamma_{-1}}{W}, & \text{for } \gamma_0 > \delta I_{max}(t) \geq \gamma_{-1} \end{cases} \tag{4}$$

where μ_i represents the membership value of variable corresponding to i^{th} linguistic function.

Similarly the membership grades can be defined for negative UOD.

Simplest FLC

The concept of simplest FLC was introduced by Ying (2000). The term simplest refers to a minimal possible configuration in terms of number of input variables, fuzzy sets and fuzzy rules for any properly functional FLC. Two triangular

Figure 1. Block diagram of a fuzzy logic controller

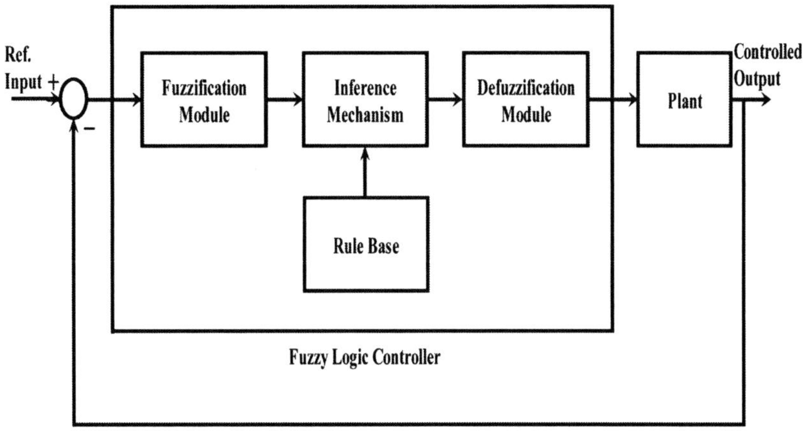

membership functions for each input variable, i.e., error (e) and change in error (ce) are used in the universe of discourse (UOD) of (-L, L) as shown in Figure 3(a) and (b), respectively. Three triangular membership functions are used for output variable δI_{max} in the UOD of (-H, H) as shown in Figure 3(c). The UOD for input and output variables is taken from –1 to 1. The centre-of-gravity defuzzification method is used to obtain the output as a crisp value.

Let x be the instantaneous value of any input variable e(t) and ce(t) at sampling instant t. The membership grades of membership functions for both the input variables can be given as:

$$\mu_{n,x} = \begin{cases} 1, & \text{for } x \leq \text{-L} \\ \dfrac{\text{L-}x}{2\text{L}}, & \text{for -L} \leq x \leq \text{L} \\ 0, & \text{for } x \geq \text{L} \end{cases} \quad (5)$$

$$\mu_{p,x} = \begin{cases} 0, & \text{for } x \leq \text{-L} \\ \dfrac{\text{L}+x}{2\text{L}}, & \text{for -L} \leq x \leq \text{L} \\ 1, & \text{for } x \geq \text{L} \end{cases} \quad (6)$$

Similarly, the membership grades of triangular output membership functions can be given as:

$$\mu_{n,\delta I_{max}} = \begin{cases} 1, & \text{for } \delta I_{max} \leq \text{-H} \\ \dfrac{-\delta I_{max}}{\text{H}}, & \text{for -H} \leq \delta I_{max} \leq 0 \\ 0, & \text{for } \delta I_{max} \geq 0 \end{cases} \quad (7)$$

$$\mu_{p,\delta I_{max}} = \begin{cases} 1, & \text{for } \delta I_{max} \geq \text{H} \\ \dfrac{\delta I_{max}}{\text{H}}, & \text{for } 0 \leq \delta I_{max} \leq \text{H} \\ 0, & \text{for } \delta I_{max} \leq 0 \end{cases} \quad (8)$$

$$\mu_{z,\delta I_{max}} = \begin{cases} \dfrac{-\delta I_{max}}{\text{H}}, & \text{for } 0 \leq \delta I_{max} \leq \text{H} \\ \dfrac{\delta I_{max}}{\text{H}}, & \text{for H} \leq \delta I_{max} \leq 0 \\ 1, & \text{for } \delta I_{max} = 0 \end{cases} \quad (9)$$

The four rules of simplest FLC are given below. These four rules can cover the entire range of system behaviour as explained by Ying (2000):

Figure 2. Membership functions for (a) each input variable error and change in error, (b) output δImax and central values of linguistic variables of δImax for 49 rules FLC

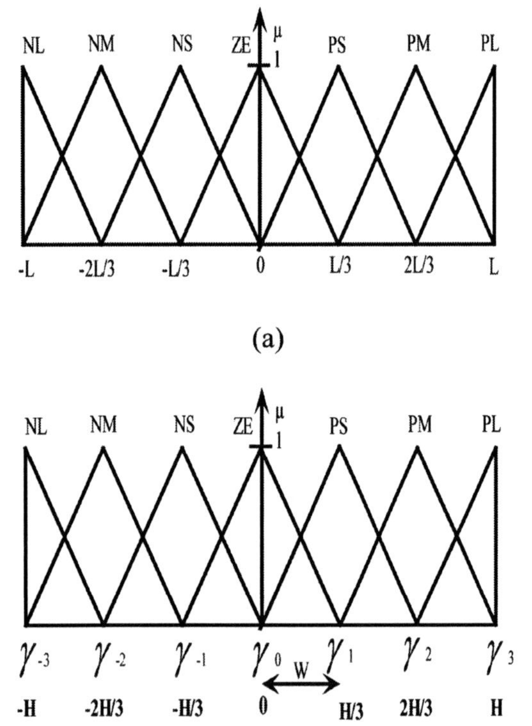

Rule 1: If e is negative (N) and ce is negative (N) then δI_{max} is negative (N).
Rule 2: If e is negative (N) and ce is positive (P) then δI_{max} is zero (ZE).
Rule 3: If e is positive (P) and ce is negative (N) then δI_{max} is zero (ZE).
Rule 4: If e is positive (P) and ce is positive (P) then δI_{max} is positive (P).

Approximation Principle

The basic principle of approximation method is based on the comparing of the outputs of both the controllers at central values of linguistic membership functions and then deriving the compensating factors to minimize the deviation, the methodology used is shown in Figure 4.

Comparison of Membership Grades of Two Controllers

When the membership grades of two controllers are compared, the following observations are obtained:

Between $\gamma_{(J-j)}$ to $\gamma_{(J-j)-1,}$ for j = 0, 1, 2 and J = 3, the two membership grades can be written as:

$$\frac{u(t)}{W_1} = \frac{u_1(t) - \gamma_{(J-j)-1}}{W} \qquad (10)$$

where $u(t)$ and $u_1(t)$ are the outputs of simplest FLC and 49 rule FLC respectively. Considering the UOD between −1 to +1, W_1 (the space between central values of two membership functions for simplest FLC) will be unity. So the above equation can be written as:

$$u_1(t) = u(t) \times W + \gamma_{(J-j)-1} \qquad (11)$$

Similarly for negative UOD between $\gamma_{-(J-j)}$ to $\gamma_{-((J-j)-1)}$, the relationship can be given as:

$$\frac{u(t)}{W_1} = \frac{u_1(t) - \gamma_{-((J-j)-1)}}{W} \qquad (12)$$

$$\frac{u(t)}{W_1} = \frac{u_1(t) + \gamma_{(J-j)-1}}{W} \qquad (13)$$

as $\gamma_{-((J-j)-1)} = -\gamma_{(J-j)-1}$. The above equation can be written as

$$u_1(t) = u(t) \times W - \gamma_{(J-j)-1} \qquad (14)$$

Figure 3. Membership functions of (a) error, (b) change in error, and (c) output δI_{max} of simplest FLC

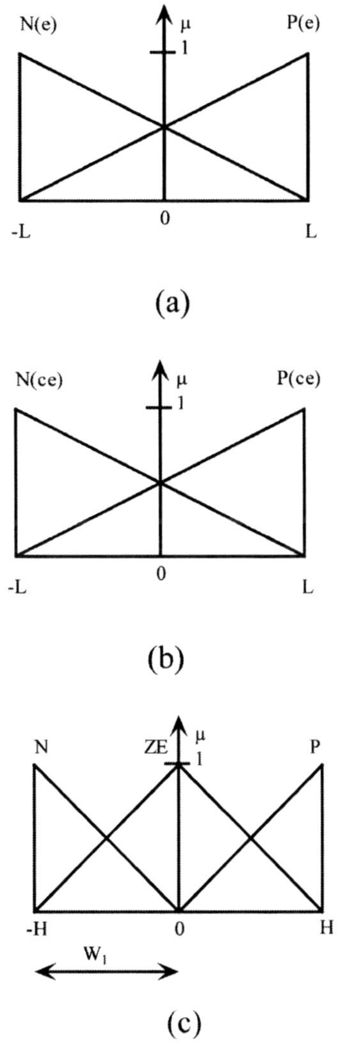

Equality sign in (11) and (14) holds well if the following conditions are satisfied:

a. The outputs of two controllers are identical for the given input variables.
b. The instantaneous output of FLC is dependent on only one membership function.

These conditions are however not satisfied because:

a. The output of 49 rules FLC is never equal to the output of 4-rule FLC for the given input variables due to nonlinear nature of FLC.
b. The instantaneous output of FLC is not dependent on one membership function, whereas it depends on all active membership functions.

Due to these reasons the equality relationship will not hold good which will affect the approximation process adversely. Therefore, the error for the corresponding range is taken care of by suitable compensating factors. Thus after introducing the compensating factor for the corresponding ranges, (11) and (14) can be written as:

$$u_2(t) = u(t) \times \mathrm{W} \pm \gamma_{(J-j)-1} \pm C_n \ ,$$

for n = J, (J-1), ...1 (15)

The positive and negative signs are used for positive and negative UODs respectively. In the above equation u2(t) is a new output variable representing the output of approximated simplest FLC.

Design of Compensating Factors

A simplest FLC is more nonlinear as compared to a 49 rule FLC Ying (2000). The variation of these non-linearities is more dominating in the middle of UOD and keeps decreasing towards the extreme values of UOD. So the compensating factors are to be designed to minimize the non-linearities and make the response of approximated simplest FLC comparable with that of a 49-rule based FLC. The compensating factors are designed based on following criterion:

Figure 4. *Methodology used in designing proposed approximated simplest FLC*

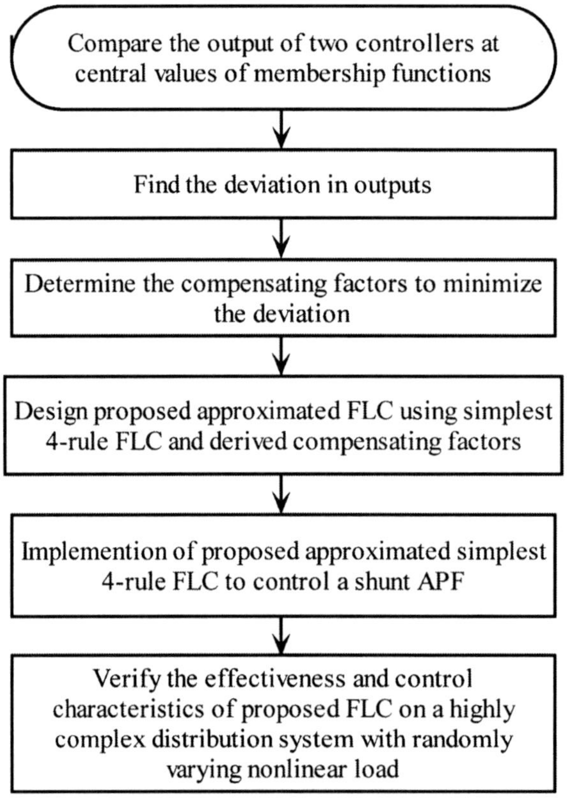

1. Two compensating factors as multiplier near zero on each side of UOD, because addition of average error near zero will provide large control values, which are not desired.
2. Four compensating factors to take care of the remaining UOD towards the extremes which will use the addition of average error between the outputs of simplest FLC and 49 rule-based FLC.

Therefore, the six compensating factors will cover the entire range of UOD. By simulating the output of simplest FLC and 49 rule-based FLC at corresponding central values, following results are obtained:

$$u_{simplest\ FLC}(1,1) = 0.673 \tag{16}$$

$$u_{49\ rule\ FLC}(1,1) = 0.895 \tag{17}$$

$$u_{simplest\ FLC}(\frac{2}{3},\frac{2}{3}) = 0.361 \tag{18}$$

$$u_{49\ rule\ FLC}(\frac{2}{3},\frac{2}{3}) = 0.895 \tag{19}$$

After comparing the outputs at central values (1, 1) and (2/3, 2/3) the average error comes out to be 0.0558. This error will be compensated by adding another compensating factor, i.e., C_3 = 0.0558. Similarly the values of other compensating factors are calculated and are shown in Table 2.

The proposed scheme of using compensating factors is shown in Figure 5.

Table 2. Values and ranges of compensating factors

Compensating Factor	Range	Value
C_3(Positive UOD)	1 to 2/3	0.0558
C_2(Positive UOD)	2/3 to 1/3	0.1948
C_1(Positive UOD)	1/3 to 0	8.3
C_{-1}(Negative UOD)	0 to –1/3	8.3
C_{-2}(Negative UOD)	-1/3 to –2/3	0.1948
C_{-3}(Negative UOD)	-2/3 to -1	0.0558

BASIC COMPENSATION PRINCIPLE AND CONTROL SCHEME OF SHUNT ACTIVE POWER FILTER (APF)

Figure 6 represents the single line diagram of basic compensation principle of a shunt APF. The compensation of current harmonics in shunt active power filters is achieved by injecting equal and opposite harmonic compensating current. A shunt active power filter can be visualized as a current source injecting the harmonic component equal in magnitude but with a phase shift of 180o to that generated by the nonlinear load. Therefore, the source currents remain sinusoidal and in phase with the respective phase voltages. The scheme is used by Grady et al. (1990), Singh, Chandra, et al. (1999), Dixon et al. (1999), Jain et al. (2002), and Singh, Singh, and Mitra (2007) in their work.

The control scheme plays a vital role in efficient operation of a shunt APF, and its effectiveness depends on the execution of basically three tasks, i.e., signal conditioning, derivation of reference signal and generation of gating signal. In signal conditioning three phase source currents and voltages of two phases are sensed for further processing. The reference source currents will be generated in such a way that they force the actual source currents to be in phase with the source voltages in order to achieve ideal compensation. The instantaneous values of reference source currents can be given as:

$$i_{sa}^{*} = i_{s,\max} \sin \omega t \tag{20}$$

$$i_{sb}^{*} = i_{s,\max} \sin (\omega t - \frac{2\pi}{3}) \tag{21}$$

$$i_{sc}^{*} = i_{s,\max} \sin (\omega t + \frac{2\pi}{3}) \tag{22}$$

where, i^*_{sa}, i^*_{sb} and i^*_{sc} are instantaneous values of reference source currents of phase a, b and c, respectively and $i_{s,max}$ is the peak value of reference source currents.

The maximum value of reference current template can be obtained by regulating the dc link voltage and the respective phase angles can be derived from unit voltage vectors obtained from source voltages. A current controlled voltage source PWM converter is used as shunt APF. The schematic diagram of shunt APF with the proposed control scheme is shown in Figure 7. The dc link voltage is sensed and compared with a set reference value. The error and change in error are of dc link voltage are used as inputs to FLC. The output of controller in both the cases will be the incremental change in peak value of reference source current. This incremental change is added to the previous peak value of reference current to generate the peak value of current sampling instant as shown by (23).

Figure 5. Block diagram of proposed approximated simplest FLC

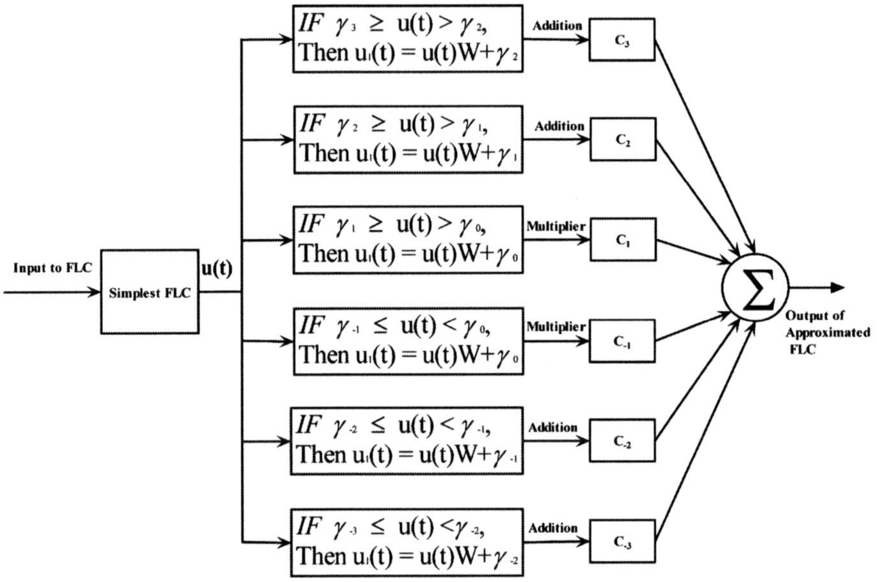

$$i_{s,\max}(t) = \delta i_{s,\max}(t) + i_{s,\max}(t-1) \tag{23}$$

The reference currents so obtained are compared with the actual source currents in the hysteresis current controller loop to generate the gate pulses for the shunt APF.

SIMULATION RESULTS

The system parameters given in Table 3 and gain values of conventional PI controller ($K_p = 0.58$, $K_i = 10$) are selected based on the discussion by Jain et al. (2002), Mishra & Bhende, (2007), Mohan, Undeland, and Robbins (2003) and Singh et al. (2007). The simulation results are obtained with conventional PI controller, 49 rules based FLC and proposed approximated simplest 4-rule FLC. Simulations are performed on MATLAB® in Simulink toolbox.

The system is simulated for both the FLCs using same normalization and denormalization factors. The performance of shunt APF is analyzed for the three randomly varying loading conditions, initially the filter is switched on at 0.05 sec to compensate the current harmonics injected by a nonlinear load. The load is varied after every 10 cycles in three steps as discussed in the following sections.

Figure 6. Single line diagram of basic compensation principle of a shunt active power filter

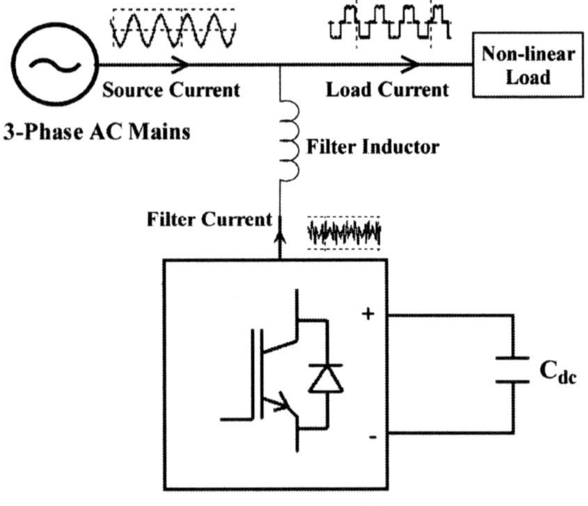

Case I

In this case the dynamic behaviour of various controllers is analysed during switch on response at 0.05 sec, with load resistance and inductance of 30 Ω and 20 mH respectively as shown in Figure 8. The peak overshoot in dc link voltage using PI controller is 4.06% as compared to 1.32% and 1.79% with 49-rule FLC and proposed FLC respectively. The 49 rules FLC takes 1.81 cycles and proposed FLC takes 2.30 cycles to settle (within 1% band of reference value) in comparison to six cycles taken by PI controller as shown in Table 4.

Case II

Here, the load resistance is changed from 30Ω to 60Ω. Due to load change, the load current is reduced from 13.83 A (rms) to 6.95 A (rms). Figure 9 shows the effect of load change and the responses of various controllers to regulate the dc link capacitor voltage. In this case, the peak overshoot in capacitor voltage due to load change with all the controllers is comparable but again when it come to settling time both the FLCs settles within two cycles whereas PI controller takes four cycles as shown in Table 4.

Case III

At t = 0.45 sec, the load resistance is changed from 60Ω to 45Ω. Due to load change, the load current is increased from 6.95 A (rms) 9.26 A (rms). Figure 10 shows the effect of load change and the responses of various controllers to regulate the dc link capacitor voltage. In this case also a similar trend of dynamic response is repeated as shown in Table 4.

The percentage THD in source current after compensation is shown in Figure 11. Although the % THD after compensation comes below 2% within one cycle after switch on as well during sudden load change for all the controllers, but most of the time the proposed FLC exhibits best compensation capabilities, among the three controllers considered here.

Figure 7. Schematic diagram of proposed simplest 4-rule FLC (approximating a 49-rule FLC) used to control shunt APF

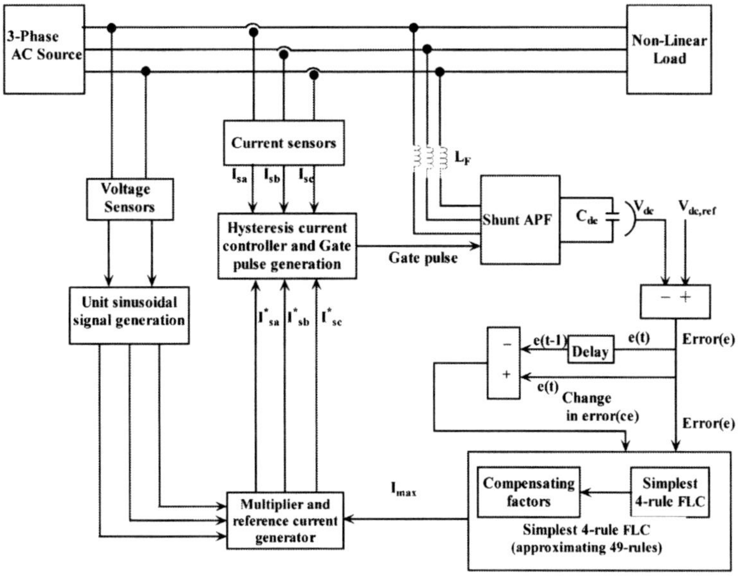

Table 3. System parameters

System Parameter	Value
Source voltage(V_s)	230 V(rms/phase)
System frequency(f)	50 Hz
Source impedance(R_s,L_s)	0.1Ω, 0.5 mH
Filter impedance (R_F,L_F)	0.4Ω, 3.35 mH
Reference DC link voltage ($V_{dc,ref}$)	680 V
DC link capacitance (C_{dc})	2000 μF
Load Resistance (R_L)	30 Ω, 60 Ω, 45Ω
Load Inductance (L_L)	20 mH

Comparison on the Basis of Computational Memory Requirement

For comparative analysis of computational memory requirement, an FLC is functionally divided into three sections, i.e., fuzzification, fuzzy inference and knowledge base and defuzzification. In approximated FLC, which is a reduced rule FLC, additional memory is required to perform approximation process. Even after this additional requirement, total memory requirement of approximated FLC is less than a 49-rule FLC due to less number of linguistic membership functions and reduced size of rule base. Table 5 provides a clear view of the comparative memory requirement for 49-rule FLC and proposed approximated FLC.

Comparison on the Basis of Performance Indices and Simulation Execution Time

A comparison on the basis of the integrated time averaged error (ITAE), integrated average error (IAE), integrated square error (ISE), integrated time square error (ITSE) and simulation time computed during the entire simulation (i.e., 325000 samples) is presented in Table 6. The proposed FLC is comparable with 49 rules FLC against all the performances indices and is quite ahead of conventional PI controller. The execution time of simulation for proposed FLC is 243.06 sec while for a 49 rule FLC it is 399.43 sec, so there is a considerable reduction in execution time due to the reduced complexity of proposed FLC. Although the simulation execution time of PI controller is minimum but it lacks on other performance indices.

Figure 8. Switch ON response with (a) PI controller, (b) 49-rule FLC, and (c) the proposed FLC

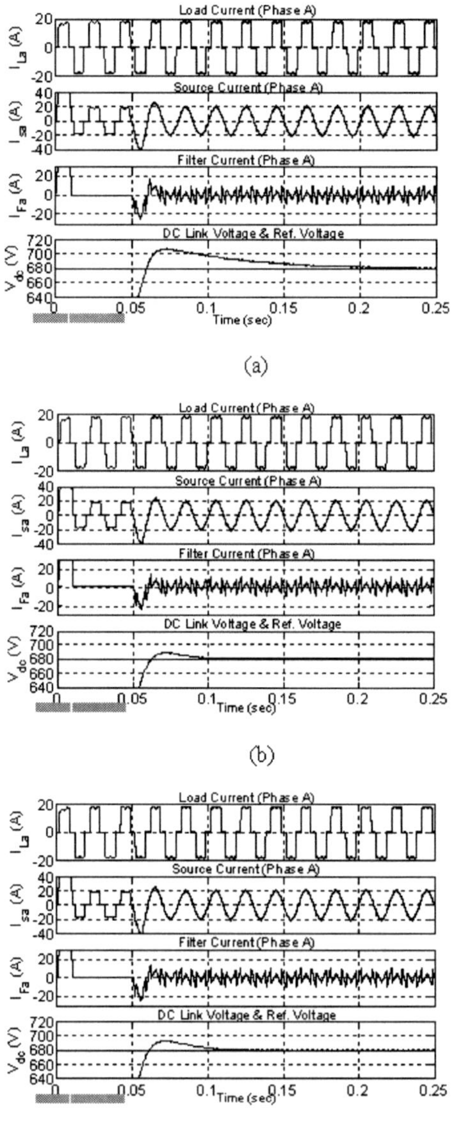

(a) (b) (c)

Table 4. Comparison of dynamic response of different controllers

Load	Peak Overshoot / Undershoot (%)			Settling Time (cycles)		
	PI	49 Rule FLC	Proposed FLC	PI	49 Rule FLC	Proposed FLC
Case 1	4.06	1.32	1.79	6.0	1.81	2.30
Case 2	2.16	2.05	2.14	4.0	1.80	2.00
Case 3	1.29	1.17	1.25	0.6	0.47	0.55

CONCLUSION

In this paper an approximated simplest 4 rule based FLC is proposed for a shunt active power filter to achieve harmonic mitigation. Traditionally conventional PI controllers are used to regulate the dc link voltage of shunt APF. They provide efficient harmonic compensation but suffer on the front of poor dynamic response. This drawback is overcome by 49-rule FLC. As the number of rules increases the control action becomes more efficient due to smooth transition from one membership function to other, this is achieved at the cost of increased complexity. The proposed FLC being a simplest 4-rule FLC is less complex as the numbers of fuzzy rules are greatly reduced and at the same time approximates the computational functionality of a 49 rule based FLC. Furthermore, with the reduction of rules the computational time of decision-making has reduced tremendously without compromising the control performance, which has been confirmed through computer simulations under varying load conditions. The proposed FLC was also proven to

Figure 11. Percentage THD in source current after filter switch on

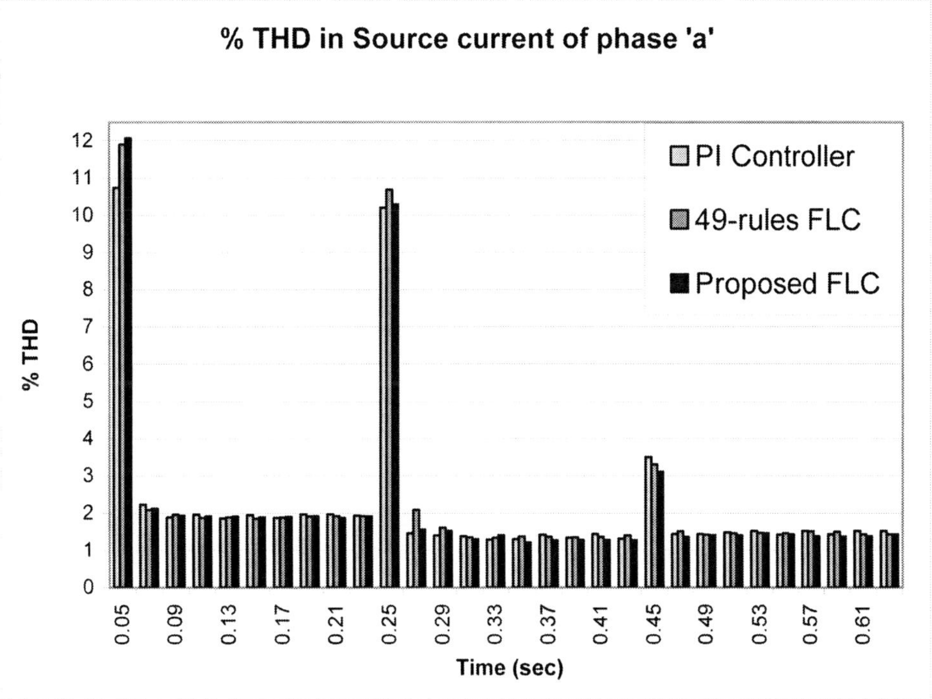

Figure 9. Load perturbation response of (a) PI controller, (b) 49-rule FLC, and (c) the proposed FLC at 0.25 sec

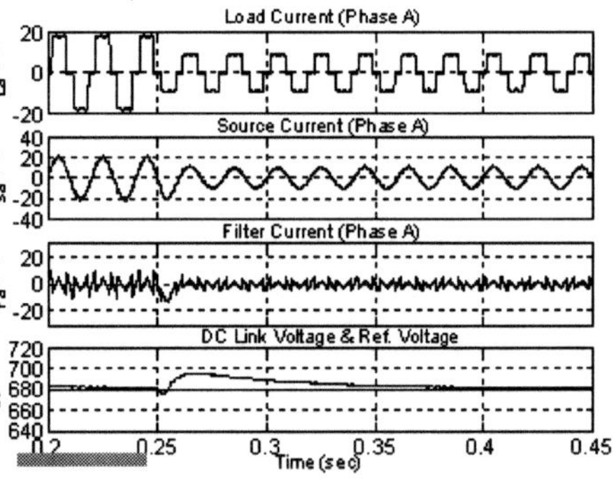

(a)

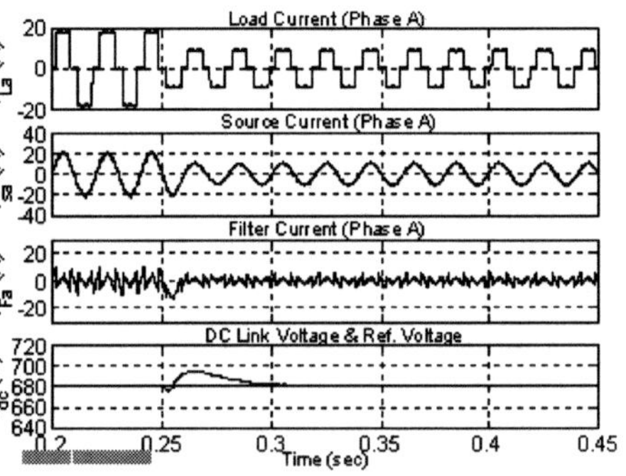

(b)

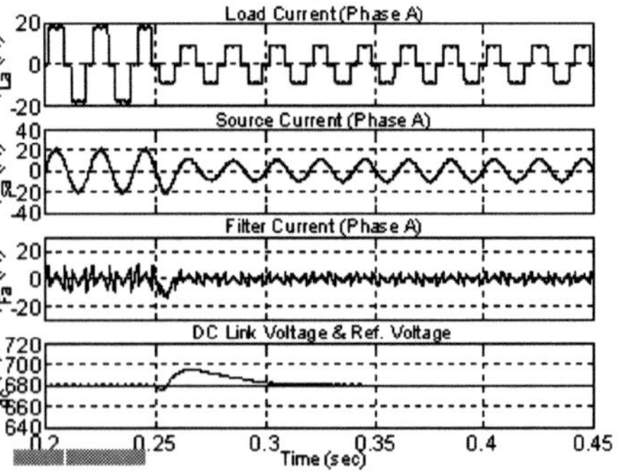

(c)

Figure 10. Load perturbation response of (a) PI controller (b) 49 rules FLC and (c) the proposed FLC at 0.45 sec

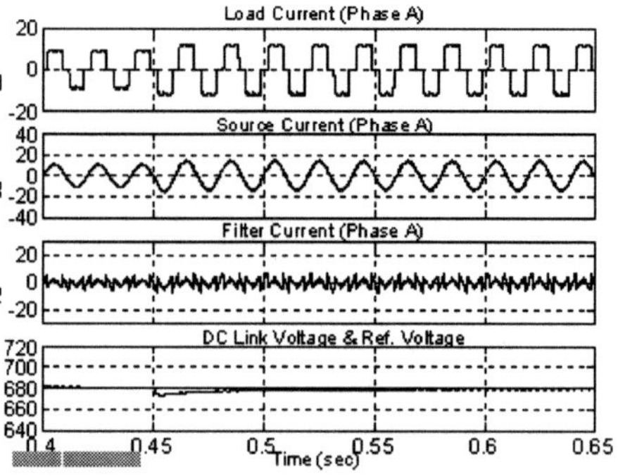

(a)

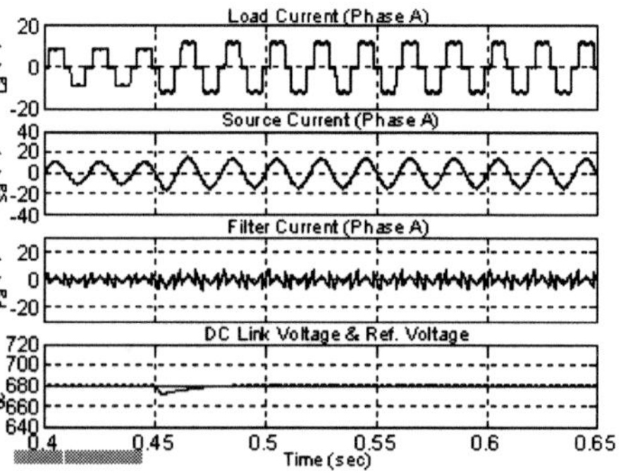

(b)

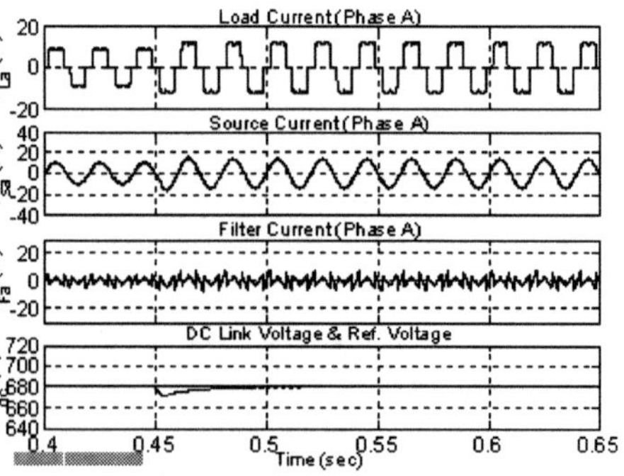

Table 5.Comparison of computational memory requirement where m is memory required to describe each membership function

Sl. No.	Module	Parameter	49-rule FLC	Approximated 4-rule FLC
1.	Fuzzification	No. of input variables	2	2
		No. of membership function for each input	7	2
		Memory units required	14 * m	4* m
2.	Fuzzy Inference and Knowledge Base	No. of rules	49	4
		No. of antecedent	7	4
		No. of consequent	7	4
		Memory units required	49*(7+7)=686	4*(4+4)=32
3.	Defuzzification	No. of output variable	1	1
		Membership function for output variable δI_{max}	7	3
		Memory units required	7 * m	3* m
4.	Approximation	No. of comparators		6
		No. of multipliers		2
		No. of adders		11
		Memory units required		19
Total Memory Requirement			**21*m+686**	**7*m+51**

Table 6. Comparison of controllers on the basis of performance indices

Performance Index	PI	49 Rule FLC	Proposed FLC
ITAE	0.74	0.36	0.36
IAE	9.60	7.62	7.64
ISE	2001.51	1970.76	1971.53
ITSE	15.76	11.88	12.01
Simulation Execution Time (Sec)	170.94	399.43	243.06

equally competitive in terms of dynamic response and compensation performance. The THD of supply current is well within 5%, the limitations imposed by IEEE-519 standards (IEEE Computer Society, 1993). A comparison of the performance of proposed FLC is made with conventional PI controller and 49-rule FLC. The proposed FLC is found to be performing best among all controllers in reducing THD as well as showing better dynamic response than conventional PI controller.

The important features of proposed scheme are that it is system independent, less complex in design, takes less memory, reduced computational time and provides efficient control action in satisfying the demand of effective compensation as well better dynamic response. The future scope of this work may include on line adaptive selection of compensating factors which are now calculated off line without increasing the system complexity, such that the basic motive of complexity reduction should not be defeated.

REFERENCES

Akagi, H., Kanajawa, Y., & Nabae, A. (1984). Instantaneous reactive power compensator comprising switching devices without energy storage components. *IEEE Transactions on Industry Applications*, *20*(3), 625–630. doi:10.1109/TIA.1984.4504460

An, L., Zhikang, S., Wenji, Z., Ruixiang, F., & Chunming, T. (2009). Development of hybrid active power filter based on adaptive fuzzy dividing frequency-control method. *IEEE Transactions on Power Delivery*, *24*(1), 424–432. doi:10.1109/TPWRD.2008.2005877

Anis Ibrahim, W. R., & Morcos, M. M. (2002). Artificial intelligence and advanced mathematical tools for power quality applications: A survey. *IEEE Transactions on Power Delivery*, *17*(2), 668–673. doi:10.1109/61.997958

Arya, R. K. (2006). Approximations of Mamdani type multi-fuzzy sets fuzzy controller by simplest fuzzy controller. *International Journal of Computational Cognition*, *4*, 35–47.

Bezine, H., Derbel, N., & Alimi, A. M. (2002). Fuzzy control of robotic manipulator: some issues on design and rule base size reduction. *Engineering Applications of Artificial Intelligence*, *15*, 401–416. doi:10.1016/S0952-1976(02)00075-1

Bose, B. K. (1994). Expert systems, fuzzy logic, and neural network applications in power electronics and motion control. *Proceedings of the IEEE*, *82*(8), 1303–1323. doi:10.1109/5.301690

Ciliz, M. K. (2005). Rule base reduction for knowledge based fuzzy controller with application to vacuum cleaner. *Expert Systems with Applications*, *28*, 175–184. doi:10.1016/j.eswa.2004.10.009

Dixon, J. W., Contrado, J. M., & Morán, L. A. (1999). A fuzzy-controlled active front-end rectifier with current harmonic filtering characteristics and minimum sensing variables. *IEEE Transactions on Power Electronics*, *14*(4), 724–729. doi:10.1109/63.774211

Grady, W. M., Samotyj, M. J., & Noyola, A. H. (1990). Survey of active power line conditioning methodologies. *IEEE Transactions on Power Delivery*, *5*(3), 1536–1542. doi:10.1109/61.57998

Hampel, R., & Chaker, N. (1998). Minimizing the variable parameters for optimizing the fuzzy controller. *Fuzzy Sets and Systems*, *100*, 131–142. doi:10.1016/S0165-0114(97)00059-6

IEEE. Computer Society. (1993). *Standard 519-1992: IEEE Recommended Practices and Requirements for Harmonic Control in Electrical Power Systems*. Washington, DC: Author.

Jain, S. K., Agrawal, P., & Gupta, H. O. (2002). Fuzzy logic controlled shunt active power filter for power quality improvement. *IEE Proceedings. Electric Power Applications*, *149*(5), 317–328. doi:10.1049/ip-epa:20020511

Karuppanan, P., & Mahapatra, K. K. (2011). PLL with fuzzy logic controller based shunt active power filter for harmonic and reactive power compensation. In *Proceedings of the International Conference on Power Electronics*, New Delhi, India (pp. 1-6).

Kumar, P., & Mahajan, A. (2009). Soft computing techniques for the control of an active power filter. *IEEE Transactions on Power Delivery*, *24*(1), 452–461. doi:10.1109/TPWRD.2008.2005881

Mishra, S., & Bhende, C. N. (2007). Bacterial foraging technique-based optimized active power filter for load compensation. *IEEE Transactions on Power Delivery*, *22*(1), 457–465. doi:10.1109/TPWRD.2006.876651

Mohan, N., Undeland, T., & Robbins, W. P. (2003). *Power Electronics-Converters, Applications and Design*. New York, NY: John Wiley & Sons.

Moser, B., & Navarata, M. (2002). Fuzzy controllers with conditionally firing rules. *IEEE Transactions on Fuzzy Systems*, *10*(3), 340–349. doi:10.1109/TFUZZ.2002.1006437

Peng, F. Z., Akagi, H., & Nabae, A. (1990). Study of active power filters using quad series voltage source PWM converters for harmonic compensation. *IEEE Transactions on Power Electronics*, *5*(1), 9–15. doi:10.1109/63.45994

Singh, B., Al-Haddad, K., & Chandra, A. (1999). A review of active power filters for power quality improvement. *IEEE Transactions on Industrial Electronics*, *46*(5), 960–969. doi:10.1109/41.793345

Singh, B., Chandra, A., & Al-Haddad, K. (1999). Computer aided modeling and simulation of active power filters. *Journal of Electric Machines & Power Systems*, *27*, 1227–1241. doi:10.1080/073135699268687

Singh, G. K., Singh, A. K., & Mitra, R. (2007). A simple fuzzy logic based robust active power filter for harmonic minimization under random load variation. *Electric Power Systems Research*, *77*, 1101–1111. doi:10.1016/j.epsr.2006.09.006

Ying, H. (2000). *Fuzzy Control and Modeling: Analytical foundations and applications*. New York, NY: Wiley-IEEE Press.

Zeng, X., & Singh, M. G. (1994). Approximation theory of fuzzy systems-SISO case. *IEEE Transactions on Fuzzy Systems*, *2*(2), 162–194. doi:10.1109/91.277964

Zeng, X., & Singh, M. G. (1995). Approximation theory of fuzzy systems-MIMO case. *IEEE Transactions on Fuzzy Systems*, *3*(2), 219–235. doi:10.1109/91.388175

This work was previously published in the International Journal of Fuzzy System Applications, Volume 1, Issue 4, edited by Toly Chen, pp. 18-36, copyright 2010 by IGI Publishing (an imprint of IGI Global).

Chapter 12
Semidefinite Programming-Based Method for Implementing Linear Fitting to Interval-Valued Data

Minghuang Li
Beijing Normal University, China

Fusheng Yu
Beijing Normal University, China

ABSTRACT

Building a linear fitting model for a given interval-valued data set is challenging since the minimization of the residue function leads to a huge combinatorial problem. To overcome such a difficulty, this article proposes a new semidefinite programming-based method for implementing linear fitting to interval-valued data. First, the fitting model is cast to a problem of quadratically constrained quadratic programming (QCQP), and then two formulae are derived to develop the lower bound on the optimal value of the nonconvex QCQP by semidefinite relaxation and Lagrangian relaxation. In many cases, this method can solve the fitting problem by giving the exact optimal solution. Even though the lower bound is not the optimal value, it is still a good approximation of the global optimal solution. Experimental studies on different fitting problems of different scales demonstrate the good performance and stability of our method. Furthermore, the proposed method performs very well in solving relatively large-scale interval-fitting problems.

INTRODUCTION

The objective of linear regression is to adjust the parameters of a linear model function so as to best fit a given data set. In the real world, the observations and measurements are usually subject to imprecision and vagueness. In order to assess such uncertainty one ideal way is to represent the data as intervals or fuzzy sets.

The linear regression analysis involving fuzzy data is the central concern of fuzzy regression. Tanaka (1982) first proposed a fuzzy linear regres-

DOI: 10.4018/978-1-4666-1870-1.ch012

sion model by minimizing the index of fuzziness of the system. Diamond (1990, 1994, 1997) introduced a metric on the set of fuzzy numbers and used this metric to define a least-sum-of-squares criterion function as in the usual sense. Further discussion can be found in (Chang, 2001; Coppi et al., 2004; D'Urso, 2003, 2006; Tanaka et al., 1982, 1987, 1996; Sakawa et al., 1992; Ma et al., 1997; Chang et al., 2001; Wu, 2003) and other references therein.

At the interval level, the similar topic has also been discussed extensively in the field of symbolic data analysis (Bock et al., 2000). Bertrand and Goupil (2000) and Billard and Diday (2000, 2003) introduced statistics methods based on covariance and correlation function suitable for interval-valued data. A centre and range method was developed in Lima Neto et al. (2008) for fitting interval data. Various forms of support vector interval regression approaches have been discussed (An et al., 2008; Jeng et al., 2003; Hong et al., 2005, 2006). Additionally, in practical context, this model has shown its potential in forecasting interval-valued time series (Maia et al., 2008).

The model studied in this article is fitting a crisp linear function to interval-valued input-output data. More precisely, the model consists of minimizing the sum of squared interval distances between observed and predicted values. The predicted values are calculated by interval arithmetic (or the extension principle).

However, as far as we know, solving this fitting problem in high dimensional case is still an open problem. The main difficulty is that 2^n different combinations of the signs of n regression coefficients have to be treated separately. Hence the complexity of the problem will increase extremely.

Such combinatorial difficulty was also mentioned in Diamond et al. (1997), where the author discussed the fitting model involving minimizing the sum of squared L_2 distances between observed and predicted values represented by fuzzy numbers.

Theoretically, the problem involving minimization of interval distance may lead to a hard problem, but we have found that this problem, in many cases, can be exactly solved in polynomial time. It thus makes it possible for this type of 'distance-based' fitting model to handle high dimensional problems.

The method proposed in this paper consists of two main steps: casting the interval fitting problem as a quadratically constrained quadratic programming (QCQP) problem, and using the relaxation theory to obtain a lower bound of the optimal value by solving a semidefinite programming problem. It is worthy of emphasizing that in many situations, the lower bound is exactly equal to the global minimum.

Considering the fact that the lower bound may not always be the global minimum, we also present a correcting step to enhance the accuracy of the solution by solving a quadratic programming problem. The combination of the two procedures performs very well in practical setting, especially in dealing with relatively large scale problems

The remainder of this paper is organized as follows. First we provide the preliminary knowledge and notations used throughout this paper. Next, we give the QCQP form of an interval fitting problem. Two formulas for solving the optimization problem based on two types of relaxations are presented. Finally, we show the experiment results of solving small scale (n=2) and large scale(n=200) fitting problems and conclude this study.

PRELIMINARIES

In this section, we give some notations used in this paper and briefly introduce the related preliminary knowledge.

Notation

Throughout this paper, the following notations are used:

For a vector x, $x \geq 0$ $(x > 0)$ means element-wise $x_k \geq 0$ $(x_k > 0)(k = 1, 2, ..., n)$.

For matrix A, its range, pseudo-inverse, transpose, and trace are denoted by $\mathcal{R}(A)$, $A^{\dagger}$, A^t and $\mathrm{tr}(A)$ respectively.

For two matrices A and B, $A \geq B$ $(A > B)$ means $A - B$ is positive semidefinite (positive definite).

The norm $||\cdot||$ refers to the Euclidean norm (l_2 norm).

All the intervals are closed and represented in midpoint-spread form $\langle a, c\rangle = [a - c, a + c]$ $c \geq 0$.

We also use some standard abbreviations such as QP (quadratic programming), SDP (semidefinite programming), QCQP (quadratically constrained quadratic programming) and KKT (Karush-Kuhn-Tucker) conditions (Bazaraa et al., 2006; Boyd et al., 2004).

Linear Combination of Intervals

Our fitting model studied in this article is based on the extension principle (or interval arithmetic). In more detail, the predicted value is calculated by the linear combination of input variables under the extension principle.

Let $f(x_1, x_2, ..., x_n) = \beta_1 x_1 + \beta_2 x_2 +, ..., +\beta_n x_n$ bethe linear function. For n fuzzy subsets $\{\widetilde{X}_1, \widetilde{X}_2, ..., \widetilde{X}_n\}$ defined on $\mathbb{R}^1$, the linear combination of these n subsets is given by the extension principle. More precisely, the membership function of $f(\widetilde{X}_1, \widetilde{X}_2, ..., \widetilde{X}_n)$ is defined by

$$\mu_{f(\widetilde{X}_1, \widetilde{X}_2, ..., \widetilde{X}_n)}(r) = \sup_{r = \beta_1 w_1 + ... + \beta_n w_n} \min\{X_1(w_1), ..., X_n(w_n)\}$$

Generally, there is no analytical expression for this function. However, when $\widetilde{X}_1, \widetilde{X}_2, ..., \widetilde{X}_n$ are intervals, the linear combination is actually the range of function $f(x_1, x_2, ..., x_n)$ on $\widetilde{X}_1 \times \widetilde{X}_2 \times ... \times \widetilde{X}_n$. Thus, it is an interval which has the following analytical form:

$$f(\widetilde{X}_1, \widetilde{X}_2, ..., \widetilde{X}_n) = \langle \sum_{k=1}^{n} \beta_k a_k, \sum_{k=1}^{n} |\beta_k| c_k \rangle$$

where $\widetilde{X}_k = \langle a_k, c_k \rangle$ $(k = 1, 2, ..., n)$ are the interval-valued input data. This expression coincides with the one defined by interval arithmetic (Moor, 1979).

Interval Distance

Given two intervals $I_1 = \langle a_1, c_1 \rangle$, $I_2 = \langle a_2, c_2 \rangle$, the interval distance (Moor, 1979) is defined as

$$\begin{aligned} d_{interval}(I_1, I_2) &= [(a_1 - c_1 - a_2 + c_2)^2 + (a_1 + c_1 - a_2 - c_2)^2]^{1/2} \\ &= \sqrt{2}\,[(a_1 - a_2)^2 + (c_1 - c_2)^2]^{1/2} \end{aligned}$$

It is easy to verify that $d_{interval}$ is a distance measurement.

Note that for two intervals I_1 and I_2, $d_{interval}(I_1, I_2)$ generally is not equal to $||I_1 - I_2||_{interval}$ which is defined by

$$||\langle a, c\rangle||_{interval} = [(a - c)^2 + (a + c)^2]^{1/2}.$$

$||I_1 - I_2||_{interval}$ is actually not a distance measurement, since

$$||\langle a, c\rangle - \langle a, c\rangle||_{interval} = ||\langle 0, 2c\rangle||_{interval} = 2\sqrt{2}c \neq 0$$

(for $c \neq 0$).

Therefore, for the fitting problem of interval-valued data, it is inappropriate to adopt $||I_1 - I_2||_{interval}$ as the measurement of deviation between two intervals.

QCQP FORM OF AN INTERVAL-VALUED DATA FITTING PROBLEM

In this section, we first derive the residue function for the interval-valued data fitting problem and point out that the minimization of this residue function will lead to a huge combinatorial problem. To overcome such difficulty, we then cast the fitting model to a problem of quadratically constrained quadratic programming (QCQP).

Formulation of Interval-Valued Data Fitting Problem

Let us consider the following linear fitting problem

$$\tilde{y}_i \cong \tilde{x}_{i1} \cdot \beta_1 + \tilde{x}_{i2} \cdot \beta_2 + \ldots + \tilde{x}_{in} \cdot \beta_n \qquad i = 1, 2, \ldots, m \tag{1}$$

where $\beta_k \in \mathbb{R}^1$ $(k = 1, 2, \ldots, n)$ are fitting (regression) coefficients. $\tilde{x}_{ik} = \langle a_{ik}, c_{ik} \rangle$ and $\tilde{y}_i = \langle d_i, f_i \rangle$ are input-output interval-valued data. m is the size of the data set.

By interval arithmetic (or extension principle), the predicted value $\tilde{y}_i^*$ calculated by Equation (1) is interpreted as

$$\tilde{y}_i^* = \langle \sum_{k=1}^{n} \beta_k a_{ik}, \sum_{k=1}^{n} | \beta_k | \, c_{ik} \rangle$$

The aim of the interval linear fitting problem proposed in this article is to find the coefficients $\beta = (\beta_1, \ldots, \beta_n)^t$ which will minimize the following residue:

$$residue(\beta) = (1/2) \sum_{i=1}^{m} d_{interval}^2 (\tilde{y}_i^*, \tilde{y}_i)$$

To simplify the expression, we rewrite the overall residue in the following compact form:

$$\begin{aligned} residue(\beta) &= \sum_{i=1}^{m} \left[(\sum_{k=1}^{n} a_{ik}\beta_k - d_i)^2 + (\sum_{k=1}^{n} c_{ik} \, | \beta_k | - f_i)^2 \right] \\ &= || M\beta - d ||^2 + || N | \beta | - f ||^2 \end{aligned} \tag{2}$$

where $| \beta | = (| \beta_1 |, \ldots, | \beta_n |)^t$, $M = (a_{ik})_{mxn}$, $N = (c_{ik})_{mxn}$, $d = (d_1, \ldots, d_m)^t$, $f = (f_1, \ldots, f_m)^t$. Thus, the interval-valued linear fitting problem is converted to the following unconstrained minimization problem:

$$minimize \qquad residue(\beta) \tag{3}$$

We have noticed that the problem of fitting a crisp model to interval or fuzzy input-output data was already studied in literature (D'Urso, 2003; Lima Neto et al., 2008; Billard et al., 2000, 2003; Bertrand et al., 2000) where the spread of predicted interval-valued data is assumed to be a linear function of β (or β^r in Lima Neto et al. (2008)). Such a simple assumption does facilitate the solving process significantly. But, in our fitting situation, the uncertainty is actually enlarged by $| \beta |$ thus no fully linear structure can be exploited.

But there is one special case in which (3) can be easily solved. If the observed values $\tilde{y}_i$ are crisp points, which means *f=0*, then (3) is convex optimization. To see this, we only need to introduce slack variables t_i, then (3) is equivalent to

$$\begin{aligned} &minimize && \sum_{i=1}^{m} (\sum_{k=1}^{n} a_{ik}\beta_k - d_i)^2 + \sum_{i=1}^{m} t_i^2 \\ &subject\ to && \sum_{k=1}^{n} c_{ik} \, | \beta_k | \le t_i \quad i = 1, 2, \ldots, m \end{aligned}$$

We further introduce slack variables z_k, and then the constraints are equivalent to

$$\sum_{k=1}^{n} c_{ik} z_k \leq t_i \qquad -z_k \leq \beta_k \leq z_k$$
$$i = 1, 2, \ldots, m \qquad k = 1, 2, \ldots, n$$

which is a convex quadratic programming problem (Bazaraa, 2006) and can be solved efficiently.

Combinatorial Problem

It is important to notice that, in general, $residue(\beta)$ is neither convex nor differentiable, which makes it hard to be handled by the traditional steepest descent algorithm directly.

The primary difficulty stems from the absolute values in (2). One straightforward method is to eliminate *n* absolute values by confining β into 2^n orthants respectively. More precisely, for each *k*, fix the sign of β_k, and then solve the optimization problem below:

$$\begin{aligned} &minimize && \| M\beta - d \|^2 + \| ND\beta - f \|^2 \\ &subject\ to && D\beta \geq 0 \end{aligned} \tag{4}$$

The diagonal matrix D contains a combination of the signs we have chosen

$$D = \begin{bmatrix} s_1 & \cdots & 0 \\ \vdots & \ddots & \vdots \\ 0 & \cdots & s_n \end{bmatrix} \qquad s_k \in \{-1, 1\}\ (k = 1, 2, \ldots, n)$$

For each *D*, the optimization is a QP and can be solved efficiently. However, in order to find the global minimum of (3), 2^n different choices of *D* need to be considered. When *n* is small, solving (4) seems to be feasible. But, when *n* is relatively large, the complexity of computation will increase extremely.

The above phenomenon was also mentioned in Diamond et al. (1997) where the coefficients are confined in various cones on Banach space containing the cone of triangular fuzzy number. In addition, most of the fuzzy models involving linear combinations induced by the extension principle have such combinatorial difficulty. Thus, most of the existing methods are only useful in low dimensional cases.

QCQP Form of An Interval-Valued Data Fitting Problem

To overcome the combinatorial difficulty, we will present a new method to handle (3) based on semidefinite programming rather than QP. The first step is to introduce slack variables and convert (3) to a new form with quadratic equality constraints.

Note that any real number z can be expressed in the form $z = z^+ - z^-$, where $z^+ z^- = 0,\ z^+ \geq 0,\ z^- \geq 0$. In this case, $|z| = z^+ + z^-$. Likewise, for vector β we can introduce two slack vectors $\beta^+,\ \beta^-$, satisfying $\beta^+ \geq 0,\ \beta^- \geq 0$ and $(\beta^+)^t \beta^- = 0$, then $|\beta| = \beta^+ + \beta^-$. Replace β by $\beta^+ - \beta^-$ in (2), we arrive at the following equivalent optimization problem:

$$\begin{aligned} &minimize && \left\| \begin{bmatrix} M & -M \\ N & N \end{bmatrix} \begin{bmatrix} \beta^+ \\ \beta^- \end{bmatrix} - \begin{bmatrix} d \\ f \end{bmatrix} \right\|^2 \\ &subject\ to && (\beta^+)^t \beta^- = 0,\ \ \beta^+ \geq 0,\ \ \beta^- \geq 0 \end{aligned}$$

By combining β^+, β^- as one variable denoted by $x \in \mathbb{R}^{2n}$ $(x^t = [(\beta^+)^t, (\beta^-)^t])$, we convert an interval-valued data fitting problem to the standard QCQP form:

$$\begin{aligned} &minimize && x^t A x + b^t x + c \\ &subject\ to && x^t P x = 0,\ \ x \geq 0 \end{aligned} \tag{5}$$

where $c = d^t d + f^t f$.

$$P = \begin{bmatrix} 0 & I \\ I & 0 \end{bmatrix} \quad b = -2\begin{bmatrix} M^t & N^t \\ -M^t & N^t \end{bmatrix}\begin{bmatrix} d \\ f \end{bmatrix}$$

$$A = \begin{bmatrix} M^tM + N^tN & N^tN - M^tM \\ N^tN - M^tM & M^tM + N^tN \end{bmatrix}$$

Although the constraints and objective function in (5) are differentiable, the problem is still hard to solve, since a nonconvex quadratic equality constrain is involved. In general sense, solving this type of problem is NP-hard (Nesterov et al., 2000; Boyd et al., 1997).

However, we can still obtain a lower bound of $residue(\beta)$ by Lagrangian duality or semidefnite relaxation (Nesterov et al., 2000) based on (5), which is shown in the next section.

SOLVING THE INTERVAL-VALUED DATA FITTING PROBLEM WITH SEMIDEFINITE PROGRAMMING

We divide this section into three parts. In the first part, we derive two relaxation formulae to develop the lower bound on the optimal value of the nonconvex QCQP (5). Since the lower bound of $residue(\beta)$ may not always be obtainable, we discuss in the second part the situation when exact solution cannot be obtained from the lower bound. After that, the overall procedure for solving the interval-valued data fitting problem with semidefinite programming is presented in the third part.

Lower Bound of the Residue

The relaxation theory guarantees that a lower bound on the optimal value of (5) can always be readily found. Two available strategies to obtain the lower bound are the well-known semidefinite relaxation and the Lagrangian relaxation based on the duality theory (Lieven et al., 1996; Boyd et al., 1997; Nesterov et al., 2000; Boyd, 2004). The formulae derived from the two strategies give the same lower bound, but have their own convenience in specific context. Therefore, for practical purposes, it is necessary to give a brief introduction to the formulae based on the two relaxation methods.

To make our discuss compact, we will give out all the propositions with their proofs presented in appendices.

Proposition 1. (Semidefinite relaxation of an interval fitting problem)

A lower bound of $residue(\beta)$ in (5) can be obtained by solving the following (primal) semidefinite programming (SDP) problem.

$$\begin{aligned} &minimize \quad && \mathrm{tr}(AX) + b^t x + c \\ &subject\ to \quad && \mathrm{tr}(PX) = 0 \\ & && \begin{bmatrix} X & x \\ x^t & 1 \end{bmatrix} \geq 0, \quad x \geq 0 \end{aligned} \tag{6}$$

where X is a $2n \times 2n$ symmetric matrix. Let p^* be the minimum value of (3). d^* be the optimal value of the objective function in (6), and the corresponding optimal solution be $(\bar{X}, \bar{x})$. If $\bar{X} = \bar{x}\bar{x}^t$, then the lower bound is tight ($d^* = p^*$) and $\bar{x}$ is the global minimum of (5).

Unlike the original problem, the SDP problem (6) is convex and can be efficiently solved by interior-point methods in polynomial time (Lieven et al., 1996; Boyd et al., 1997; Nesterov et al., 2000).

Proposition 1 guarantees $d^* \leq p^*$, thereby d^* can sever as a certificate for the global minimum. That is, suppose we can find $\bar{\beta}$ in (3) such that $residue(\bar{\beta}) = d^*$ then $\bar{\beta}$ is the global optimal solution of (3).

Similarly, the same lower bound can also be obtained for the (dual) SDP problem which is based on the Lagrangian relaxation.

Proposition 2. (Lagrangian relaxation of an interval fitting problem)

A lower bound of $residue(\beta)$ in (5) can be obtained by solving the following (dual) SDP:

$$\begin{array}{ll} maximize & t \\ subject\ to & \begin{bmatrix} 4(A+\mu P) & b-\lambda \\ (b-\lambda)^t & c-t \end{bmatrix} \geq 0, \ \lambda \geq 0 \end{array} \tag{7}$$

The corresponding optimal solution is denoted by $(\overline{t}, \overline{\lambda}, \overline{\mu})$. To simplify notation, we still denote the optimal value of (7) as d^* (which means $d^* = \overline{t}$). If $A + \overline{\mu}P$ is nonsingular, then $d^* = p^*$ holds and the global minimum of (5), denoted as $\overline{x}$, can be calculated by

$$\overline{x} = -(1/2)(A + \overline{\mu}P)^{-1}(b - \overline{\lambda}).$$

The Lagrangian relaxation techniques provide the same lower bounds on the optimal value of (5). In general sense, they don't give particular way to compute the feasible point $\overline{x}$. But in this interval fitting problem, the global minimum can be, in many cases, recovered from $\overline{\lambda}$ and $\overline{\mu}$.

To illustrate the relation between problem (6) and (7), we describe the well-known strong duality theory below. The proof can be found in Lieven et al. (1996).

Proposition 3. Problems (6) and (7) are duals of each other. If they are strictly feasible, they will give exactly the same lower bound on p^*.

The reason we provide two relaxation forms is that different SDP solvers may have different performances in dealing with the same problem. For instance, the problem structure of (6) is more suitable to cone solvers such as SeDuMi, while (7) is minimizing a linear function subject to Linear Matrix Inequality (LMI) (Nesterov et al., 2000) and ordinary linear inequality, thus it can be handled conveniently by many LMI solvers. In this regard, it is advisable to choose a proper problem description according to the specific context.

Proposition 4. (Global optimality conditions)

The following two conditions appeared in proposition 1 and proposition 2 are called global optimality conditions with respect to the two relaxation formulas:

$X = \overline{x}\overline{x}^t$ (1)
$A + \overline{\mu}P$ (2) is nonsingular

If either of the conditions holds, we have $d^* = p^*$ and problem (3) can be exactly solved by letting $\overline{\beta} = (\overline{x}_1 - \overline{x}_{n+1}, ..., \overline{x}_n - \overline{x}_{2n})$ according to the definition of x in (5).

Theoretically, for this nonconvex QCQP, the equality $d^* = p^*$ doesn't necessarily hold in general (Jeyakumar et al., 2007, Polik & Terlaky, 2007), which implies the global optimality conditions are not satisfied. However, for our interval fitting problem, numerical experiments suggest that the equality $d^* = p^*$ holds in most cases. Even though it doesn't hold, the negative effect can be significantly alleviated by solving a QP, which is shown in the next part.

Discussion When Optimality Conditions Not Satisfied

According to proposition 1 and proposition 2, if $X \neq \overline{x}\overline{x}^t$ or $A + \overline{\mu}P$ is singular, we can't obtain the exact p^* from (6) or (7). It implies the lower bound might not be tight. In this situation, it is not advisable to adopt $\overline{x}$ directly. This phenomenon may come from the round-off errors in SDP solver or the difficulty in the problem structure.

But we notice that in (3), what really important is actually the sign of $\overline{\beta}$. When the optimality conditions are not satisfied, the $\overline{\beta}$ is usually not the global minimum but close to it, so it can provide a proper hint on the sign of the global minimum.

To enhance accuracy, we therefore only adopt the sign of $\overline{\beta}$ and then solve a QP in (4). This simple heuristic method works well in many contexts and can give a good approximation of the global optimal solution.

Procedure of Solving an Interval-Valued Data Linear Fitting Problem

We summarize the overall solving procedure as follows:

Step 1: Given a sufficiently small $\varepsilon > 0$, midpoint matrix *M,d* and spread matrix *N,f*, form the new matrix *A,b,c,P* according to (5).

Step 2: Solve the SDP (6) or (7), obtain the optimal valued $\overline{x}$ and the corresponding lower bound d^*.

Step 3: Construct the optimal solution in (3) by letting $\overline{\beta} = (\overline{x}_1 - \overline{x}_{n+1}, ..., \overline{x}_n - \overline{x}_{2n})$. If $residue(\overline{\beta}) - d^* < \varepsilon$ then $\overline{\beta}$ is global optimal and the problem has been exactly solved, otherwise go to step 4.

Step 4: Create the matrix *D* by

$$D = \begin{bmatrix} \text{sgn}(\overline{\beta}_1) & \cdots & 0 \\ \vdots & \ddots & \vdots \\ 0 & \cdots & \text{sgn}(\overline{\beta}_n) \end{bmatrix}$$

Step 5: Solve the QP in (4), obtain an approximate solution $\overline{\beta}^*$, and update $\overline{\beta} = \overline{\beta}^*$.

Note that if $residue(\overline{\beta}) \approx d^*$, we can reasonably consider $\overline{\beta}$ as a satisfactory solution. When $residue(\overline{\beta}) \neq d^*$, solving (3) may be a hard problem. Such a difficulty can be alleviated, to some extent, by modifying the original problem (2). Roughly speaking, if we reduce *f* sufficiently, (2) will approach to a convex optimization problem thus can be completed solved. Empirical evidence suggests that our method works well in practice and in many cases $residue(\overline{\beta}) = d^*$ without any modification. We will illustrate this in the next section.

EXPERIMENTAL STUDIES

In this section, we design two experiments on the interval-valued data fitting problems of different scales. In the small scale case, we will demonstrate the solving procedure in detail, and show how the combinatorial problem (4) is avoided by solving a SDP. In the large-scale case, the good performance and stability of our method are revealed sufficiently. Meanwhile, several fitting problems that don't satisfy the optimality conditions will also be discussed. All the computation is implemented by SeDuMi (a Matlab toolbox).

Experiment on a Small-Scale Fitting Problem

To show the solving procedure in detail, we study here a small scale problem with *n=2, m=6*. The midpoint and spread of each interval-valued data are generated randomly, the corresponding matrices are:

$$M = \begin{bmatrix} 1.476 & 0.222 \\ -0.813 & 1.871 \\ 0.645 & 0.110 \\ -1.309 & -0.411 \\ -0.867 & 0.511 \\ -0.474 & -1.199 \end{bmatrix} \quad N = \begin{bmatrix} 0.130 & 0.454 \\ 0.254 & 0.904 \\ 0.803 & 0.282 \\ 0.667 & 0.065 \\ 0.013 & 0.476 \\ 0.561 & 0.984 \end{bmatrix} \quad d = \begin{bmatrix} -0.096 \\ 0.445 \\ -0.295 \\ -0.167 \\ 0.179 \\ 0.421 \end{bmatrix} \quad f = \begin{bmatrix} 0.922 \\ 0.561 \\ 0.652 \\ 0.772 \\ 0.106 \\ 0.001 \end{bmatrix}$$

The surface of $residue(\beta)$ is illustrated in Figure 1. Obviously, the residue function is neither differentiable nor convex, and has four local minimums. Traditional steepest descent algo-

Figure 1. The surface of the function $residue(\beta)$ near (0,0). The global minimum $\overline{\beta} = [-0.253, 0.161]^t$ can be located directly by solving a SDP

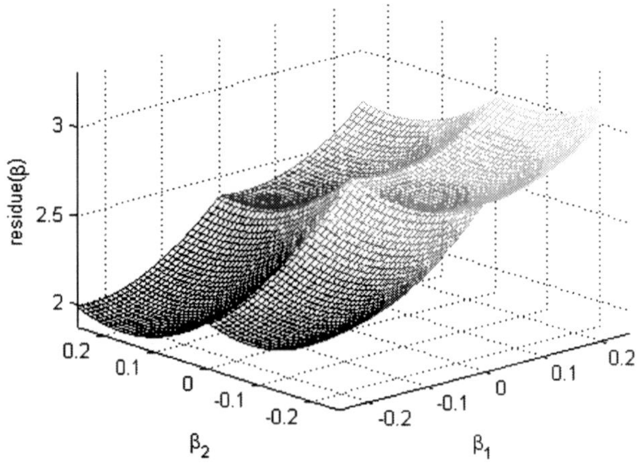

rithms might fail in dealing with this type of function, because the line search process is quite easy to be trapped into one of the four local minimums.

Next, we implement the method presented in this article. Firstly, we form A, b, c and P according to (5) and solve SDP (6). Then, we obtain the optimal result: $\overline{x} = [0, 0.161, 0.253, 0]^t$ and $d^* = 1.924$. The norm of $\overline{X} - \overline{x}\overline{x}^t$ is 1.638e-009, thus the global optimality condition (1) holds. Finally, we obtain

$$\overline{\beta} = [\overline{x}_1 - \overline{x}_3, \overline{x}_2 - \overline{x}_4] = [-0.253, 0.161]^t \text{ and } residue(\overline{\beta}) = d^* = 1.924.$$

This global minimum is illustrated clearly in Figure 1.

This simple experiment says that the global minimum can be located directly by solving a SDP. Interestingly, it might not be obvious that we do not need to eliminate absolute values and solve 4 QPs in (4) at all. The benefit is significant especially when n is relatively large, say 200.

Experiment on a Large Scale Fitting Problem

In this part, we will present computational results on the fitting problem with 200 regression coefficients (n=200) and 400 data (m=400). The elements of matrix M and d are drawn from a standard normal distribution with mean 0 and standard deviation 1, each entry of the interval spread matrix N is generated from 0.1ξ, where ξ is a random variable following a uniform distribution in [0,1].

The elements of f are generated by $L\xi$ with L being a positive number. The performance of our method will be investigated by solving a series of problems with different choices of L. For each L, we execute the algorithm corresponding to our method 3 times based on different initial data. These experiments will also be carried by the dual SDP (7) to show the impact of the singularity of $A + \overline{\mu}P$.

For such large scale fitting problem, we can reasonably guess that there could be as much as 2^{200} local minimums. In this case, the approach based on solving 2^{200} QPs in (4) will definitely fail.

Table 1. Experiment results under different choices of L

L	d^*	$residue(\bar{\beta})$	$\|\| \bar{X} - \bar{x}\bar{x}^t \|\|$	$\lambda_{min}(A + \bar{\mu}P)$
0.1	241.70	241.70	1.7e-9	0.974
	220.84	220.84	1.2e-9	0.923
	242.66	242.66	1.9e-9	1.137
0.5	213.49	213.49	1.0e-9	1.887
	226.96	226.96	2.8e-9	0.948
	212.02	212.02	5.5e-9	1.034
1.0	219.76	219.76	1.8e-8	0.669
	247.69	247.69	7.3e-9	0.752
	231.21	231.21	4.3e-9	0.577
1.5	259.12	259.65	0.002	1.2e-6
	279.37	280.11	0.006	2.5e-7
	255.66	256.26	0.005	7.0e-7
2.0	396.95	401.20	0.028	2.8e-7
	380.47	387.09	0.040	3.5e-8
	371.59	378.88	0.022	6.3e-8
2.5	464.03	474.66	0.067	1.2e-7
	441.76	457.06	0.090	2.4e-8
	463.23	477.50	0.064	6.8e-8

To bypass the combinatorial issue, we turn to the method presented in this article. The experiment results are summarized in Table 1. Where, d^* refers to the lower bound of $residue(\beta)$, $|| \bar{X} - \bar{x}\bar{x}^t ||$ is the matrix l_2 norm of $\bar{X} - \bar{x}\bar{x}^t$. $\lambda_{min}(A + \bar{\mu}P)$ refers to the minimum eigenvalue of $A + \bar{\mu}P$. Note that $A + \bar{\mu}P \geq 0$ holds, and $A + \bar{\mu}P$ is nonsingular if and only if $\lambda_{min}(A + \bar{\mu}P) \neq 0$.

It can be seen that when L varies from 0.1 to 1.0, the fitting problems have been completely solved, since the lower bound of $residue(\beta)$ has been attained by some $\bar{\beta}$. This can also be verified by the values of $|| \bar{X} - \bar{x}\bar{x}^t ||$ or $\lambda_{min}(A + \bar{\mu}P)$. Both of them indicate that the global optimality conditions hold. Actually, this range of L, for this problem, has already met the requirements of real world applications.

When L is greater than 1, the norm $|| \bar{X} - \bar{x}\bar{x}^t ||$ increases rapidly and $A + \bar{\mu}P$ tends to a singular matrix, which means the global optimality conditions no longer hold from the numerical perspective. In this regard, the lower bound d^* might not be attainable, but satisfactory solutions can still be obtained by the procedure in step 4. It is unknown whether there exists other β such that $residue(\beta) < residue(\bar{\beta})$, but observe that $d^* \approx residue(\bar{\beta})$ which suggests $\bar{\beta}$ is approximately global optimal.

If we further increase L, the gap between d^* and $residue(\bar{\beta})$ will be enlarged dramatically. But such a large value of L makes little sense, because the variation of spreads even dominates the distribution of data midpoints. So it might be a bad strategy to fit such data set by a linear model.

It can be seen from the experiments that in many cases the nonconvex nondifferential problem (3) can be exactly solved by the SDP (6) or (7) even if it seems to be a huge combinatorial problem. Moreover, we usually encounter trouble in finding a feasible solution based on the relaxed problem. In more detail, if the global optimality conditions do not hold, finding a good feasible point of (5) from the SDP will turn out to be a hard problem. However, in the fitting problem, the QCQP is only an auxiliary procedure and the original problem (3) is actually unconstrained, so it doesn't really matter whether we can find the exact feasible solution of (5).

CONCLUSION

In this paper, we proposed a new method to fit interval-valued data by a linear model function. The central procedure consists of two steps: firstly, solve a semidefinite programming and secondly, if the global optimality conditions are violated, solve a quadratic programming to enhance the accuracy. Although the fitting problem might be NP-hard, experimental studies show that our method can give a good approximation of the global optimal solution and in many situations, exactly solve the fitting problem.

It is well-known that many linear models involving linear combinations of intervals or fuzzy sets induced by the extension principle may lead to hard problems in high dimensional cases. The method presented in this article provides a feasible way to overcome such difficulty to some extent. Moreover, this fitting method can also be generalized further to fit fuzzy-number data. The reason is that each fuzzy number can be represented by its α level sets which are closed intervals.

ACKNOWLEDGMENT

Support from Project 60775032 by National Natural Science Foundation of China (NSFC) is gratefully acknowledged. This study is also sponsored by the priority discipline of Beijing Normal University.

REFERENCES

An, W., Angulo, C., & Sun, Y. (2008). Support vector regression with interval-input interval-output. *International Journal of Computational Intelligence Systems, 1*(4), 299–303. doi:10.2991/ijcis.2008.1.4.2

Bazaraa, M. S. (2006). *Nonlinear programming-theory and algorithms*. Chichester, UK: John Wiley & Sons.

Bertrand, P., & Goupil, F. (2000). Descriptive statistic for symbolic data. In Bock, H.-H., & Diday, E. (Eds.), *Analysis of symbolic data* (pp. 106–124). Heidelberg, Germany: Springer-Verlag.

Billard, L., & Diday, E. (2000). Regression analysis for interval-valued data. In *Proceedings of the 7th Conference of the International Federation of Classification Societies* (pp. 369-374).

Billard, L., & Diday, E. (2003). From the statistics of data to the statistics of knowledge: Symbolic data analysis. *Journal of the American Statistical Association, 98*(462), 470–487. doi:10.1198/016214503000242

Bock, H. H., & Diday, E. (2000). *Analysis of symbolic data*. Heidelberg, Germany: Springer-Verlag.

Boyd, S., & Vandenberghe, L. (1997). Semidefinite programming relaxations of non-convex problems in control and combinatorial optimization. In Paulraj, A., Roychowdhury, V., & Schaper, C. D. (Eds.), *Communications, computation, control and signal processing: A tribute to Thomas Kailath*. Amsterdam, The Netherlands: Kluwer Academic.

Boyd, S., & Vandenberghe, L. (2004). *Convex optimization*. Cambridge, UK: Cambridge University Press.

Chang, Y. H. O. (2001). Hybrid fuzzy least-squares regression analysis and its reliability measures. *Fuzzy Sets and Systems*, *119*, 225–246. doi:10.1016/S0165-0114(99)00092-5

Chang, Y. H. O., & Ayyub, B. M. (2001). Fuzzy regression methods-a comparative assessment. *Fuzzy Sets and Systems*, *119*, 187–203. doi:10.1016/S0165-0114(99)00091-3

Coppi, R., D'Urso, P., Giordani, P., & Santoro, A. (2006). Least squares estimation of a linear regression model with LR fuzzy response. *Computational Statistics & Data Analysis*, *51*(1), 267–286. doi:10.1016/j.csda.2006.04.036

D' Urso, P. (2003). Linear regression analysis for fuzzy/crisp input and fuzzy/crisp output data. *Computational Statistics & Data Analysis*, *42*(1-2), 47–72. doi:10.1016/S0167-9473(02)00117-2

D'Urso, P., & Santoro, A. (2006). Goodness of fit and variable selection in the fuzzy multiple linear regression. *Fuzzy Sets and Systems*, *157*, 2627–2647. doi:10.1016/j.fss.2005.03.015

Diamond, P. (1990). Least squares fitting of compact set-valued data. *Journal of Mathematical Analysis and Applications*, *14*, 531–544.

Diamond, P., & Kloeden, P. (1994). *Metric spaces of fuzzy sets*. Singapore: World Scientific.

Diamond, P., & Korner, R. (1997). Extended fuzzy linear models and least squares estimates. *Computers & Mathematics with Applications (Oxford, England)*, *33*(9), 15–32. doi:10.1016/S0898-1221(97)00063-1

Hong, D. H., & Hwang, C. (2005). Interval regression analysis using quadratic loss support vector machine. *IEEE Transactions on Fuzzy Systems*, *13*, 229–237. doi:10.1109/TFUZZ.2004.840133

Hong, D. H., & Hwang, C. (2006). Support vector interval regression machine for crisp input and output data. *Fuzzy Sets and Systems*, *157*, 1114–1125. doi:10.1016/j.fss.2005.09.008

Jeng, J. T., Chuang, C. C., & Su, S. F. (2003). Support vector interval regression networks for interval regression analysis. *Fuzzy Sets and Systems*, *138*, 283–300. doi:10.1016/S0165-0114(02)00570-5

Jeyakumar, V., Rubinov, A. M., & Wu, Z. Y. (2007). Non-convex quadratic minimization problems with quadratic constraints: Global optimality conditions. *Mathematics Programming Series A*, *110*, 521–541. doi:10.1007/s10107-006-0012-5

Lima Neto, E. A., & de Carvalho, F. A. T. (2008). Centre and range method for fitting a linear regression model to symbolic interval data. *Computational Statistics & Data Analysis*, *52*, 1500–1515. doi:10.1016/j.csda.2007.04.014

Ma, M., Friedman, M., & Kandel, A. (1997). General fuzzy least squares. *Fuzzy Sets and Systems*, *88*, 107–118. doi:10.1016/S0165-0114(96)00051-6

Maia, A. L. S., de Carvalho, F. A. T., & Ludermir, T. B. (2008). Forecasting models for interval-valued time series. *Neurocomputing*, *71*, 3344–3352. doi:10.1016/j.neucom.2008.02.022

Moor, R. E. (1979). *Methods and applications of interval analysis*. Philadelphia, PA: SIAM.

Nesterov, Y. E., & Todd, M. J. (1998). Primal-dual interior-point methods for self-scaled cones. *SIAM Journal on Optimization*, *8*(2), 324–364. doi:10.1137/S1052623495290209

Nesterov, Y. E., Wolkowicz, H., & Ye, Y. (2000). Semidefinite programming relaxations of non-convex quadratic optimization. In Wolkowicz, H., Saigal, R., & Vandenberghe, L. (Eds.), *Handbook of semidefinite programming*. Amsterdam, The Netherlands: Kluwer Academic.

Polik, I., & Terlaky, T. A. (2007). Survey of the S-Lemma. *SIAM, 49*(3), 371-418.

Sakawa, M., & Yano, H. (1992). Fuzzy linear regression analysis for fuzzy input-output data. *Information Sciences*, *63*, 191–206. doi:10.1016/0020-0255(92)90069-K

Schon, S., & Kutterer, H. (2005). Using zonotopes for overestimation-free interval least-squares -some geodetic applications. *Reliable Computing*, *11*, 137–155. doi:10.1007/s11155-005-3034-4

Tanaka, H. (1987). Fuzzy data analysis by possibility linear model. *Fuzzy Sets and Systems*, *24*,363–375. doi:10.1016/0165-0114(87)90033-9

Tanaka, H., & Koyama, K. (1996). Interval regression analysis by quadratic programming approach. *IEEE Transactions on Fuzzy Systems*, *6*(4), 473–481. doi:10.1109/91.728436

Tanaka, H., Uejima, S., & Asia, K. (1982). Linear regression analysis with fuzzy model. *IEEE Transactions on Systems, Man, and Cybernetics*, *12*, 903–907. doi:10.1109/TSMC.1982.4308925

Vandenberghe, L., & Boyd, S. (1996). Semidefinite programming. *SIAM Review*, *38*, 49–95. doi:10.1137/1038003

Wu, H.-C. (2003). Linear regression analysis for fuzzy input and output data using the extension principle. *Computers & Mathematics with Applications (Oxford, England)*, *45*(12), 1849–1859. doi:10.1016/S0898-1221(03)90006-X

APPENDIX A

Proof of Proposition 1

Using $x^t Ax = \mathrm{tr}(Axx^t)$, we can transform (5) into an equivalent form

$$\begin{aligned} &\textit{minimize} && \mathrm{tr}(AX) + b^t x + c \\ &\textit{subject to} && \mathrm{tr}(PX) = 0 \\ & && X = xx^t, \quad x \geq 0 \end{aligned}$$

Note that the equality constraint $X = xx^t$ is not a convex constraint, so it can lead to a hard problem. The idea of the semidefinite relaxation approach is replacing the constraint $X = xx^t$ by $X \geq xx^t$. We can thus transform (6) to a convex problem.

$$\begin{aligned} &\textit{minimize} && \mathrm{tr}(AX) + b^t x + c \\ &\textit{subject to} && \mathrm{tr}(PX) = 0 \\ & && X \geq xx^t \quad x \geq 0 \end{aligned}$$

Apparently, the optimal value of this problem is a lower bound of (5). Furthermore, the inequality $X \geq xx^t$ is actually a convex constraint, using the schur complement (Nesterov et al., 2000), it can be reformulated as the semidefinite cone constraint

$$\begin{bmatrix} X & x \\ x^t & 1 \end{bmatrix} \geq 0$$

Thus we have the SDP relaxation formula (6). Obviously, if we obtain $\bar{X} = \bar{x}\bar{x}^t$, then $\bar{x}$ is the global minimum of (5).

APPENDIX B

Proof of Proposition 2

Let $\lambda \geq 0$, μ be dual variables and the Lagrangian associated with (5) be defined as

$$L(x, \lambda, \mu) = x^t Ax + b^t x + c + \sum_{i=1}^{2n} \lambda_i(-x_i) + \mu x^t Px$$

The Lagrange dual function can be expressed as

$$\begin{aligned} g(\lambda,\mu) &= \inf_{x\in\mathbf{R}^{2n}} L(x,\lambda,\mu) \\ &= \inf_{x\in\mathbf{R}^{2n}} x^t(A+\mu P)x + (b-\lambda)^t x + c \end{aligned}$$

$g(\lambda,\mu)$ can be written explicitly, see Boyd et al. (2004).

$$g(\lambda,\mu) = \begin{cases} -(1/4)(b-\lambda)^t(A+\mu P)^{\dagger}(b-\lambda)+c & A+\mu P \geq 0, b-\lambda \in \mathcal{R}(A+\mu P) \\ -\infty & otherwise \end{cases}$$

Consider the dual problem

$$\begin{array}{ll} maximize & g(\lambda,\mu) \\ subject\ to & \lambda \geq 0 \end{array} \quad (*)$$

Using the property of schur complement (Boyd et al., 2004), it is equivalent to

$$\begin{array}{ll} maximize & t \\ subject\ to & \begin{bmatrix} 4(A+\mu P) & b-\lambda \\ (b-\lambda)^t & c-t \end{bmatrix} \geq 0,\ \lambda \geq 0 \end{array}$$

From weak duality $d^* = \sup_{\lambda\geq 0} g(\lambda,\mu) \leq p^*$, we have proved the first statement.

Furthermore, note that $(\bar{\lambda},\bar{\mu})$ is the global maximum of (*). If $A+\bar{\mu}P$ is nonsingular, we can fix $\bar{\mu}$, then $\bar{\lambda}$ is the global minimum of the following optimization problem:

$$\begin{array}{ll} minimize & (1/4)(b-\lambda)^t(A+\bar{\mu}P)^{-1}(b-\lambda) \\ subject\ to & \lambda \geq 0 \end{array}$$

Then, there exists a dual variable $\bar{v} \geq 0$ associated with λ such that KKT conditions at $(\bar{\lambda},\bar{\mu})$ are satisfied, that is

$$\begin{aligned} -(A+\bar{\mu}P)^{-1}(b-\bar{\lambda}) - 2\bar{v} &= 0 \\ \bar{\lambda}^t\bar{v} &= 0 \\ \bar{v} \geq 0, \bar{\lambda} &\geq 0 \end{aligned}$$

On the other hand, using $A+\bar{\mu}P > 0$, then $\bar{\mu}$ is the minimum of the following optimization problem

$$\begin{array}{ll} minimize & y(\mu) = (1/4)(b-\bar{\lambda})^t(A+\mu P)^{-1}(b-\bar{\lambda}) \\ subject\ to & A+\mu P > 0 \end{array}$$

The problem is minimizing a convex function subject to an open convex domain, thus $y(\mu)$ achieves minimum if and only if the derivative vanishes at $\bar{\mu}$. Choosing $\Delta\mu$ sufficiently small, and we have

$$\begin{aligned}
y(\bar{\mu}+\Delta\mu) &= (1/4)(b-\bar{\lambda})^t(A+\bar{\mu}P+\Delta\mu P)^{-1}(b-\bar{\lambda}) \\
&= (1/4)(b-\bar{\lambda})^t[(A+\bar{\mu}P)^{1/2}(I+(A+\bar{\mu}P)^{-1/2}\Delta\mu P(A+\bar{\mu}P)^{-1/2})(A+\bar{\mu}P)^{1/2}]^{-1}(b-\bar{\lambda}) \\
&= (1/4)(b-\bar{\lambda})^t(A+\bar{\mu}P)^{-1/2}[I+(A+\bar{\mu}P)^{-1/2}\Delta\mu P(A+\bar{\mu}P)^{-1/2}]^{-1}(A+\bar{\mu}P)^{-1/2}(b-\bar{\lambda}) \\
&\approx (1/4)(b-\bar{\lambda})^t(A+\bar{\mu}P)^{-1/2}[I-(A+\bar{\mu}P)^{-1/2}\Delta\mu P(A+\bar{\mu}P)^{-1/2}](A+\bar{\mu}P)^{-1/2}(b-\bar{\lambda}) \\
&= y(\bar{\mu})-\Delta\mu(\bar{v}^tP\bar{v})
\end{aligned}$$

Here we have used the first-order approximation $(I+A)^{-1}\approx I-A$, valid when A is small. Hence, we obtain $y'(\bar{\mu})=-\bar{v}^tP\bar{v}=0$. Put $\bar{v}$ into the objective function of (5) and use the complementary slackness $\bar{\lambda}^t\bar{v}=0$, we obtain

$$\begin{aligned}
&\bar{v}^tA\bar{v}+b^t\bar{v}+c \\
&= \bar{v}^t(A+\bar{\mu}P)\bar{v}+b^t\bar{v}+c \\
&= -(1/2)\bar{v}^t(b-\bar{\lambda})+b^t\bar{v}+c \\
&= (1/2)(b-\bar{\lambda})^t\bar{v}+c \\
&= -(1/4)(b-\bar{\lambda})^t(A+\bar{\mu}P)^{-1}(b-\bar{\lambda})+c=d^*
\end{aligned}$$

On the other hand, note that $\bar{v}$ satisfies the constraints in (5), which means the lower bound d^* can be achieved by a feasible $\bar{v}$ .We thus complete the proof.

This work was previously published in the International Journal of Fuzzy System Applications, Volume 1, Issue 3, edited by Toly Chen, pp. 32-46, copyright 2010 by IGI Publishing (an imprint of IGI Global).

Chapter 13
System Identification Based on Dynamical Training for Recurrent Interval Type-2 Fuzzy Neural Network

Tsung-Chih Lin
Feng Chia University, Taiwan

Yi-Ming Chang
Feng Chia University, Taiwan

Tun-Yuan Lee
Feng Chia University, Taiwan

ABSTRACT

This paper proposes a novel fuzzy modeling approach for identification of dynamic systems. A fuzzy model, recurrent interval type-2 fuzzy neural network (RIT2FNN), is constructed by using a recurrent neural network which recurrent weights, mean and standard deviation of the membership functions are updated. The complete back propagation (BP) algorithm tuning equations used to tune the antecedent and consequent parameters for the interval type-2 fuzzy neural networks (IT2FNNs) are developed to handle the training data corrupted by noise or rule uncertainties for nonlinear system identification involving external disturbances. Only by using the current inputs and most recent outputs of the input layers, the system can be completely identified based on RIT2FNNs. In order to show that the interval IT2FNNs can handle the measurement uncertainties, training data are corrupted by white Gaussian noise with signal-to-noise ratio (SNR) 20 dB. Simulation results are obtained for the identification of nonlinear system, which yield more improved performance than those using recurrent type-1 fuzzy neural networks (RT1FNNs).

INTRODUCTION

In the past decades, fuzzy sets and their associated fuzzy logic have supplanted conventional technologies in many scientific applications and engineering systems, especially in control systems, pattern recognition and system identification. We have also witnessed a rapid growth in the use of fuzzy logic in a wide variety of consumer products and industrial systems. Since 1985, there

DOI: 10.4018/978-1-4666-1870-1.ch013

has been a strong growth in their use for dealing with the control of, especially nonlinear, time varying systems. For instance, fuzzy controllers have generated a great deal of excitement in various scientific and engineering areas, because they allow for ill-defined and complex systems rather than requiring exact mathematical models (Narendra & Parthasarathy, 1990; Wang 1993; Wang, 1994; Rovithakis & Christodoulou, 1994; Castro et al., 1995; Chen et al., 1996). The most important issue for fuzzy control systems is how to deal with the guarantee of stability and control performance, and recently there have been significant research efforts on the issue of stability in fuzzy control systems (Spooner & Passino, 1996; Ma & Sun, 2000).

Quite often, the information that is used to construct the rules in a fuzzy logic system (FLS) is uncertain. At least, there are four possible ways of rule uncertainties in type-1 FLS (Mendel 2004; Liang & Mendel, 2000): (1) Being words mean different things to different people, the meanings of the words that are used in antecedents and consequents of rules can be uncertain; (2) consequents may have a histogram of values associated with them, especially when knowledge is extracted from a group of experts who do not all agree; (3) measurements that activate a type-1 FLS may be noisy and therefore uncertain; (4) finally, the data that are used to tune the parameters of a type-1 FLS may also be noisy. Therefore, antecedent or consequent uncertainties translate into uncertain antecedent or consequent membership functions (MFs). Type-1 FLSs are unable to directly handle rule uncertainties, since their membership functions are type-1 fuzzy sets. On the other hand, type-2 FLSs involved in this paper whose antecedent or consequent membership functions are type-2 fuzzy sets can handle rule uncertainties (Hagras, 2004; Hagras, 2007; Martinez et al., 2009; Sepulveda et al., 2009; Castro et al., 2009). A type-2 FLS is characterized by IF-THEN rules, but its antecedent or consequent sets are type-2. Hence, type-2 FLSs can be used when the circumstances are too uncertain to determine exact membership grades such as when training data is corrupted by noise.

Type-2 FLSs have been applied successfully to deal with decision making (Yager, 1980), time-series forecasting (Karnik & Mendel, 1999), time varying channel equalization (Liang & Mendel, 2000), fuzzy controller designs (Wang, 1994; Lin et al., 2009, 2010; Lin, 2010b), VLSI fault diagnosis (Lin, 2010a) and control of mobile robots (Wu, 1996), due to the type-2 FLSs ability to handle uncertainties. Further, genetic algorithm (GA) was adopted to fine tune the Gaussian MFs in the antecedent part of type-1 FNN (Wang et al., 2001). The dynamical optimal training algorithm for the two-layer consequent part of interval TFNN (Wang et al., 2004), was proposed to learn the parameters of the antecedent type-2 MFs as well as of the consequent weighting factors of the consequent part of the T2FNN. The back propagation (BP) equations proposed by Wang et al. (2004) are not correct and were modified by Hagras (2006).

However, the BP algorithm presented in Hagras (2006) was the only one of the sixteen possible combinations for the different permutations what active branch x_k and the status of rule i compared to numbers R and L used to compute y_r and y_l, respectively. Based on the presented generalized analysis for the computing derivatives in interval type-2 FLSs (Mendel, 2004), in this paper, the all sixteen possible combinations of the specific BP equation used to tune the antecedent and consequent parameters of the interval T2FNN will be developed.

This paper is organized as follows. First, a brief description of interval type-2 fuzzy logic system is introduced. The construction of RIT2FNN is described and the BP algorithm for RIT2FNN is presented. Finally, a simulation example to demonstrate the performance of the proposed method is provided and the conclusions of the advocated design methodology are given.

BRIEF DESCRIPTION OF INTERVAL TYPE-2 FUZZY LOGIC SYSTEM

Due to the complexity of type reduction, the general type-2 FLS becomes computationally intensive. In order to make things simpler and easier to compute meet and join operations, the secondary MFs of an interval type-2 FLS are all unity which leads finally to simplify type reduction. The 2-D interval type-2 Gaussian membership function (MF) with uncertain mean $m \in [m_1,\ m_2]$ and a fixed deviation σ is shown in Figure 1.

$$\mu_{\tilde{A}}(x) = \exp\left[-\frac{1}{2}\left(\frac{x-m}{\sigma}\right)^2\right], m \in [m_1, m_2] \tag{1}$$

It is obvious that the type-2 fuzzy set is in a region bounded by an upper MF and a lower MF denoted as $\bar{\mu}_{\tilde{A}}(x)$ and $\underline{\mu}_{\tilde{A}}(x)$, respectively, and is called a footprint of uncertainty (FOU). In the meantime, the firing strength F^i for the *i*th rule can be an interval type-1 set (Liang & Mendel, 2000) expressed as

$$F^i = \begin{bmatrix} \underline{f}^i \\ \bar{f}^i \end{bmatrix} \tag{2}$$

where

$$\underline{f}^i = \underline{\mu}_{\tilde{F}_1^i}(x_1) * \cdots * \underline{\mu}_{\tilde{F}_n^i}(x_n) = \Pi_{j=1}^{n}\underline{\mu}_{\tilde{F}_j^i}(x_j) \tag{3}$$

$$\bar{f}^i = \bar{\mu}_{\tilde{F}_1^i}(x_1) * \cdots * \bar{\mu}_{\tilde{F}_n^i}(x_n) = \Pi_{j=1}^{n}\bar{\mu}_{\tilde{F}_j^i}(x_j) \tag{4}$$

Based on the center of set type reduction, the defuzzified crisp output from an interval type-2 FLS is the average of y_l and y_r, i.e.

$$y\ (\underline{x}) = \frac{y_l + y_r}{2} \tag{5}$$

where y_l and y_r are the left most and right most points of the interval type-1 set which can be obtained as

$$y_l = \frac{\sum_{i}^{M} f_l^i w_l^i}{\sum_{i}^{M} f_l^i} = \frac{\sum_{i=1}^{L} \bar{f}^i w_l^i + \sum_{i=L+1}^{M} \underline{f}^i w_l^i}{\sum_{i=1}^{L} \bar{f}^i + \sum_{i=L+1}^{M} \underline{f}^i} \tag{6}$$

Figure 1. Interval type-2 fuzzy set with uncertain mean

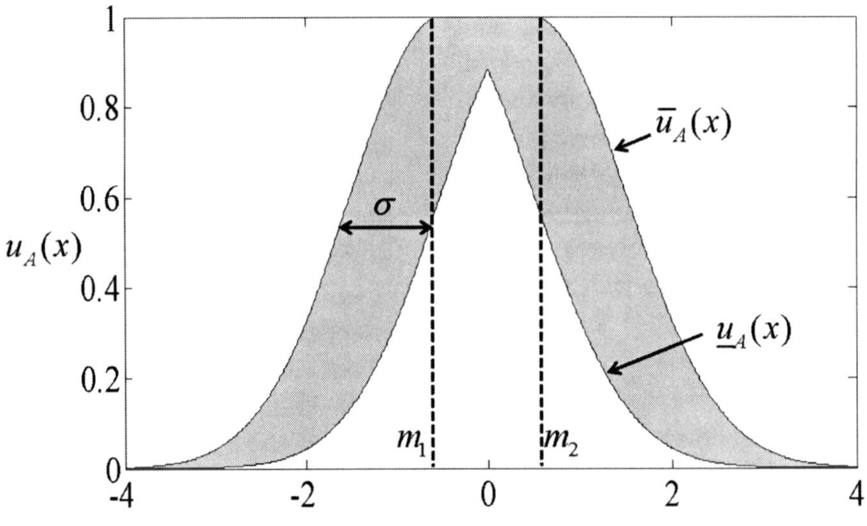

and

$$y_r = \frac{\sum_{i}^{M} f_r^i w_r^i}{\sum_{i}^{M} f_r^i} = \frac{\sum_{i=1}^{R} \underline{f}^i w_r^i + \sum_{i=R+1}^{M} \overline{f}^i w_r^i}{\sum_{i=1}^{R} \underline{f}^i + \sum_{i=R+1}^{M} \overline{f}^i} \qquad (7)$$

where M is the total number of rules in the rule base of the RIT2FNN. The weighting factors w_l^i and w_r^i of the consequent part represent the centroid interval set of the consequent type-2 fuzzy set of the *i*-th rule. In the meantime, R and L can be determined by using the iterative Karnik-Mendel procedure (Wang et al., 2004).

A type-2 FLS is very similar to a type-1 FLS as shown in Figure 2 (Wang et al., 2004), the major structure difference being that the defuzzifier block of a type-1 FLS is replaced by the output processing block in a type-2 FLS which consists of type-reduction followed by defuzzification.

There are five main parts in a type-2 FLS: fuzzifier, rule base, inference engine, type-reducer and defuzzifier. A type-2 FLS is a mapping $f : \Re^p \to \Re^1$. After defuzzification, fuzzy inference, type-reduction and defuzzification, a crisp output can be obtained.

Consider a type-2 FLS having p inputs $x_1 \in X_1$, $\cdots$, $x_p \in X_p$ and one output $y \in Y$. The type-2 fuzzy rule base consists of a collection of IF-THEN rules. As in the type-1 case, we assume there are M rules and the rule of a type-2 relation between the input space $X_1 \times X_2 \times \cdots \times X_p$ and the output space Y can be expressed as

$$R^l : \text{IF } x_1 \text{ is } \tilde{F}_1^l \text{ and } \cdots \text{ and } x_p \text{ is } \tilde{F}_p^l, \qquad (8)$$

where $\tilde{F}_j^l$s are antecedent type-2 sets ($j = 1, 2, \cdots, p$) and $\tilde{G}^l s$ are consequent type-2 sets.

The inference engine combines rules and gives a mapping from input type-2 fuzzy sets to output type-2 fuzzy sets. To achieve this process, we have to compute the union and intersection of type-2 sets, as well as the composition of type-2 relations. The output of the inference engine block is a type-2 set. By using the extension principle of type-1 defuzzification method, type-reduction takes us from type-2 output sets of the FLS to a type-1 set called the "type-reduced set". This set may then be defuzzified to obtain a single crisp value.

Figure 2. The structure of a type-2 fuzzy logic system

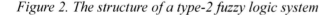

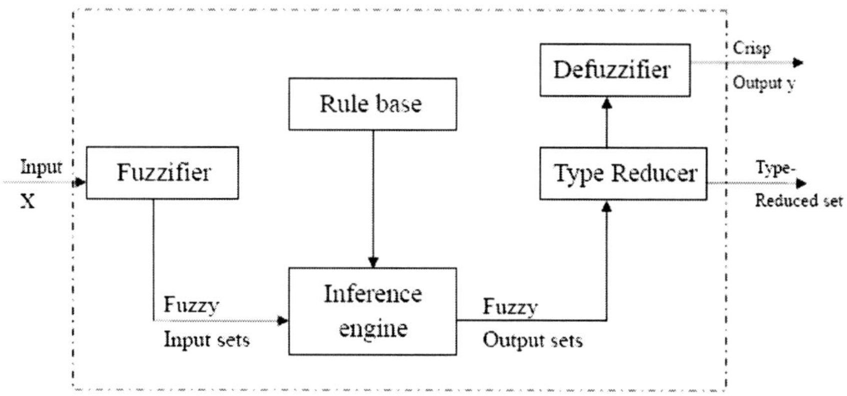

1. CONSTRUCTION OF RIT2FNN

The structure of the proposed RIT2FNN as shown in Figure 3, consists of four layers which includes the input, the membership function, the rule and the output layers. By adding feedback connection through delay in the input layer, the recurrent feedback is embedded in the network. All notations are denoted as in Table 1.

The signal propagation and the interaction between each layer are given as follows:

Layer 1 - Input Layer: *The nodes in this layer transmit input values to the next layer directly. By using the recurrent weights* w_{d_j}, *feedback connections are added in this input layer to embed temporal relations. For* j *th node in this layer, the input and output are described as:*

Input variable of the j th node: $x_j^1(p)$
Output variable of the j th node:

$$O_j^1(p) = x_j^1(p) + w_{d_j} O_j^1(p-1) \ , \quad j = 1....N \tag{9}$$

where p is the number of iterations.

Layer 2 - Membership Layer: *In this layer, each node represents the terms of respective linguistic variables and the interval type-2 Gaussian function is adopted as the membership function in Equation (10).*

$$N(m_j^i, \sigma_j^i, O_j^1) = \exp\left(-\frac{1}{2}\left(\frac{O_j^1 - m_j^i}{\sigma_j^i}\right)^2\right),$$

$$m_j^i \in \left[m_{j1}^i \ \ m_{j2}^i\right] \ , \ i=1,\cdots,M \ , \ j = 1,....,N \tag{10}$$

Meanwhile, for network input $u_{ij}^2 = O_j^1$, the output is

$$O_{ij}^2 = N(m_j^i, \sigma_j^i, u_{ij}^2) = \exp\left(-\frac{1}{2}\left(\frac{u_{ij}^2 - m_j^i}{\sigma_j^i}\right)^2\right) \tag{11}$$

Figure 3. Structure of four-layer RIT2FNN

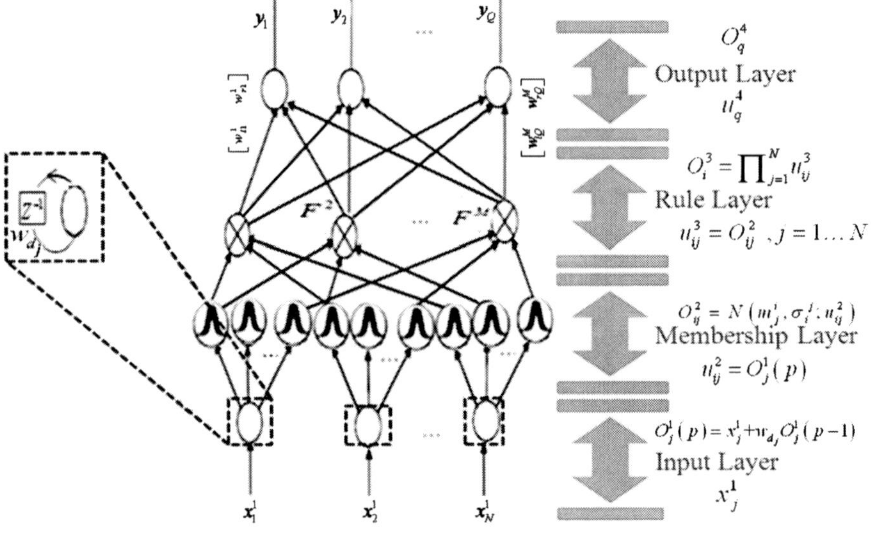

Table 1. The notations of the construction of RIT2FNN

Layer	Input variable	Output variable
1	$x_j^1,\ j = 1,2,\cdots,N$	O_j^1
2	$u_{ij}^2, i = 1,2,\cdots,M, j = 1,2,\cdots,N$	O_{ij}^2
3	$u_i^3 = \Pi_{j=1}^N O_{ij}^2$	$O_i^3 = u_i^3$
4	u_q^4	O_q^4

where $j = 1, 2, \cdots, N$ and $i = 1, 2, \cdots, M$ are used to index the i-th term of the j-th input variable. Furthermore, $O_{ij}^2 \in \left[\underline{O}_{ij}^2, \overline{O}_{ij}^2\right]$ and $\underline{O}_{ij}^2$ and $\overline{O}_{ij}^2$ are lower and upper membership function, respectively and can be expressed as

$$\underline{O}_{ij}^2 = \begin{cases} N\left(m_{j2}^i, \sigma_i^j; u_{ij}^2\right), & u_{ij}^2 \le \dfrac{m_{j1}^i + m_{j2}^i}{2} \\ N\left(m_{j1}^i, \sigma_i^j; u_{ij}^2\right), & u_{ij}^2 > \dfrac{m_{j1}^i + m_{j2}^i}{2} \end{cases} \tag{12}$$

$$\overline{O}_{ij}^2 = \begin{cases} N\left(m_{j1}^i, \sigma_i^j; u_{ij}^2\right), & u_{ij}^2 \le m_{j1}^i \\ 1, & m_{j1}^i \le u_{ij}^2 \le m_{j2}^i \\ N\left(m_{j2}^i, \sigma_i^j; u_{ij}^2\right), & u_{ij}^2 \ge m_{j1}^i \end{cases} \tag{13}$$

Layer 3 - Rule Layer: *Nodes consist of the preconditions of the rule, i.e., the firing strength F^i from (2) as shown in the following context. The IF-THEN rule for the interval RT2FNN can be expressed as:*

R^i: IF O_1^1 is $\widetilde{F}_1^i$, and IF O_2^1 is $\widetilde{F}_2^i$, and $\cdots$, and IF O_N^1 is $\widetilde{F}_N^i$

THEN y_1 is $[w_{l1}^i \quad w_{r1}^i]$, and y_2 is $[w_{l2}^i \quad w_{r2}^i]\cdots$, and y_q is $[w_{lQ}^i \quad w_{rQ}^i]$ (14)

where $i = 1,2,\cdots,M$ is rule number; $\widetilde{F}_N^i$ is the interval type-2 fuzzy set of antecedent part; w_{lq}^i and w_{rq}^i ($q = 1,2,\cdots,Q$) are left and right output weights, respectively.

Layer 4 - Output Layer: *In this layer, the left most and right most points are evaluated using Karnik-Mendel type-reduction algorithm and after defuzzification the system's output (Wang et al., 2004; Mendel, 2004; Liang & Mendel, 2000) is obtained as:*

$$y_q(\underline{x}) = O_q^4 = \frac{y_{lq} + y_{rq}}{2} \tag{15}$$

where y_{lq} and y_{rq} are the left most and right most points of the interval type-2 set which can be expressed as

$$y_{lq} = \frac{\sum_i^M f_l^i w_{lq}^i}{\sum_i^M f_l^i} = \frac{\sum_{i=1}^L \overline{f}^i w_{lq}^i + \sum_{i=L+1}^M \underline{f}^i w_{lq}^i}{\sum_{i=1}^L \overline{f}^i + \sum_{i=L+1}^M \underline{f}^i} \tag{16}$$

and

$$y_{rq} = \frac{\sum_{i}^{M} f_r^i w_{rq}^i}{\sum_{i}^{M} f_r^i} = \frac{\sum_{i=1}^{R} \underline{f}^i w_{rq}^i + \sum_{i=R+1}^{M} \overline{f}^i w_{rq}^i}{\sum_{i=1}^{R} \underline{f}^i + \sum_{i=R+1}^{M} \overline{f}^i} \quad (17)$$

where

$$\underline{f}^i = \underline{O}_{i1}^2 * \cdots * \underline{O}_{iN}^2 = \Pi_{j=1}^{N} \underline{O}_{ij}^2 \quad (18)$$

$$\overline{f}^i = \overline{O}_{i1}^2 * \cdots * \overline{O}_{iN}^2 = \Pi_{j=1}^{N} \overline{O}_{ij}^2 \quad (19)$$

and M is the total number of rules in the rule base of the RIT2FNN.

1. BP ALGORITHM FOR THE RIT2FNN

Based on BP algorithm, for P input/output training data pairs($\underline{x}^p ; d^p$), $p = 1,2,\cdots,P$ the weights of the RIT2FNN can be adjusted such that the following error function must be minimized:

$$e_q^p = \frac{1}{2}\left[y_q(\underline{x}^p) - d_q^p\right]^2, \quad p = 1,2,\cdots,P \quad (20)$$

Therefore, the BP algorithm may be written briefly as:

$$W(p+1) = W(p) + \Delta W(p) = W(p) - \eta\left(-\frac{\partial e_q^p(p)}{\partial W(p)}\right) \quad (21)$$

where η is the learning rate and W represents the tuning weights, in this case, which are means m_1, m_2, deviation σ, recurrent weight w_{d_j} and output weighting factors w_{lq}^i and w_{rq}^i.

Let's reconsider k-th input and i-th rule, a Gaussian primary MF with a fixed standard deviation σ_k and an uncertain mean with interval value $[m_{k1}^i, m_{k2}^i]$, the upper and lower MFs can be described as

$$\underline{O}_{ik}^2 = \begin{cases} N\left(m_{k2}^i, \sigma_k^i; O_k^1\right), & O_k^1 \le \frac{m_{k1}^i + m_{k2}^i}{2} \\ N\left(m_{k1}^i, \sigma_k^i; O_k^1\right), & O_k^1 > \frac{m_{k1}^i + m_{k2}^i}{2} \end{cases} \quad (22)$$

and

$$\overline{O}_{ik}^2 = \begin{cases} N\left(m_{k1}^i, \sigma_k^i; O_k^1\right), & O_k^1 < m_{k1}^i \\ 1, & m_{k1}^i \le O_k^1 \le m_{k2}^i \\ N\left(m_{k2}^i, \sigma_k^i; O_k^1\right), & O_k^1 > m_{k2}^i \end{cases} \quad (23)$$

where

$$N\left(m_k^i, \sigma_k^i; O_k^1\right) = \exp\left(-\frac{1}{2}\left(\frac{O_k^1 - m_k^i}{\sigma_k^i}\right)^2\right),$$

$$m_k^i \in \left[m_{k1}^i \ \ m_{k2}^i\right] \quad (24)$$

The detail tuning procedure for all training weights can be described as follows:

- Tuning output weighting factors w_l^i and w_r^i:

Using Equation (21) and the chain rule, we can obtain

$$w_{lq}^i(p+1) = w_{lq}^i(p) - \eta\frac{\partial e_q^p}{\partial w_{lq}^i}$$

$$= w_{lq}^{i}(p) - \eta \left[\frac{\partial e_q^p}{\partial y_q} \frac{\partial y_q}{\partial y_{lq}} \frac{\partial y_{lq}}{\partial w_{lq}^{i}} \right] \tag{25}$$

$$w_{rq}^{i}(p+1) = w_{rq}^{i}(p) - \eta \frac{\partial e_q^p}{\partial w_{rq}^{i}}$$

$$= w_{rq}^{i}(p) - \eta \left[\frac{\partial e_q^p}{\partial y_q} \frac{\partial y_q}{\partial y_{rq}} \frac{\partial y_{rq}}{\partial w_{rq}^{i}} \right] \tag{26}$$

Furthermore,

$$\frac{\partial e_q^p}{\partial y_q} = y(x^p) - d^p, \frac{\partial y_q}{\partial y_{lq}} = \frac{1}{2}, \frac{\partial y_q}{\partial y_{rq}} = \frac{1}{2} \tag{27}$$

and

$$\frac{\partial y_l}{\partial w_{lq}^{i}} = \begin{cases} \dfrac{\overline{f}^i}{\sum_{i=1}^{L} \overline{f}^i + \sum_{i=L+1}^{M} \underline{f}^i}, & i \le L \\ \dfrac{\underline{f}^i}{\sum_{i=1}^{L} \overline{f}^i + \sum_{i=L+1}^{M} \underline{f}^i}, & i > L \end{cases} \tag{28}$$

$$\frac{\partial y_r}{\partial w_{rq}^{i}} = \begin{cases} \dfrac{\underline{f}^i}{\sum_{i=1}^{R} \underline{f}^i + \sum_{i=R+1}^{M} \overline{f}^i}, & i \le R \\ \dfrac{\overline{f}^i}{\sum_{i=1}^{R} \overline{f}^i + \sum_{i=R+1}^{M} \underline{f}^i}, & i > R \end{cases} \tag{29}$$

Thus, from Equations (25)-(29) the output weighting factors w_{lq}^{i} and w_{rq}^{i} can be adjusted by

$$w_{lq}^{i}(p+1) = \begin{cases} w_{lq}^{i}(p) - \eta \left(\frac{1}{2}\left(y_q - d_q^p\right) \right), \\ \qquad \dfrac{\overline{f}^i}{\sum_{i=1}^{L} \overline{f}^i + \sum_{i=L+1}^{M} \underline{f}^i}, & i \le L \\ w_{lq}^{i}(p) - \eta \left(\frac{1}{2}\left(y_q - d_q^p\right) \right), \\ \qquad \dfrac{\underline{f}^i}{\sum_{i=1}^{L} \overline{f}^i + \sum_{i=L+1}^{M} \underline{f}^i}, & i > L \end{cases} \tag{30}$$

and

$$w_{rq}^{i}(p+1) = \begin{cases} w_{rq}^{i}(p) - \eta \left(\frac{1}{2}\left(y_q - d_q^p\right) \right), \\ \qquad \dfrac{\underline{f}^i}{\sum_{j=1}^{R} \underline{f}^i + \sum_{j=R+1}^{M} \overline{f}^i}, & i \le R \\ w_{rq}^{i}(p) - \eta \left(\frac{1}{2}\left(y_q - d_q^p\right) \right), \\ \qquad \dfrac{\overline{f}^i}{\sum_{i=1}^{R} \underline{f}^i + \sum_{i=R+1}^{M} \overline{f}^i}, & i > R \end{cases} \tag{31}$$

- Tuning the mean m_{k1}^{i} and m_{k2}^{i}:

Using Equation (21) and the chain rule, the update laws for m_{k1}^{i} can be derived as

$$m_{k1}^{i}(p+1) = m_{k1}^{i}(p) - \eta \frac{\partial e^p}{\partial m_{k1}^{i}}$$

$$= m_{k1}^i(p) - \eta\left[\frac{\partial e^p}{\partial y_q}\frac{\partial y_q}{\partial y_{lq}}\frac{\partial y_{lq}}{\partial m_{k1}^i} + \frac{\partial e^p}{\partial y_q}\frac{\partial y_q}{\partial y_{rq}}\frac{\partial y_{rq}}{\partial m_{k1}^i}\right]$$

$$= m_{k1}^i(p) - \eta\left[\frac{\partial e^p}{\partial y_q}\frac{\partial y_q}{\partial y_{lq}}\left(\frac{\partial y_{lq}}{\partial \overline{O}_{ij}^2}\frac{\partial \overline{O}_{ij}^2}{\partial m_{k1}^i} + \frac{\partial y_{lq}}{\partial \underline{O}_{ij}^2}\frac{\partial \underline{O}_{ij}^2}{\partial m_{k1}^i}\right)\right.$$
$$\left. + \frac{\partial e^p}{\partial y_q}\frac{\partial y_q}{\partial y_{rq}}\left(\frac{\partial y_{rq}}{\partial \overline{O}_{ij}^2}\frac{\partial \overline{O}_{ij}^2}{\partial m_{k1}^i} + \frac{\partial y_{rq}}{\partial \underline{O}_{ij}^2}\frac{\partial \underline{O}_{ij}^2}{\partial m_{k1}^i}\right)\right] \tag{32}$$

and

$$\frac{\partial y_{lq}}{\partial \underline{O}_{ij}^2} = \begin{cases} 0 & , \ i \le L \\ \dfrac{\left(\prod_{\substack{j=1\\ j\ne k}}^{n} \underline{O}_{ij}^2\right)\left(w_{lq}^i - y_{lq}\right)}{\sum_{i=1}^{L}\overline{f}^i + \sum_{i=L+1}^{M}\underline{f}^i} & , \ i > L \end{cases} \tag{33}$$

$$\frac{\partial y_{lq}}{\partial \overline{O}_{ij}^2} = \begin{cases} \dfrac{\left(\prod_{\substack{j=1\\ j\ne k}}^{n} \overline{O}_{ij}^2\right)\left(w_{lq}^i - y_{lq}\right)}{\sum_{i=1}^{L}\overline{f}^i + \sum_{i=L+1}^{M}\underline{f}^i} & , \ i \le L \\ 0 & , \ i > L \end{cases} \tag{34}$$

$$\frac{\partial y_{rq}}{\partial \underline{O}_{ij}^2} = \begin{cases} \dfrac{\left(\prod_{\substack{j=1\\ j\ne k}}^{n} \underline{O}_{ij}^2\right)\left(w_{rq}^i - y_{rq}\right)}{\sum_{i=1}^{R}\underline{f}^i + \sum_{i=R+1}^{M}\overline{f}^i} & , \ i \le R \\ 0 & , \ i > R \end{cases} \tag{35}$$

$$\frac{\partial y_{rq}}{\partial \overline{O}_{ij}^2} = \begin{cases} 0 & , \ i \le R \\ \dfrac{\left(\prod_{\substack{j=1\\ j\ne k}}^{n} \overline{O}_{ij}^2\right)\left(w_{rq}^i - y_{rq}\right)}{\sum_{i=1}^{R}\underline{f}^i + \sum_{i=R+1}^{M}\overline{f}^i} & , \ i > R \end{cases} \tag{36}$$

Moreover,

$$\frac{\partial \overline{O}_{ij}^2}{\partial m_{k1}^i} = \begin{cases} \dfrac{\left(O_k^1 - m_{k1}^i\right)N\left(m_{k1}^i, \sigma_k^i; O_k^1\right)}{\left(\sigma_k^i\right)^2} & , O_k^1 < m_{k1}^i \\ 0 & , m_{k1}^i \le O_k^1 \le m_{k2}^i \\ 0 & , \ O_k^1 > m_{k2}^i \end{cases} \tag{37}$$

$$\frac{\partial \underline{O}_{ij}^2}{\partial m_{k1}^i} = \begin{cases} 0 & O_k^1 \le \dfrac{m_{k1}^i + m_{k2}^i}{2} \\ \dfrac{\left(O_k^1 - m_{k1}^i\right)N\left(m_{k1}^i, \sigma_k^i; O_k^1\right)}{\left(\sigma_k^i\right)^2}, & O_k^1 > \dfrac{m_{k1}^i + m_{k2}^i}{2} \end{cases} \tag{38}$$

$$\frac{\partial \overline{O}_{ij}^2}{\partial m_{k2}^i} = \begin{cases} 0 & , \ O_k^1 < m_{k1}^i \\ 0 & , \ m_{k1}^i \le O_k^1 \le m_{k2}^i \\ \dfrac{\left(O_k^1 - m_{k2}^i\right)N\left(m_{k2}^i, \sigma_k^i; O_k^1\right)}{\left(\sigma_k^i\right)^2}, & O_k^1 > m_{k2}^i \end{cases} \tag{39}$$

and

$$\frac{\partial \underline{O}_{ij}^{2}}{\partial m_{k2}^{i}} = \begin{cases} \dfrac{\left(O_k^1 - m_{k2}^{i}\right) N\left(m_{k2}^{i}, \sigma_k^{i}; O_k^1\right)}{\left(\sigma_k^{i}\right)^2}, & O_k^1 \le \dfrac{m_{k1}^{i} + m_{k2}^{i}}{2} \\ 0, & O_k^1 > \dfrac{m_{k1}^{i} + m_{k2}^{i}}{2} \end{cases} \tag{40}$$

Therefore, from the above analysis, there are sixteen possible combinations of $\eta \dfrac{\partial e_q^p}{\partial m_{k1}^{i}}$ for different values O_k^1 and *i* as shown in Table 2. Similarly, the update laws for m_{k2}^{i} can be derived from Equation (31) by replacing m_{k1}^{i} by m_{k2}^{i} and there are sixteen possible combinations of $\eta \dfrac{\partial e_q^p}{\partial m_{k2}^{i}}$ for different values O_k^1 and *i* as shown in Table 3.

- Tuning standard deviation σ_k^{i}:

Using Equation (21) and the chain rule, the standard deviation σ_k^{i} can be adjusted as

$$\sigma_k^{i}(p+1) = \sigma_k^{i}(p) - \eta \frac{\partial e^p}{\partial \sigma_k^{i}}$$

$$= \sigma_k^{i}(p) - \eta \left[\frac{\partial e_q^p}{\partial y_q} \frac{\partial y_q}{\partial y_{lq}} \left(\frac{\partial y_{lq}}{\partial \overline{O}_{ij}^{2}} \frac{\partial \overline{O}_{ij}^{2}}{\partial \sigma_k^{i}} + \frac{\partial y_{lq}}{\partial \underline{O}_{ij}^{2}} \frac{\partial \underline{O}_{ij}^{2}}{\partial \sigma_k^{i}} \right) \right] \tag{41}$$

and

$$\frac{\partial \overline{O}_{ij}^{2}}{\partial \sigma_k^{i}} = \begin{cases} \dfrac{\left(O_k^1 - m_{k1}^{i}\right)^2 N\left(m_{k1}^{i}, \sigma_k^{i}; O_k^1\right)}{\left(\sigma_k^{i}\right)^3}, & O_k^1 < m_{k1}^{i} \\ 0, & m_{k1}^{i} \le O_k^1 \le m_{k2}^{i} \\ \dfrac{\left(O_k^1 - m_{k2}^{i}\right)^2 N\left(m_{k2}^{i}, \sigma_k^{i}; O_k^1\right)}{\left(\sigma_k^{i}\right)^3}, & O_k^1 > m_{k2}^{i} \end{cases} \tag{42}$$

$$\frac{\partial \underline{O}_{ij}^{2}}{\partial \sigma_k^{i}} = \begin{cases} \dfrac{\left(O_k^1 - m_{k2}^{i}\right)^2 N\left(m_{k2}^{i}, \sigma_k^{i}; O_k^1\right)}{\left(\sigma_k^{i}\right)^3}, & O_k^1 \le \dfrac{m_{k1}^{i} + m_{k2}^{i}}{2} \\ \dfrac{\left(O_k^1 - m_{k1}^{i}\right)^2 N\left(m_{k1}^{i}, \sigma_k^{i}; O_k^1\right)}{\left(\sigma_k^{i}\right)^3}, & O_k^1 > \dfrac{m_{k1}^{i} + m_{k2}^{i}}{2} \end{cases} \tag{43}$$

Finally, from the above analysis, there are sixteen possible combinations of $\eta \dfrac{\partial e_q^p}{\partial \sigma_k^{i}}$ for different values O_k^1 and *i* as shown in Table 4.

- Tuning recurrent weight w_{d_k}:

Using Equation (21) and the chain rule, the recurrent weight w_{d_k} can be adjusted as

$$w_{d_k}(p+1) = w_{d_k}(p) - \eta \frac{\partial e_q^p}{\partial w_{d_k}}$$

$$= w_{d_k}(p) - \eta \left[\frac{\partial e_q^p}{\partial y_q} \frac{\partial y_q}{\partial y_{lq}} \frac{\partial y_{lq}}{\partial w_{d_k}} + \frac{\partial e_q^p}{\partial y_q} \frac{\partial y_q}{\partial y_{rq}} \frac{\partial y_{rq}}{\partial w_{d_k}} \right]$$

Table 2. Sixteen possible combinations of $\eta\frac{\partial e_q^p}{\partial m_{k1}^i}$ for different values of O_k^1 and i

	$i \le L \,\&\, i \le R$	$i \le L \,\&\, i > R$	$i > L \,\&\, i \le R$	$i > L \,\&\, i > R$
$O_k^1 < m_{k1}^i$ $O_k^1 \le \frac{m_{k1}^i + m_{k2}^i}{2}$	$\eta\frac{\partial e_q^p}{\partial y_q}\frac{\partial y_q}{\partial y_{lq}}\frac{\partial y_{lq}}{\partial \overline{O}_{ij}^2}\frac{\partial \overline{O}_{ij}^2}{\partial m_{k1}^i}$	$\eta\left(\frac{\partial e_q^p}{\partial y_q}\frac{\partial y}{\partial y_{lq}}\frac{\partial y_{lq}}{\partial \overline{O}_{ij}^2}\frac{\partial \overline{O}_{ij}^2}{\partial m_{k1}^i} + \frac{\partial e_q^p}{\partial y_q}\frac{\partial y_q}{\partial y_{rq}}\frac{\partial y_{rq}}{\partial \overline{O}_{ij}^2}\frac{\partial \overline{O}_{ij}^2}{\partial m_{k1}^i}\right)$	0	$\eta\frac{\partial e_q^p}{\partial y_q}\frac{\partial y_q}{\partial y_{rq}}\frac{\partial y_{rq}}{\partial \overline{O}_{ij}^2}\frac{\partial \overline{O}_{ij}^2}{\partial m_{k1}^i}$
$O_k^1 < m_{k1}^i$ $O_k^1 > \frac{m_{k1}^i + m_{k2}^i}{2}$	$\eta\left(\frac{\partial e_q^p}{\partial y_q}\frac{\partial y_q}{\partial y_{lq}}\frac{\partial y_{lq}}{\partial \overline{O}_{ij}^2}\frac{\partial \overline{O}_{ij}^2}{\partial m_{k1}^i} + \frac{\partial e_q^p}{\partial y_q}\frac{\partial y_q}{\partial y_{rq}}\frac{\partial y_{rq}}{\partial \underline{O}_{ij}^2}\frac{\partial \underline{O}_{ij}^2}{\partial m_{k1}^i}\right)$	$\eta\left(\frac{\partial e_q^p}{\partial y_q}\frac{\partial y_q}{\partial y_{lq}}\frac{\partial y_{lq}}{\partial \overline{O}_{ij}^2}\frac{\partial \overline{O}_{ij}^2}{\partial m_{k1}^i} + \frac{\partial e_q^p}{\partial y_q}\frac{\partial y_q}{\partial y_{rq}}\frac{\partial y_{rq}}{\partial \overline{O}_{ij}^2}\frac{\partial \overline{O}_{ij}^2}{\partial m_{k1}^i}\right)$	$\eta\left(\frac{\partial e_q^p}{\partial y_q}\frac{\partial y_q}{\partial y_{lq}}\frac{\partial y_{lq}}{\partial \underline{O}_{ij}^2}\frac{\partial \underline{O}_{ij}^2}{\partial m_{k1}^i} + \frac{\partial e_q^p}{\partial y_q}\frac{\partial y_q}{\partial y_{rq}}\frac{\partial y_{rq}}{\partial \underline{O}_{ij}^2}\frac{\partial \underline{O}_{ij}^2}{\partial m_{k1}^i}\right)$	$\eta\left(\frac{\partial e_q^p}{\partial y_q}\frac{\partial y_q}{\partial y_{lq}}\frac{\partial y_{lq}}{\partial \underline{O}_{ij}^2}\frac{\partial \underline{O}_{ij}^2}{\partial m_{k1}^i} + \frac{\partial e_q^p}{\partial y_q}\frac{\partial y_q}{\partial y_{rq}}\frac{\partial y_{rq}}{\partial \overline{u}}\frac{\partial \overline{u}}{\partial m_{k1}^i}\right)$
$O_k^1 > m_{k1}^i$ $O_k^1 \le \frac{m_{k1}^i + m_{k2}^i}{2}$	0	0	0	0
$O_k^1 > m_{k1}^i$ $O_k^1 > \frac{m_{k1}^i + m_{k2}^i}{2}$	$\eta\frac{\partial e_q^p}{\partial y_q}\frac{\partial y_q}{\partial y_{rq}}\frac{\partial y_{rq}}{\partial \underline{O}_{ij}^2}\frac{\partial \underline{O}_{ij}^2}{\partial m_{k1}^i}$	0	$\eta\left(\frac{\partial e_q^p}{\partial y_q}\frac{\partial y_q}{\partial y_{lq}}\frac{\partial y_{lq}}{\partial \underline{O}_{ij}^2}\frac{\partial \underline{O}_{ij}^2}{\partial m_{k1}^i} + \frac{\partial e_q^p}{\partial y_q}\frac{\partial y_q}{\partial y_{rq}}\frac{\partial y_{rq}}{\partial \underline{O}_{ij}^2}\frac{\partial \underline{O}_{ij}^2}{\partial m_{k1}^i}\right)$	$\eta\frac{\partial e_q^p}{\partial y_q}\frac{\partial y_q}{\partial y_{lq}}\frac{\partial y_{lq}}{\partial \underline{O}_{ij}^2}\frac{\partial \underline{O}_{ij}^2}{\partial m_{k1}^i}$

Table 3. Sixteen possible combinations of $\eta\frac{\partial e_q^p}{\partial m_{k2}^i}$ for different values of O_k^1 and i

	$i \le L \,\&\, i \le R$	$i \le L \,\&\, i > R$	$i > L \,\&\, i \le R$	$i > L \,\&\, i > R$
$O_k^1 < m_{k1}^i$ $O_k^1 \le \frac{m_{k1}^i + m_{k2}^i}{2}$	$\eta\frac{\partial e_q^p}{\partial y_q}\frac{\partial y_q}{\partial y_{rq}}\frac{\partial y_{rq}}{\partial \underline{O}_{ij}^2}\frac{\partial \underline{O}_{ij}^2}{\partial m_{k2}^i}$	0	$\eta\left(\frac{\partial e_q^p}{\partial y_q}\frac{\partial y_q}{\partial y_{lq}}\frac{\partial y_{lq}}{\partial \underline{O}_{ij}^2}\frac{\partial \underline{O}_{ij}^2}{\partial m_{k2}^i} + \frac{\partial e_q^p}{\partial y_q}\frac{\partial y_q}{\partial y_{rq}}\frac{\partial y_{rq}}{\partial \underline{O}_{ij}^2}\frac{\partial \underline{O}_{ij}^2}{\partial m_{k2}^i}\right)$	$\eta\frac{\partial e_q^p}{\partial y_q}\frac{\partial y_q}{\partial y_{lq}}\frac{\partial y_{lq}}{\partial \underline{O}_{ij}^2}\frac{\partial \underline{O}_{ij}^2}{\partial m_{k2}^i}$
$O_k^1 < m_{k1}^i$ $O_k^1 > \frac{m_{k1}^i + m_{k2}^i}{2}$	0	0	0	0
$O_k^1 > m_{k1}^i$ $O_k^1 \le \frac{m_{k1}^i + m_{k2}^i}{2}$	$\eta\left(\frac{\partial e_q^p}{\partial y_q}\frac{\partial y_q}{\partial y_{lq}}\frac{\partial y_{lq}}{\partial \overline{O}_{ij}^2}\frac{\partial \overline{O}_{ij}^2}{\partial m_{k2}^i} + \frac{\partial e_q^p}{\partial y_q}\frac{\partial y_q}{\partial y_{rq}}\frac{\partial y_{rq}}{\partial \underline{O}_{ij}^2}\frac{\partial \underline{O}_{ij}^2}{\partial m_{k2}^i}\right)$	$\eta\left(\frac{\partial e_q^p}{\partial y_q}\frac{\partial y_q}{\partial y_{lq}}\frac{\partial y_{lq}}{\partial \overline{O}_{ij}^2}\frac{\partial \overline{O}_{ij}^2}{\partial m_{k2}^i} + \frac{\partial e_q^p}{\partial y_q}\frac{\partial y_q}{\partial y_{rq}}\frac{\partial y_{rq}}{\partial \overline{O}_{ij}^2}\frac{\partial \overline{O}_{ij}^2}{\partial m_{k2}^i}\right)$	$\eta\left(\frac{\partial e_q^p}{\partial y_q}\frac{\partial y_q}{\partial y_{lq}}\frac{\partial y_{lq}}{\partial \underline{O}_{ij}^2}\frac{\partial \underline{O}_{ij}^2}{\partial m_{i2}^j} + \frac{\partial e_q^p}{\partial y_q}\frac{\partial y_q}{\partial y_{rq}}\frac{\partial y_{rq}}{\partial \underline{O}_{ij}^2}\frac{\partial \underline{O}_{ij}^2}{\partial m_{i2}^j}\right)$	$\eta\left(\frac{\partial e_q^p}{\partial y_q}\frac{\partial y_q}{\partial y_{lq}}\frac{\partial y_{lq}}{\partial \underline{O}_{ij}^2}\frac{\partial \underline{O}_{ij}^2}{\partial m_{k2}^i} + \frac{\partial e_q^p}{\partial y_q}\frac{\partial y_q}{\partial y_{rq}}\frac{\partial y_{rq}}{\partial \overline{O}_{ij}^2}\frac{\partial \overline{O}_{ij}^2}{\partial m_{k2}^i}\right)$
$O_k^1 > m_{k1}^i$ $O_k^1 > \frac{m_{k1}^i + m_{k2}^i}{2}$	$\eta\frac{\partial e_q^p}{\partial y_q}\frac{\partial y_q}{\partial y_{lq}}\frac{\partial y_{lq}}{\partial \overline{O}_{ij}^2}\frac{\partial \overline{O}_{ij}^2}{\partial m_{k2}^i}$	$\eta\left(\frac{\partial e_q^p}{\partial y_q}\frac{\partial y_q}{\partial y_{lq}}\frac{\partial y_{lq}}{\partial \overline{O}_{ij}^2}\frac{\partial \overline{O}_{ij}^2}{\partial m_{k2}^i} + \frac{\partial e_q^p}{\partial y_q}\frac{\partial y_q}{\partial y_{rq}}\frac{\partial y_{rq}}{\partial \overline{O}_{ij}^2}\frac{\partial \overline{O}_{ij}^2}{\partial m_{k2}^i}\right)$	0	$\eta\frac{\partial e_q^p}{\partial y_q}\frac{\partial y_q}{\partial y_{rq}}\frac{\partial y_{rq}}{\partial \overline{O}_{ij}^2}\frac{\partial \overline{O}_{ij}^2}{\partial m_{k2}^i}$

Table 4. Sixteen possible combinations of $\eta\dfrac{\partial e_q^p}{\partial\sigma_k^i}$ *for different values of* O_k^1 *and i*

	$i \le L\ \&\ i \le R$	$i \le L\ \&\ i > R$	$i > L\ \&\ i \le R$	$i > L\ \&\ i > R$
$O_k^1 < m_{k1}^i$ $O_k^1 \le \frac{m_{k1}^i + m_{k2}^i}{2}$	$\eta\left(\frac{\partial e_q^p}{\partial y_q}\frac{\partial y_q}{\partial y_{lq}}\frac{\partial y_{lq}}{\partial\overline{O}_{ij}^2}\frac{\partial\overline{O}_{ij}^2}{\partial\sigma_k^i(m_{k1}^i)}+\frac{\partial e_q^p}{\partial y_q}\frac{\partial y_q}{\partial y_{rq}}\frac{\partial y_{rq}}{\partial\underline{O}_{ij}^2}\frac{\partial\underline{O}_{ij}^2}{\partial\sigma_k^i(m_{k2}^i)}\right)$	$\eta\left(\frac{\partial e_q^p}{\partial y_q}\frac{\partial y_q}{\partial y_{lq}}\frac{\partial y_{lq}}{\partial\overline{O}_{ij}^2}\frac{\partial\overline{O}_{ij}^2}{\partial\sigma_k^i(m_{k1}^i)}+\frac{\partial e_q^p}{\partial y_q}\frac{\partial y_q}{\partial y_{rq}}\frac{\partial y_{rq}}{\partial\overline{O}_{ij}^2}\frac{\partial\overline{O}_{ij}^2}{\partial\sigma_k^i(m_{k1}^i)}\right)$	$\eta\left(\frac{\partial e_q^p}{\partial y_q}\frac{\partial y_q}{\partial y_{lq}}\frac{\partial y_{lq}}{\partial\underline{O}_{ij}^2}\frac{\partial\underline{O}_{ij}^2}{\partial\sigma_k^i(m_{k2}^i)}+\frac{\partial e_q^p}{\partial y_q^p}\frac{\partial y_q}{\partial y_{rq}}\frac{\partial y_{rq}}{\partial\underline{O}_{ij}^2}\frac{\partial\underline{O}_{ij}^2}{\partial\sigma_k^i(m_{k2}^i)}\right)$	$\eta\left(\frac{\partial e_q^p}{\partial y_q}\frac{\partial y_q}{\partial y_{lq}}\frac{\partial y_{lq}}{\partial\underline{O}_{ij}^2}\frac{\partial\underline{O}_{ij}^2}{\partial\sigma_k^i(m_{k2}^i)}+\frac{\partial e_q^p}{\partial y_q}\frac{\partial y_q}{\partial y_{rq}}\frac{\partial y_{rq}}{\partial\overline{O}_{ij}^2}\frac{\partial\overline{O}_{ij}^2}{\partial\sigma_k^i(m_{k1}^i)}\right)$
$O_k^1 < m_{k1}^i$ $O_k^1 > \frac{m_{k1}^i + m_{k2}^i}{2}$	$\eta\left(\frac{\partial e_q^p}{\partial y_q}\frac{\partial y_q}{\partial y_{lq}}\frac{\partial y_{lq}}{\partial\overline{O}_{ij}^2}\frac{\partial\overline{O}_{ij}^2}{\partial\sigma_k^i(m_{k1}^i)}+\frac{\partial e_q^p}{\partial y_q}\frac{\partial y_q}{\partial y_{rq}}\frac{\partial y_{rq}}{\partial\underline{O}_{ij}^2}\frac{\partial\underline{O}_{ij}^2}{\partial\sigma_k^i(m_{k1}^i)}\right)$	$\eta\left(\frac{\partial e_q^p}{\partial y_q}\frac{\partial y_q}{\partial y_{lq}}\frac{\partial y_{lq}}{\partial\overline{O}_{ij}^2}\frac{\partial\overline{O}_{ij}^2}{\partial\sigma_k^i(m_{k1}^i)}+\frac{\partial e_q^p}{\partial y_q}\frac{\partial y_q}{\partial y_{rq}}\frac{\partial y_{rq}}{\partial\overline{O}_{ij}^2}\frac{\partial\overline{O}_{ij}^2}{\partial\sigma_k^i(m_{k1}^i)}\right)$	$\eta\left(\frac{\partial e_q^p}{\partial y_q}\frac{\partial y_q}{\partial y_{lq}}\frac{\partial y_{lq}}{\partial\underline{O}_{ij}^2}\frac{\partial\underline{O}_{ij}^2}{\partial\sigma_k^i(m_{k1}^i)}+\frac{\partial e_q^p}{\partial y_q}\frac{\partial y_q}{\partial y_{rq}}\frac{\partial y_{rq}}{\partial\underline{O}_{ij}^2}\frac{\partial\underline{O}_{ij}^2}{\partial\sigma_k^i(m_{k1}^i)}\right)$	$\eta\left(\frac{\partial e_q^p}{\partial y_q}\frac{\partial y_q}{\partial y_{lq}}\frac{\partial y_{lq}}{\partial\underline{O}_{ij}^2}\frac{\partial\underline{O}_{ij}^2}{\partial\sigma_k^i(m_{k1}^i)}+\frac{\partial e_q^p}{\partial y_q^p}\frac{\partial y_q}{\partial y_{rq}}\frac{\partial y_{rq}}{\partial\overline{O}_{ij}^2}\frac{\partial\overline{O}_{ij}^2}{\partial\sigma_k^i(m_{k1}^i)}\right)$
$O_k^1 > m_{k1}^i$ $O_k^1 \le \frac{m_{k1}^i + m_{k2}^i}{2}$	$\eta\left(\frac{\partial e_q^p}{\partial y_q}\frac{\partial y_q}{\partial y_{lq}}\frac{\partial y_{lq}}{\partial\overline{O}_{ij}^2}\frac{\partial\overline{O}_{ij}^2}{\partial\sigma_k^i(m_{k2}^i)}+\frac{\partial e_q^p}{\partial y_q}\frac{\partial y_q}{\partial y_{rq}}\frac{\partial y_{rq}}{\partial\underline{O}_{ij}^2}\frac{\partial\underline{O}_{ij}^2}{\partial\sigma_k^i(m_{k2}^i)}\right)$	$\eta\left(\frac{\partial e_q^p}{\partial y_q}\frac{\partial y_q}{\partial y_{lq}}\frac{\partial y_{lq}}{\partial\overline{O}_{ij}^2}\frac{\partial\overline{O}_{ij}^2}{\partial\sigma_k^i(m_{k2}^i)}+\frac{\partial e_q^p}{\partial y_q}\frac{\partial y_q}{\partial y_{rq}}\frac{\partial y_{rq}}{\partial\overline{O}_{ij}^2}\frac{\partial\overline{O}_{ij}^2}{\partial\sigma_k^i(m_{k2}^i)}\right)$	$\eta\left(\frac{\partial e_q^p}{\partial y_q}\frac{\partial y_q}{\partial y_{lq}}\frac{\partial y_{lq}}{\partial\underline{O}_{ij}^2}\frac{\partial\underline{O}_{ij}^2}{\partial\sigma_k^i(m_{k2}^i)}+\frac{\partial e_q^p}{\partial y_q}\frac{\partial y_q}{\partial y_{rq}}\frac{\partial y_{rq}}{\partial\underline{O}_{ij}^2}\frac{\partial\underline{O}_{ij}^2}{\partial\sigma_k^i(m_{k2}^i)}\right)$	$\eta\left(\frac{\partial e_q^p}{\partial y_q}\frac{\partial y_q}{\partial y_{lq}}\frac{\partial y_{lq}}{\partial\underline{O}_{ij}^2}\frac{\partial\underline{O}_{ij}^2}{\partial\sigma_k^i(m_{k2}^i)}+\frac{\partial e_q^p}{\partial y_q}\frac{\partial y_q}{\partial y_{rq}}\frac{\partial y_{rq}}{\partial\overline{O}_{ij}^2}\frac{\partial\overline{O}_{ij}^2}{\partial\sigma_k^i(m_{k2}^i)}\right)$
$O_k^1 > m_{k1}^i$ $O_k^1 > \frac{m_{k1}^i + m_{k2}^i}{2}$	$\eta\left(\frac{\partial e_q^p}{\partial y_q}\frac{\partial y_q}{\partial y_{lq}}\frac{\partial y_{lq}}{\partial\overline{O}_{ij}^2}\frac{\partial\overline{O}_{ij}^2}{\partial\sigma_k^i(m_{k2}^i)}+\frac{\partial e_q^p}{\partial y_q}\frac{\partial y_q}{\partial y_{rq}}\frac{\partial y_{rq}}{\partial\underline{u}}\frac{\partial\underline{u}}{\partial\sigma_k^i(m_{k1}^i)}\right)$	$\eta\left(\frac{\partial e_q^p}{\partial y_q}\frac{\partial y_q}{\partial y_{lq}}\frac{\partial y_{lq}}{\partial\overline{O}_{ij}^2}\frac{\partial\overline{O}_{ij}^2}{\partial\sigma_k^i(m_{k2}^i)}+\frac{\partial e_q^p}{\partial y_q}\frac{\partial y_q}{\partial y_{rq}}\frac{\partial y_{rq}}{\partial\overline{O}_{ij}^2}\frac{\partial\overline{O}_{ij}^2}{\partial\sigma_k^i(m_{k2}^i)}\right)$	$\eta\left(\frac{\partial e_q^p}{\partial y_q}\frac{\partial y_q}{\partial y_{lq}}\frac{\partial y_{lq}}{\partial\underline{O}_{ij}^2}\frac{\partial\underline{O}_{ij}^2}{\partial\sigma_k^i(m_{k1}^i)}+\frac{\partial e_q^p}{\partial y_q}\frac{\partial y_q}{\partial y_{rq}}\frac{\partial y_{rq}}{\partial\underline{O}_{ij}^2}\frac{\partial\underline{O}_{ij}^2}{\partial\sigma_k^i(m_{k1}^i)}\right)$	$\eta\left(\frac{\partial e_q^p}{\partial y_q}\frac{\partial y_q}{\partial y_{lq}}\frac{\partial y_{lq}}{\partial\underline{O}_{ij}^2}\frac{\partial\underline{O}_{ij}^2}{\partial\sigma_k^i(m_{k1}^i)}+\frac{\partial e_q^p}{\partial y_q}\frac{\partial y_q}{\partial y_{rq}}\frac{\partial y_{rq}}{\partial\overline{O}_{ij}^2}\frac{\partial\overline{O}_{ij}^2}{\partial\sigma_k^i(m_{k2}^i)}\right)$

$$= w_{d_k}(p) - \eta\left[\frac{\partial e_q^p}{\partial y_q}\frac{\partial y_q}{\partial y_{lq}}\left(\frac{\partial y_{lq}}{\partial\overline{O}_{ik}^2}\frac{\partial\overline{O}_{ik}^2}{\partial w_{d_k}} + \frac{\partial y_{lq}}{\partial\underline{O}_{ik}^2}\frac{\partial\underline{O}_{ik}^2}{\partial w_{d_k}}\right)\right] \tag{44}$$

and

$$\frac{\partial\overline{O}_{ij}^2}{\partial w_{d_k}} = \begin{cases} \dfrac{O_{k(P-1)}^1\left(m_{k1}^j - O_k^1\right)N\left(m_{k1}^j, \sigma_k^j; O_k^1\right)}{\left(\sigma_k^j\right)^2}, & O_k^1 < m_{k1}^i \\ 0 & , m_{k1}^i \le O_k^1 \le m_{k2}^i \\ \dfrac{O_{i(k-1)}^1\left(m_{k2}^j - O_k^1\right)N\left(m_{k2}^j, \sigma_k^j; O_k^1\right)}{\left(\sigma_k^j\right)^2}, & O_k^1 > m_{k2}^i \end{cases} \tag{45}$$

$$\frac{\partial\underline{O}_{ij}^2}{\partial w_{d_k}} = \begin{cases} \dfrac{O_{k(P-1)}^1\left(m_{k2}^j - O_k^1\right)N\left(m_{k2}^j, \sigma_k^j; O_k^1\right)}{\left(\sigma_k^j\right)^2}, & O_k^1 \le \dfrac{m_{k1}^i + m_{k2}^i}{2} \\ \dfrac{O_{k(P-1)}^1\left(m_{k1}^j - O_k^1\right)N\left(m_{k1}^j, \sigma_k^j; O_k^1\right)}{\left(\sigma_k^j\right)^2}, & O_k^1 > \dfrac{m_{k1}^i + m_{k2}^i}{2} \end{cases} \tag{46}$$

Finally, from the above analysis, there are sixteen possible combinations of $\eta\dfrac{\partial e_q^p}{\partial w_{d_k}}$ for different values O_k^1 and *i* as shown in Table 5.

Table 5. Sixteen possible combinations of $\eta \frac{\partial e_q^p}{\partial w_{d_k}}$ *for different values of* O_k^1 *and i*

	$i \le L \,\&\, i \le R$	$i \le L \,\&\, i > R$	$i > L \,\&\, i \le R$	$i > L \,\&\, i > R$
$O_k^1 < m_{k1}^i$ $O_k^1 \le \frac{m_{k1}^i + m_{k2}^i}{2}$	$\eta\left(\frac{\partial e_q^p}{\partial y_q}\frac{\partial y_q}{\partial y_{lq}}\frac{\partial y_{lq}}{\partial \overline{O}_{ij}^2}\frac{\partial \overline{O}_{ij}^2}{\partial w_{d_k}(m_{k1}^i)}+\frac{\partial e_q^p}{\partial y_q}\frac{\partial y_q}{\partial y_{rq}}\frac{\partial y_{rq}}{\partial \underline{O}_{ij}^2}\frac{\partial \underline{O}_{ij}^2}{\partial w_{d_k}(m_{k2}^i)}\right)$			$\eta\left(\frac{\partial e_q^p}{\partial y_q}\frac{\partial y_q}{\partial y_{lq}}\frac{\partial y_{lq}}{\partial \underline{O}_{ij}^2}\frac{\partial \underline{O}_{ij}^2}{\partial w_{d_k}(m_{k2}^i)}+\frac{\partial e_q^p}{\partial y_q}\frac{\partial y_q}{\partial y_{rq}}\frac{\partial y_{rq}}{\partial \overline{O}_{ij}^2}\frac{\partial \overline{O}_{ij}^2}{\partial w_{d_k}(m_{k1}^i)}\right)$
$O_k^1 < m_{k1}^i$ $O_k^1 > \frac{m_{k1}^i + m_{k2}^i}{2}$			$\eta\left(\frac{\partial e_q^p}{\partial y_q}\frac{\partial y_q}{\partial y_{lq}}\frac{\partial y_{lq}}{\partial \underline{O}_{ij}^2}\frac{\partial \underline{O}_{ij}^2}{\partial w_{d_k}(m_{k1}^i)}+\frac{\partial e^p}{\partial y_q}\frac{\partial y_q}{\partial y_{rq}}\frac{\partial y_{rq}}{\partial \underline{O}_{ij}^2}\frac{\partial \underline{O}_{ij}^2}{\partial w_{d_k}(m_{k1}^i)}\right)$	$\eta\left(\frac{\partial e_q^p}{\partial y_q}\frac{\partial y_q}{\partial y_{lq}}\frac{\partial y_{lq}}{\partial \underline{O}_{ij}^2}\frac{\partial \underline{O}_{ij}^2}{\partial w_{d_k}(m_{k1}^i)}+\frac{\partial e_q^p}{\partial y_q}\frac{\partial y_q}{\partial y_{rq}}\frac{\partial y_{rq}}{\partial \overline{O}_{ij}^2}\frac{\partial \overline{O}_{ij}^2}{\partial w_{d_k}(m_{k1}^i)}\right)$
		$\eta\left(\frac{\partial e_q^p}{\partial y_q}\frac{\partial y_q}{\partial y_{lq}}\frac{\partial y_{lq}}{\partial \overline{O}_{ij}^2}\frac{\partial \overline{O}_{ij}^2}{\partial w_{d_k}(m_{k2}^i)}+\frac{\partial e_q^p}{\partial y_q}\frac{\partial y_q}{\partial y_{rq}}\frac{\partial y_{rq}}{\partial \overline{O}_{ij}^2}\frac{\partial \overline{O}_{ij}^2}{\partial w_{d_k}(m_{k2}^i)}\right)$	$\eta\left(\frac{\partial e_q^p}{\partial y_q}\frac{\partial y_q}{\partial y_{lq}}\frac{\partial y_{lq}}{\partial \underline{O}_{ij}^2}\frac{\partial \underline{O}_{ij}^2}{\partial w_{d_k}(m_{k2}^i)}+\frac{\partial e_q^p}{\partial y_q}\frac{\partial y_q}{\partial y_{rq}}\frac{\partial y_{rq}}{\partial \underline{O}_{ij}^2}\frac{\partial \underline{O}_{ij}^2}{\partial w_{d_k}(m_{k2}^i)}\right)$	
	$\eta\left(\frac{\partial e_q^p}{\partial y_q}\frac{\partial y_q}{\partial y_{lq}}\frac{\partial y_{lq}}{\partial \overline{O}_{ij}^2}\frac{\partial \overline{O}_{ij}^2}{\partial w_{d_k}(m_{k2}^i)}+\frac{\partial e_q^p}{\partial y_q}\frac{\partial y_q}{\partial y_{rq}}\frac{\partial y_{rq}}{\partial \underline{O}_{ij}^2}\frac{\partial \underline{O}_{ij}^2}{\partial w_{d_k}(m_{k1}^i)}\right)$	$\eta\left(\frac{\partial e_q^p}{\partial y_q}\frac{\partial y_q}{\partial y_{lq}}\frac{\partial y_{lq}}{\partial \overline{O}_{ij}^2}\frac{\partial \overline{O}_{ij}^2}{\partial w_{d_k}(m_{k2}^i)}+\frac{\partial e_q^p}{\partial y_q}\frac{\partial y_q}{\partial y_{rq}}\frac{\partial y_{rq}}{\partial \overline{O}_{ij}^2}\frac{\partial \overline{O}_{ij}^2}{\partial w_{d_k}(m_{k2}^i)}\right)$		

Figure 4. RIT2FNNs system identification scheme

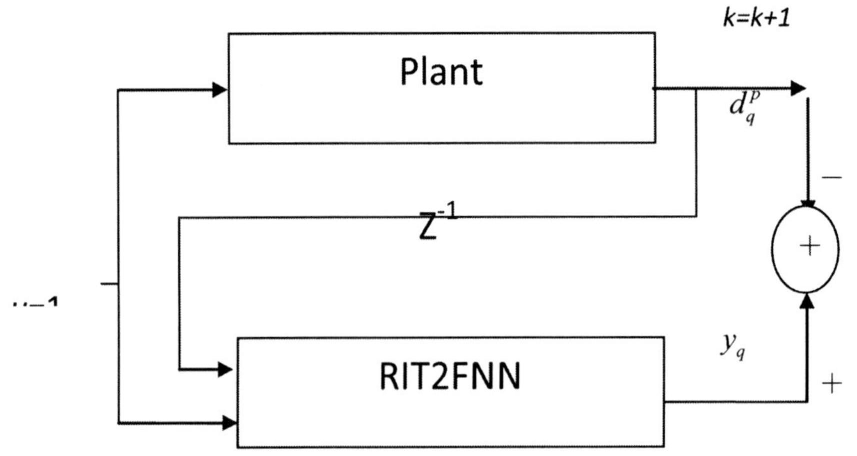

SIMULATION EXAMPLE

Based on the above system identification training for the interval T2FNNs, the block diagram of the identifier is shown in Figure 4. Consider the Duffing forced oscillation system (Wang, 1994).

$$\dot{x}_1 = x_2$$
$$\dot{x}_2 = -0.1x_2 - x_1^3 + 12\cos t + u + d$$
$$y = x_1$$

Table 6. Initial means and standard deviations of the type-1 and interval type-2 of Gaussian membership functions

	Input 1				Input 2			
	Deviation	T2 Center		T1 Center	Deviation	T2 Center		T1 Center
	σ_1^i	m_{11}^i	m_{12}^i	m_1^i	σ_2^i	m_{21}^i	m_{22}^i	m_2^i
$\mu_{F_j^1}(x_j)$	1	-6.6	-5.4	-6	1.5	-9.9	-8.1	-9
$\mu_{F_j^2}(x_j)$	1	-4.6	-3.4	-4	1.5	-6.9	-5.1	-6
$\mu_{F_j^3}(x_j)$	1	-2.6	-1.4	-2	1.5	-3.9	-2.1	-3
$\mu_{F_j^4}(x_j)$	1	-0.6	0.6	0	1.5	-0.9	0.9	0
$\mu_{F_j^5}(x_j)$	1	1.4	2.6	2	1.5	2.1	3.9	3
$\mu_{F_j^6}(x_j)$	1	3.4	4.6	4	1.5	5.1	6.9	6
$\mu_{F_j^7}(x_j)$	1	5.4	6.6	6	1.5	8.1	9.9	9

The simulation time $t = 10$ seconds and the step size h=0.01. For states x_1 and x_2, the initial means and standard deviations of the type-1 and interval type-2 FNN Gaussian MFs are given in Table 6 and as shown in Figures 5 and 6, respectively.

Figures 7, 8, and 9 show the identification performance for noise-free Duffing forced oscillation system for IT2RFNN, IT2FNN and T1RFNN, respectively. All three figures indicate

Figure 5. Interval type-2 Gaussian MFs of state x_1

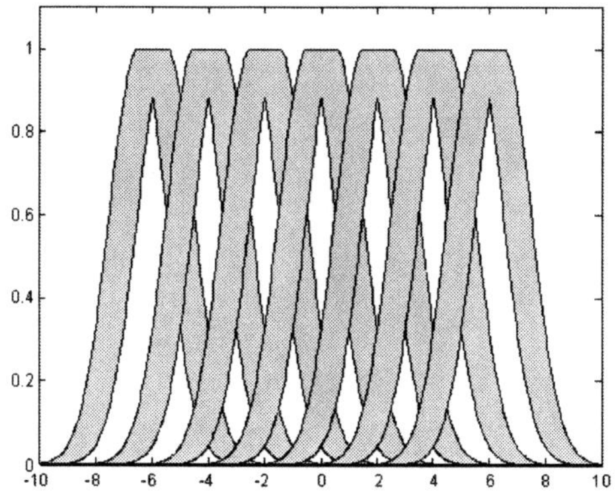

Figure 6. Interval type-2 Gaussian MFs of state x_2

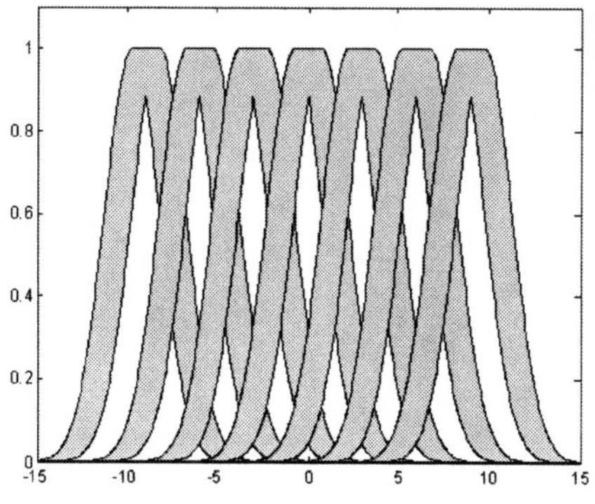

Figure 7. Identification performance of the RIT-2FNN for noise free

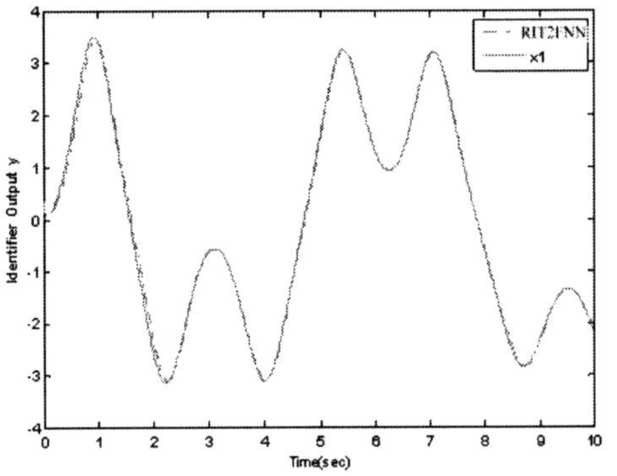

Figure 8. Identification performance of the IT-2FNN for noise free

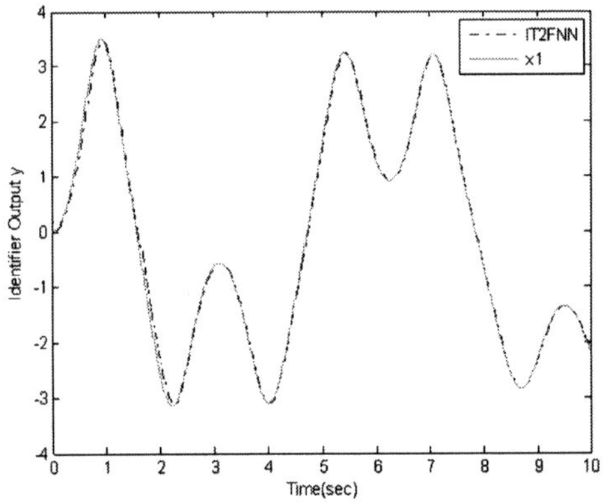

Figure 9. Identification performance of the RIT-1FNN for noise free

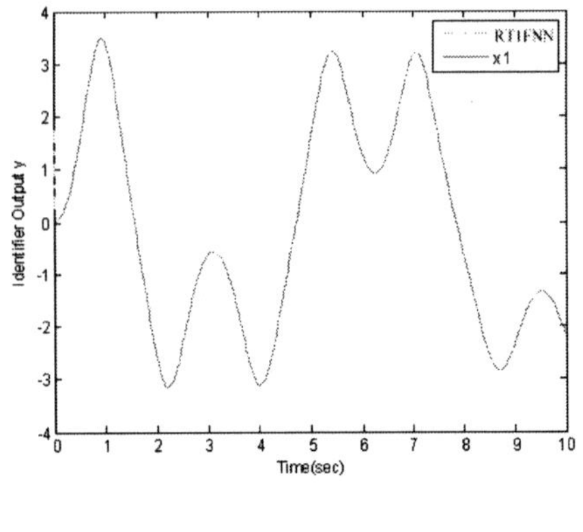

Figure 10. Error function e_q^p of the RIT2FNNIT-2FNN and RT1FNN noise free

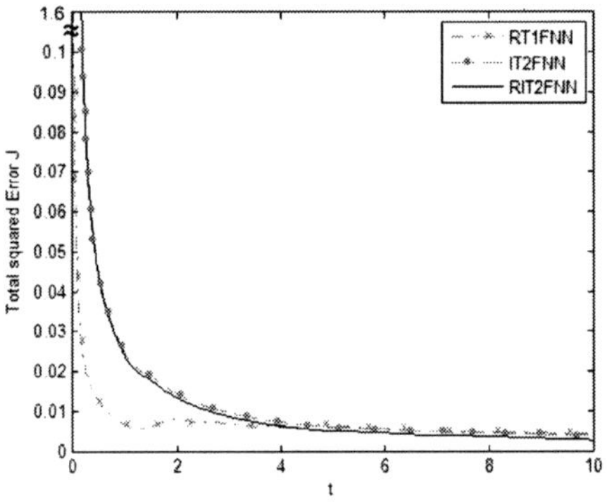

Figure 11. Results of prediction using the RIT2FNN with noise SNR= 20dB

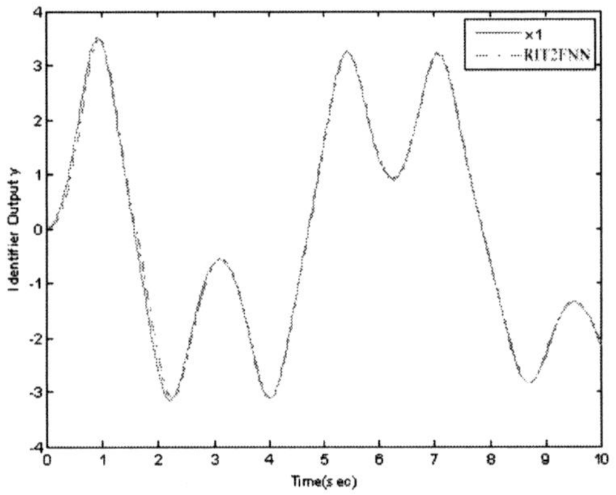

Figure 12. Results of prediction using the IT2FNN with noise SNR= 20dB

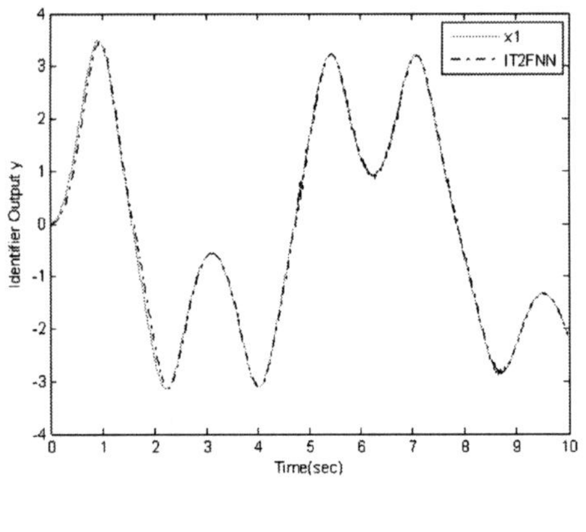

Figure 13. Results of prediction using the RIT-1FNN with noise SNR=20dB

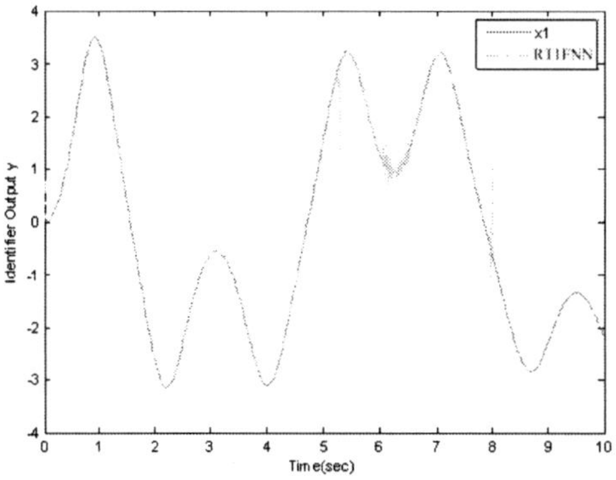

the fact that the identification performances can be guaranteed. Figure 10 shows the error function e_q^p of RIT2FNN, IT2FNN and RIT1FNN, respectively.

In order to show that the interval type-2 can handle measurement uncertainties, training data are corrupted by white Gaussian noise (WGN) with signal-to-noise ratio (SNR) as 20 dB. Figures 11, 12 and 13 show the identification performance with WGN SNR=20 dB of the Duffing forced oscillation system by IT2RFNN, IT2FNN and T1RFNN, respectively. The error functions e_q^p of RIT2FNN, IT2FNN and RT1FNN, respectively, are shown in Figure 14. From Figure 14, it is obvious that after 8 seconds the T1FNN is divergent. The mean square errors for noise-free and

Table 7. Performance comparison for RIT2FNN, IT2FNN and RITIFNN

Type	RIT2FNN	IT2FNN	RT1FNN
No. of linguistic variables	7	7	7
Learning rate	0.5	0.5	0.5
MSE of Noise free	0.0079	0.0082	0.029
MSE with Noise	0.0097	0.0107	inf

noisy training data corrupted by WGN with SNR= 20dB cases are given in Table 7.

CONCLUSION

A novel fuzzy modeling approach for identification of dynamic systems is constructed based on a recurrent interval type-2 fuzzy neural network. The antecedent and consequent parameters for the IT2FNNs are tuned by using the complete BP algorithm tuning equations. Furthermore, a new dynamical training algorithm for RIT2FNN is developed to handle the training data corrupted by noise or rule uncertainties for nonlinear system identification involving external disturbances. The system can be completely identified based on IT2FNNs only by using the current inputs and most recent outputs of the system. Furthermore,

Figure 14. Squared errors of the RIT2FNN, IT2FNN and RT1FNN with noise SNR=20dB

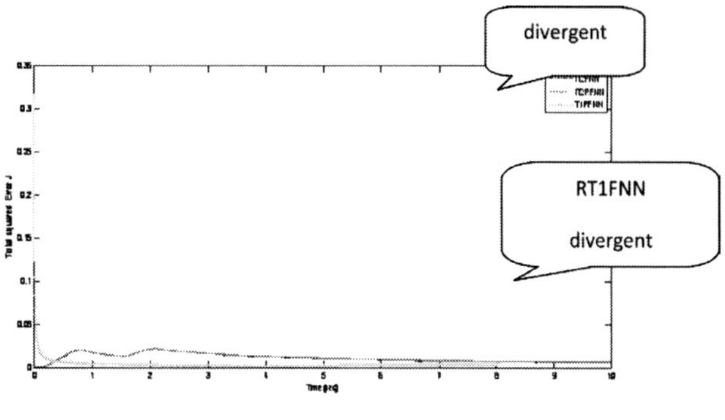

in order to show that the interval IT2FNNs can handle measurement uncertainties, training data are corrupted by white Gaussian noise with SNR as 20 dB. Simulation results confirm the validity and performance of the advocated identification methodology.

However, the description of some systems is more accurate when the fractional order derivative is used. Fractional order system identification will be considered in future research.

REFERENCES

Castro, J. L. (1995). Fuzzy logical controllers are universal approximators. *IEEE Transactions on Systems, Man, and Cybernetics. Part B, Cybernetics*, *25*, 629–635.

Castro, J. R., Castillo, O., Melin, P., & Rodriguez-Diaz, A. (2009). A hybrid learning algorithm for a class of interval type-2 fuzzy neural networks. *Information Sciences*, *179*, 2175–2193. doi:10.1016/j.ins.2008.10.016

Chen, B. S., Lee, C. H., & Chang, Y. C. (1996). tracking design of uncertain nonlinear SISO systems: Adaptive fuzzy approach. *IEEE Transactions on Fuzzy Systems*, *4*, 32–43.H^∞ doi:10.1109/91.481843

Hagras, H. A. (2006). Comments on dynamical optimal training for interval type-2 fuzzy neural network (T2FNN). *IEEE Transactions on Systems, Man, and Cybernetics. Part B, Cybernetics*, *36*(5), 1206–1209. doi:10.1109/TCSI.2006.873184

Hagras, H. A. (2007). Type-2 FLCs: A new generation of fuzzy controllers. *IEEE Computational Intelligent Magazine*, *2*, 30–43. doi:10.1109/MCI.2007.357192

Hagras, H. A., & Hierarchical, A. A. (2004). Type-2 fuzzy logic control architecture for autonomous mobile robots. *IEEE Transactions on Fuzzy Systems*, *12*(4), 524–539. doi:10.1109/TFUZZ.2004.832538

Karnik, N. N., & Mendel, J. M. (1999). Applications of type-2 fuzzy logic systems to forecasting of time-series. *Information Science*, *120*, 89–111. doi:10.1016/S0020-0255(99)00067-5

Liang, Q., & Mendel, J. M. (2000). Equalization of nonlinear time-varying channels using type-2 fuzzy adaptive filters. *IEEE Transactions on Fuzzy Systems*, *8*, 551–563. doi:10.1109/91.873578

Liang, Q., & Mendel, J. M. (2000). Interval type-2 logic systems: Theory and design. *IEEE Transactions on Fuzzy Systems*, *8*, 535–550. doi:10.1109/91.873577

Lin, T. C. (2010a). Analog circuits fault diagnosis under parameter variations based on type-2 fuzzy logic systems. *International Journal of Innovative Computing. Information and Control*, *6*(5), 2137–2158.

Lin, T. C. (2010b). Observer-based robust adaptive interval type-2 fuzzy tracking control of multivariable nonlinear systems. *Engineering Applications of Artificial Intelligence*, *23*(3), 386–399. doi:10.1016/j.engappai.2009.11.007

Lin, T. C., Kuo, M. J., & Hsu, C. H. (2010). Robust adaptive tracking control of multivariable nonlinear systems based on interval type-2 fuzzy approach. *International Journal of Innovative Computing. Information and Control*, *6*(3), 941–961.

Lin, T. C., Liu, H. L., & Kuo, M. J. (2009). Direct adaptive interval type-2 fuzzy control of multivariable nonlinear systems. *Engineering Applications of Artificial Intelligence*, *22*, 420–430. doi:10.1016/j.engappai.2008.10.024

Ma, X. J., & Sun, Z. Q. (2000). Output tracking and regulation of nonlinear system based on Takagi-Sugeno fuzzy model. *IEEE Transactions on Systems, Man, and Cybernetics. Part B, Cybernetics*, *30*, 47–59. doi:10.1109/3477.826946

Martinez, R., Castillo, O., & Aguilar, L. T. (2009). Optimization of interval type-2 fuzzy logic controllers for a perturbed autonomous wheeled mobile robot using genetic algorithm. *Information Sciences*, *179*, 2157–2174. doi:10.1016/j.ins.2008.12.028

Mendel, J. M. (2004). Computing derivatives in interval type-2 fuzzy logic systems. *IEEE Transactions on Fuzzy Systems*, *12*(1), 84–98. doi:10.1109/TFUZZ.2003.822681

Narendra, K. S., & Parthasarathy, K. (1990). Identification and control of dynamical systems using neural networks. *IEEE Transactions on Neural Networks*, *1*, 4–27. doi:10.1109/72.80202

Rovithakis, G. A., & Christodoulou, M. A. (1994). Adaptive control of unknown plants using dynamical neural networks. *IEEE Transactions on Systems, Man, and Cybernetics. Part B, Cybernetics*, *24*, 400–412.

Sepulveda, R., Castillo, O., Melin, P., Rodriguez-Diaz, A., & Montiel, O. (2009). Experimental study of intelligent controller under uncertainty using type-1 and type-2 fuzzy logic. *Information Sciences*, *177*, 2023–2048. doi:10.1016/j.ins.2006.10.004

Spooner, J. T., & Passino, K. M. (1996). Stable adaptive control using fuzzy systems and neural networks. *IEEE Transactions on Fuzzy Systems*, *4*, 339–359. doi:10.1109/91.531775

Wang, C. H., Cheng, C., & Lee, T. (2004). Dynamical optimal training for interval type-2 fuzzy neural network (T2FNN). *IEEE Transactions on Systems, Man, and Cybernetics. Part B, Cybernetics*, *34*(3), 1472–1477. doi:10.1109/TSMCB.2004.825927

Wang, C. H., Liu, H. L., & Lin, C. T. (2001). Dynamic learning rate optimalization of the back propagation algorithm. *IEEE Transactions on Systems, Man, and Cybernetics. Part B, Cybernetics*, *31*, 669–677.

Wang, L. X. (1993). Stable adaptive fuzzy control of nonlinear systems. *IEEE Transactions on Fuzzy Systems*, *1*, 146–155. doi:10.1109/91.227383

Wang, L. X. (1994). *Adaptive fuzzy systems and control: Design and stability analysis*. Upper Saddle River, NJ: Prentice-Hall.

Wu, K. C. (1996). Fuzzy interval control of mobile robots. *Computers & Electrical Engineering*, *22*(3), 211–229. doi:10.1016/0045-7906(95)00038-0

Yager, R. R. (1980). Fuzzy subsets of type II in decisions. *Journal of Cybernetics*, *10*, 137–159. doi:10.1080/01969728008927629

This work was previously published in the International Journal of Fuzzy System Applications, Volume 1, Issue 3, edited by Toly Chen, pp. 66-85, copyright 2010 by IGI Publishing (an imprint of IGI Global).

Chapter 14
Integrated Circuit Emission Model Extraction with a Fuzzy Logic System

Tsung-Chih Lin
Feng Chia University, Taiwan

Ming-Jen Kuo
Feng Chia University, Taiwan

Alexandre Boyer
University of Toulouse, France

ABSTRACT

This paper describes a novel technique for multiple parameter extraction of the S12X TEM cell model using a fuzzy logic system (FLS). The FLS is utilized to capture the circuit information and to extract the circuit parameters based on experiential knowledge. The proposed extraction technique uses both linguistic information (i.e., human-like knowledge and experience) and numerical data of measurement to construct the fuzzy macromodel. The simulation results confirm the validity and estimation performance of the equivalent circuit by the advocated design methodology.

INTRODUCTION

Nowadays, the electromagnetic compatibility (EMC) analysis of integrated circuit (IC) design has become an important issue due to the rapid increasing of electromagnetic interference (EMI) and the tendency of adopting tremendous technologies such as higher operating frequency, higher dissipation and lower supply voltage (Bendhiaet al., 2005). On the other hand, the progress in the macromodeling technology (Leontaris & Billings, 1987) extends its applications to provide representations of various kinds of ICs under measurement. It has also been integrated with more intelligent algorithms, such as fuzzy neural networks (FNNs), vector fitting (VF) and so on (Mutnury, 2005). However, the primary purpose of macromodeling is to search for a mathematical model that approximately describes input-output mapping in terms of equations. Furthermore, both

DOI: 10.4018/978-1-4666-1870-1.ch014

IC vendors and designers need to extract the accurate parameters of the models.

According to its inventor, Zadeh, fuzzy logic systems (FLSs) were particularly developed as a methodology for numerical processing and other inaccurate models which have no other ways to solve this difficulty (Zadeh, 1965; Lin et al. 2007a, 2007b). Undoubtedly, fuzzy logic systems have been applied successfully in many engineering territories. Applications, such as control engines, fault tolerance, parallelism, have exhibited that fuzzy systems can produce better performance than conventional techniques. In the field of circuit design, the major benefit of an FLS is that it provides mathematical strength to the emulation of certain circuit behavior which can be substituted by simple "IF-THEN" relations instead of complicated descriptions. Then, the circuit model can infer measurement information properly by using linguistic attributes associated with human-like cognition and adoptive capacity.

According to Lin et al. (2008), an FLS was able to model high frequency effects of parasitic elements of a circuit by extracting values of symmetric parasitic capacitors. For the consideration of multiple components, organizing fuzzy sets, deciding the shape of membership functions (MFs) are required to improve the fitting performance due to the correlation among each element of models. The fuzzy set theory provides interconnection between logic and intuition by acquiring the quantitative and appropriate experiential knowledge. It forms a series of reasonable regulations or connections by means of a fuzzy inference engine (FIE) that combine fuzzy IF-THEN rules into a mapping from fuzzy sets. After aggregating the results from these fuzzy IF-THEN rules, the fuzzy defuzzifier converts the aggregation result into a crisp quantity for further processing. A suitable modeling methodology is required to increase the information of measurement through the modeler's judgment. To this purpose, we propose an effective approach with software assistance, which has been developed to help building an integrated circuit emission model (ICEM) (IC-EMC, 2006), by using SPICE transient simulation.

The main objective of this paper is to demonstrate the extraction capability of a single model to predict the electromagnetic emission of a digital circuit by using a FLS. The remainder of this paper is organized as follows. First we describe the model of the test chip. Next we present the framework of the FLS. Then the extracting procedure of modeling circuits with the FLS is proposed. Experimental results of the extracting procedure are given in and we conclude the findings of this study.

MODELING OF THE CIRCUIT UNDER TEST

This paper works on the model of a commercial component, which is named S12X 16-bit microcontroller and has been widely embedded in automotive electronic systems. Various conducted and radiated emission measurements were performed and a single ICEM model was proposed to predict the emission spectrum level (Labussiere et al., 2008). Radiated measurements were done in a gigahertz transverse electromagnetic (GTEM) cell designed to extend the frequency range of measurement (typically up to 18 GHz) higher than a TEM cell in the frequency domain due to its match termination and tapered structure (IC-EMC, 2002). The protocol for IC emission measurement in the GTEM cell is shown in Figure 1. For each program running in the microcontroller and the different orientations of the test board in the cell, we select the maximum radiated spectrum as the desired data. Frequency of interest is limited to be up to 2GHz.

From these measurements and the basic physical information about the circuit (technology, power supply pin number, size of the die and so on), the SPICE model shown in Figure 2 was proposed to predict the emission level of the test chip (Labussiere et al., 2008). The model of S12X

Figure 1. GTEM cell measurement setup (IC-EMC, 2002)

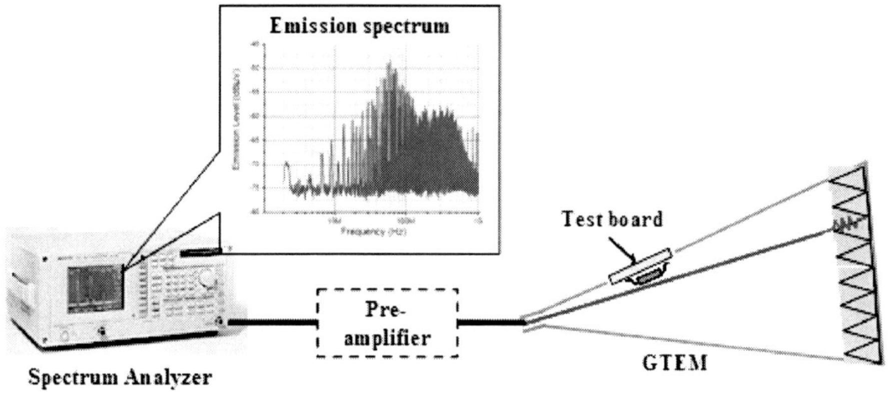

for GTEM cell measurements consists of five sub-models which are established as follows:

1. **Core model**: According to the estimation of the current source based on logic cell description and the information of passive distribution network (PDN), a symmetric ICEM model is built by using analytic formulas based on the physical dimensions of the associated signal tracks.
2. **Regulator model**: A passive equivalent model for the on-chip regulator is proposed in order to decrease the external supply voltage of the core model.
3. **Supply model**: A simplified passive model for the power supply network is proposed from power supply pin placement and connections between power supply and ground pins. Power supply model was confirmed by S parameter measurements.

Figure 2. Model of S12X for GTEM cell measurements (Labussiere et al., 2008)

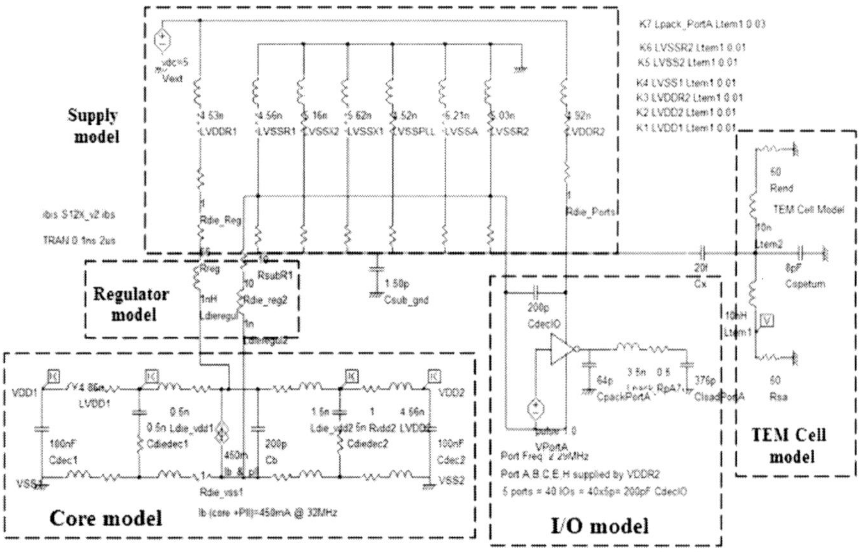

4. **Input/Output (I/O) model**: I/O model is built from IBIS file data.
5. **TEM Cell model**: TEM Cell model is constructed based on the observation of magnetic and electric coupling effects between IC and TEM cell mechanism. The simulated output is defined here for representing the connection to the spectrum analyzer through an optional pre- amplifier.

From the simulation with IC-EMC software (Sicard et al., 2009), Figure 3 presents the comparison between the radiated emission simulation and measurement of the GTEM cell. The spectrum of the predicted model only fits accurately within the frequency range from 10MHz to 1GHz. The parameters of this model may be redesigned to fit the measurements more precisely.

As we described before, an FLS provides an effective and systematic architecture to integrate human experience and empirical knowledge by means of capturing the approximated and inexact information of the measurements. In addition, it is useful to deal with a complex analysis without any complicated mathematical descriptions. Therefore, the process information of modeling will be transformed into associated fuzzy conditions for describing the relation between practical measurement and behavioral model. For this reason, we will optimize the current value of the multiple components within the model by using an FLS. The brief description of the FLS is proposed in the next section.

DESIGN OF THE FUZZY LOGIC SYSTEM

In this section, we propose an FLS, which consists of the dependent linguistic knowledge of system variables (for instance, low, middle, high), to extract multiple output variables. The configuration of the FLS is described in Figure 4. It includes multiple fuzzy rule-based systems which are built from a collection of linguistic information, and it performs a four-part nonlinear mapping from inputs (stimuli) to outputs (responses).

First, the purpose of the fuzzifier is to map the set of sensor (such as the amplitude of the harmonics of the spectrum) to values ranging from 0 to 1, which is denoted as fuzzy sets, by using a set of MFs. Second, the fuzzy rule-based system uses

Figure 3. Comparison between measurement and simulation

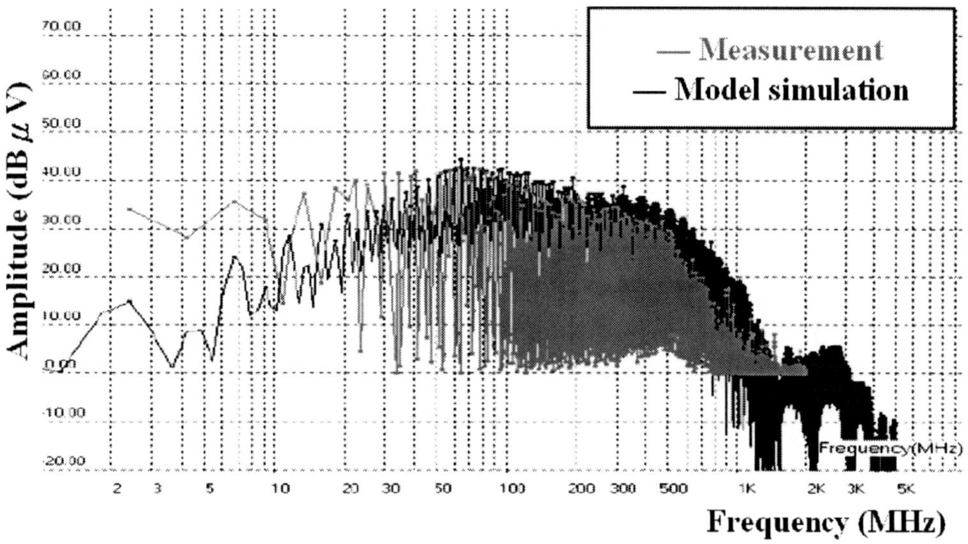

Figure 4. The configuration of the FLS

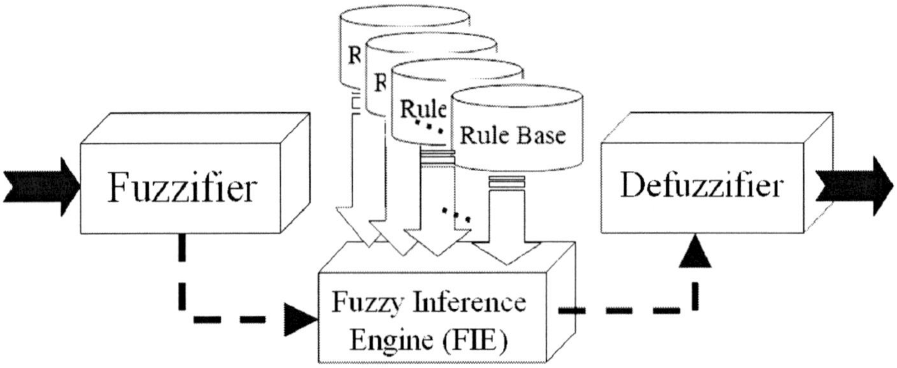

"IF-THEN" description of the *l*th rule as shown in equation (1):

Rule *l*:

IF x_1 is F_1^l, and $\cdots$, and x_n is F_n^l,

THEN

$$y^l = q_0^l + q_1^l x_1 + \cdots + q_n^l x_n = \underline{\theta}_l^T \cdot [1\ \underline{x}^T]^T \tag{1}$$

where F_i^l are fuzzy sets and $\underline{\theta}_l^T = [q_0^l, q_1^l, \cdots, q_n^l]$ is a vector of the adjustable factors of the consequence part of the fuzzy rule. y^l is the output linguistic variable, and a fuzzy inference engine is used to combine the fuzzy IF-THEN rules (Takagi & Sugeno, 1985) of the fuzzy rule base into a mapping from an input linguistic vector $\underline{x}^T = [x_1, x_2, \ldots, x_n] \in R^n$ to an output variable $y \in R$. Let l=1, 2, …, L be the number of fuzzy IF-THEN rules. For instance, if three levels are given as; low, middle, high, then L is equal to three.

As a result, each component has been considered as L rules. In order to verify the capability of the FLS, we try to give identical conditions for multiple components (i.e., M components), so that the total number of the fuzzy rules will be L^M. Third, the fuzzy inference engine (FIE) formulates suitable connections between input data and rule-based information based on the concepts of fuzzy set theory. Finally, it is desired to come up with the set of crisp outputs by using the central average defuzzifier, which takes the output distribution and finds its average (Abdulghafour, 1994). The output of the fuzzy logic system with the central average defuzzifier, product inference and singleton fuzzifier can be expressed as

$$y(\underline{x}) = \frac{\sum_{l=1}^{M} \upsilon^l \cdot y_l}{\sum_{l=1}^{M} \upsilon^l} = \frac{\sum_{l=1}^{M} \upsilon^l \cdot \underline{\theta}_l^T [1\ \underline{x}^T]}{\sum_{l=1}^{M} \upsilon^l} \tag{2}$$

where $\mu_{F_i^l}(x_i)$ is the MF value of the fuzzy variable x_i and $\upsilon^l = \prod_{i=1}^{n} \mu_{F_i^l(x_i)}$ is the true value of the lth implication. Equation (2) can be rewritten as

$$y(\underline{x}) = \underline{\theta}^T \underline{\psi}(\underline{x}) \tag{3}$$

where $\underline{\theta}^T = \left[\underline{\theta}_1^T\ \ \underline{\theta}_2^T\ \ \cdots\ \ \underline{\theta}_M^T\right]$ is an adjustable parameter vector and $\underline{\psi}^T(\underline{x}) = [\psi^1(\underline{x}), \psi^2(\underline{x}), \ldots, \psi^M(\underline{x})]$ is a fuzzy basis function (Wang, 1994); vector $\psi^l(x)$ is defined as

$$\psi^l(\underline{x}) = \frac{\upsilon^l [1\ \underline{x}^T]}{\sum_{l=1}^{M} \upsilon^l} \tag{4}$$

Due to the mapping between crisp values and fuzzy operations, an advanced methodology based on the FLS was successfully adopted to solve the extraction of multiple components. The FLS applied in this section belongs to TS fuzzy systems. There are other kinds of fuzzy inference systems like Mamdani and Assilina's (1975) system.

THE BRIEF DESCRIPTION OF CIRCUIT EXTRACTION

In general, the process of circuit extraction consists in obtaining the information that includes the adequate fuzzy input variables of the FLS, the number of rule-based inferences and the estimated parameters of the model under test (MUT). An FLS, developed to imitate human inference in its capability to make accurate judgments from obscure and uncertain information (Jang et al., 1997), can offer convincing performance in the determination of parameters which consist of their conditional influence on the behavior of model (Leontaris & Billings, 1987; Scott & Ray, 1993; Yager & Filev, 1994; Sjoberg et al., 1995; Simoes et al., 1997). Therefore, to summarize the above analysis, the extraction algorithm for the critical parameters, such as bypass capacitors, decoupling capacitors and parasitic capacitors, is proposed as follows:

- **Fuzzy Variables**: According to the responses (spectrums) of the MUT with the adjustments, we define adequate sets as the associated fuzzy variables between inputs and outputs of the FLS to determine the best testing frequency where the maximum difference among those waveforms is observed.
- **MFs**: In order to extract reasonable values from the FLS, the suitable MFs must be constructed with respect to the fuzzy sets of the FLS. By observing the variation of testing spectrums, we define three different levels of MFs (Low, Mid, High) for each component to be extracted as shown in Figure 5. They are Gaussian-type fuzzy numbers.
- **Logical Rule Base**: Because of the variety of multiple components, we apply logical rule-based conditions which define causal phenomena, and make proper connections for the FIE.
- **Defuzzification Output**: The process of mapping from fuzzy sets to the crisp output (the values of the components to be extracted) is to be performed via the central average defuzzifier (2). The are other defuzzification approaches such as first of maxima (FOM), middle of maxima (MOM), mean of support (MOS), basic defuzzification distribution (BADD), height defuzzification, etc.

An FLS has been validated to be an accurate estimator for a class of nonlinear systems. For this reason, an advanced methodology, which relies on the extraction capability of multiple components with the logical rule base, is described above. We will give an experimental example in the next section.

EXPERIMENTAL EXAMPLE

In order to confirm the validity and performance of the advocated methodology, the experimental example of model S12X is given in the following. By adjusting each component constantly, we focus on the critical parameters that offer the most influence on the ICEM. The values of indicated components which will be estimated as shown in Figure 2 are given as:

$C_b = 200pF$, $C_{diedec1} = C_{diedec2} = 2nF$, $C_{decIO} = 200pF$,

Figure 5. Three different levels of MFs (Low, Mid, High)

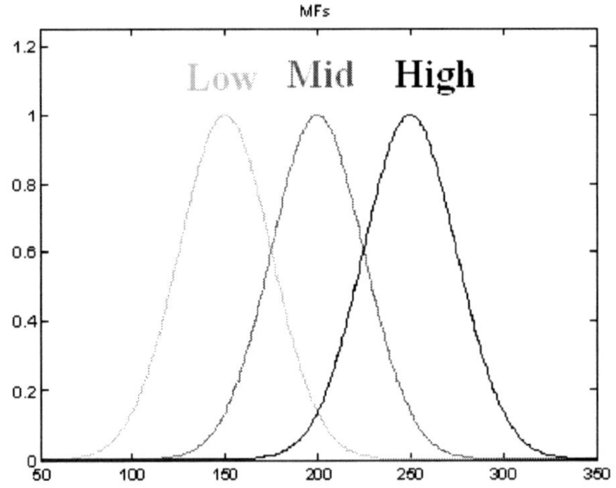

$C_x = 20fF$.

It is assumed that $C_{diedec1}$ and $C_{diedec2}$ change equally because of the symmetric structure of the ICEM model. The flowchart of extracting multiple components by using the FLS is described in Figure 6.

- **Define MFs by circuit information:** As we observe the response by changing the value of each component, we indicate the suitable circuit information of each component for establishing the fuzzy membership functions that are recorded in Table 1. Bounds given to the values of these different elements are set based on the modeler's experience about the circuit under test. Be sure that the levels of MFs are defined by the amplitude of the spectrum, not the value of the component (*Cdiedec1* and *Cdiedec2*).
- **Construct rule-based systems:** In order to represent the variety of the output response, all the circuit information are connected as shown in Figure 7. Meanwhile, we also establish a rule-based system which is a collection of fuzzy linguistic information, and it performs nonlinear mapping between circuit information and the output response. Furthermore, the reasonable maximum and minimum of the rule-based system is defined which will produce the highest and the lowest boundaries of the output responses. The total rule num-

Figure 6. The flowchart of extracting multiple components by using FLS

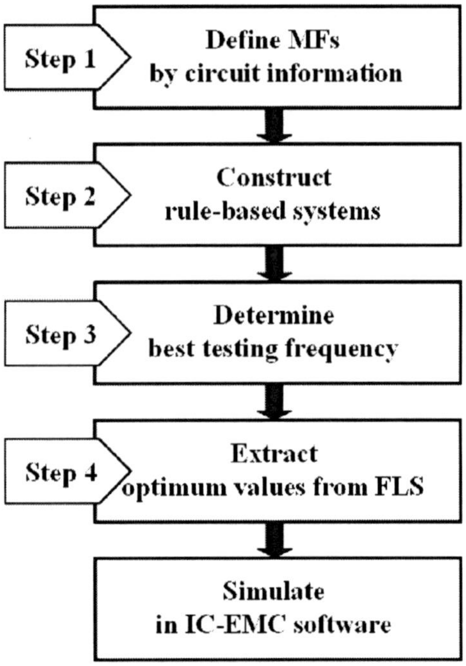

bers of rules in the rule base is equal to $3\times3\times3\times3=3^4=81$.

- **Determine the best testing frequency:** The testing frequencies are determined from the maximum differences between the highest and lowest boundaries of the output responses (Liu, 1991; Catelani and Giraldi, 1998; Dai and Xu, 1999; Li, 2004). In Figure 8, the best testing frequency can be determined as 7.99e8Hz and the measured responses are located inside the defined region.
- **Extract optimum values from the FLS:** After defining the associated fuzzy sets and rule bases, the outputs of the FLS can be derived from the FIE that combines fuzzy rules into a mapping from fuzzy sets. According to the central average method, we indicate the information of each component as adjustable parameter vectors in Equation (2) and the crisp outputs can be defuzzified from Equation (3). The extracted results of the FLS are listed in Table 2.

Table 1. Circuit information of each component used as fuzzy output variables

Var. MFtypes	Cb	Cdidee 1 & 2	CdecIO	Cx
High	250p	1.5n	350p	35f
Mid	200p	2n	300p	30f
Low	150p	2.5n	250p	25f

Figure 7. The constructed rule-based system

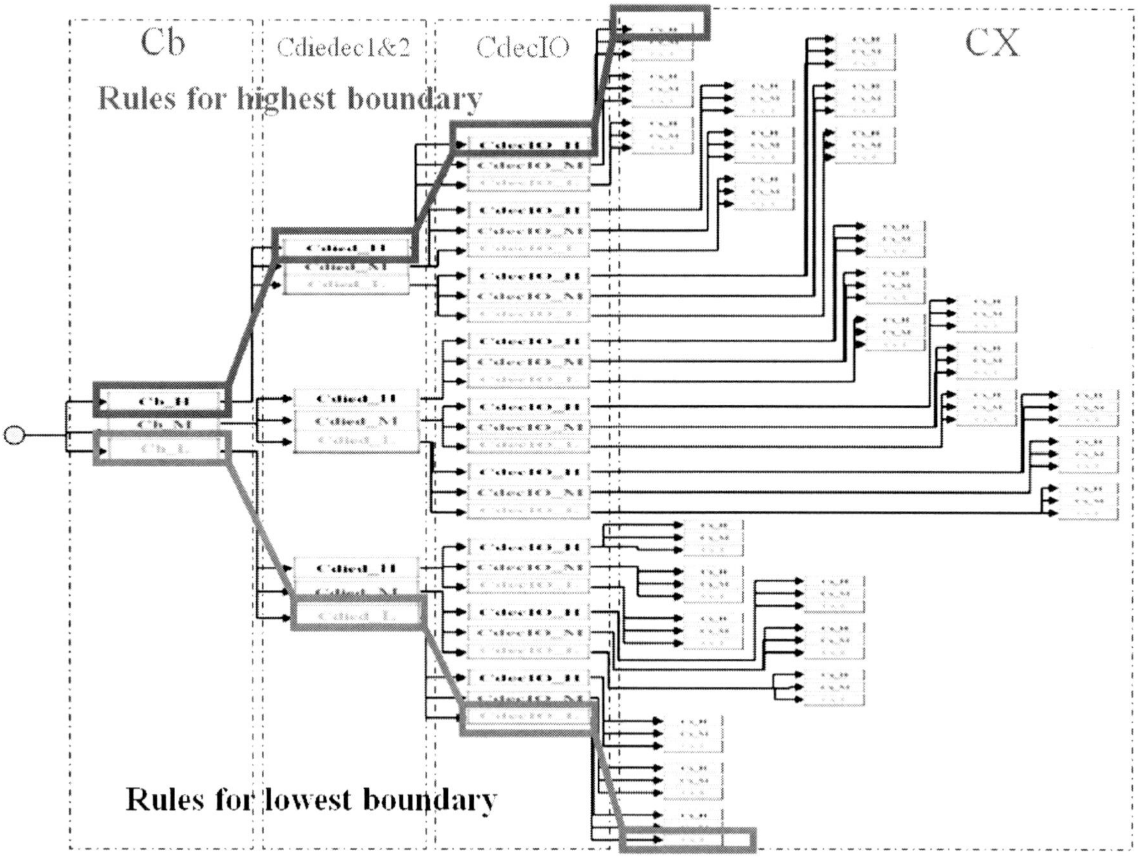

Table 2. Results of FLS

Components	Cb	Cdiedec 1 & 2	CdecIO	Cx
Fuzzy results	231.3762 *pF*	1.8932 *nF*	85.4381 *pF*	46.6429 *fF*

After extracting multiple parameters based on the proposed FLS model, spectrum envelopes are compared with our previous work (Lin et al., 2008) as shown in Figures 9 and 10, respectively. The estimation performance is given in Table 3. According to the deviation between simulation results and measurements, the FLS method improves the fitting performance significantly. For example, the deviation is reduced from 4.8892 dBμV to 2.8829 dBμV. However, it takes almost two hours to extract the optimum values by using the FLS.

CONCLUSION

This paper has proposed a novel and advanced approach for extracting multiple parameters based on the extending fuzzy rule base of an FLS. It evaluates the overall variation of the output response in the circuit and automatically extracts suitable values for the critical parameters of the ICEM. A case study is used in the paper to confirm the extraordinary modeling technique, and the simulation outcome validates the better accuracy for fitting the measurement waveform. However,

Figure 8. The best testing frequency

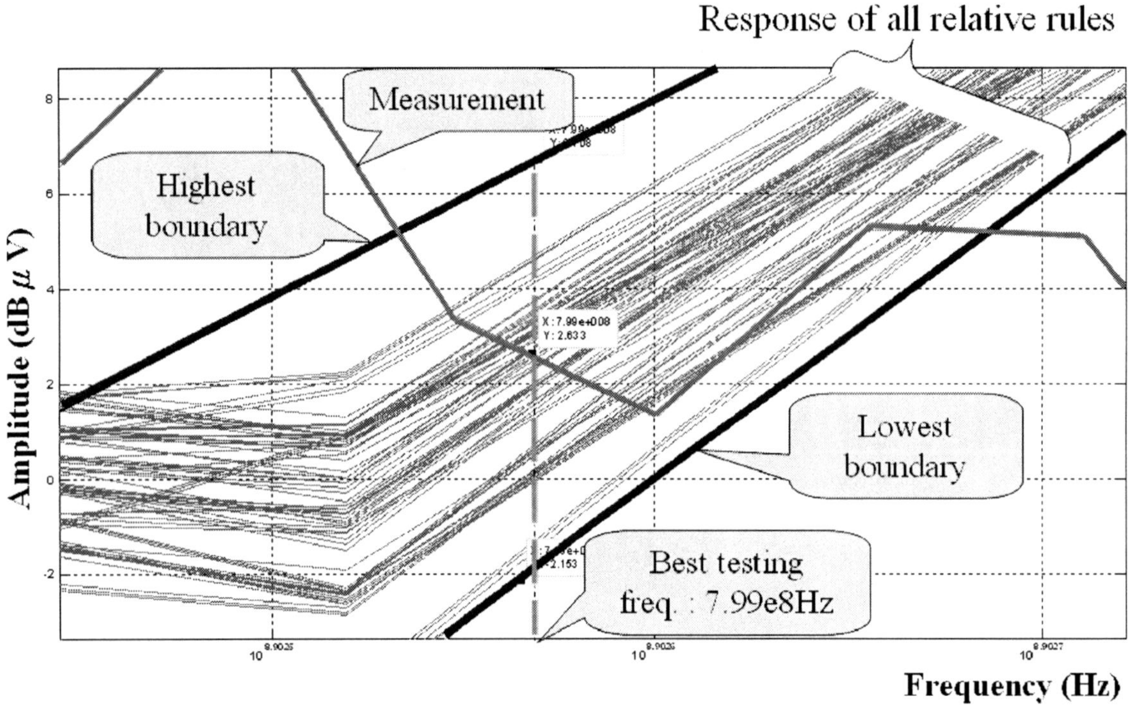

Figure 9. Envelop of Lin et al.'s model (Lin et al., 2008)

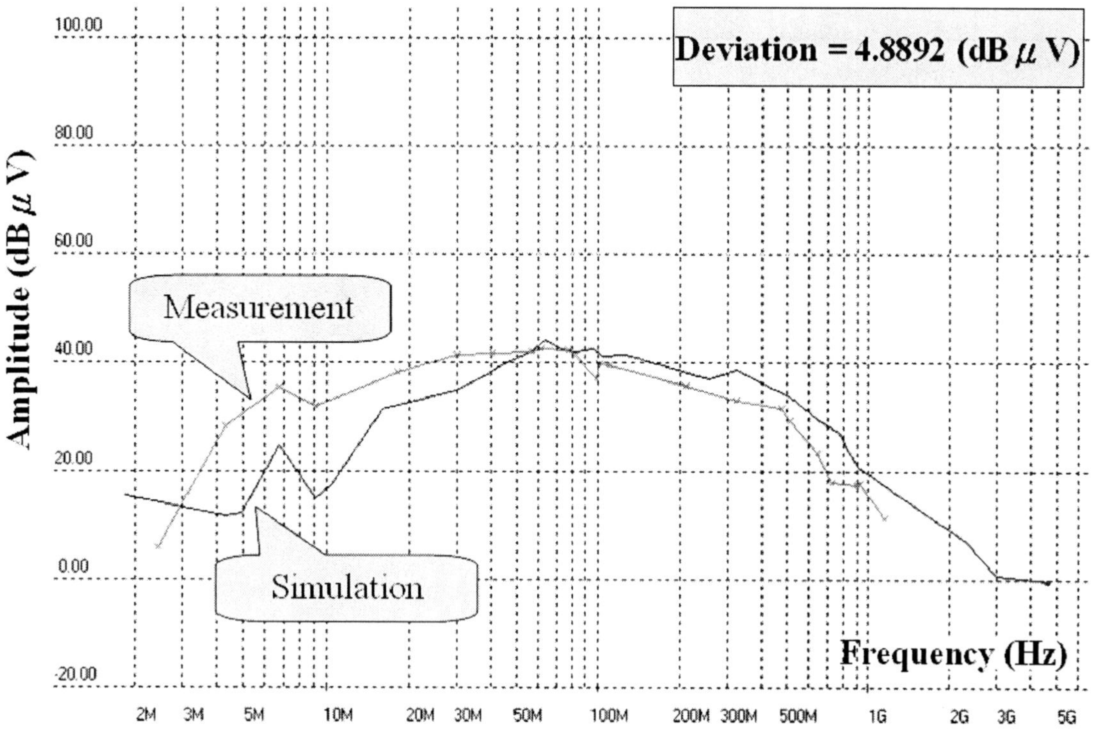

Figure 10. Envelop of the proposed FLS model

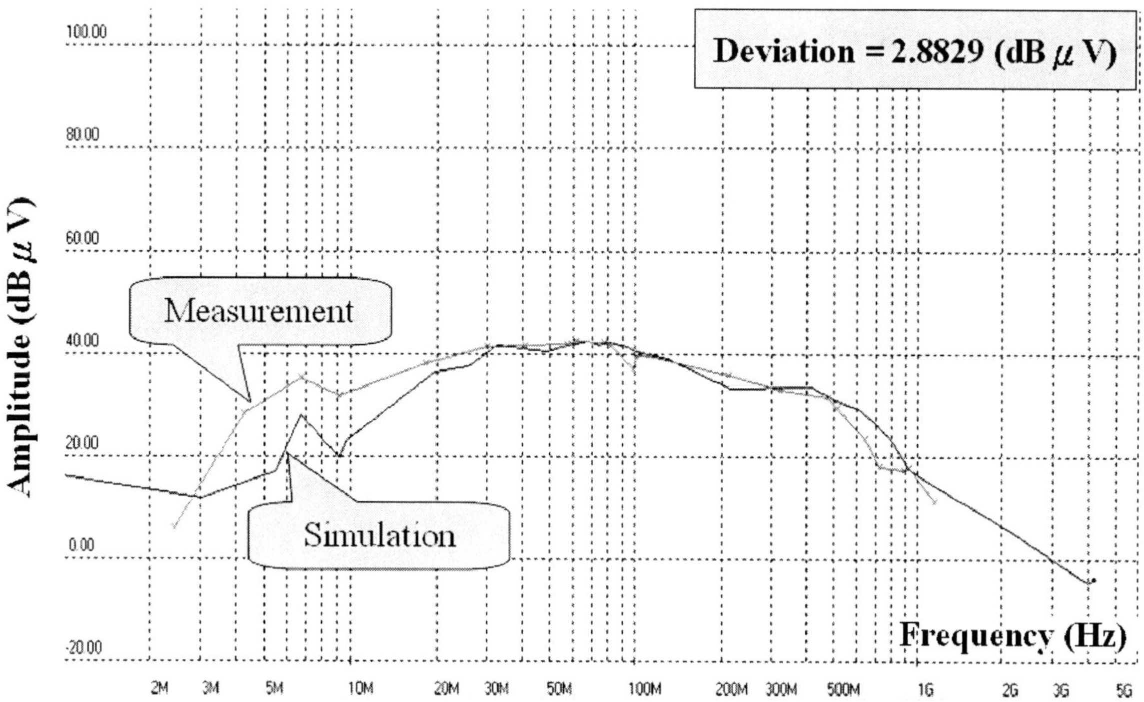

Table 3. Comparison the performances of two models

Model names	Deviation (dB μ V)
Previous model	4.8892
Fuzzy model	2.8829

a type-1 FLS cannot fully handle linguistic and numerical uncertainties associated with dynamic unstructured environments. In order to overcome the limitation of a type-1 FLS, in the future research a type-2 FLS can be used to establish a more precise integrated circuit emission model.

REFERENCES

Abdulghafour, M., Fellah, A., & Abidi, M. A. (1994). Fuzzy logic-based data integration: theory and application. In. *Proceedings of the IEEE International Conference in Multisensor Fusion and Integration for Intelligent Systems, 1*, 151–160.

Bendhia, S., Ramdani, M., & Sicard, E. (2005). *EMC of ICs: Techniques for low emission and susceptibility*. New York, NY: Springer.

Catelani, M., & Giraldi, S. (1998). Fault diagnosis of analog circuits with model-based techniques. In. *Proceedings of the IEEE Conference on Instrumentation and Measurement Technology, 1*, 501–504.

Dai, Y., & Xu, J. (1999). Analog circuit fault diagnosis based on noise measurement. *Microelectronics and Reliability, 39*(8), 1293–1298. doi:10.1016/S0026-2714(99)00029-3

IC-EMC. (2002). *61967-2: Methods of radiated emission – TEM cell method and wideband TEM cell method 150 KHz to 8 GHz*. Geneva, Switzerland: International Electrotechnical Commission.

IC-EMC. (2006). *62433-2: Models of integrated circuits for EMI behavioral simulation – ICEM-CE. ICEM conducted emission model*. Geneva, Switzerland: International Electrotechnical Commission.

Jang, J. S., Sun, C. T., & Mizutani, E. (1997). *Neuro-fuzzy and soft computing: A computational approach to learning and machine intelligence*. Upper Saddle River, NJ: Prentice-Hall.

Labussiere-Dorgan, C., Bendhia, S., Sicard, E., Junwu, T., Quaresma, H. J., & Lochot, C. (2008). Modeling the electromagnetic emission of a microcontroller using a single model. *IEEE Transactions on Electromagnetic Compatibility, 50*, 22–34. doi:10.1109/TEMC.2007.911918

Leontaris, I. J., & Billings, S. A. (1987). Model selection and validation methods for nonlinear systems. *International Journal of Control, 45*, 311–341. doi:10.1080/00207178708933730

Li, M. (2004). *Test frequency selection for analog circuits based on bode diagrams and equivalent fault grouping*. Unpublished master's thesis, University of Cincinnati, OH.

Lin, T. C., Chang, Y. M., & Kuo, M. J. (2008). Circuit extraction for ICEM model based on fuzzy logic system. In *Proceedings of the EMC Technology and Practice Symposium*.

Lin, T. C., Chen, Y. C., & Kuo, M. J. (2007). Analog circuits fault diagnosis under parameter variations based on fuzzy logic system. In *Proceedings of the VLSI Design/CAD Symposium*, Hualien, Taiwan.

Lin, T. C., Kuo, M. J., & Chen, Y. C. (2007). Frequency domain analog circuit fault diagnosis based on radial basis function neural network. In *Proceedings of the VLSI Design/CAD Symposium*, Hualien, Taiwan.

Liu, R.-W. (1991). *Testing and diagnosis of analog circuits and systems*. New York, NY: Van Nostrand Reinhold.

Mamdani, E. H., & Assilina, S. (1975). An experiment in linguistic synthesis with a fuzzy logic controller. *International Journal of Man-Machine Studies*, *7*(1), 1–13. doi:10.1016/S0020-7373(75)80002-2

Mutnury, B. (2005). *Macromodeling of nonlinear driver and receiver circuits*. Unpublished doctoral dissertation, Georgia Institute of Technology, Atlanta, GA.

Scott, G. M., & Ray, W. H. (1993). Creating efficient nonlinear network process models that allow model interpretation. *Journal of Process Control*, *3*(3), 163–178. doi:10.1016/0959-1524(93)80022-4

Sicard, E., & Boyer, A. (2009). *IC-EMC – user's manual*. Retrieved from http://www.ic-emc.org

Simoes, M. G., Bose, B. K., & Spiegel, R. J. (1997). Fuzzy logic based intelligent control of a variable speed cage machine wind generation system. *IEEE Transactions on Power Electronics*, *12*(1), 87–95. doi:10.1109/63.554173

Sjoberg, J., Zhang, Q., Ljung, L., Benveniste, A., Delyon, B., & Glorennec, P. (1995). Nonlinear black-box modeling in system identification: a unified overview. *Automatica*, *31*, 1691–1724. doi:10.1016/0005-1098(95)00120-8

Takagi, T., & Sugeno, M. (1985). Fuzzy identification of systems and its applications to modeling and control. *IEEE Transactions on Systems, Man, and Cybernetics*, *15*, 116–132.

Wang, L. X. (1994). *Adaptive fuzzy systems and control: Design and stability analysis*. Upper Saddle River, NJ: Prentice-Hall.

Yager, R. R., & Filev, D. P. (1994). *Essentials of fuzzy modeling and control*. New York, NY: John Wiley & Sons.

Zadeh, L. A. (1965). Fuzzy sets. *Information and Control*, *8*, 338–353. doi:10.1016/S0019-9958(65)90241-X

This work was previously published in the International Journal of Fuzzy System Applications, Volume 1, Issue 2, edited by Toly Chen, pp. 17-28, copyright 2010 by IGI Publishing (an imprint of IGI Global).

Chapter 15
Optimizing the Performance of Plastic Injection Molding Using Weighted Additive Model in Goal Programming

Abbas Al-Refaie
University of Jordan, Jordan

Ming-Hsien Li
Feng Chia University, Taiwan

ABSTRACT

Injection molding process is increasingly more significant in today's plastic production industries because it provides high-quality product, short product cycles, and light weight. This research optimizes the performance of this process with three main quality responses: defect count, cycle time, and spoon weight, using the weighted additive goal programming model. The three quality responses and process factors are described by appropriate membership functions. The Taguchi's orthogonal array is then utilized to provide experimental layout. A linear optimization based on the weighted additive model in goal programming model is built to minimize the deviations of the product/process targets from their corresponding imprecise fuzzy values specified by the process engineer's preferences. The results show that the average defect count is reduced from an average of 0.75 to 0.16. Moreover, the average cycle time becomes 13.06 seconds, which is significantly smaller than that obtained at initial factor settings (= 15.10 seconds). Finally, the average spoon weight is exactly on its target value of 2.0 gm.

INTRODUCTION

Plastic injection molding process (Yang & Gao, 2006; Oktem et al., 2007; Huang & Lin, 2008) is one of the most important polymer processing operations in today's plastic production industries for providing short product cycles, high-quality part surfaces, good mechanical properties, low costs, and light weight.

In practice, the plastic injection molding process consists of four phases, including plastication, injection, packing, and cooling. Recently, investigating the factors affect the performance of this process has gained significant research attention.

DOI: 10.4018/978-1-4666-1870-1.ch015

For example, Ong and Koh (2005) investigated the effects of the mold temperature, injection pressure, injection rate, and injection time on the weights of the plastic parts in the micro injection molding process. Song et al. (2007) studied the effects of injection rate, injection pressure, melt temperature, metering size, and part thickness on ultra-thin wall plastic parts in the injection molding process using Taguchi method. Tang et al. (2007) investigated the effects of the melt temperature, filling time, packing pressure, and packing time on reducing warpage in the plastic injection molding process using Taguchi method. Deng et al. (2008) studied the effects of injection time, velocity pressure switch, packing pressure, and injection velocity on product's weight in the injection process by integrating Taguchi method, regression analysis, and the Davidon-Fletcher-Powell method. Chen et al. (2008) employed self-organizing map with back-propagation neural network and Taguchi method to investigate the effects of injection time, packing pressure, injection velocity, and packing time on part's weight. Altan (2010) reduced shrinkage in injection moldings using a hybrid Taguchi and neural network approach by optimizing the weight of the produced part.

However, the above-mentioned research only considered optimizing a single quality response. Recently, optimizing multiple quality responses of the injection molding process has received significant research attention. Among them, Chiang (2007) integrated Taguchi method, back-propagation neural network, genetic algorithm and engineering optimization concepts to optimize the multiple-input and multiple-output plastic injection molding process. Five process factors were investigated including melt temperature, mold temperature, injection velocity, injection pressure, and velocity pressure switch. Three quality responses were studied, including the strength of welding line, shrinkage, and the differences in distributive temperature. Kamoun et al. (2009) used the simplex method to the on-line optimization of the parameters of an injection molding process. Eight process factors were studied including back pressure, injection rate, injection pressure, cooling time, holding pressure, holding time, nozzle temperature, and opening stroke. The quality responses were cycle time and the percentage of defective.

Typically, determining precise targets for multiple quality responses is often a difficult task for a process/product engineer in the real world. The conventional goal programming (GP) models usually consider the response goals as precise and deterministic. Therefore, several formulations of GP models were introduced for solving the fuzzy GP (FGP) problems taking into account the decision maker's preferences. An effective approach in dealing with FGP is the weighted additive model which takes into account all shapes of membership functions (Yaghoobi et al., 2008). Therefore, this research utilizes the weighted additive model to optimize the performance of an injection molding process for plastic spoons with three quality responses, including defect count, process cycle time and spoon weight. Each quality response is described by a suitable membership function. Also, the process factor levels, which are assigned by process knowledge, are described by a desired operational interval, and hence they are represented by a trapezoid membership function. An optimization model is then formulated to minimize the weighted deviations from the imprecise fuzzy values for all quality responses and process factors. Finally, the optimal factor settings are obtained then the anticipated and confirmation improvements are calculated.

INJECTION MOLDING

Identifying Main Quality Responses

The main idea of the injection molding process is based on forcing molten plastic resins through a nozzle under pressure into a mold having two basic parts. Plastic material fills the space in between these two major parts called the mold cavity-core chamber. Figure 1 shows the basic components of the injection molding process. Plastic powder is molten though the use of some proper heating technique applied to a barrel containing a screw that is free to rotate inside the barrel. Injection process takes place under proper pressure and velocity to force the molten plastic material, which are important to the quality of the injected product. That is, pressure and velocity besides the volume of shot material all together play a basic role in the filling of the material injected into the cavity-core chamber of the mould. The injection molding process used in producing the plastic products is equipped with a computerized controller capable of setting all the controllable process factors easily.

In this research, one of the highly demanded plastic products, which is the medical dosing spoon with an annual demand of at least six million spoons is considered. The spoon is usually manufactured from a whitened plastic material. Its major content is Polypropylene which is semi-transparent while the minor content is a white coloring additive. In general, the spoon is composed of 97.5% Polypropylene and 2.5% whitening additive. Process knowledge indicated that the spoon weight, which belongs to the nominal-the-best (NTB) type response with specification ranges as 2.0±0.07 gm, is a critical quality response and can be considered a good indication of the stability of the injection molding process. Two other quality responses are of main interest, including the defect count and cycle time, which are both smaller-the-better (STB) type responses.

Deciding Key Process Factors

Based on the process knowledge of injection molding setup requirements, thirteen controllable factors of the injection molding process that might affect the defect count, cycle time and spoon weight will be investigated concurrently. These factors are denoted by $x_1, x_2, \ldots, x_{13}$ that are divided into 5 categories as follows:

1. The factors x_1, x_2, and x_3 represent the factors controlling temperature settings in the preparation stage of molten plastic. Their physical values range from 150 to 200 C°.
2. The factors x_4, x_5, x_6, and x_7 denote the control factors of injection pressure. Increasing the injection pressure might cause over-dose or flash, whereas reducing the injection causes insufficient dose. The controllable pressure range is between 1~140 bars.

Figure 1. Schematic illustration of Injection molding process

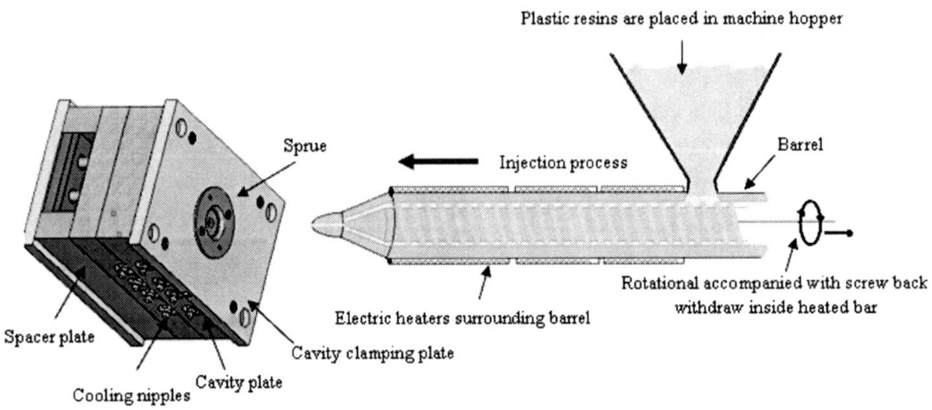

3. The factors x_8, x_9, x_{10}, and x_{11} represent injection velocities, which are controlled by means of changing the flow of the hydraulic oil passing through the actuator related to that movement. Changing the flow is performed through varying the useful flow of the pump. The flow proportional valve is computer-controlled and it ranges between 1-100% of the maximum pump discharge rate in liters per minutes. The involved velocities affect cycle time as well as part quality.
4. The factor x_{12} codes cooling time – the time period between the end of the injection time and the moment the mold starts to open. Increasing this time lengthens the cycle time and consequently reduces productivity. However, reducing this time much below the allowable limit could result in defective parts.
5. The factor x_{13} is injection time – the time period during which the injection piston moves forward forcing the molten plastic material through the nozzle of the machine barrel and via the sprue of the mold to fill the cavity within the mold. Reducing this time period might result in insufficient shots or parts partially produced and actually wasted. Whereas, increasing this time period over an optimal value lengthens the cycle time of production and hence reduces the productivity of the machine. Table 1 lists these factors with their corresponding level ranges. In order to conduct experimental work, the range of each factor level is divided into three levels: the minimum, medium, and maximum. For example, the three levels of Temperature-1 are assigned 175, 185, and 195 C° respectively. An appropriate orthogonal array to investigate thirteen three-level factors concurrently is the L_{27} (13^3) array as shown in Table 2. Let y_1, y_2 and y_3 denote the replicate value of defect count, cycle time and spoon weight, respectively. Each experiment is repeated four times. Let $\bar{y}_1$, $\bar{y}_2$ and $\bar{y}_3$ denote the corresponding average values. The values of $\bar{y}_1$, $\bar{y}_2$ and $\bar{y}_3$ are computed and shown respectively in the last three columns of Table 2.

THE PROPOSED PROCEDURE

The proposed procedure for optimizing the performance of the injection molding process is described as follows. Let g_{y1}, g_{y2}, and g_{y3} be the imprecise fuzzy values (aspiration levels) of y_1, y_2 and y_3, respectively. These values are usually specified by the process engineer's knowledge. Also, let Δ^+_{yi} and Δ^-_{yi}; where $i = 1,.., 3$, represent the chosen values of the right and the left maximum admissible violation from the imprecise fuzzy value, respectively. Then, the weighted additive model is outlined in the following steps:

Step 1: In Taguchi's method, the relationship between the controllable process factors and each response is unknown. Let x_j denotes the jth process factor; j=1,2,...,13. The relationship between these factors and the kth response, y_k, can be expressed as

$$y_k = f\left(x_1,\ x_2,\ ....,\ x_{13}\right), \qquad k=1,...,3 \tag{1}$$

Step 2: A suitable membership function is selected to represent each response as follows:

For defect count and cycle time, which are the STB type responses, the membership functions, $MF(y_1)$ and $MF(y_2)$ depicted in Figure 2, are both defined as follows:

$$MF(y_k) = \begin{cases} 1, & y_k \le g_{y_k} \\ 1 - \dfrac{y_k - g_{y_k}}{\Delta^+_{y_k}}, & g_{y_k} \le y_k \le g_{y_k} + \Delta^+_{y_k} \\ 0, & y_k \ge g_{y_k} + \Delta^+_{y_k} \end{cases} \quad \text{, where } k=1,\ 2 \tag{2}$$

Table 2. The experimental results with L_{27} array

Exp. No.	Controllable factors													defect count	cycle time	Spoon weight
	1	2	3	4	5	6	7	8	9	10	11	12	13			
	x_1	x_2	x_3	x_4	x_5	x_6	x_7	x_8	x_9	x_{10}	x_{11}	x_{12}	x_{13}	$\bar{y}_1$	$\bar{y}_2$	$\bar{y}_3$
1	1	1	1	1	1	1	1	1	1	1	1	1	1	0	13.1	1.9825
2	1	1	1	1	2	2	2	2	2	2	2	2	2	0	15.9	2.0125
3	1	1	1	1	3	3	3	3	3	3	3	3	3	2	17.5	2.12
4	1	2	2	2	1	1	1	2	2	2	3	3	3	0	17.3	2.0075
5	1	2	2	2	2	2	2	3	3	3	1	1	1	1.25	12.9	2.1125
6	1	2	2	2	3	3	3	1	1	1	2	2	2	0.75	15.1	2.07
7	1	3	3	3	1	1	1	3	3	3	2	2	2	0	16	2.01
8	1	3	3	3	2	2	2	1	1	1	3	3	3	0	17.5	2.0275
9	1	3	3	3	3	3	3	2	2	2	1	1	1	0.25	13.5	2.02
10	2	1	2	3	1	2	3	1	2	3	1	2	3	1.25	15.3	2.135
11	2	1	2	3	2	3	1	2	3	1	2	3	1	1.25	16.3	2.1
12	2	1	2	3	3	1	2	3	1	2	3	1	2	2	13.9	2.175
13	2	2	3	1	1	2	3	2	3	1	3	1	2	0.5	13.2	2.015
14	2	2	3	1	2	3	1	3	1	2	1	2	3	0	15.2	2.0275
15	2	2	3	1	3	1	2	1	2	3	2	3	1	0.75	16.6	2.015
16	2	3	1	2	1	2	3	3	1	2	2	3	1	1.75	15.5	2.08
17	2	3	1	2	2	3	1	1	2	3	3	1	2	0.25	13	2.03
18	2	3	1	2	3	1	2	2	3	1	1	2	3	0	15.2	2.03
19	3	1	3	2	1	3	2	1	3	2	1	3	2	0.25	16.9	2.0225
20	3	1	3	2	2	1	3	2	1	3	2	1	3	1	14.7	2.0725
21	3	1	3	2	3	2	1	3	2	1	3	2	1	1	14.9	2.085
22	3	2	1	3	1	3	2	2	1	3	3	2	1	0	13.9	2.02
23	3	2	1	3	2	1	3	3	2	1	1	3	2	0.5	16.1	2.09
24	3	2	1	3	3	2	1	1	3	2	2	1	3	0	13.9	2.0325
25	3	3	2	1	1	3	2	3	2	1	2	1	3	0.75	13.9	2.065
26	3	3	2	1	2	1	3	1	3	2	3	2	1	0	14.3	2.06
27	3	3	2	1	3	2	1	2	1	3	1	3	2	0	17	2.015

Generally, the larger $MF(y_k)$ is, the better is the performance. Consequently, the process engineer is interested of region I in Figure 2, in which $MF(y_k)$ is equal to one. When y_s falls within region II, $MF(y_k)$ decreases and hence the performance worsens. In this context, the following three constraints will be considered, including:

$$y_k - \delta^+_{y_k} \le g_{y_k}, \quad k=1, 2 \tag{3-a}$$

$$MF(y_k) + \frac{\delta^+_{y_k}}{\Delta^+_{y_k}} = 1, \quad k=1, 2 \tag{3-b}$$

$$0 \le \delta^+_{y_k} \le \Delta^+_{y_k}, \quad k=1, 2 \tag{3-c}$$

where $\delta^+_{y_k}$ represents the positive deviation from the fuzzy value. From Eq. (3), when $\delta^+_{y_k}$ equals zero, the first constraint ensures that y_k is in region I and hence provides $MF(y_k)$ equal to one.

For spoon wei ght which is the NTB type response, the triangular membership function, $MF(y_3)$ shown in Figure 3, is used and can be expressed as:

$$MF(y_3) = \begin{cases} 0, & y_3 \le g_{y_3} - \Delta^-_{y_3} \\ 1 - \dfrac{g_{y_3} - y_3}{\Delta^-_{y_3}}, & g_{y_3} - \Delta^-_{y_3} \le y_{y_3} \le g_{y_3} \\ 1 - \dfrac{y_3 - g_{y3}}{\Delta^+_{y_3}}, & g_{y_3} \le y_3 \le g_{y_3} + \Delta^+_{y_3} \\ 0, & y_3 \ge g_{y_3} + \Delta^+_{y_3} \end{cases} \tag{4}$$

Figure 2. The general membership functions of three responses

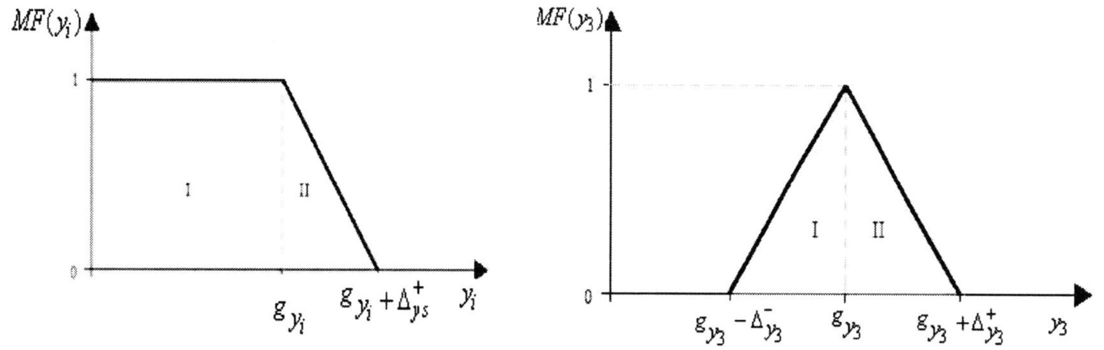

(a) Defect count and cycle time

(b) spoon weight

In Figure 2, the process engineer seeks a target value of y_3 equals exactly g_{y3} and hence provides the best performance. However, y_3 may deviate to the left or right; *i.e.*, regions I and II, depending on the negative and positive deviation variables $\delta^-_{y_3}$ and $\delta^+_{y_3}$, respectively. Thus, the following two constraints are considered:

$$y_3 + \delta^-_{y_3} - \delta^+_{y_3} = g_{y_3} \qquad \text{(5-a)}$$

$$MF(y_3) + \frac{\delta^-_{y_3}}{\Delta^-_{y_3}} + \frac{\delta^+_{y_3}}{\Delta^+_{y_3}} = 1 \qquad \text{(5-b)}$$

Notice that when $\delta^-_{y_3}$ and $\delta^+_{y_3}$ have zero values, y_3 exactly equals to g_{y3} with $MF(y_3)$ value of one, which is the desired target.

Step 3: In Taguchi's method, the controllable process factors are defined by operating discrete levels. In injection molding process, a process factor can be set at any level within a continuous interval. This region represents the preference of the process engineer. Hence, the trapezoidal membership function, $MF(x_j)$; where $j = 1,\ldots, 13$, is considered as an appropriate selection of describing the settings of a process factor. The $MF(x_j)$ is shown in Figure 4 and is defined as follows:

$$MF(x_j) = \begin{cases} 0, & x_j \le g^l_{x_j} - \Delta^-_{x_j} \\ 1 - \dfrac{g^l_{x_j} - x_j}{\Delta^-_{x_j}}, & g^l_{x_j} - \Delta^-_{x_j} \le x_j \le g^l_{x_j} \\ 1, & g^l_{x_j} \le x_j \le g^u_{x_j} \\ 1 - \dfrac{x_j - g^u_{x_j}}{\Delta^+_{x_j}}, & g^u_{x_j} \le x_j \le g^u_j + \Delta^+_{x_j} \\ 0, & x_j \ge g^u_{x_j} + \Delta^+_{x_j} \end{cases} \text{, where } j=1,\ldots, 13 \qquad (6)$$

where $g^l_{x_j}$ and $g^u_{x_j}$ are the lower and upper limits of the *j*th process factor x_j, respectively. $\Delta^-_{x_j}$ and $\Delta^+_{x_j}$ represent the left and right maximal admissible violations from the imprecise fuzzy value, respectively.

From Figure 3, the process engineer is interested mostly in region II, which corresponds to the maximal $MF(x_j)$ value of one. While, if the level of a process factor lies below $g^l_{x_j}$ or above $g^u_{x_j}$, as shown in regions I and III, respectively, the process performance decreases linearly. Finally, the factor setting outside these two regions is unacceptable. Hence, the following constraints can be formulated:

$$x_j + \delta^-_{x_j} \ge g^l_{x_j}, \qquad j=1,\ldots, 13 \qquad \text{(7-a)}$$

$$0 \le \delta^-_{x_j} \le \Delta^-_{x_j}, \qquad j=1,\ldots, 13 \tag{7-b}$$

$$x_j - \delta^+_{x_j} \le g^u_{x_j}, \qquad j=1,\ldots, 13 \tag{7-c}$$

$$0 \le \delta^+_{x_j} \le \Delta^+_{x_j}, \qquad j=1,\ldots, 13 \tag{7-d}$$

$$MF(x_j) + \frac{\delta^-_{x_j}}{\Delta^-_{x_j}} + \frac{\delta^+_{x_j}}{\Delta^+_{x_j}} = 1, \qquad j=1,\ldots, 13 \tag{7-e}$$

where $\delta^-_{x_j}$ and $\delta^+_{x_j}$ represent the negative and positive deviations from the lower and upper limits of x_j, respectively, where only one of these two deviation variables is strictly positive. Note that when both are equal to zero, x_j will fall in region II, which is preferred as the operating interval for a process factor.

Step 4: The overall objective function, Z, which is the sum of the weighted deviations of the three responses and the thirteen controllable process variables, is then formulated as:

$$Z = \sum_{k=1}^{2} w_{y_k}\left(\frac{\delta^+_{y_k}}{\Delta^+_{y_k}}\right) + w_{y_3}\left(\frac{\delta^-_{y_3}}{\Delta^-_{y_3}} + \frac{\delta^+_{y_3}}{\Delta^+_{y_3}}\right) + \sum_{j=1}^{13} w_{x_j}\left(\frac{\delta^-_{x_j}}{\Delta^-_{x_j}} + \frac{\delta^+_{x_j}}{\Delta^+_{x_j}}\right) \tag{8}$$

where the first, second, and third terms represent the sum of the weighted deviations corresponding to the defect count and cycle time, spoon weight, and process factors, respectively. Also, w_{y1}, w_{y2}, w_{y3}, and w_{xj} denote the weights assigned by the process engineer to y_1, y_2, y_3, and x_j. These weights are not necessarily equal and are set by considering customer's priorities and/or the process engineer's preferences.

Step 5: Typically, the process engineer aims at minimizing the total sum of the weighted deviations. Using the formulated equations in steps 2 through 4, the complete linear programming model will be:

$$\text{Minimize } Z = \sum_{k=1}^{2} w_{y_k}\left(\frac{\delta^+_{y_k}}{\Delta^+_{y_k}}\right) + w_{y_3}\left(\frac{\delta^-_{y_3}}{\Delta^-_{y_3}} + \frac{\delta^+_{y_3}}{\Delta^+_{y_3}}\right) + \sum_{j=1}^{13} w_{x_j}\left(\frac{\delta^-_{x_j}}{\Delta^-_{x_j}} + \frac{\delta^+_{x_j}}{\Delta^+_{x_j}}\right)$$

Subject to:

Figure 3. The membership function of each process factor

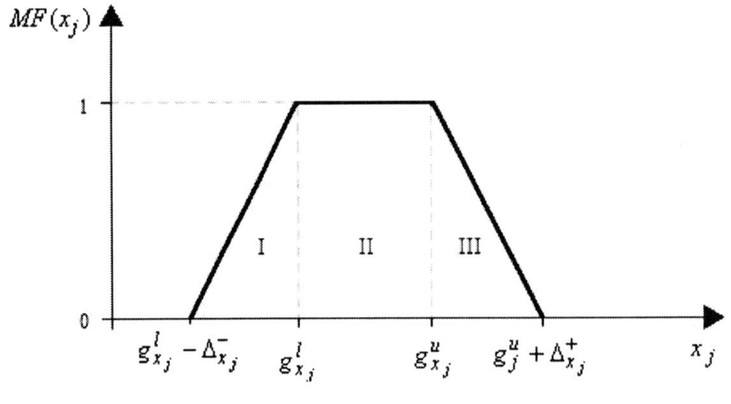

Figure 4. The membership function values of the three quality responses

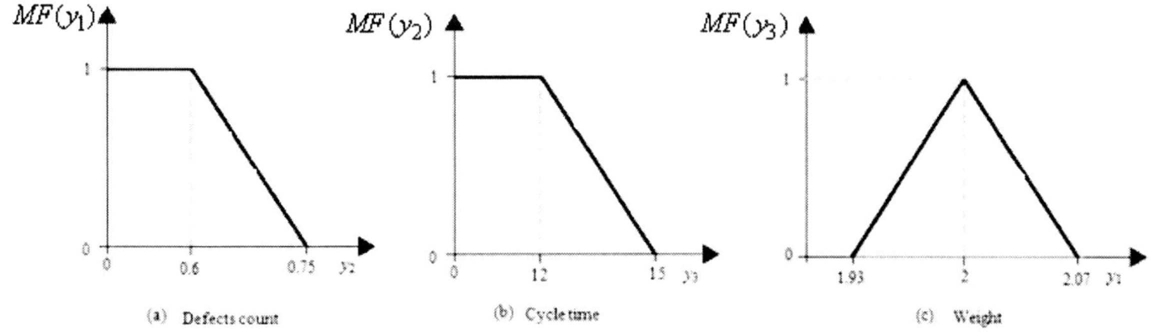

$$y_k = f(x_1,\ x_2,\ ....,\ x_{13}), \qquad k = 1,..., 3$$

$$y_k - \delta^{+}_{y_k} \le g_{y_k}, \qquad k = 1,2$$

$$MF(y_k) + \frac{\delta^{+}_{y_k}}{\Delta^{+}_{y_k}} = 1, \qquad k = 1,2$$

$$\delta^{+}_{y_k} \le \Delta^{+}_{y_k}, \qquad k = 1,2$$

$$y_3 + \delta^{-}_{y_3} - \delta^{+}_{y_3} = g_{y_3},$$

$$MF(y_3) + \frac{\delta^{-}_{y_3}}{\Delta^{-}_{y_3}} + \frac{\delta^{+}_{y_3}}{\Delta^{+}_{y_3}} = 1,$$

$$\delta^{+}_{y_3} \le \Delta^{+}_{y_3},$$

$$x_j + \delta^{-}_{x_j} \ge g^{l}_{x_j}, \qquad j=1,..., 13$$

$$x_j - \delta^{+}_{x_j} \le g^{u}_{x_j}, \qquad j=1,..., 13$$

$$MF(x_j) + \frac{\delta^{-}_{x_j}}{\Delta^{-}_{x_j}} + \frac{\delta^{+}_{x_j}}{\Delta^{+}_{x_j}} = 1, \qquad j=1,..., 13$$

$$0 \le \delta^{-}_{x_j} \le \Delta^{-}_{x_j}, \qquad j=1,..., 13$$

$$0 \le \delta^{+}_{x_j} \le \Delta^{+}_{x_j}, \qquad j=1,..., 13$$

IMPLEMENTATION OF OPTIMIZATION PROCEDURE

The following procedure was proposed to optimize the performance of the plastic injection molding process as follows:

Step 1: The multiple linear regression equations that relate the coded process factors $x_1, x_2, \ldots, x_{13}$ to each of the three responses are obtained and are expressed respectively as follows:

$$y_1 = -\ 4.05833 - 0.00694\ x_1 + 0.00278\ x_2 + 0.01250\ x_3 + 0.00389\ x_4 - 0.00056\ x_5 + 0.01667\ x_6 + 0.01222\ x_7 + 0.00463\ x_8 + 0.00833\ x_9 + 0.00463\ x_{10} - 0.00278\ x_{11} - 0.21296\ x_{12} - 0.08333\ x_{13},$$

$$y_2 = -22.56889 + 0.05278\ x_1 + 0.15889\ x_2 + 0.00167\ x_3 + 0.00356\ x_4 + 0.00067\ x_5 + 0.00056\ x_6 - 0.00333\ x_7 - 0.00704\ x_8 + 0.00926\ x_9 - 0.00111\ x_{10} + 0.01630\ x_{11} - 0.09630\ x_{12} - 0.35556\ x_{13},$$

$$y_3 = 1.63231 + 0.00024\ x_1 - 0.00015\ x_2 + 0.00058\ x_3 + 0.00014\ x_4 + 0.00007\ x_5 + 0.00108\ x_6 + 0.00083\ x_7 + 0.00012\ x_8 + 0.00083\ x_9 + 0.00110\ x_{10} - 0.00038\ x_{11} - 0.01361\ x_{12} + 0.01111\ x_{13},$$

Step 2: Based on process knowledge and customer requirements, the membership functions for the three responses are built as shown in Figure 4. For the average defect count, g_{y1} and $\Delta^{+}_{y_1}$ are determined as 0.6 and 0.15, respectively. That is, the average defect count is mostly preferred when its value falls within the interval [0, 0.6]. However, the process engineer will be dissatisfied when it falls in the interval (0.6, 0.75]. For cycle time, the values of g_{y2} and $\Delta^{+}_{y_2}$ are decided as 12 and 3, respectively. Finally, for the spoon weight, the target value, g_{y3}, is 2 gm with $\Delta^{-}_{y_1}$ and $\Delta^{+}_{y_1}$ of 0.07 gm.

Step 3: Based on process knowledge, the values of $\Delta^{-}_{x_j}$ and $\Delta^{+}_{x_j}$ are both defined equal to the value of 0.05 for each of the thirteen

Figure 5. The membership function of x_1 (Temperature-1)

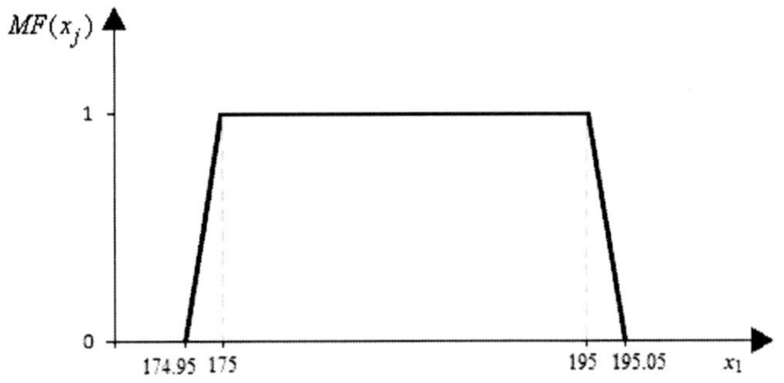

process factors. That is, the positive or negative deviation from the upper or lower level range is 0.05 to ensure a low probability of process engineer's dissatisfaction. For example, the membership function of the factor Temperature-1 is constructed as shown in Figure 5.

The membership functions of the other twelve factors in Table 1 are constructed in the same manner.

Step 4: The negative and positive deviations of responses y_1, y_2, and y_3 are assigned with weights of 0.27, 0.20, and 0.4, respectively, whereas the process variables are assigned with equal weights of 0.01. The total weight equals one (= 0.27 + 0.2 + 0.4 + 0.13).

Step 5: Combining the equations in steps 1 to 4, the complete model is shown in the Appendix. After solving this model, the optimized Z value is 0.0, while the values of μ_{y1}, μ_{y2}, μ_{y3}, and μ_{yj} are all equal to one for all j values. Interestingly, by adopting the optimal factor settings the average defect count is reduced

Table 1. The physical values of factor levels

Category	Control factors	Coded variable	Level range	Level		
				1	2	3
1	Temperature-1	x_1	[175, 195]	175	185	195
	Temperature-2	x_2	[165, 185]	165	175	185
	Temperature-3	x_3	[155, 175]	155	165	175
2	Pressure-1	x_4	[75, 125]	75	90	125
	Pressure-2	x_5	[95, 145]	95	120	145
	Pressure-3	x_6	[110, 150]	110	130	150
	Pressure-4	x_7	[100, 150]	100	125	150
3	Velocity-1	x_8	[35, 65]	35	50	65
	Velocity -2	x_9	[35, 65]	35	50	65
	Velocity -3	x_{10}	[45, 75]	45	60	75
	Velocity -4	x_{11}	[55, 85]	55	70	85
4	Cooling time	x_{12}	[4, 7]	4	5.5	7
5	Injection time	x_{13}	[2, 3]	2	2.5	3

Table 3. Improvement analysis using the weighted additive model

Response (average)	Initial Condition (I)	Optimal Condition (II)	Anticipated Improvement (II)-(I)	Confirmation Improvement
Defects count	0.75	0.16	0.59	0.53
Cycle time (seconds)	15.10	13.06	2.04	2.05
Weight (gm)	2.07	2.00	0.07	0.065

from an average of 0.75 to 0.16. Moreover, the average cycle time is 13.06 seconds, which is significantly smaller than that obtained at initial factor settings (= 15.10 seconds). Finally, the average spoon weight is exactly on its target value of 2.0 gm. Confirmation experiments are conducted with five repetitions, and then the anticipated and confirmation improvements are both summarized in Table 3. It is obvious that the confirmation improvement is very close to the anticipated improvement for the three quality responses, which indicates the efficiency of the proposed model in optimizing the performance of the injection molding process.

In summary, the proposed approach can provide reliable results when (1) the quality responses and process factors can be described by membership functions rather than deterministic targets, (2) the process factors are described with continuous intervals rather than discrete factor levels, and (3) priority or preferences should be considered in the decision making process.

CONCLUSION

This research efficiently optimized the performance of the injection molding process with three quality response, including the average defect count, cycle time, and spoon weight using the weighted additive fuzzy model. These responses and thirteen process factors are described by various membership functions. A complete fuzzy model is then formulated and solved to determine the optimal factor settings. Confirmation results showed that the proposed approach effectively optimized the performance of the injection molding process for a plastic part while incorporating customer/process engineer's preferences.

ACKNOWLEDGMENT

This research was supported in part by the National Science Council of the Republic of China, Grant No. NSC 99-2221-E-035-070-.

REFERENCES

Altan, M. (2010). Reducing shrinkage in injection moldings via the Taguchi, ANOVA and neural network methods. *Materials & Design*, *31*, 599–604. doi:10.1016/j.matdes.2009.06.049

Chen, W. C., Tai, P. H., Wang, M. W., Deng, W. J., & Chen, C. T. (2008). A neural network-based approach for dynamic quality prediction in a plastic injection molding process. *Expert Systems with Applications*, *35*, 843–849. doi:10.1016/j.eswa.2007.07.037

Chiang, K. T. (2007). The optimal process conditions of an injection-molded thermoplastic part with a thin shell feature using grey-fuzzy logic: A case study on machining the PC/ABS cell phone shell. *Materials & Design*, *28*, 1851–1860. doi:10.1016/j.matdes.2006.04.008

Deng, W. J., Chen, C. T., Sun, C. H., Chen, W. C., & Chen, C. P. (2008). An effective approach for process parameter optimization in injection molding of plastic housing components. *Polymer-Plastics Technology and Engineering*, *47*, 910–919. doi:10.1080/03602550802189142

Huang, M. S., & Lin, T. Y. (2008). An innovative regression model-based searching method for setting the robust injection molding parameters. *Journal of Materials Processing Technology*, *198*, 436–444. doi:10.1016/j.jmatprotec.2007.07.022

Kamoun, A., Jaziri, M., & Chaabouni, M. (2009). The use of the simplex method and its derivatives to the on-line optimization of the parameters of an injection moulding process. *Chemometrics and Intelligent Laboratory Systems*, *96*, 117–122. doi:10.1016/j.chemolab.2008.04.010

Oktem, H., Erzurumlu, T., & Uzman, I. (2007). Application of Taguchi optimization technique in determining plastic injection molding process parameters for a thin-shell part. *Materials & Design*, *28*, 1271–1278. doi:10.1016/j.matdes.2005.12.013

Ong, N. S., & Koh, Y. H. (2005). Experimental investigation into micro injection molding of plastic parts. *Materials and Manufacturing Processes*, *20*, 245–253. doi:10.1081/AMP-200042004

Song, M. C., Liu, Z., Wang, M. J., Yu, T. M., & Zhao, D. Y. (2007). Research on effects of injection process parameters on the molding process for ultra-thin wall plastic parts. *Journal of Materials Processing Technology*, *187-188*, 668–671. doi:10.1016/j.jmatprotec.2006.11.103

Tang, S. H., Tan, Y. J., Sapuan, S. M., Sulaiman, S., Ismail, N., & Samin, R. (2007). The use of Taguchi method in the design of plastic injection mould for reducing warpage. *Journal of Materials Processing Technology*, *182*, 418–426. doi:10.1016/j.jmatprotec.2006.08.025

Yaghoobi, M. A., Jones, D. F., & Tamiz, M. (2008). Weighted additive models for solving fuzzy goal programming problems. *Asia-Pacific Journal of Operational Research*, *25*(5), 715–733. doi:10.1142/S0217595908001973

Yang, Y., & Gao, F. (2006). Injection molding product weight: Online prediction and control based on a nonlinear principal component regression model. *Polymer Engineering and Science*, *46*(4), 540–548. doi:10.1002/pen.20522

APPENDIX

The complete optimization model is formulated as follows

$$\text{Minimize } Z = \frac{27}{100} \times \frac{\delta^+_{y_1}}{0.15} + \frac{20}{100} \times \frac{\delta^+_{y_2}}{3} + \frac{40}{100} \times (\frac{\delta^-_{y_3}}{0.07} + \frac{\delta^+_{y_3}}{0.07}) + \frac{1}{100} \times \sum_{j=1}^{13} \left(\frac{\delta^-_{x_j}}{0.05} + \frac{\delta^+_{x_j}}{0.05} \right)$$

subject to

$f(y_k)$, $\quad k=1,\ldots,3$

$y_1 - \delta^+_{y_1} \le 0.6; \quad MF(y_1) + \dfrac{\delta^+_{y_1}}{0.15} = 1,$

$y_2 - \delta^+_{y_2} \le 12; \quad MF(y_2) + \dfrac{\delta^+_{y_2}}{3} = 1,$

$y_3 + \delta^-_{y_3} - \delta^+_{y_3} = 2; \quad MF(y_3) + \dfrac{\delta^-_{y_3}}{0.07} + \dfrac{\delta^+_{y_3}}{0.07} = 1,$

$x_1 + \delta^-_{x_1} \ge 175;$ $x_1 - \delta^+_{x_1} \le 195;$ $x_2 + \delta^-_{x_2} \ge 95;$ $x_2 - \delta^+_{x_2} \le 145;$ $x_3 + \delta^-_{x_3} \ge 155;$
$x_3 - \delta^+_{x_3} \le 175;$ $x_4 + \delta^-_{x_4} \ge 75;$ $x_4 - \delta^+_{x_4} \le 125;$ $x_5 + \delta^-_{x_5} \ge 95;$ $x_5 - \delta^+_{x_5} \le 145;$
$x_6 + \delta^-_{x_6} \ge 110;$ $x_6 - \delta^+_{x_6} \le 150;$ $x_7 + \delta^-_{x_7} \ge 100;$ $x_7 - \delta^+_{x_7} \le 150;$ $x_8 + \delta^-_{x_8} \ge 35,$
$x_8 - \delta^+_{x_8} \le 65;$ $x_9 + \delta^-_{x_9} \ge 35;$ $x_9 - \delta^+_{x_9} \le 65;$ $x_{10} + \delta^-_{x_{10}} \ge 45;$ $x_{10} - \delta^+_{x_{10}} \le 75;$
$x_{11} + \delta^-_{x_{11}} \ge 55;$ $x_{11} - \delta^+_{x_{11}} \le 85;$ $x_{12} + \delta^-_{x_{12}} \ge 4;$ $x_{12} - \delta^+_{x_{12}} \le 7;$ $x_{13} + \delta^-_{x_{13}} \ge 2;$
$x_{13} - \delta^+_{x_{13}} \le 3,$

$\mu_{x_j} + \dfrac{\delta^-_{x_j}}{0.05} + \dfrac{\delta^+_{x_j}}{0.05} = 1, \quad j = 1, \ldots, 13$

$y_k, \mu_{y_k}, \delta^-_{y_k}, \delta^+_{y_k} \ge 0, \quad k = 1, \ldots, 3$

$x_j, \mu_{x_j}, \delta^-_{x_j}, \delta^+_{x_j} \ge 0, \quad j = 1, \ldots, 13$

After solving this model, the optimal factor settings are obtained as follows:

$x_1 = 175;$ $x_2 = 165;$ $x_3 = 175;$ $x_4 = 75;$ $x_5 = 95;$ $x_6 = 110;$ $x_7 = 100;$
$x_8 = 35;$ $x_9 = 35;$ $x_{10} = 45.04;$ $x_{11} = 55;$ $x_{12} = 4;$ $x_{13} = 2.0$

This work was previously published in the International Journal of Fuzzy System Applications, Volume 1, Issue 2, edited by Toly Chen, pp. 43-54, copyright 2010 by IGI Publishing (an imprint of IGI Global).

Chapter 16
Modelling the Long-Term Cost Competitiveness of a Semiconductor Product with a Fuzzy Approach

Toly Chen
Feng Chia University, Taiwan

ABSTRACT

Unit cost is undoubtedly the most critical factor for the competitiveness of a product. Therefore, evaluating the competitiveness of a product according to its unit cost is a reasonable idea. The current practice assesses the competitiveness at a number of check points in the product life cycle, and then averages the results. This approach is computationally simple but theoretically doubtful. This study evaluates the long-term cost competitiveness of a semiconductor product based on its cost learning model from a new viewpoint – the trend in the mid-term competitiveness. Using a fuzzy value to express the long-term cost competitiveness, the flexibility in the interpretation and implementation of the evaluation result is increased. A practical example is used to illustrate the proposed methodology.

INTRODUCTION

Competitiveness is the ability and performance of a firm, sub-sector or country to sell and supply goods and/or services in a given market. Competitiveness engineering is a systematic procedure, including a series of activities of assessing and enhancing competitiveness. Competitiveness assessment is a one of the major tasks. There have been many relevant references in this field, but most of them focused on exploring the factors affecting competitiveness (such as cost, quality, customer satisfaction, technical competence, etc.) and ways to improve competitiveness (such as balanced scorecard, blue ocean strategy, lean production, green supply chain, learning organization, etc.). How to assess competitiveness in a quantitative way is rarely discussed.

DOI: 10.4018/978-1-4666-1870-1.ch016

Competition in the semiconductor manufacturing industry has reached an unprecedented level of intensity. Typical products of this industry include dynamic random access memory (DRAM), flash, application-specific integrated circuits, etc. For sustainable development under such a difficult environment, semiconductor manufacturers come up with various ways to improve their competitiveness, such as alliances, becoming fabless, outsourcing, and developing next-generation technologies.

Porter (1979), is a pioneering study in this field. He came to the five forces that influence the competitiveness of an enterprise such as the threat of substitute products, the threat of established rivals, the threat of new entrants, the bargaining power of suppliers, and the bargaining power of customers. However, these invisible forces mostly come from outside the company, and it is also difficult to assess their impact. Armstrong (1989) advocated that a competitive semiconductor manufacturer should meet four principles, including continuous measurable improvement, statistical thinking, constraint-focus and people development/empowerment. Jenkins et al. (1990) stressed the importance of quality, and described how to design quality into products and processes. Dr. Robert Helms, CEO of International SEMATECH, remarked, "In our industry, it used to be that the big companies eat the small. Today, the fast run over the slow" (Helms, 2001). Leachman (2002) benchmarked ten semiconductor manufacturing facilities to identify the factors that influence competitive semiconductor manufacturing (CSM). Peng and Chien (2003) focused on how to create values for customers. Shortening cycle time, producing high-quality products, on-time delivery of orders, continual reducing costs, and improving efficiency were considered as the most direct and effective ways. Recently, Walsh et al. (2005) observed that the competitiveness and long-term success of an enterprise is closely related. Although Liao and Hu (2007) claimed that knowledge transfer is an important factor for the competitiveness of semiconductor manufacturers. However, according to Crowder's view, the leadership of manufacturing science and technology will not necessarily deliver a competitive advantage (Crowder, 1989).

The traditional way to measure the competitiveness of a semiconductor manufacturer is to interview stakeholders, such as its past and present key management personnel, marketing and technological consultants, professional analysts, major capital equipment suppliers, and even competitors (Leachman, 2002; Walsh et al., 2005). This process is subjective. It may lead to imprecise assessment that may not be suitable for quantitative analyses. In practice, another frequently used approach is the hierarchical assessment approach, in which several aspects of competitiveness are to be evaluated, and then a simple (weighted) average method is used to integrate the assessment results. However, such a treatment is unfounded in practice and questionable. Moreover, competitiveness is a subjective and uncertain concept, but the existing methods cannot maintain this flexibility. On the other hand, quantitative measures such as market share (Walsh et al., 2005; Defree, 2007) and revenues (Walsh et al., 2005; Defree, 2007) has been very sensitive to market conditions that are beyond our control, and need to be compared with those of the competitors or with the average levels in the whole industry. In addition, these measures are not the sources but rather the outcomes of competitiveness. Moreover, the mid-term financial performance is difficult to predict.

The competitiveness of a manufacturer comes from all of its products. A product is competitive because it is trendy, has good quality, can be manufactured with low costs, etc. Based on this belief, Chen (2007) proposed a systematic procedure to assess the yield competitiveness of a product. However, even if a product is competitive in some ways, it may not be competitive in other ways. For example, some semiconductor products have high yields at the later stages of their product life cycles, but their costs are still too high to generate profits. On the competitiveness of a semiconductor product, product cost has always been regarded as one of the most basic and important factors. Although semiconductor manufacturers may have relevant technologies to meet future growth in the years, the ability to implement these technologies must be cost-effective (Dance et al., 1996). Competitive products are therefore essential. Reducing the unit cost of each product type was considered the most important goal to a factory (Carnes, 1991). Foxconn's pay increase in 2010 led to the mainland continued to improve human wage level, so that to obtain low-cost human resources becomes difficult. Many companies started as nomads, migrating to regions with lower wage levels. Labor cost is undoubtedly one of the major product cost. Besides, there is a continuing trend in the semiconductor industry's disintegration. The aim is to improve the cost effectiveness at each stage of the semiconductor value chain (Hwang et al., 2008). In addition, during the financial crisis in 2009, many semiconductor-manufacturing factories closed down mainly because the sales were far insufficient to cover the costs. For these reasons, to assess the competitiveness of a semiconductor manufacturing factory, considering the costs of all its products is a reasonable idea. In addition, compared with other financial indicators, cost is more predictable, because in the long term cost reduction has a learning feature. For these reasons, Chen (2010) proposed a systematic procedure to evaluate the mid-term cost competitiveness. The time period considered in a short-term or mid-term evaluation is usually one or two stages in the product life cycle (Chen & Wang, 2009), and in a long-term evaluation, this will be the entire product life cycle. In this study, we propose a flexible way to assess the long-term cost competitiveness of a semiconductor product. The idea is that the trend in the mid-tem competitiveness can be used to evaluate the long-term one. Since the mid-term cost competitiveness can be expressed with a linguistic term such as "very competitive" or "somewhat uncompetitive" that is flexible in interpretation and implementation, the long-term cost competitiveness becomes a fuzzy correlation coefficient with the same advantages (Chen, 2010). The motives for this study include:

1. To the best of our knowledge, a quantitative way of evaluating the long-term cost competitiveness has yet been proposed in the past studies.
2. Some well-known recent events have a profound impact on the cost competitiveness of related firms, and also need to be explained from a theoretical point of view.
3. Unlike yield, cost can be decomposed into multiple sub-items, for each of which an independent evaluation of the competitiveness can be conducted. Hence there is a need for a hierarchical structure to aggregate the results of these evaluations.

The rest of this paper is organized as follows. Before evaluating the long-term cost competitiveness of a semiconductor product, the future unit cost has to be forecasted in advance from a learning point of view. Then, the long-term cost competitiveness of the semiconductor product can be evaluated based on the forecasted unit cost. A practical example with data collected from a real semiconductor manufacturing factory is used to demonstrate the application of the proposed methodology. Finally, the concluding remarks with a view to the future are given.

FORECASTING THE FUTURE UNIT COST

Parameters used in the proposed methodology are defined as follows.

a_t: the normalized unit cost at time period t.
b: the learning constant.
C: wafer cost.

$c^*_{m(k)}$: cost target at check point k.
$\tilde{c}_t$: unit cost forecast at period t. $\tilde{c}_t = (c_{t1}, c_{t2}, c_{t3})$.
c_t: actual unit cost at time period t.
G: gross die.
K: the number of check points.
$m(k)$: the k-th check point.
$P(k)$: the mid-term competitiveness at check point k.
$r(t)$: homoscedastical, serially non-correlated error term.
$\tilde{R}$: fuzzy correlation coefficient.
t: time.
T: the current time.
Y_t: yield at time period t.
Y_0 : the asymptotic/final yield.

Before evaluating the long-term cost competitiveness of a semiconductor product, the future unit cost has to be forecasted in advance. It would be very easy if the reduction in unit cost follows a learning process. We first describe the relationship between yield and cost. According to Gruber (1994), the yield of a semiconductor product follows a learning process:

$$Y_t = Y_0 e^{-b/t + r(t)} \tag{1}$$

The unit cost can be calculated as

$$c_t = C / (Y_t \cdot G) \tag{2}$$

Obviously, the change in the unit cost is also a learning process, not a usual time-series. After converting to logarithms,

$$\ln c_t = \ln C - \ln Y_0 - \ln G + b / t - r(t) \tag{3}$$

where $a = \ln C - \ln Y_0 - \ln G$. To consider the uncertainty in the unit cost, parameters in Equation (3) are expressed by asymmetric triangular fuzzy numbers (TFNs) as follows:

$$\tilde{a} = (a_1, \tag{4}$$

$$\tilde{b} = (b_1, \tag{5}$$

Therefore,

$$\ln \tilde{c}_t \cong (\ln c_{t1}, \ln c_{t2}, \ln c_{t3}) = \tilde{a}(+)\tilde{b} / t - r(t) = (a_1 + b_1 / t, a_2 + b_2 / t, a_3 + b_3 / t) - r(t) \tag{6}$$

where (+) represents fuzzy addition. Equation (6) is obviously a fuzzy linear regression equation that can be fitted by solving the following linear programming problem according to Tanaka and Watada's method (Tanaka & Watada, 1988):

$$\text{Min } Z = \sum_{t=1}^{T} (\ln c_{t3} - \ln c_{t1}) \tag{7}$$

subject to

$$\ln c_t \geq \ln c_{t1} + s(\ln c_{t2} - \ln c_{t1}) \tag{8}$$

$$\ln c_t \leq \ln c_{t3} + s(\ln c_{t2} - \ln c_{t3}) \tag{9}$$

$$c_{t1} = a_1 + b_1 / t \tag{10}$$

$$c_{t2} = a_2 + b_2 / t \tag{11}$$

$$c_{t3} = a_3 + b_3 / t \tag{12}$$

$$0 \leq a_1 \leq a_2 \leq a_3 \tag{13}$$

$$0 \leq b_1 \leq b_2 \leq b_3 \tag{14}$$

$$t = 1 \sim T \tag{15}$$

In this way, all the actual values fall within the ranges of the fuzzy forecasts. Other methods are also applicable to this purpose, such as Peters's linear programming method, the quadratic non-possibilistic method (Donoso et al., 2006), the nonlinear programming method (Chen & Lin, 2008), and so on.

EVALUATING THE LONG-TERM COST COMPETITIVENESS

Subsequently, the assessment of the mid-term cost competitiveness of a semiconductor product on the basis of the projected cost is as follows. Assume there are k competitive regions specified as

$$\text{CR}(k)\text{: } c_t \leq c^*_{m(k)} \text{ when } t \geq m(k) \tag{16}$$

$k = 1{\sim}K$

The mid-term cost competitiveness of the product at each check point can be assessed using the procedure introduced in Chen (2010):

$$\tilde{P}(k) = \int_{m(k)}^{m(k)+\Delta t} c^*_{m(k)} dt - \int_{m(k)}^{m(k)+\Delta t} \tilde{c}_t dt \tag{17}$$

$k = 1{\sim}K$

$$c^*_{m(k)}\Delta t - (a_3 + b_3)\Delta t \leq \tilde{P}(k) \leq c^*_{m(k)}\Delta t \tag{18}$$

To evaluate the long-term cost competitiveness, the trend in the mid-term one is an innovative idea. For this purpose, the fuzzy correlation coefficient between the mid-term cost competitiveness and time is calculated:

$$\tilde{R} = \tilde{S}_{xy} / \sqrt{\tilde{S}_{xx}\tilde{S}_{yy}} = (\sum_{\text{all } k} m(k)\tilde{P}(k)(-)K\bar{m}\bar{\tilde{P}}) / \sqrt{\sum_{\text{all } k} m^2(k) - K\bar{m}^2}\sqrt{\sum_{\text{all } k} \tilde{P}^2(k)(-)K\bar{\tilde{P}}} \tag{19}$$

where

$$\bar{\tilde{P}} = \sum_{k=1}^{K} \tilde{P}(k) \Big/ K \tag{20}$$

$$\bar{m} = \sum_{k=1}^{K} m(k) \Big/ K \tag{21}$$

To derive $\tilde{R}$, the arithmetic for TFNs cannot be directly applied. On the contrary, the α-cut operations are used:

$$A_\alpha = \{x \mid x \in R,\ \mu_{\tilde{A}}(x) \geq \alpha\} \equiv \text{the } \alpha\text{-cut of } \tilde{A} \tag{22}$$

$$B_\alpha = \{x \mid x \in R,\ \mu_{\tilde{B}}(x) \geq \alpha\} \equiv \text{the } \alpha\text{-cut of } \tilde{B} \tag{23}$$

Then

$$A_\alpha(+)B_\alpha = [a_1^\alpha,\ a_2^\alpha](+)[b_1^\alpha,\ b_2^\alpha] = [a_1^\alpha + b_1^\alpha,\ a_2^\alpha + b_2^\alpha] \tag{24}$$

$$A_\alpha(-)B_\alpha = [a_1^\alpha,\ a_2^\alpha](-)[b_1^\alpha,\ b_2^\alpha] = [a_1^\alpha - b_2^\alpha,\ a_2^\alpha - b_1^\alpha] \tag{25}$$

$$A_\alpha(\times)B_\alpha = [a_1^\alpha,\ a_2^\alpha](\times)[b_1^\alpha,\ b_2^\alpha] = [a_1^\alpha \cdot b_1^\alpha,\ a_2^\alpha \cdot b_2^\alpha]$$

$a_1^\alpha \geq 0,$

$$b_1^\alpha \geq 0 \tag{26}$$

$$A_\alpha(/)B_\alpha = [a_1^\alpha,\ a_2^\alpha](/)[b_1^\alpha,\ b_2^\alpha] = [a_1^\alpha / b_2^\alpha,\ a_2^\alpha / b_1^\alpha],$$

$a_1^\alpha \geq 0,$

$$b_1^{\alpha} > 0 \tag{27}$$

where (+), (−), (×), and (/) denote fuzzy addition, subtraction, multiplication and division, respectively. $\tilde{R}$ ranges from -1 to 1. The semiconductor product does not have long-term competitiveness if $\tilde{R}$ is negative.

The long-term competitiveness measured in terms of $\tilde{R}$ satisfies the following properties:

1. A product is mid-term competitive does not guarantee that it is long-term competitive. On the contrary, a long-term competitive product might not have mid-term competitiveness (Figure 1). Mid-term competitiveness and long-term competitiveness should be different concepts. Otherwise, there is no need to measure the long-term competitiveness.
2. The long-term competitiveness should be measured in a manner that provides some theoretical explanation for the related competitive actions and stimulates the continuing growth of the firm. For example, Foxconn's pay increase seems to reduce its competitiveness, but from a theoretical point of view it may not be so. As the first company to increase wages, Foxconn was therefore entitled to set the cost targets for evaluating the competitiveness in the industry. By setting a cost target high enough, Foxconn skillfully weakened the competitiveness of its rivals.
3. Obviously, the uncertainty in the long-term competitiveness is higher than that in the mid-term one. From this point of view, using the simple or weighted average of the mid-term competitiveness (Chen, 2009) to measure the long-term competitiveness is problematic.

Theorem 1. *The uncertainty in the average of some fuzzy values is not higher than those in the original values.*

Proof. The uncertainty in a fuzzy value can be measured with the spread of the fuzzy value. Assume there are M fuzzy values indicated with $\tilde{A}_m = (A_{m1}, A_{m2}, A_{m3})$, $m = 1\sim M$. The average of these fuzzy values is calculated as

$$\bar{\tilde{A}} = \sum_{m=1}^{M} \tilde{A}_m / M = \left(\sum_{m=1}^{M} A_{m1} / M, \sum_{m=1}^{M} A_{m2} / M, \sum_{m=1}^{M} A_{m3} / M\right) \tag{28}$$

The uncertainty in the average is measured as

$$\begin{aligned} &\sum_{m=1}^{M} A_{m3} / M - \sum_{m=1}^{M} A_{m1} / M \\ &= \sum_{m=1}^{M} (A_{m3} - A_{m1}) / M \\ &\le \sum_{m=1}^{M} \max_h (A_{h3} - A_{h1}) / M = \\ &M \cdot \max_h (A_{h3} - A_{h1}) / M = \\ &\max_h (A_{h3} - A_{h1}) \end{aligned} \tag{29}$$

Theorem 1 is proven.

A DEMONSTRATIVE CASE

To illustrate the applicability of the proposed methodology, a real case from a semiconductor manufacturing factory is used (Table 1). The fitted cost learning model is shown in Figure 2. There are three check points for which the following competitive regions are specified:

CR(1): $c_t \ge 1.15$ when $t \ge 12$,

CR(2): $c_t \ge 1.04$ when $t \ge 15$,

CR(3): $c_t \ge 0.95$ when $t \ge 21$.

Figure 1. Mid-term competitiveness and long-term competitiveness

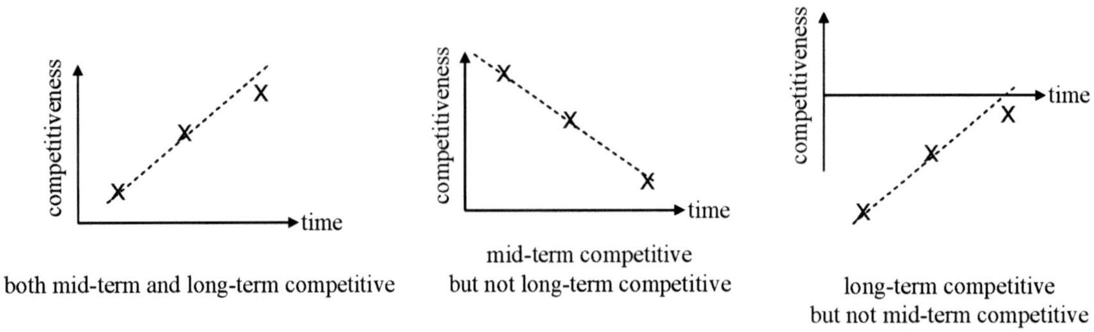

The mid-term cost competitiveness at each check point is assessed. The results are summarized as follows:

$\tilde{c}_{(1)} = (-0.19, 0.31, 0.35)$,

$\tilde{c}_{(2)} = (-0.07, 0.43, 0.46)$,

$\tilde{c}_{(3)} = (0.03, 0.53, 0.56)$.

After applying the α-cut operations, the fuzzy correlation coefficient between the mid-term competitiveness and time is obtained as shown in Figure 3, which is a fuzzy value with such a complex shape that cannot be approximated with a TFN.

Subsequently, the fuzzy correlation coefficient is compared with a number of linguistic terms, including "very competitive (VC)", "somewhat competitive (SC)", "moderate (M)", "somewhat uncompetitive (SU)", and "very uncompetitive (VU)":

VU = (-1, -1, -0.3)

SU = (-1, -0.3, 0)

M = (-0.3, 0, 0.3)

SC = (0, 0.3, 1)

VC = (0.3, 1, 1)

According to the experimental results:

1. Obviously, the uncertainty in the long-term competitiveness is higher than those in the mid-term ones.
2. It seems that the long-term cost competitiveness can be best represented with "SU" ~ "VC" (Figure 4).

1. Experimental results also show the difficulty of estimating the future competitiveness of the product.

Table 1. The collected data

t	1	2	3	4	5	6	7	8	9	10
c_t (US$)	2.57	1.61	1.76	1.28	1.53	1.19	1.32	1.32	1.61	1.32

Figure 2. The fitted cost learning model

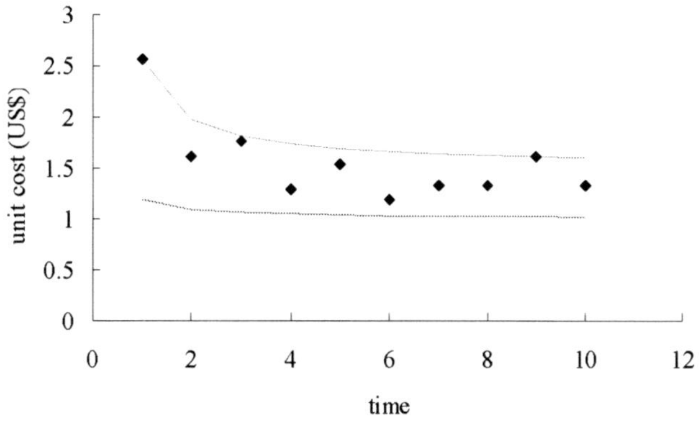

To further elaborate the performance of the proposed methodology, the fuzzy correlation coefficient is first defuzzified with Wrather and Yu's approach:

$$D(\tilde{R}) = \int R\mu_{\tilde{R}}(R)dR \;/\; \int \mu_{\tilde{R}}(R)dR \qquad (30)$$

The defuzzification result is -0.005, and therefore the product can be summarized as slightly lack of competitiveness from the crisp point of view. Parametric analyses have also been completed to assess the impact of some parameters on the long-term cost competitiveness. For example, the advance or delay of a check point changes the long-term competitiveness. Taking the first check point as an example, the results are shown in Figure 5. Obviously, any delay to reach the cost target is certainly not conducive to the long-term cost competitiveness. On the other hand, establishing a higher cost goal (Figure 6) seems to help to improve the long-term cost competitiveness. The reason is that the first mid-term cost competitiveness is pulled down and the following ones have a lower basis, which improves their trend.

To further elaborate the effectiveness of the proposed methodology, five domain experts (including three product managers and two executive officers) were requested to evaluate the cost competitiveness of the product in the long term.

Figure 3. The fuzzy correlation coefficient

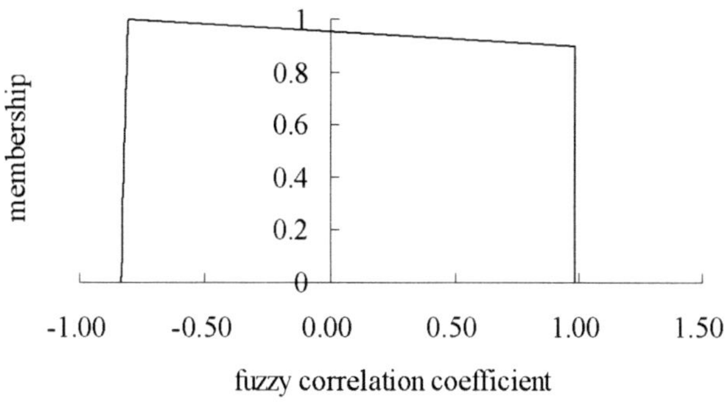

Figure 4. The interpretation of the fuzzy correlation coefficient

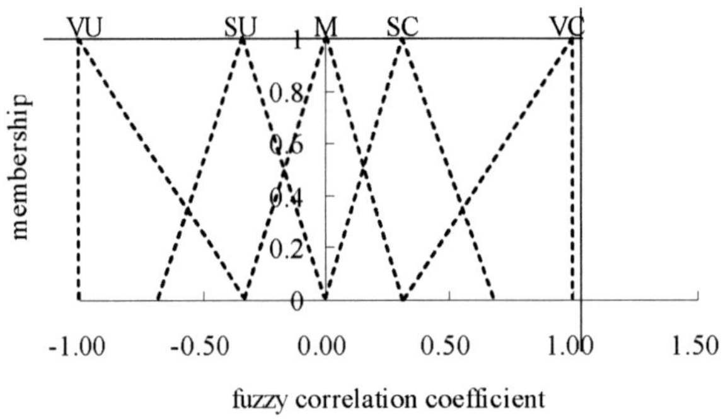

We showed these experts the forecasted costs of this product during 18 periods. They also considered the product life cycles of similar products, the substitutions for this product, capacity limitation, future prices, and other issues. Finally, they gave the following answers: {"SU", "U", "SU", "M", "U"}. After aggregation by the arithmetic of triangular fuzzy numbers, their consensus could be expressed with a triangular fuzzy numbers (0.05, 0.2, 0.45), approximately equal to "SU", which happened to be the same with that made by the proposed methodology.

CONCLUSION

Recently, some papers have been published about competitiveness engineering in manufacturing; however, there are several issues that have not yet been addressed. For example, How to assess competitiveness in a quantitative way is rarely discussed. Product cost is undoubtedly the most critical factor for the competitiveness of a product in a semiconductor fabrication plant. Therefore, evaluating the competitiveness of a product with its unit cost is a reasonable idea. To this end, a flexible approach is proposed in this study. Initially, the fuzzy linear regression approach is applied to estimate the unit cost of a semiconductor product.

Figure 5. The effects of changing the first check point

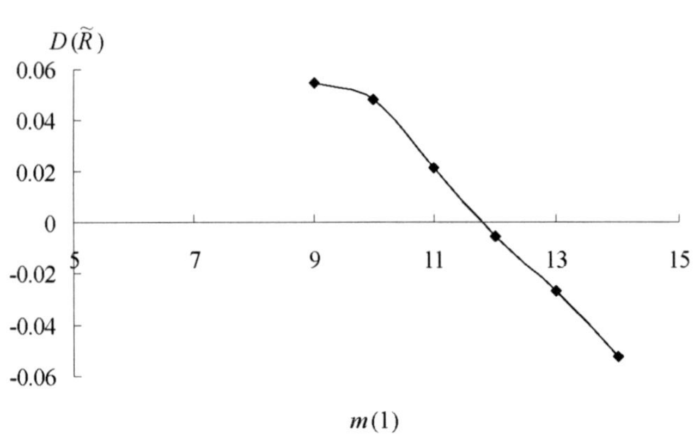

Figure 6. The effects of changing the first cost goal

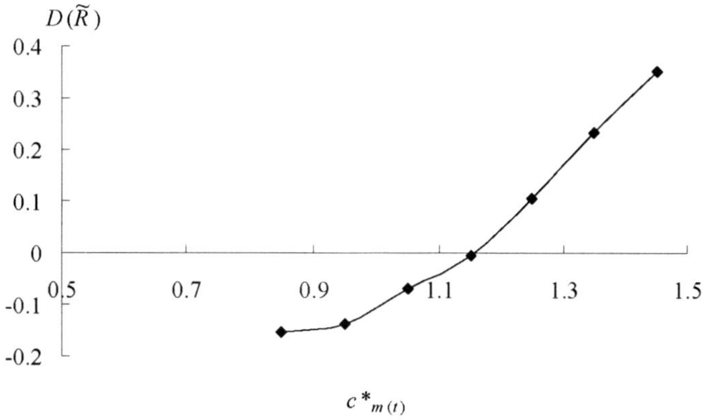

Subsequently, the cost region for the semiconductor product to be competitive is specified. The mid-term cost competitiveness of the semiconductor product is assessed by comparing the estimated unit cost and the competitive region. To obtain the long-term cost competitiveness, the trend in the mid-term competitiveness is a good index. A practical example with data collected from a real semiconductor-manufacturing factory is used to demonstrate the applicability of the proposed methodology. According to experimental results, the evaluated long-term cost competitiveness of the product was low. However, considering the uncertainty in cost learning, the long-term competitiveness was highly uncertain.

However, to further evaluate the advantages and disadvantages of the proposed methodology, it should be applied to more different types of products. Besides this, more sophisticated approaches of evaluating cost competitiveness can be developed in future studies.

ACKNOWLEDGMENT

This study is financially supported by National Science Council of Taiwan.

REFERENCES

Armstrong, E. *(1989). Principles for competitive semiconductor manufacturing. In* Proceedings of the IEEE/SEMI International Semiconductor Manufacturing Science Symposium

Carnes, R. *(1991). Long-term cost of ownership: Beyond purchase price. In* Proceedings of the IEEE/SEMI International Semiconductor Science Symposium *(pp. 39-43).*

Chen, T. (2007). Evaluating the mid-term competitiveness of a product in a semiconductor fabrication factory with a systematic procedure. *Computers & Industrial Engineering*, *53*, 499–513. doi:10.1016/j.cie.2007.05.008

Chen, T. (2009). A FNP approach for evaluating and enhancing the long-term competitiveness of a semiconductor fabrication factory through yield learning modeling. *International Journal of Advanced Manufacturing Technology*, *40*, 993–1003. doi:10.1007/s00170-008-1414-8

Chen, T. (2010). A FNP approach for establishing the optimal and efficient capacity re-allocation plans for enhancing the long-term competitiveness of a semiconductor product. *International Journal of Fuzzy Systems*, *12*(2), 158–169.

Chen, T., & Lin, Y. C. (2008). A fuzzy-neural system incorporating unequally important expert opinions for semiconductor yield forecasting. *International Journal of Uncertainty, Fuzziness, and Knowledge-based Systems*, *16*(1), 35–58. doi:10.1142/S0218488508005030

Chen, T., & Wang, Y. C. (2009). A fuzzy set approach for evaluating and enhancing the mid-term competitiveness of a semiconductor factory. *Fuzzy Sets and Systems*, *160*, 569–585. doi:10.1016/j.fss.2008.06.006

Crowder, B. C. *(1989). Manufacturing science and manufacturing competitiveness. In* Proceedings of the IEEE/SEMI International Semiconductor Manufacturing Science Symposium *(pp. 32-33).*

Deffree, S. *(2007).* Semiconductor winners and losers split by competitiveness. *Retrieved from* http://www.electronicsweekly.com/

Donoso, S., Marin, N., & Vila, M. A. *(2006). Quadratic programming models for fuzzy regression. In* Proceedings of the International Conference on Mathematical and Statistical Modeling in Honor of Enrique Castillo.

Gruber, H. (1994). *Learning and strategic product innovation: Theory and evidence for the semiconductor industry*. Boston, MA: Elsevier.

Helms, R. (2001). Chips: The fastest downturn in history has its optimist. *Business Week.*

Hwang, B. N., Chang, S. C., Yu, H. C., & Chang, C. W. (2008). Pioneering e-supply chain integration in semiconductor industry: A case study. *International Journal of Advanced Manufacturing Technology*, *36*, 825–832. doi:10.1007/s00170-006-0886-7

Jenkins, T., Phail, F., & Sackman, S. (1990). Semiconductor competitiveness in the 1990s. *Proceedings of the Society of Automotive Engineers*, *223*, 249–255.

Leachman, R. C. (2002). *Competitive semiconductor manufacturing: Final report on findings from benchmarking eight-inch, sub-350nm wafer fabrication lines*. Berkley, CA: University of California.

Liao, S. H., & Hu, T. C. (2007). Knowledge transfer and competitive advantage on environmental uncertainty: An empirical study of the Taiwan semiconductor industry. *Technovation*, *27*(6-7), 402–411. doi:10.1016/j.technovation.2007.02.005

Peng, C. Y., & Chien, C. F. (2003). *Data value development to enhance competitive advantage: A retrospective study of EDA systems for semiconductor fabrication.* International Journal of Services. *Technology and Management*, *4*(4-6), 365–383.

Porter, M. E. (1979). How competitive forces shape strategy. Harvard Business Review.

Tanaka, H., & Watada, J. (1988). Possibilistic linear systems and their application to the linear regression model. *Fuzzy Sets and Systems*, *272*, 275–289. doi:10.1016/0165-0114(88)90054-1

Walsh, S. T., Boylan, R. L., McDermott, C., & Paulson, A. (2005). The semiconductor silicon industry roadmap: Epochs driven by the dynamics between disruptive technologies and core competencies. *Technological Forecasting and Social Change*, *72*, 213–236. doi:10.1016/S0040-1625(03)00066-0

This work was previously published in the International Journal of Fuzzy System Applications, Volume 1, Issue 4, edited by Toly Chen, pp. 62-72, copyright 2010 by IGI Publishing (an imprint of IGI Global).

Chapter 17
Extension of the TOPSIS for Multi-Attribute Group Decision Making under Atanassov IFS Environments

Deng-Feng Li
Fuzhou University, China

Jiang-Xia Nan
Guilin University of Electronic Technology, China

ABSTRACT

This paper extends the technique for order preference by similarity to ideal solution (TOPSIS) for solving multi-attribute group decision making (MAGDM) problems under Atanassov intuitionistic fuzzy set (IFS) environments. In this methodology, weights of attributes and ratings of alternatives on attributes are extracted from fuzziness inherent in decision data and making process and described using Atanassov IFSs. An Euclidean distance measure is developed to calculate the differences between alternatives for each decision maker and an Atanassov IFS positive ideal solution (IFSPIS) as well as an Atanassov IFS negative ideal-solution (IFSNIS). Degrees of relative closeness to the Atanassov IFSPIS for all alternatives with respect to each decision maker in the group are calculated. Then all decision makers in the group may be regarded as "attributes" and a corresponding classical MADM problem is generated and hereby solved by the TOPSIS. The proposed methodology is validated and compared with other similar methods. A numerical example is examined to demonstrate the implementation process of the methodology proposed in this paper.

INTRODUCTION

Multi-attribute group decision making (MAGDM) problems are of importance in most kinds of fields such as engineering, economics and management. It is obvious that much knowledge in the real world is fuzzy rather than precise. Imprecision comes from a variety of sources such as unquantifiable information (Chen & Hwang, 1992; Triantaphyllou & Lin, 1996; Chen, 2000; Chu, 2002a, 2002b; Braglia et al., 2003; Chu & Lin, 2003; Li, 2003; Chen & Tzeng, 2004; Zhao et al., 2006).

DOI: 10.4018/978-1-4666-1870-1.ch017

In many situations decision makers have imprecise/vague information about alternatives with respect to attributes. One of the methods which describe imprecise cases is the fuzzy set (FS) introduced by Zadeh (1965). The main characteristic of an FS $\tilde{A}$ is that a membership degree $\mu_{\tilde{A}}(u) \in [0,1]$ is assigned to each element u in a universe of discourse and a non-membership degree automatically is equal to $1-\mu_{\tilde{A}}(u)$, i.e., this membership degree combines the evidence for u and the evidence against u. However, a human being who expresses the degree of membership of a given element in an FS very often does not express the corresponding degree of non-membership as the complement to 1. This reflects a well-known psychological fact that the linguistic negation is not always identified with the logical negation. Thus in 1983 Atanassov (1986, 1999) introduced the concept of an intuitionistic fuzzy set (IFS), which is called the Atanassov IFS for short in this paper. The Atanassov IFS is characterized by two functions expressing the degree of belongingness and the degree of non-belongingness, respectively. This idea is useful in decision making processes. Especially, the Atanassov IFS may give us a very natural tool for modeling preferences. Sometimes it seems to be more natural to describe imprecise and uncertain opinions not only by membership functions due to the fact that in some situations it is easier to describe our negative feeling than positive attitude. Even more, quite often one can easily specify alternatives (or objects) one dislikes, but simultaneously cannot specify clearly what are really wanted. Let us consider a situation observed in a real estate agency. Very often an interested customer looking for an apartment is not completely convinced on the location and considers several variants. It is obvious that some districts are more preferable than others whereas there are some districts that the customer dislikes. It seems to be much suitable for modeling such situations using the Atanassov IFSs. For example, it may happen that the customer asked about his/her favorite district in Warsaw cannot definitely determine whether it is Ochota, Mokotów or Zoliborz, but he/she feels sure that he/she hates Wola. Thus the preferences of the customer may be modeled using the Atanassov IFS, where the membership function expresses the degree of a given district being preferred by the customer while the non-membership function indicates the degree of the given district which is not preferred (Grzegorzewski & Mrówka, 2005). People maybe meet a similar situation when comparing preferences expressed by means of orderings which admit uncertainty due to imprecision, vagueness and hesitance. In this case the Atanassov IFS may give us a natural tool for modeling such improper orderings (Grzegorzewski & Mrówka, 2005; Deschrijver & Kerre, 2007).

Obviously, the Atanassov IFS is a natural generalization of the FS. Gau and Buehrer (1993) introduced the concept of a vague set characterized by a truth-membership function and a false-membership function. But Bustince and Burillo (1996) showed that the vague set is the Atanassov IFS. Over the last decades, Atanassov's IFS has been applied to many different fields such as decision making (Chen & Tan, 1994; Szmidt & Kacprzyk, 2002, 2004; Gabriella et al., 2004; Atanassov et al., 2005; Herrera et al., 2005; Li, 2005, 2010a, 2010b; Pankowska & Wygralak, 2006; Lin et al., 2007; Liu & Wang, 2007; Li et al., 2010; Xu, 2007) and logic programming (Atanassov & Georgiev, 1993) as well as pattern recognition (Li & Cheng, 2002; Mitchell, 2005). Chen and Tan (1994) presented a technique for handling multi-criteria fuzzy decision making problems based on vague sets. They provided a score function to measure the degree of suitability of each alternative with respect to a set of criteria presented by vague values. Lin et al. (2007) and Liu and Wang (2007) discussed similar decision making problems using the Atanassov IFSs. Szmidt and Kacprzyk (2002, 2004) considered the use of the Atanassov IFSs for building soft decision making models with imprecise information and proposed two solution concepts about the Atanassov IFS core and the consensus winner for group decision making using the Atanassov IFSs. Atanassov et al. (2005) proposed an intuitionistic fuzzy interpretation of multi-person multi-criteria decision making. Gabriella et al. (2004) constructed a generalized net model of multi-person multi-criteria decision making model based on intuitionistic fuzzy graphs. Atanassov and Georgiev (1993) presented a logic programming system, which usesd Atanassov's IFSs to model various forms of uncertainty and considered the problem of propagating uncertainty through logical inference and various models of interpretation. The discussed framework allows knowledge representation and inference under uncertainty in the form of rules suitable for expert systems. Li and Cheng (2002) and Mitchell (2005) introduced some similarity measures of the Atanassov IFSs. Li (2005) constructed a series of linear programming models for MADM problems using Atanassov's IFSs and hereby proposed a closeness degree method. Pankowska and Wygralak (2006) studied general Atanassov IFSs with triangular norms and their application to group decision making. Xu (2007) discussed intuitionistic fuzzy preference relations and their applications to group decision making based on the order weighted averaging (OWA) operator. However, there exists little investigation on MAGDM problems in which multiple attributes are considered explicitly and both weights of all attributes and ratings of alternatives on attributes are expressed with the Atanassov IFSs. Li et al. (2009) developed a fractional programming methodology for solving this kind of MAGDM problems. However, in the methodology developed by Li et al. (2009), a series of fractional programming problems need to be solved for constructing closeness coefficient Atanassov IFSs, which are used to rank the alternatives. The computation is tedious and ranking the closeness coefficient Atanassov IFSs is not easy to be implemented.

The technique for order preference by similarity to ideal solution (TOPSIS) developed by Hwang and Yoon (1981) is one of the well-known classical MADM methods and has been widely used in real life decision situations. In this paper, TOPSIS is further extended to solve MAGDM problems in which all weights of attributes and ratings of alternatives on attributes are expressed with Atanassov's IFSs. In this methodology, an Atanassov IFS positive ideal-solution (IFSPIS) and an Atanassov IFS negative ideal-solution (IFSNIS) are defined and an Euclidean distance is defined to calculate the differences between each alternative for each decision maker and Atanassov IFSPIS as well as Atanassov IFSNIS. Furthermore, the degrees of the relative closeness to Atanassov IFSPIS for alternatives with respect to decision makers are calculated. In total, $K\times m$ relative closeness degrees are obtained since there are K decision makers in the group and m alternatives. All K decision makers may be regarded as "attributes". Then a corresponding classical MADM problem is generated and solved using the TOPSIS.

The remainder of this paper is organized as follows. We briefly review the concept, operations, and distance measures of Atanassov IFSs and TOPSIS (Hwang & Yoon, 1981). Next we present the process and algorithm of the extended TOPSIS for MAGDM problems under Atanassov IFS environments. Finally, a numerical example and short conclusion are given.

PRELIMINARIES

Atanassov IFSs and Operations

The concept of Atanassov IFS was firstly introduced by Atanassov (1986, 1999).

Definition 1 *(Atanassov, 1986, 1999). Let* $X=\{x_1, x_2, \ldots, x_m\}$ *be a finite universal set. An Atanassov IFS A in X is an object having the following form:*

$$A = \{< x_l, \mu_A(x_l), \upsilon_A(x_l) > | \, x_l \in X\},$$

where the functions

$$\mu_A : X \mapsto [0,1]$$
$$x_l \to \mu_A(x_l)$$

and

$$\upsilon_A : X \mapsto [0,1]$$
$$x_l \to \upsilon_A(x_l)$$

define the degree of membership and the degree of non-membership of an element $x_l \in X$ to the set $A \subseteq X$, respectively, and such that they satisfy the condition: $0 \leq \mu_A(x_l) + \upsilon_A(x_l) \leq 1$ for every $x_l \in X$.

Let

$$\pi_A(x_l) = 1 - \mu_A(x_l) - \upsilon_A(x_l),$$

which is called Atanassov intuitionistic index of element x_l in set A. It is the degree of indeterminacy membership of element x_l to set A. Obviously, $0 \leq \pi_A(x_l) \leq 1$.

Some operations and distances of the Atanassov IFSs, which will be employed, are introduced in the following.

Let A and B be two Atanassov IFSs in the set X. Namely,

$$A = \{< x_l, \mu_A(x_l), \upsilon_A(x_l) > | \, x_l \in X\}$$

and

$$B = \{< x_l, \mu_B(x_l), \upsilon_B(x_l) > | \, x_l \in X\},$$

where

$$\pi_A(x_l) = 1 - \mu_A(x_l) - \upsilon_A(x_l)$$

and

$$\pi_B(x_l) = 1 - \mu_B(x_l) - \upsilon_B(x_l).$$

Definition 2 *(Atanassov, 1986, 1999). Let A and B be two Atanassov IFSs in the set X.*

1. $A \subset B$ if and only if for any $x_l \in X$, $\mu_A(x_l) \leq \mu_B(x_l)$ and $\upsilon_A(x_l) \geq \upsilon_B(x_l)$;
2. $A = B$ if and only if for any $x_l \in X$, $\mu_A(x_l) = \mu_B(x_l)$ and $\upsilon_A(x_l) = \upsilon_B(x_l)$;
3. $A + B = \{< x_l, \mu_A(x_l) + \mu_B(x_l) - \mu_A(x_l)\mu_B(x_l), \upsilon_A(x_l)\upsilon_B(x_l) > | \, x_l \in X\}$;
4. $AB = \{< x_l, \mu_A(x_l)\mu_B(x_l), \upsilon_A(x_l) + \upsilon_B(x_l) - \upsilon_A(x_l)\upsilon_B(x_l) > | \, x_l \in X\}$.

The concept of the distance between the Atanassov IFSs was firstly introduced by Atanassov (1999). Some deeper discussion of the distance was given by Szmidt and Kacprzyk (2001) and Li (2004). Here an Euclidean distance between two Atanassov IFSs A and B is defined as in Box 1.

TOPSIS

TOPSIS developed by Hwang and Yoon (1981), with reference to Chen and Hwang (1992), is used to select the best one of ranking from a finite set of feasible alternatives. The basic principle is that the chosen alternative should have the shortest distance from the positive ideal-solution (PIS) and the farthest distance from the negative ideal-solution (NIS). Suppose that a MADM problem has m alternatives X_i (i=1, 2,..., m) evaluated with respect to n attributes A_j (j=1, 2,..., n), both quantitatively and qualitatively. Denote $X=\{X_1, X_2,..., X_m\}$ and $A=\{A_1, A_2,..., A_m\}$. Let f_{ij} be the value of each alternative X_i (i=1, 2,..., m) on every attribute A_j (j=1, 2,..., n). All these values are expressed concisely in the matrix format as $\boldsymbol{F}=(f_{ij})_{m \times n}$, which is referred to as a decision matrix usually represented a MADM problem.

Box 1. Euclidean distance between two Atanassov IFSs A and B

$$D_2(A,B)=\sqrt{\frac{1}{2}\sum_{l=1}^{m}[(\mu_A(x_l)-\mu_B(x_l))^2+(\upsilon_A(x_l)-\upsilon_B(x_l))^2+(\pi_A(x_l)-\pi_B(x_l))^2]} \quad (1)$$

Different attributes may have different importance. Assume that the relative weights of attributes A_j are ω_j (j=1, 2,..., n), which satisfy the normalization conditions: $\omega_j \in[0,1]$ and $\sum_{j=1}^{n}\omega_j=1$. Let $\boldsymbol{\omega}=(\omega_1, \omega_2, ..., \omega_n)^T$ be the relative weight vector of all attributes.

The main process of TOPSIS can be summarized as follows:

a. Normalize the decision matrix $\boldsymbol{F}=(f_{ij})_{m\times n}$ using the following formula: $r_{ij}=f_{ij}/\sqrt{\sum_{i=1}^{m}(f_{ij})^2}\,(i=1,2,\cdots,m;\,j=1,2\cdots,n)$, (2) which are called the normalized values.
b. Calculate the weighted normalized decision matrix $\boldsymbol{Y}=(y_{ij})_{m\times n}$ using the following formula:
$y_{ij}=\omega_j r_{ij}(i=1,2,\cdots,m;\,j=1,2\cdots,n)$. (3)
c. Determine the PIS X^* and the NIS X^- whose normalized values are defined as follows: $\mathbf{Y}^*=(y_1^*,y_2^*,\cdots,y_n^*)$ (4) and $\mathbf{Y}^-=(y_1^-,y_2^-,\cdots,y_n^-)$, (5) respectively, where $y_j^*=\max_{1\le i\le m}\{y_{ij}\}$ and $y_j^-=\min_{1\le i\le m}\{y_{ij}\}$ for $j\in\Omega_b$, $y_j^*=\min_{1\le i\le m}\{y_{ij}\}$ and $y_j^-=\max_{1\le i\le m}\{y_{ij}\}$ for $j\in\Omega_c$. Here Ω_b and Ω_c are subscript sets of the benefit attributes and the cost attributes in A, respectively.
d. Calculate the Euclidean distances of each alternative from the PIS and the NIS as follows: $D_i^*=\sqrt{\sum_{j=1}^{n}(y_{ij}-y_j^*)^2}\,(i=1,2,\cdots,m)$ (6) and $D_i^-=\sqrt{\sum_{j=1}^{n}(y_{ij}-y_j^-)^2}\,(i=1,2,\cdots,m)$, (7) respectively.
e. Calculate the relative closeness degree of each alternative to the PIS X^* as follows:
$$C_i=\frac{D_i^-}{D_i^*+D_i^-}\,(i=1,2,\cdots,m). \quad (8)$$
Obviously, $C_i\in[0,1]$ and the larger C_i the better X_i.
f. Rank the alternatives according to the relative closeness degrees to the PIS. The best alternative is the one with the largest relative closeness degree to the PIS.

THE EXTENDED TOPSIS FOR MAGDM PROBLEMS UNDER ATANASSOV IFS ENVIRONMENTS

Presentation of MAGDM Problems under Atanassov IFS Environments

Suppose that there exists an alternative set $X=\{X_1, X_2,..., X_m\}$, which consists of m non-inferior/feasible alternatives from which the most preferred alternative is to be selected by a group of K decision makers P_k (k=1,2,...,K). Assume decision maker P_k (k=1,2,...,K) constructs an Atanassov IFS $X_{ij}^k=\{<X_i,\mu_{ij}^k,\upsilon_{ij}^k>\}$, usually denoted by $X_{ij}^k=<\mu_{ij}^k,\upsilon_{ij}^k>$ for short, where μ_{ij}^k and υ_{ij}^k are the degree of membership/satisfaction and the degree of non-membership/non-satisfaction of alternative $X_i\in X$ (i=1, 2,..., m) with respect to attribute $A_j\in A$ (j=1,2,...,n) on the fuzzy concept "excellence" given by P_k, respectively; $0\le\mu_{ij}^k\le1$, $0\le\upsilon_{ij}^k\le1$ and $0\le\mu_{ij}^k+\upsilon_{ij}^k\le1$. In other words, the evaluation of alternative $X_i \in X$ on $A_j \in A$ given by decision maker P_k is an Atanassov IFS. Thus, evaluation of an alternative $X_i \in X$ given by P_k may be expressed as follows:

$$X_i = (X_{i1}^k, X_{i2}^k, \cdots, X_{in}^k) = (<\mu_{i1}^k, \upsilon_{i1}^k>, <\mu_{i2}^k, \upsilon_{i2}^k>, \cdots, <\mu_{in}^k, \upsilon_{in}^k>) \quad (9)$$

Then, a MAGDM problem under the Atanassov IFS environment can be expressed concisely in the matrix format as follows:

$$\mathbf{F}^k = (<\mu_{ij}^k, \upsilon_{ij}^k>)_{m\times n} = \begin{matrix} & A_1 & A_2 & \cdots & A_n \\ X_1 & <\mu_{11}^k, \upsilon_{11}^k> & <\mu_{12}^k, \upsilon_{12}^k> & \cdots & <\mu_{1n}^k, \upsilon_{1n}^k> \\ X_2 & <\mu_{21}^k, \upsilon_{21}^k> & <\mu_{22}^k, \upsilon_{22}^k> & \cdots & <\mu_{2n}^k, \upsilon_{2n}^k> \\ \vdots & \vdots & \vdots & & \vdots \\ X_m & <\mu_{m1}^k, \upsilon_{m1}^k> & <\mu_{m2}^k, \upsilon_{m2}^k> & \cdots & <\mu_{mn}^k, \upsilon_{mn}^k> \end{matrix} \quad (10)$$

for k=1, 2,…, K, which are usually referred to Atanassov IFS decision matrixes representing the MAGDM problem under the Atanassov IFS environment.

Similarly, assume that decision maker P_k (k=1, 2,…, K) constructs an Atanassov IFS $\omega_j^k = \{<A_j, \eta_j^k, \tau_j^k>\}$, usually denoted by $\omega_j^k = <\eta_j^k, \tau_j^k>$ for short, where η_j^k and τ_j^k are the degree of membership and the degree of non-membership of attribute $A_j \in A$ on the fuzzy concept "importance" given by P_k, respectively; $0 \le \eta_j^k \le 1, 0 \le \tau_j^k \le 1$ and $0 \le \eta_j^k + \tau_j^k \le 1$. Then, a weight vector of all attributes can be concisely expressed in the following format:

$$\boldsymbol{\omega}^k = (<\eta_1^k, \tau_1^k>, <\eta_2^k, \tau_2^k>, \cdots, <\eta_n^k, \tau_n^k>)^{\mathrm{T}} \quad (k = 1, 2, \cdots, K) \quad (11)$$

Process and Algorithm of the Extended TOPSIS for MAGDM Problems under Atanassov IFS Environments

According to (4) of Definition 2, using Equations (10) and (11), the weighted normalized Atanassov IFS decision matrix for decision maker P_k (k=1, 2,…, K) can be calculated as follows:

$$\bar{\mathbf{F}}^k = (<\bar{\mu}_{ij}^k, \bar{\upsilon}_{ij}^k>)_{m\times n} = \begin{matrix} & A_1 & A_2 & \cdots & A_n \\ X_1 & <\bar{\mu}_{11}^k, \bar{\upsilon}_{11}^k> & <\bar{\mu}_{12}^k, \bar{\upsilon}_{12}^k> & \cdots & <\bar{\mu}_{1n}^k, \bar{\upsilon}_{1n}^k> \\ X_2 & <\bar{\mu}_{21}^k, \bar{\upsilon}_{21}^k> & <\bar{\mu}_{22}^k, \bar{\upsilon}_{22}^k> & \cdots & <\bar{\mu}_{2n}^k, \bar{\upsilon}_{2n}^k> \\ \vdots & \vdots & \vdots & & \vdots \\ X_m & <\bar{\mu}_{m1}^k, \bar{\upsilon}_{m1}^k> & <\bar{\mu}_{m2}^k, \bar{\upsilon}_{m2}^k> & \cdots & <\bar{\mu}_{mn}^k, \bar{\upsilon}_{mn}^k> \end{matrix}, \quad (12)$$

where

$$<\bar{\mu}_{ij}^k, \bar{\upsilon}_{ij}^k> = \omega_j^k X_{ij}^k = <\eta_j^k, \tau_j^k><\mu_{ij}^k, \upsilon_{ij}^k> = <\eta_j^k \mu_{ij}^k, \tau_j^k + \upsilon_{ij}^k - \tau_j^k \upsilon_{ij}^k> \quad (13)$$

for i=1, 2,…, m and j=1, 2,…, n.

Using (1) of Definition 2 and Equation (12), an Atanassov IFSPIS and an Atanassov IFSNIS denoted as X^{k^+} and X^{k^-} are defined as follows:

$$X^{k^+} = (<\mu_1^{k^+}, \upsilon_1^{k^+}>, <\mu_2^{k^+}, \upsilon_2^{k^+}>, \cdots, <\mu_n^{k^+}, \upsilon_n^{k^+}>) \quad (14)$$

and

$$X^{k^-} = (<\mu_1^{k^-}, \upsilon_1^{k^-}>, <\mu_2^{k^-}, \upsilon_2^{k^-}>, \cdots, <\mu_n^{k^-}, \upsilon_n^{k^-}>), \quad (15)$$

respectively, where

$$\mu_j^{k^+} = \max_{1\le i\le m}\{\bar{\mu}_{ij}^k\},$$

$$\upsilon_j^{k^+} = \min_{1\le i\le m}\{\bar{\upsilon}_{ij}^k\},$$

$$\mu_j^{k^-} = \min_{1\le i\le m}\{\bar{\mu}_{ij}^{\ k}\},$$

$$\upsilon_j^{k^-} = \max_{1\le i\le m}\{\bar{\upsilon}_{ij}^{\ k}\}. \tag{16}$$

The Euclidean distances of an alternative $X_i \in X$ from Atanassov IFSPIS X^{k^+} and Atanassov IFSNIS X^{k^-} are defined as follows:

$$D_2^k(X_i, X^{k^+}) = \sqrt{\frac{1}{2}\sum_{j=1}^{n}[(\bar{\mu}_{ij}^{\ k} - \mu_j^{k^+})^2 + (\bar{\upsilon}_{ij}^{\ k} - \upsilon_j^{k^+})^2 + (\bar{\pi}_{ij}^{\ k} - \pi_j^{k^+})^2]} \tag{17}$$

and

$$D_2^k(X_i, X^{k^-}) = \sqrt{\frac{1}{2}\sum_{j=1}^{n}[(\bar{\mu}_{ij}^{\ k} - \mu_j^{k^-})^2 + (\bar{\upsilon}_{ij}^{\ k} - \upsilon_j^{k^-})^2 + (\bar{\pi}_{ij}^{\ k} - \pi_j^{k^-})^2]}, \tag{18}$$

respectively, where

$$\bar{\pi}_{ij}^{\ k} = 1 - \bar{\mu}_{ij}^{\ k} - \bar{\upsilon}_{ij}^{\ k},$$

$$\pi_j^{k^+} = 1 - \mu_j^{k^+} - \upsilon_j^{k^-},$$

$$\pi_j^{k^-} = 1 - \mu_j^{k^-} - \upsilon_j^{k^-}. \tag{19}$$

According to Equation (8), the relative closeness degree of an alternative X_i to Atanassov IFSPIS for each decision maker P_k is calculated as follows:

$$\rho_i^k = \frac{D_2^k(X_i, X^{k^-})}{D_2^k(X_i, X^{k^+}) + D_2^k(X_i, X^{k^-})}$$

$$(i = 1, 2, \cdots, m; k = 1, 2, \cdots, K). \tag{20}$$

Obviously, $0 \le \rho_i^k \le 1$ for i=1, 2,..., m and k=1, 2,..., K.

In total, $K\times m$ relative closeness degrees are obtained since there are K decision makers in the group and m alternatives. Each decision maker P_k (k=1, 2,..., K) may be regarded as an "attribute", which is still denoted by P_k. Hence, ρ_i^k given in Equation (20) is the value of each alternative X_i (i=1, 2,..., m) with respect to every "attribute" P_k (k=1, 2,..., K) (i.e., the decision maker). Then, we can construct a classical MADM problem with the decision matrix as follows:

$$\boldsymbol{B} = (\rho_i^k)_{m\times K} = \begin{matrix} & P_1 & P_2 & \cdots & P_K \\ X_1 & \rho_1^1 & \rho_1^2 & \cdots & \rho_1^K \\ X_2 & \rho_2^1 & \rho_2^2 & \cdots & \rho_2^K \\ \vdots & \vdots & \vdots & \vdots & \vdots \\ X_m & \rho_m^1 & \rho_m^2 & \cdots & \rho_m^K \end{matrix}. \tag{21}$$

In the group decision making problem, the group consists of different decision makers (or human beings) such as executive managers and domain experts. Usually different decision makers may have different importance, which should be taken into consideration in the aggregation procedure. Assume that the relative weights of decision makers P_k (k=1, 2,..., K) is w_k which satisfy the normalization conditions: $w_k \in [0,1]$ and $\sum_{k=1}^{K} w_k = 1$. Let $w=(w_1, w_2, ..., w_k)^T$ be the relative weight vector of the decision makers.

Combining Equation (3) with Equation (21), the weighted normalized decision matrix is calculated as follows:

$$\bar{\boldsymbol{B}} = (\bar{\rho}_i^k)_{m\times K} = \begin{matrix} & P_1 & P_2 & \cdots & P_K \\ X_1 & \bar{\rho}_1^1 & \bar{\rho}_1^2 & \cdots & \bar{\rho}_1^K \\ X_2 & \bar{\rho}_2^1 & \bar{\rho}_2^2 & \cdots & \bar{\rho}_2^K \\ \vdots & \vdots & \vdots & \vdots & \vdots \\ X_m & \bar{\rho}_m^1 & \bar{\rho}_m^2 & \cdots & \bar{\rho}_m^K \end{matrix}, \tag{22}$$

where $\bar{\rho}_i^k = w_k \rho_i^k$ (i=1, 2,..., m; k=1, 2,..., K).

Using Equations (4) and (5) with Equation (22), a PIS and a NIS denoted by X^+ and X^- are defined as follows:

$$X^+ = (\rho_1^+, \rho_2^+, \cdots \rho_K^+) \tag{23}$$

and

$$X^- = (\rho_1^-, \rho_2^-, \cdots \rho_K^-), \tag{24}$$

respectively, where $\rho_k^+ = \max_{1 \le i \le m}\{\bar{\rho}_i^k\}$ and $\rho_k^- = \min_{1 \le i \le m}\{\bar{\rho}_i^k\}$ (k=1,2,...,K).

The Euclidean distances of each alternative X_i from the PIS X^+ and the NIS X^- are defined using Equations (6) and (7) as follows:

$$D_i^+ = \sqrt{\sum_{k=1}^{K} (\bar{\rho}_i^k - \rho_k^+)^2} \tag{25}$$

and

$$D_i^- = \sqrt{\sum_{k=1}^{K} (\bar{\rho}_i^k - \rho_k^-)^2}, \tag{26}$$

respectively.

The relative closeness degree of each alternative $X_i \in X$ with respect to the PIS X^+ is defined as follows:

$$C_i = \frac{D_i^-}{D_i^+ + D_i^-} \quad . \tag{27}$$

It is easily seen that the larger C_i the better X_i. Hence the ranking order of all alternatives can be generated according to the descending order of the relative closeness degrees, and the best alternative is the one with the largest relative closeness degree.

A NUMERICAL EXAMPLE

In this section, a numerical example is worked out to illustrate the method proposed in this paper. Assume that there exists a case study of three experts (i.e., decision makers) P_1, P_2 and P_3 forming a group which assesses three C[3]I systems (i.e., alternatives) in service, denoted by the alternative set { X_1, X_2, X_3}. Four operational readiness indexes such as information accuracy A_1, information consistency A_2, system availability A_3 as well as picture completeness A_4 are identified as the evaluation attributes for the C[3]I systems. Using statistical methods, for each expert P_k (k=1,2,3), the degree of membership/satisfaction μ_{ij}^k and the degree of non-membership/on-satisfaction υ_{ij}^k for alternative X_i (i=1,2,3) with respect to attribute A_j (j=1,2,3,4) on the fuzzy concept "excellence" can be obtained and summarized in the Atanassov IFS matrix format as follows:

$$\mathbf{F}^1 = (<\mu_{ij}^1, \upsilon_{ij}^1>)_{3\times 4} = \begin{matrix} & A_1 & A_2 & A_3 & A_4 \\ X_1 & <0.75,\ 0.10> & <0.80,\ 0.15> & <0.40,\ 0.45> & <0.62,\ 0.18> \\ X_2 & <0.60,\ 0.25> & <0.68,\ 0.20> & <0.75,\ 0.05> & <0.49,\ 0.08> \\ X_3 & <0.80,\ 0.20> & <0.45,\ 0.50> & <0.60,\ 0.30> & <0.76,\ 0.06> \end{matrix}$$

$$\mathbf{F}^2 = (<\mu_{ij}^2, \upsilon_{ij}^2>)_{3\times 4} = \begin{matrix} & A_1 & A_2 & A_3 & A_4 \\ X_1 & <0.71,\ 0.15> & <0.82,\ 0.11> & <0.31,\ 0.48> & <0.40,\ 0.35> \\ X_2 & <0.58,\ 0.35> & <0.58,\ 0.30> & <0.81,\ 0.15> & <0.65,\ 0.12> \\ X_3 & <0.84,\ 0.05> & <0.61,\ 0.30> & <0.65,\ 0.20> & <0.74,\ 0.20> \end{matrix}$$

and

$$\mathbf{F}^3 = (<\mu_{ij}^3, \upsilon_{ij}^3>)_{3\times 4} = \begin{matrix} & A_1 & A_2 & A_3 & A_4 \\ X_1 & <0.85,\ 0.10> & <0.75,\ 0.10> & <0.48,\ 0.32> & <0.60,\ 0.35> \\ X_2 & <0.75,\ 0.05> & <0.70,\ 0.15> & <0.65,\ 0.15> & <0.55,\ 0.05> \\ X_3 & <0.60,\ 0.30> & <0.56,\ 0.20> & <0.70,\ 0.16> & <0.52,\ 0.20> \end{matrix}$$

respectively.

In a similar way, the degree of membership/importance η_j^k and the degree of non-membership/non-importance τ_j^k for the attribute A_j (j=1,2,3,4) given by each expert P_k (k=1,2,3) can be obtained and expressed in the Atanassov IFS vector format as follows:

$$\acute{\boldsymbol{E}}^1 = (<\eta_j^1, \tau_j^1>)_{4\times 1} = (<0.35,\ 0.25>\ <0.25,\ 0.40>\ <0.30,\ 0.55>\ <0.38,\ 0.56>)^{\mathrm{T}},$$

$$\acute{\boldsymbol{E}}^2 = (<\eta_j^2, \tau_j^2>)_{4\times 1} = (<0.25,\ 0.25>\ <0.30,\ 0.65>\ <0.35,\ 0.40>\ <0.30,\ 0.65>)^{\mathrm{T}},$$

and

$$\acute{\boldsymbol{E}}^3 = (<\eta_j^3, \tau_j^3>)_{4\times 1} = (<0.31,\ 0.45>\ <0.22,\ 0.50>\ <0.28,\ 0.59>\ <0.30,\ 0.52>)^{\mathrm{T}},$$

respectively.

According to Equation (12), three weighted normalized Atanassov IFS decision matrixes can be obtained as follows:

$$\bar{\mathbf{F}}^1 = (<\bar{\mu}_{ij}^1, \bar{\upsilon}_{ij}^1>)_{3\times 4} = \begin{matrix} & A_1 & A_2 & A_3 & A_4 \\ X_1 & <0.26,\ 0.33> & <0.20,\ 0.49> & <0.12,\ 0.75> & <0.24,\ 0.64> \\ X_2 & <0.21,\ 0.44> & <0.17,\ 0.52> & <0.23,\ 0.57> & <0.19,\ 0.60> \\ X_3 & <0.28,\ 0.40> & <0.11,\ 0.70> & <0.18,\ 0.69> & <0.29,\ 0.59> \end{matrix}$$

$$\bar{\mathbf{F}}^2 = (<\bar{\mu}_{ij}^2, \bar{\upsilon}_{ij}^2>)_{3\times 4} = \begin{matrix} & A_1 & A_2 & A_3 & A_4 \\ X_1 & <0.18,\ 0.36> & <0.25,\ 0.69> & <0.11,\ 0.69> & <0.12,\ 0.77> \\ X_2 & <0.15,\ 0.51> & <0.17,\ 0.76> & <0.28,\ 0.49> & <0.20,\ 0.69> \\ X_3 & <0.21,\ 0.29> & <0.18,\ 0.76> & <0.23,\ 0.52> & <0.22,\ 0.72> \end{matrix}$$

and

$$\bar{\mathbf{F}}^3 = (<\bar{\mu}_{ij}^3, \bar{\upsilon}_{ij}^3>)_{3\times 4} = \begin{matrix} & A_1 & A_2 & A_3 & A_4 \\ X_1 & <0.26,\ 0.51> & <0.17,\ 0.55> & <0.13,\ 0.72> & <0.18,\ 0.69> \\ X_2 & <0.23,\ 0.48> & <0.15,\ 0.58> & <0.18,\ 0.65> & <0.17,\ 0.54> \\ X_3 & <0.19,\ 0.61> & <0.12,\ 0.60> & <0.20,\ 0.66> & <0.16,\ 0.62> \end{matrix}$$

respectively.

Using Equations (14) and (15), an Atanassov IFSPIS X^{k^+} (k=1,2,3) and an Atanassov IFSNIS X^{k^-} $(k = 1,2,3)$ can be determined as follows:

$$X^{1^+} = (<0.28, 0.33>, <0.20, 0.49>, <0.23, 0.57>, <0.29, 0.59>)$$

$$X^{1^-} = (<0.21, 0.44>, <0.11, 0.70>, <0.12, 0.75>, <0.19, 0.64>)$$

$$X^{2^+} = (<0.21, 0.29>, <0.25, 0.69>, <0.28, 0.49>, <0.22, 0.69>)$$

$$X^{2^-} = (<0.15, 0.36>, <0.17, 0.76>, <0.11, 0.69>, <0.12, 0.77>),$$

$$X^{3^+} = (<0.26, 0.48>, <0.17, 0.55>, <0.20, 0.65>, <0.18, 0.54>),$$

and

$$X^{3^-} = (<0.19, 0.61>, <0.12, 0.60>, <0.13, 0.72>, <0.16, 0.69>),$$

respectively.

Using Equations (17)-(19), the distances of the alternative X_i (i=1,2,3) from X^{k^+} (k=1,2,3) and from X^{k^-} are calculated as in Table 1.

Using Equation (20), the relative closeness degrees ρ_i^k $(i = 1,2,3)$ to X^{k^+} $(k = 1,2,3)$ are calculated as in Table 2.

Using Equation (21), the decision matrix presenting relative closeness degrees can be obtained as follows:

$$\boldsymbol{B} = \begin{matrix} & P_1 & P_2 & P_3 \\ X_1 & 0.561 & 0.272 & 0.382 \\ X_2 & 0.620 & 0.544 & 0.811 \\ X_3 & 0.355 & 0.680 & 0.402 \end{matrix}.$$

Taking the weight vector of the experts P_k (k=1,2,3) as $\boldsymbol{w}=(0.35,0.35,0.3)^{\mathrm{T}}$ and using Equation (22), the weighted normalized decision matrix is calculated as follows:

$$\bar{\boldsymbol{B}} = \begin{matrix} & P_1 & P_2 & P_3 \\ X_1 & 0.196 & 0.220 & 0.124 \\ X_2 & 0.095 & 0.190 & 0.238 \\ X_3 & 0.115 & 0.243 & 0.121 \end{matrix}.$$

Using Equations (23) and (24), the PIS X^+ and the NIS X^- can be obtained as follows:

Table 1. Distances of X_i from X^{k^+} and X^{k^-}

P_k	$D_2(X_1, X^{k^+})$	$D_2(X_2, X^{k^+})$	$D_2(X_3, X^{k^+})$	$D_2(X_1, X^{k^-})$	$D_2(X_2, X^{k^-})$	$D_2(X_3, X^{k^-})$
P_1	0.166	0.139	0.222	0.212	0.227	0.122
P_2	0.217	0.212	0.088	0.081	0.253	0.187
P_3	0.168	0.046	0.143	0.104	0.198	0.096

$X^+ = (0.220, 0.238, 0.243)$

and

$X^- = (0.124, 0.095, 0.115)$,

respectively.

Using Equations (25) and (26), the distances of the alternatives X_i (i=1,2,3) from X^+ and X^- can be obtained as follows:

$D_1^+ = 0.193$,

$D_2^+ = 0.048$,

$D_3^+ = 0.155$,

$D_1^- = 0.072$,

$D_2^- = 0.186$,

$D_3^- = 0.143$,

respectively.

According to Equation (27), the relative closeness degrees of the alternatives X_i (i=1,2,3) with respect to the PIS X^+ can be calculated as follows:

$C_1 = 0.272$,

$C_2 = 0.765$,

$C_3 = 0.480$,

respectively. Hence, the best alternative is X_2, and the ranking order of the three alternatives is $X_2 \succ X_3 \succ X_1$.

To make comparison between the proposed method in this paper and the fractional programming method developed by Li et al. (2009), using the method proposed by Li et al. (2009), the relative closeness coefficient intervals of the alternatives X_i ($i=1,2,3$) through solving six

Table 2. Relative closeness degrees ρ_i^k to X^{k^+}

P_k	ρ_1^k	ρ_2^k	ρ_3^k
P_1	0.561	0.620	0.355
P_2	0.272	0.544	0.680
P_3	0.382	0.811	0.402

fractional programming problems for the expert P_1 are obtained as follows:

$$[C_1^{1l}, C_1^{1u}] = [0.5792, 0.8273],$$

$$[C_2^{1l}, C_2^{1u}] = [0.5902, 0.8609],$$

$$[C_3^{1l}, C_3^{1u}] = [0.5759, 0.7801],$$

respectively.

Similarly, the relative closeness coefficient intervals of the alternatives X_i $(i=1,2,3)$ for the experts P_2 and P_3 can be obtained by solving twelve fractional programming problems as follows:

$$[C_1^{2l}, C_1^{2u}] = [0.4471, 0.7664],$$

$$[C_2^{2l}, C_2^{2u}] = [0.6154, 0.8030],$$

$$[C_3^{2l}, C_3^{2u}] = [0.6703, 0.8607]$$

and

$$[C_1^{3l}, C_1^{3u}] = [0.6123, 0.8217],$$

$$[C_2^{3l}, C_2^{3u}] = [0.6221, 0.9212],$$

$$[C_3^{3l}, C_3^{3u}] = [0.5677, 0.7932],$$

respectively.

According to Equation (17) given by Li et al. (2009), the relative closeness coefficient intervals of the alternatives X_i $(i=1,2,3)$ for the group can be calculated as follows:

$$\xi_1 = \sum_{k=1}^{3} w_k [C_1^{kl}, C_1^{ku}] = [0.5429,\ 0.8043],$$

$$\xi_2 = \sum_{k=1}^{3} w_k [C_2^{kl}, C_2^{ku}] = [0.6086,\ 0.8588]$$

and

$$\xi_3 = \sum_{k=1}^{3} w_k [C_3^{kl}, C_3^{ku}] = [0.6065,\ 0.8122],$$

respectively.

Using Equation (18) given by Li et al. (2009), likelihoods $p(\xi_1 \geq \xi_2)$, $p(\xi_2 \geq \xi_3)$ and $p(\xi_3 \geq \xi_1)$ can be calculated as follows:

$p(\xi_1 \geq \xi_2)$=0.3825, $p(\xi_2 \geq \xi_3)$=0.5534, $p(\xi_3 \geq \xi_1)$=0.5765,

respectively. Thus, likelihoods of pair-wise comparisons of the three alternatives X_1, X_2 and X_3 can be obtained in the matrix format as follows:

$$\boldsymbol{P} = \begin{array}{c} \\ X_1 \\ X_2 \\ X_3 \end{array} \begin{array}{c} \begin{matrix} X_1 & X_2 & X_3 \end{matrix} \\ \begin{pmatrix} 0.5 & 0.3825 & 0.4235 \\ 0.6175 & 0.5 & 0.5534 \\ 0.5765 & 0.4466 & 0.5 \end{pmatrix} \end{array}.$$

Using Equation (19) given by Li et al. (2009), optimal degrees of membership for the alternatives X_1, X_2 and X_3 can be calculated as follows:

θ_1=0.3010, θ_2=0.3618, θ_3=0.3372,

respectively. Then, the ranking order of the three alternatives is $X_2 \succ X_3 \succ X_1$, which is the same as that obtained by the proposed method in this paper. From the above process of calculation, it is easy to see that the method proposed by Li et al. (2009) has tedious calculation. However, the proposed method in this paper has small calculation and is easy to be implemented. On the other hand, the method given by Li et al. (2009) calculates the closeness coefficient intervals, which are mathematically equivalent to the closeness coefficient Atanassov IFSs. It is well-known that ranking the Atanassov IFSs is a difficult task. Moreover, different ranking methods may result in different ranking orders of the closeness coefficient intervals/Atanassov IFSs. In this case, it is intractable for the decision maker to make the final decision.

CONCLUSION

The TOPSIS is further extended to develop the extended TOPSIS under Atanassov IFS settings for solving MAGDM problems in which the weights of attributes and the ratings of alternatives on attributes are express with the Atanassov IFSs. The Euclidean distance measure is developed to calculate the differences between each alternative for each decision maker and the Atanassov IFSPIS as well as the Atanassov IFSNIS. The relative closeness degrees of all alternatives to the Atanassov IFSPIS for the decision makers in the group are calculated and all K decision makers may be regarded as "attributes". Then a corresponding classical MADM problem is generated and solved by the traditional TOPSIS.

Mathematically the Atanassov IFS and the interval-value fuzzy set (IVFS) are equivalent (Atanassov & Gargov, 1989; Grzegorzewski & Mrówka, 2005; Herrera et al., 2005). Therefore, the method proposed in this paper can be applied to MAGDM problems with IVFSs.

The method proposed in this paper allows us to use flexible ways to simulate real decision situations, which may build more realistic scenarios describing possible future events. Our study remarkably differs from those existing works (Szmidt & Kacprzyk, 2002, 2004; Li, 2005, 2010a, 2010b; Pankowska & Wygralak, 2006; Lin et al., 2007; Liu & Wang, 2007; Xu, 2007; Li et al., 2009, 2010) in that the method proposed in this paper is developed on the concept that the chosen alternative should have the shortest distance from the Atanassov IFSPIS and the farthest distance from the Atanassov IFSNIS. The defined Euclidean distances are directly used to measure differences between alternatives which are expressed with the Atanassov IFSs. The computation process is simple and easy to be implemented.

Although the method proposed in this paper is illustrated by the C^3I system selection problem, it can also be applied to similar problems under Atanassov IFS environments with a wide spectrum of possibilities, which may enable the explicit consideration of the best and the worst results one can expect.

ACKNOWLEDGMENT

The authors would like to thank the valuable comments and also appreciate the constructive suggestions from the anonymous referees and the Editor-in-Chief Toly Chen for International Journal of Fuzzy System Applications. This research was sponsored by the Natural Science Foundation of China (Nos. 70871117, 70902041 and 71001015) and the Humanities and Social Sciences research project of the Ministry of Education of China (08JC630072).

REFERENCES

Atanassov, K. T. (1986). Intuitionistic fuzzy sets. *Fuzzy Sets and Systems*, *50*, 87–96. doi:10.1016/S0165-0114(86)80034-3

Atanassov, K. T. (1999). *Intuitionistic fuzzy sets*. Heidelberg, Germany: Springer-Verlag.

Atanassov, K. T., & Gargov, G. (1989). Interval-valued intuitionistic fuzzy sets. *Fuzzy Sets and Systems*, *31*, 343–349. doi:10.1016/0165-0114(89)90205-4

Atanassov, K. T., & Georgiev, C. (1993). Intuitionistic fuzzy prolog. *Fuzzy Sets and Systems, 53*, 121–128. doi:10.1016/0165-0114(93)90166-F

Atanassov, K. T., Pasi, G., & Yager, R. R. (2005). Intuitionistic fuzzy interpretations of multi-criteria multi-person and multi-measurement tool decision making. *International Journal of Systems Science, 36*(14), 859–868. doi:10.1080/00207720500382365

Braglia, M., Frosolini, M., & Montanari, R. (2003). Fuzzy TOPSIS approach for failure mode, effects and criticality analysis. *Quality and Reliability Engineering, 19*, 425–443. doi:10.1002/qre.528

Burillo, P., & Bustince, H. (1996). Vague sets are intuitionistic fuzzy sets. *Fuzzy Sets and Systems, 79*, 403–405. doi:10.1016/0165-0114(95)00154-9

Chen, C. T. (2000). Extension of the TOPSIS for group decision-making under fuzzy environment. *Fuzzy Sets and Systems, 114*, 1–9. doi:10.1016/S0165-0114(97)00377-1

Chen, M. F., & Tzeng, G. H. (2004). Combining grey relation and TOPSIS concepts for selecting an expatriate host country. *Mathematical and Computer Modelling, 40*, 1473–1490. doi:10.1016/j.mcm.2005.01.006

Chen, S. J., & Hwang, C. L. (1992). *Fuzzy multiple attribute decision making: Methods and applications*. Berlin, Germany: Springer-Verlag.

Chen, S. M., & Tan, J. M. (1994). Handling multicriteria fuzzy decision-making problems based on vague set theory. *Fuzzy Sets and Systems, 67*, 163–172. doi:10.1016/0165-0114(94)90084-1

Chu, T. C. (2002a). Facility location selection using fuzzy TOPSIS under group decisions. *International Journal of Uncertainty. Fuzziness and Knowledge-Based Systems, 10*, 687–701. doi:10.1142/S0218488502001739

Chu, T. C. (2002b). Selecting plant location via a fuzzy TOPSIS approach. *International Journal of Advanced Manufacturing Technology, 20*, 859–864. doi:10.1007/s001700200227

Chu, T. C., & Lin, Y. C. (2003). A fuzzy TOPSIS method for robot selection. *International Journal of Advanced Manufacturing Technology, 21*, 284–290. doi:10.1007/s001700300033

Deschrijver, G., & Kerre, E. E. (2007). On the position of intuitionistic fuzzy set theory in the framework of theories modelling imprecision. *Information Sciences, 177*, 1860–1866. doi:10.1016/j.ins.2006.11.005

Gabriella, P., Yager, R. R., & Atanassov, K. T. (2004). Intuitionistic fuzzy graph interpretations of multi-person multi-criteria decision making: Generalized net approach. In *Proceedings of the Second IEEE International Conference on Intelligent Systems* (pp. 434-439).

Gau, W. L., & Buehrer, D. J. (1993). Vague sets. *IEEE Transactions on Systems, Man, and Cybernetics, 23*, 610–614. doi:10.1109/21.229476

Grzegorzewski, P., & Mrówka, E. (2005). Some notes on (Atanassov's) intuitionistic fuzzy sets. *Fuzzy Sets and Systems, 156*(3), 492–495. doi:10.1016/j.fss.2005.06.002

Herrera, F., Martinez, L., & Sanchez, P. J. (2005). Managing non-homogeneous information in group decision making. *European Journal of Operational Research, 166*, 115–132. doi:10.1016/j.ejor.2003.11.031

Hwang, C. L., & Yoon, K. (1981). *Multiple attribute decision making: Methods and applications: A state of the art survey*. Berlin, Germany: Springer-Verlag.

Li, D. F. (2003). *Fuzzy multiobjective many-person decision makings and games*. Beijing, China: National Defense Industry Press.

Li, D. F. (2004). Some measures of dissimilarity in intuitionistic fuzzy structures. *Journal of Computer and System Sciences*, *68*, 115–122. doi:10.1016/j.jcss.2003.07.006

Li, D. F. (2005). Multiattribute decision making models and methods using intuitionistic fuzzy sets. *Journal of Computer and System Sciences*, *70*, 73–85. doi:10.1016/j.jcss.2004.06.002

Li, D. F. (2010a). TOPSIS-based nonlinear-programming methodology for multiattribute decision making with interval-valued intuitionistic fuzzy sets. *IEEE Transactions on Fuzzy Systems*, *18*(2), 299–311.

Li, D. F. (2010b). Linear programming method for MADM with interval-valued intuitionistic fuzzy sets. *Expert Systems with Applications*, *37*(8), 5939–5945. doi:10.1016/j.eswa.2010.02.011

Li, D. F., Chen, G. H., & Huang, Z. G. (2010). Linear programming method for multiattribute group decision making using IF sets. *Information Sciences*, *180*(9), 1591–1609. doi:10.1016/j.ins.2010.01.017

Li, D. F., & Cheng, C. T. (2002). New similarity measures of intuitionistic fuzzy sets and application to pattern recognitions. *Pattern Recognition Letters*, *23*, 221–225. doi:10.1016/S0167-8655(01)00110-6

Li, D. F., Wang, Y. C., Liu, S., & Shan, F. (2009). Fractional programming methodology for multi-attribute group decision-making using IFS. *Applied Soft Computing*, *9*, 219–225. doi:10.1016/j.asoc.2008.04.006

Lin, L., Yuan, X. H., & Xia, Z. Q. (2007). Multicriteria fuzzy decision-making methods based on intuitionistic fuzzy sets. *Journal of Computer and System Sciences*, *73*, 84–88. doi:10.1016/j.jcss.2006.03.004

Liu, H. W., & Wang, G. J. (2007). Multi-criteria decision-making methods based on intuitionistic fuzzy sets. *European Journal of Operational Research*, *179*, 220–233. doi:10.1016/j.ejor.2006.04.009

Mitchell, H. B. (2005). Pattern recognition using type-II fuzzy sets. *Information Sciences*, *170*, 409–418. doi:10.1016/j.ins.2004.02.027

Pankowska, A., & Wygralak, M. (2006). General IF-sets with triangular norms and their applications to group decision making. *Information Sciences*, *176*, 2713–2754. doi:10.1016/j.ins.2005.11.011

Szmidt, E., & Kacprzyk, J. (2001). Distances between intuitionistic fuzzy sets. *Fuzzy Sets and Systems*, *114*, 505–518. doi:10.1016/S0165-0114(98)00244-9

Szmidt, E., & Kacprzyk, J. (2002). Using intuitionistic fuzzy sets in group decision making. *Control and Cybernetics*, *31*, 1037–1053.

Szmidt, E., & Kacprzyk, J. (2004). A concept of similarity for intuitionistic fuzzy sets and its use in group decision making. In *Proceedings of the International Joint Conference on Neural Networks and the IEEE International Conference on Fuzzy Systems* (pp. 25-29).

Triantaphyllou, E., & Lin, C. T. (1996). Development and evaluation of five fuzzy multiattribute decision-making methods. *International Journal of Approximate Reasoning*, *14*, 281–310. doi:10.1016/0888-613X(95)00119-2

Xu, Z. S. (2007). Intuitionistic preference relations and their application in group decision making. *Information Sciences*, *177*, 2363–2379. doi:10.1016/j.ins.2006.12.019

Zadeh, L. A. (1965). Fuzzy sets. *Information and Control*, *18*, 338–356. doi:10.1016/S0019-9958(65)90241-X

Zhao, M. Y., Cheng, C. T., Chau, K. W., & Li, G. (2006). Multiple criteria data envelopment analysis for full ranking units associated to environment impact assessment. *International Journal of Environment and Pollution*, *28*(3-4), 448–464. doi:10.1504/IJEP.2006.011222

This work was previously published in the International Journal of Fuzzy System Applications, Volume 1, Issue 4, edited by Toly Chen, pp. 47-61, copyright 2010 by IGI Publishing (an imprint of IGI Global).

Chapter 18
Applying the Linguistic Strategy-Oriented Aggregation Approach to Determine the Supplier Performance with Ordinal and Cardinal Data Forms

Shih-Yuan Wang
Jinwen University of Science and Technology, Taiwan

Sheng-Lin Chang
China University of Science and Technology, Taiwan

Reay-Chen Wang
Tungnan University, Taiwan

ABSTRACT

Supply chain management is a new and evolving paradigm for enterprises to cope with international competition and to improve global logistics efficiency. The suppliers' performances affect not only supply chain execution results but also the profit capability and business survivability. However, suppliers' performance assessment always involves a large dimension of supplier behaviors. Information on supplier behaviors is often difficult to be accurately demonstrated as quantitative data. For this reason, the study employs a 2-tuple linguistic variable to perform the initial evaluation and final assessment while keeping track of both linguistic information and data, which can avoid a tied result. Additionally, the modified linguistic ordered weighted averaging (M-LOWA) operator with maximum entropy is used to derive the maximum aggregation value under the current business strategy to reflect on the criteria. The focal company can then rapidly rely on the assessment results to represent the performance of suppliers and provide integrated information to decision makers. This study draws the complete framework for the issue of supplier performance assessment without limitations on categories of variables and scales.

DOI: 10.4018/978-1-4666-1870-1.ch018

INTRODUCTION

Today, enterprises must develop a new perspective toward suppliers from being competitive to cooperative in order to not only cope with the challenges of international competition and global logistics but also to establish their efficient supply chain systems. Hence, the mechanism of supplier assessment becomes critical for developing the supply chain. The assessment criteria must consider supplier ability, performance and potential capability. More importantly, the criteria must include enterprise strategy (Krause et al., 2001), product position (Aitken et al., 2003) and supplier performance (Carbonara et al., 2002) to improve the compatibility of the assessment results and to demonstrate the advantages of the supply chain.

Due to the broad range of content for assessing supplier performance, supplier behavior information should be analyzed in various ways according to the characteristics of each criterion that increases the difficulty of the assessment process. Choi and Hartley (1996) evaluated supplier performance based on consistency, reliability, relationship, flexibility, price, service, technological capability and finances, and also addressed 26 supplier selection criteria. Verma and Pullman (1998) ranked the importance of supplier attributes including quality, on-time delivery, cost, lead-time and flexibility. Vonderembse and Tracey (1999) discussed supplier and manufacturing performances that could be determined by supplier selection criteria and supplier involvement. Furthermore, they concluded that supplier selection could be evaluated by quality, availability, reliability and performance, while supplier involvement could be evaluated by product research and development (R&D) and improvement, and supplier performance could be evaluated by stoppage, delivery, damage and quality. Additionally, manufacturing performance could be evaluated by cost, quality, inventory and delivery.

Krause et al. (2001) devised a purchasing strategy based on the competitiveness in cost, quality, delivery, flexibility and innovation. Tracey and Tan (2001) developed supplier selection criteria, including quality, delivery, reliability, performance and price, and assessed customer satisfaction based on price, quality, variety and delivery. Furthermore, Kannan and Tan (2002) made supplier selection based on commitment, needs, capability, fit and honesty, and developed a system for supplier evaluation based on delivery, quality, responsiveness and information sharing. Kannan and Tan (2002) also made supplier selection and evaluated supplier performance based on the weights of evaluation attributes or criteria with crisp values that depend on subjective individual judgments.

Muralidharan et al. (2002) compared the advantages and limitations of nine previously developed methods of supplier rating, combined multiple criteria decision-making and analytic hierarchy processes to construct multi-criteria group decision-making model for supplier rating. The attributes of quality, delivery, price, technique capability, finance, attitude, facility, flexibility and service were used for supplier evaluation, and the attributes of knowledge, skill, attitude and experience were used for individual assessments. Sarkis and Talluri (2002) suggested that purchasing function has been attracting growing interest as a critical component of supply chain management, and multiple factors have been considered in supplier selection and evaluation, including strategic, operational, tangible and intangible measures within planning horizon, culture, technology, relationship, cost, quality, time and flexibility.

Chan (2003) discriminated between quantitative (cost, resource utilization) and qualitative (quality, flexibility, visibility, trust, innovativeness) performance measurements from the supply chain, and defined the belonging dimension and scale. Sharland et al. (2003) made supplier selection based on cycle time, proximity, manufacturing quality, comparative price and ease of qualifying to evaluate supplier performance and relationship. Moreover, Otto and Kotzab (2003) derived the goals of supply chain management from six perspectives, and described standard problems, solutions and performance metrics. Additionally, Gunasekaran et al. (2004) proposed a framework for supply chain performance measurement based on order planning, supplier, production and delivery performance, and defined the related activities into three layers, i.e., strategic, tactical and operational. Furthermore, Talluri and Narasimhan (2004) believed strategic sourcing to be critical for firms implementing supply chain management, and grouped supplier capability and performance assessment into six and five categories, respectively. Talluri and Narasimhan (2004) also demonstrated 15 proposed vendor evaluation techniques.

Due to the fact that supplier behaviors have the properties of dynamic and continuity, they therefore resemble fuzzy properties with uncertainty and inaccuracy. Thus, the assessment mechanism can neither be objectively nor completely based on crisp value. Aggregated criteria by crisp values are unable to sufficiently express fuzzy properties and trade-off among attributes of decision makers. Therefore, to express the level of importance using linguistic values is much simpler than using numerals; coinciding with human decision processes (Zhang et al., 2003). Subsequently, the original problem can be transformed into fuzzy multiple attribute decision making (FMADM) problem. Chen and Hwang (1992) classified the existing techniques of FMADM into three categories. In the first category, five methods (α-cut, fuzzy arithmetic, eigenvector method, weight assessing arithmetic operation, max & min operators) were introduced to deal with fuzzy information. In the second category, four methods (possibility & necessity measures, human intuition, ranking methods, linguistic transformation) were introduced to convert fuzzy information into crisp. Only one method (fuzzy outranking relation) in the third category was introduced to deal with crisp information. Aouam et al. (2003) proposed a modified method for fuzzy outranking relation that could lead the third category into the second category; hence techniques of FMADM could converge on two categories. However, this study is distinct from the techniques described in Chen and Hwang (1992) with the following characteristics:

1. This study can deal with linguistic information directly without transforming it into crisp.
2. This study applies linguistic guided weighted operator instead of pre-assigned weighted operator.
3. This study reveals the final result via linguistic instead of crisp.

The next section outlines the research method and purpose of this study. Subsequent sections describe the procedure involved in this approach and give a numerical example detailing how to apply this approach. Finally, some points are made and conclusions obtained using the proposed approach are given.

RESEARCH METHOD AND PURPOSE

This study employs a 2-tuple linguistic variable (Herrera & Martinez, 2000) to evaluate five criteria (R&D, cost, quality, service and response) and the associated supplier behaviors to each criterion. The modified linguistic ordered weighted averaging (M-LOWA) operator (Herrera et al., 1996) with maximum entropy (Filev & Yager, 1995), which is guided by a fuzzy linguistic quantifier (Herrera et al., 2000), is then employed to aggregate the evaluation results. The fuzzy linguistic quantifier expresses the fuzzy majority concept (Kacprzyk, 1986) employed to reflect the importance of the criteria based on fitting different strategies. Thus, the aggregation results are sufficient to realize the practical supplier performance and can be used to imitate human decision making. Figure 1 illustrates this approach with a flow chart.

THE MULTIPLE CRITERIA MATRIX FOR SUPPLIER PERFORMANCE EVALUATION

Generally, supplier performance is related to contract content and realization. Contract attributes and the associated supplier behaviors are defined by separating the performance criteria mentioned in previous investigations. The multiple criteria matrix for supplier performance $A = [a_{ijk}]$ is constructed on these perspectives. Table 1 lists the descriptions of these criteria that are directly modified from Chang et al. (2006, 2007), Wang (2008), and Wang et al. (2007, 2009), where suppliers $i = 1,2,\ldots,m$, criteria $j = 1,2,\ldots,5$ and behaviors $k = 1,2,\ldots,n$ belong to criteria j. The positive direction, '+', indicates that a behavior attribute is the more the better, while the negative direction, '-', represents a the-less-the-preferable attribute.

EVALUATION WITH A 2-TUPLE LINGUISTIC VARIABLE

To achieve uniform aggregation, all 2-tuple linguistic evaluated results must be converted into the positive direction before aggregation. This section discusses how to employ the 2-tuple linguistic variable and direction adjustment.

The 2-Tuple Linguistic Variable

This study employs the 2-tuple linguistic variable (s_t, α) (Herrera & Martinez, 2000) to evaluate the behaviors in Table 1, where s_t denotes the semantic element defined by linguistic term set (LTS) $S = \{s_0, s_1, \ldots, s_8\}$ and $t \in \{0,1,\ldots,8\}$ (Herrera et al., 2000) which is shown in Table 2, and $\alpha \in [-0.5, +0.5)$ is the numeric value representing how the realistic behavior matches s_t. During the following aggregation procedure, the 2-tuple linguistic variable (s_t, α) is replaced by the transformation function $\Delta(\theta)$ for performing the calculations, where $\theta = t + \alpha$ and $\theta \in [0, 8]$.

Figure 1. The procedure of the proposed approach

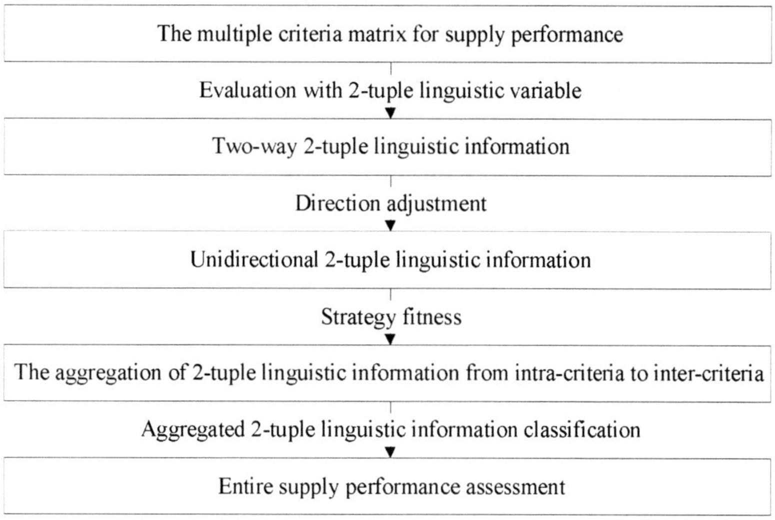

Table 1. The multiple criteria matrix for supplier performance

Criteria	Behavior		Description	Direction
R&D (supplier) a_{i1k}	Design	a_{i11}	Upgrading ability on existing design	+
	Technique	a_{i12}	Upgrading ability on existing manufacturing	+
	Odds	a_{i13}	Surpassing in trade on existing character	+
	Customization	a_{i14}	Breadth and depth variety on supply	+
	Innovation	a_{i15}	Innovating ability on the future	+
Cost (contract) a_{i2k}	Price	a_{i21}	Normal unit price	-
	Quantity	a_{i22}	Normal order quantity	-
	Discount	a_{i23}	Average discount ratio on increasing quantity	+
	Decrement	a_{i24}	Average premium ratio on decreasing quantity	-
	Rush	a_{i25}	Average premium ratio on shortening delivery	-
Quality (supplies) a_{i3k}	Import	a_{i31}	Defect ratio on incoming inspection	-
	On-line	a_{i32}	Defect ratio on in-process inspection	-
	Reliability	a_{i33}	Maintenance ratio on after-sales warrant	-
	Stability	a_{i34}	Standard deviation on incoming inspection	-
Service (supplier) a_{i4k}	Delivery	a_{i41}	Match ratio on arrangement delivery	+
	Accuracy	a_{i42}	Match ratio on arrangement quantity	+
	Assurance	a_{i43}	Duration on assurance	+
	Stockout	a_{i44}	Annual stockout ratio	-
Response (contract) a_{i5k}	Regular	a_{i51}	Normal delivery lead-time	-
	Emergency	a_{i52}	Minimum delivery lead-time	-
	Volume	a_{i53}	Requiring lead-time on changing volume	-
	Specification	a_{i54}	Requiring lead-time on changing specification	-
	Modification	a_{i55}	Requiring lead-time on changing design	-

Table 2. Semantic elements of linguistic term set S

s_t	s_0	s_1	s_2	s_3	s_4	s_5	s_6	s_7	s_8
Semantic Element	None	Very Low	Low	Almost Low	Medium	Almost High	High	Very High	Perfect

Direction Adjustment

The LTS in this study is constructed by finite and totally ordered semantic elements that satisfy the following properties [13,15] when $(s_t, \alpha_1) = \Delta(\theta_1)$ and $(s_u, \alpha_2) = \Delta(\theta_2)$ be two 2-tuples:

1. The set is ordered: $\Delta(\theta_1) \geq \Delta(\theta_2)$ if $\theta_1 \geq \theta_2$
2. The negative operator is defined: $Neg\Delta(\theta_1) = \Delta(\theta_2)$ such that $\theta_2 = (8 - \theta_1)$
3. Maximization operator: $\max\{\Delta(\theta_1), \Delta(\theta_2)\} = \Delta(\theta_1)$ if $\Delta(\theta_1) > \Delta(\theta_2)$
4. Minimization operator: $\min\{\Delta(\theta_1), \Delta(\theta_2)\} = \Delta(\theta_1)$ if $\Delta(\theta_1) < \Delta(\theta_2)$

Therefore, the evaluated results of behaviors with the negative direction in Table 1 shall be transformed into a positive direction.

THE AGGREGATION OF 2-TUPLE LINGUISTIC INFORMATION

This section explains the process of aggregation, which includes meeting the selected product strategy, obtaining the maximum entropy aggregation-weighted vector, and proceeding with the aggregation of 2-tuple linguistic information by the M-LOWA operator with maximum entropy. The philosophy of product strategy based decision criteria and entropy maximization has also been demonstrated in Chang et al. (2006, 2007), Wang (2008), and Wang et al. (2007, 2009).

Selecting the Strategy-Oriented Fuzzy Linguistic Quantifier

The aggregation-weighted vector W defined by n elements is a mapping to membership function $Q(r)$ guided by a monotonically non-decreasing fuzzy linguistic quantifier, Q, as shown in Equations (1) to (2). Besides, W_k is the kth weight for vector W. The membership function $Q(r)$ represents the membership grade on r that belongs to Q. The membership function also differs from Q (Herrera et al., 2000). This study employs three quantifiers to fit the product strategy depending on the importance of criteria, as illustrated in Figure 2.

$$w_k = Q\left(\frac{k}{n}\right) - Q\left(\frac{k-1}{n}\right) k = 1, \ldots, n \qquad (1)$$

$$Q(r) = \begin{cases} 0 & if \quad r < a \\ \dfrac{r-a}{b-a} & if \quad a \leq r \leq b \quad a, b, r \in [0,1] \\ 1 & if \quad r > b \end{cases} \qquad (2)$$

As listed in Table 3, the focal company will adopt different product strategies to meet the market demand, in gaining the competitive advantage during different phases of product life cycle (Aitken et al., 2003). Thus, it is necessary for the focal company to apply a different selection criterion for different product development strategies to enable the use of different fuzzy linguistic quantifiers in aggregating behaviors to produce the fuzzy majority rule. This study adopts

Figure 2. Monotonically non-decreasing fuzzy linguistic quantifier

Quantifier: Most Fit strategy: Major factor	Quantifier: At least half Fit strategy: Critical factor	Quantifier: As many as possible Fit strategy: Basic factor
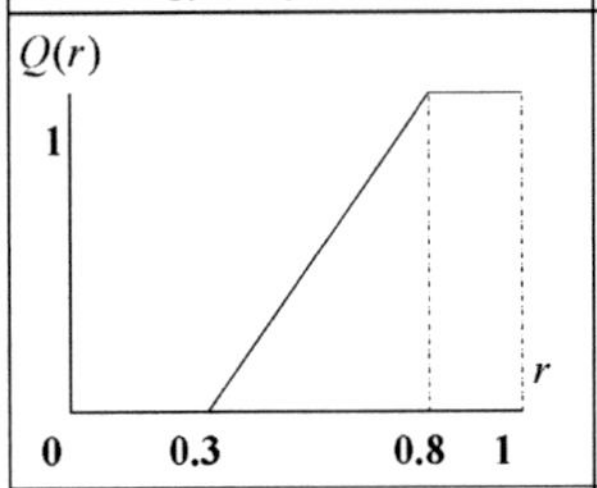	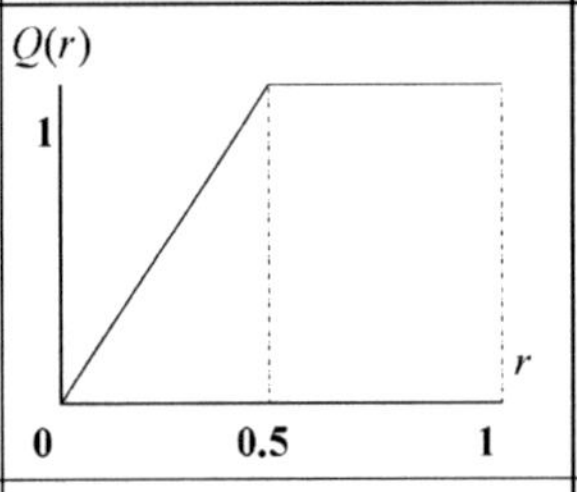	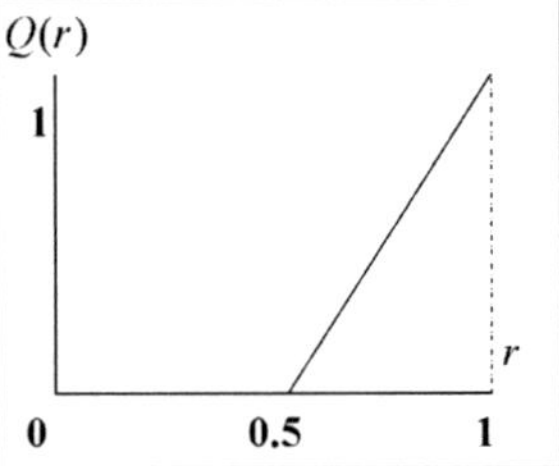
$Q(r)=\begin{cases} 0 & if \quad r<.3 \\ \frac{r-.3}{.8-.3} & if \quad .3\le r\le .8 \\ 1 & if \quad r>.8 \end{cases}$ $a,b,r\in[0,1]$	$Q(r)=\begin{cases} 0 & if \quad r<0 \\ \frac{r-0}{.5-0} & if \quad 0\le r\le .5 \\ 1 & if \quad r>.5 \end{cases}$ $a,b,r\in[0,1]$	$Q(r)=\begin{cases} 0 & if \quad r<.5 \\ \frac{r-.5}{1-.5} & if \quad .5\le r\le 1 \\ 1 & if \quad r>1 \end{cases}$ $a,b,r\in[0,1]$

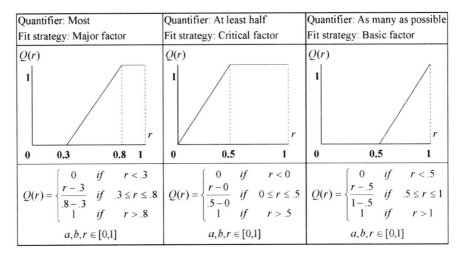

three perspectives on meeting product strategy for criteria. "Critical" factor is used for fuzzy linguistic quantifier, "At least half" to emphasize the strong influence of aggregating on results; "Major" factor is used for fuzzy linguistic quantifier, "Most" to emphasize the medium influence of aggregation on the results. Finally, "Basic" factor is used for fuzzy linguistic quantifier, "As many as possible" to represent the degree to which essential requests are satisfied.

Obtaining the Maximum Entropy Aggregation-Weighted Vector

Maximizing the entropy of the aggregation-weighted vector requires calculating the degree of "orness" and "entropy" (dispersion). The calculation is based on the aggregation-weighted vector W, as shown in Equations (3) to (4). Orness, which lies in the unit interval, is a good measurement for characterizing the degree to which the aggregation is an or-like (max-like) or and-like (min-like) operation. When orness is 1, the aggregation is equal to the maximum operation. When orness is 0, the aggregation is equal to the minimum operation. When orness is 0.5, the aggregation is equal to the arithmetic mean operation. Simultaneously, entropy represents the measurement for characterizing the degree to which the information on the individual behaviors in the aggregation process is employed (Yager, 1988).

Table 3. Strategic factor or criteria during various stages of the product life cycle

Phase	Introduction	Growth	Maturity	Saturation	Decline
Characteristic	Short life cycle Infant stage	Low volume	High volume Low variety	High volume High variety	Low volume
Critical factor	R&D	Service	Cost	Cost	Service
Major factor	Quality Cost Response	Cost Quality Response	Quality Service Response	Quality Response R&D	Cost Quality Response
Basic factor	Service	R&D	R&D	Service	R&D

$$Orness(W) = \frac{1}{n-1}\sum_{k=1}^{n}(n-k)w_k \quad (3)$$

$$Entropy(W) = -\sum_{k=1}^{n} w_k \ln w_k \quad (4)$$

The concept and purpose of maximization is based on the premise that the current orness should be kept constant to implement an amendment process for maximizing the entropy (Filev & Yager, 1995):

$$Max \quad (5)$$

$$\text{Subject to } Orness(W) = \frac{1}{n-1}\sum_{k=1}^{n}(n-k)w_k \quad (6)$$

$$\sum_{k=1}^{n} w_k = 1 \quad (7)$$

Furthermore, the Lagrange multiplier method can be used to obtain the maximum entropy aggregation-weighted vector W^*, which can aggregate the maximum information from behaviors. Filev and Yager (1995) presented the detailed information. Equation (5) can be further simplified as Equations (6) and (7). In addition, the numerical analysis approach can also be utilized to obtain h from Equation (6), and h can be substituted into Equation (7) to obtain W^*. The initial vector of W thus is replaced by the new W^*, thus maximizing the aggregation-weighted vector.

$$\sum_{k=1}^{n}\left(\frac{n-k}{n-1} - Orness(W)\right)h^{n-k} = 0 \quad (8)$$

$$w_k^* = \frac{h^{n-k}}{\sum_{k=1}^{n} h^{n-k}} \quad (9)$$

Aggregating 2-Tuple Linguistic Information by the M-LOWA Operator with Maximum Entropy

The aggregation of 2-tuple linguistic information is performed using the M-LOWA operator based on the maximum entropy aggregation-weighted vector W^*. Furthermore, the original aggregation function F_Q (Herrera et al., 1996) has to be modified to fit for 2-tuple linguistic information. Let $E = \{\Delta(\theta_1), \Delta(\theta_2), \dots, \Delta(\theta_m)\}$ denote a set of 2-tuple linguistic information to be aggregated, then the M-LOWA operator F_Q is defined as follows:

$$F_Q(\Delta(\theta_1), \Delta(\theta_2), \dots, \Delta(\theta_m)) =$$
$$W^* \cdot B^T = C^m\{w_k^*, b_k, k = 1, 2, \dots, m\}$$
$$= w_1^* \odot b_1 \oplus (1 - w_1^*) \odot C^{m-1}\{\beta_h, b_h, h = 2, 3, \dots, m\} \quad (10)$$

where $W^* = [w_1^*, w_2^*, \dots, w_m^*]$ is a maximum entropy aggregation-weighted vector, such that, $w_i^* \in [0,1]$ and $\sum_i w_i^* = 1$; $\beta_h = \frac{w_h^*}{\sum_{k=2}^{m} w_k^*}, h = 2, 3, \dots, m$, and B is the associated ordered vector of set E. Each element $b_i \in B$ is the i th largest element in the collection $\Delta(\theta_1), \Delta(\theta_2), \dots, \Delta(\theta_m)$. C^m is the convex combination operator of m elements, $\odot$ is the general product of an element by a positive real number and $\oplus$ is the general addition of elements (Delgado et al., 1992). If $m = 2$, then F_Q is defined as below:

$$F_Q(\Delta(\theta_1), \Delta(\theta_2)) = W^* \cdot B^T = C^2\{w_i^*, b_i, i = 1, 2\} = w_1^* \odot \Delta(\theta_j) \oplus (1 - w_1^*) \odot \Delta(\theta_i) = \Delta(\theta)$$

where $\theta_j, \theta_i \in [0,8]$ and $\Delta(\theta_j) \geq \Delta(\theta_i)$ (11)such that $\theta = \min\{8, i + w_1^* \cdot (j - i)\}$, where

$b_1 = \Delta(\theta_j)$, $b_2 = \Delta(\theta_i)$. If $w_j = 1$ and $w_i = 0$ with $i \neq j \ \forall i$, then the convex combination is defined as $C^m = \{w_i^*, b_i, i = 1,2,\ldots,m\} = b_j$. Finally, the aggregated information $\Delta(\theta)$ can be reversed into 2-tuple linguistic information (s_t, α) by $t = round(\theta)$ and $\alpha = \theta - t$, where $round$ is the usual roundup operation (Herrera et al., 2000).

A NUMERICAL EXAMPLE

Considering the case of a focal company specialized in manufacturing notebook computers, it has to select a supplier from three possible candidates to create a local supply chain. Supplier A possesses an advantage in R&D, Supplier B in manufacturing, and Supplier C in distribution. However, the product life cycle of notebook computers is considered in the "Maturity" phase. All behaviors are evaluated using the 2-tuple linguistic variable. Table 4 lists the multiple criteria matrix for supplier performance and the uniformity results from negative behaviors. An example of the uniformity process of a_{121}, a_{221}, a_{321} by the negative operator is displayed below:

Adjusting the result s_7 assessed from negative behavior $a_{121} = s_{8-7} = s_1$

Adjusting the result s_2 assessed from negative behavior $a_{221} = s_{8-2} = s_6$

Adjusting the result s_1 assessed from negative behavior $a_{321} = s_{8-1} = s_7$

The aggregation-weighted vector W and the maximum entropy aggregation-weighted vector W^* are calculated according to the number of behaviors within an criterion. An appropriate product strategy is determined, as listed in Tables 5 and 6. The computing process dealing with the fuzzy linguistic quantifier "Most" involves four items, $k = 1,2,\ldots,4$, as displayed below:

$$w_1 = Q\left(\frac{1}{4}\right) - Q\left(\frac{0}{4}\right) = 0 - 0 = 0$$

$$w_2 = Q\left(\frac{2}{4}\right) - Q\left(\frac{1}{4}\right) = \frac{0.5 - 0.3}{0.8 - 0.3} - 0 = 0.4$$

$$w_3 = Q\left(\frac{3}{4}\right) - Q\left(\frac{2}{4}\right) = \frac{0.75 - 0.3}{0.8 - 0.3} - \frac{0.5 - 0.3}{0.8 - 0.3} = 0.5$$

$$w_4 = Q\left(\frac{4}{4}\right) - Q\left(\frac{3}{4}\right) = 1 - \frac{0.75 - 0.3}{0.8 - 0.3} = 0.1$$

$$Orness(W) = \frac{1}{4-1}\sum_{j=1}^{4}(4-j)w_j = \frac{1}{3}(3w_1 + 2w_2 + w_3) = 0.4333333$$

$$\sum_{j=1}^{4}\left(\frac{4-j}{4-1} - 0.4333\right)h^{4-j} =$$

$$(1 - 0.4333)h^3 + (\frac{2}{3} - 0.4333)h^2 + (\frac{1}{3} - 0.4333)h - 0.4333 = 0$$

$$h = 0.8511435$$

$$w_1^* = \frac{h^3}{\sum_{j=1}^{4} h^{4-j}} = \frac{h^3}{h^3 + h^2 + h + 1} = 0.1931607$$

$$w_2^* = \frac{h^2}{\sum_{j=1}^{4} h^{4-j}} = \frac{h^2}{h^3 + h^2 + h + 1} = 0.2269426$$

$$w_3^* = \frac{h}{\sum_{j=1}^{4} h^{4-j}} = \frac{h}{h^3 + h^2 + h + 1} = 0.2666326$$

Table 4. 2-tuple linguistic evaluation and direction adjustment for candidates

Criteria	Behavior	Direction	Supplier A a_{1jk}			Supplier B a_{2jk}			Supplier C a_{3jk}		
			Assessed	Adjusted	$\Delta(\theta)$	Assessed	Adjusted	$\Delta(\theta)$	Assessed	Adjusted	$\Delta(\theta)$
R&D	a_{i11}	+	$(s_7,0)$	$(s_7,0)$	$\Delta(7)$	$(s_5,0)$	$(s_5,0)$	$\Delta(5)$	$(s_3,0)$	$(s_3,0)$	$\Delta(3)$
	a_{i12}	+	$(s_5,0)$	$(s_5,0)$	$\Delta(5)$	$(s_7,0)$	$(s_7,0)$	$\Delta(7)$	$(s_4,0)$	$(s_4,0)$	$\Delta(4)$
	a_{i13}	+	$(s_8,0)$	$(s_8,0)$	$\Delta(8)$	$(s_7,0)$	$(s_7,0)$	$\Delta(7)$	$(s_4,0)$	$(s_4,0)$	$\Delta(4)$
	a_{i14}	+	$(s_7,0)$	$(s_7,0)$	$\Delta(7)$	$(s_5,0)$	$(s_5,0)$	$\Delta(5)$	$(s_3,0)$	$(s_3,0)$	$\Delta(3)$
	a_{i15}	+	$(s_8,0)$	$(s_8,0)$	$\Delta(8)$	$(s_5,0)$	$(s_5,0)$	$\Delta(5)$	$(s_3,0)$	$(s_3,0)$	$\Delta(3)$
Cost	a_{i21}	-	$(s_7,0)$	$(s_1,0)$	$\Delta(1)$	$(s_2,0)$	$(s_6,0)$	$\Delta(6)$	$(s_1,0)$	$(s_7,0)$	$\Delta(7)$
	a_{i22}	-	$(s_7,0)$	$(s_1,0)$	$\Delta(1)$	$(s_7,0)$	$(s_1,0)$	$\Delta(1)$	$(s_2,0)$	$(s_6,0)$	$\Delta(6)$
	a_{i23}	+	$(s_1,0)$	$(s_1,0)$	$\Delta(1)$	$(s_5,0)$	$(s_5,0)$	$\Delta(5)$	$(s_7,0)$	$(s_7,0)$	$\Delta(7)$
	a_{i24}	-	$(s_6,0)$	$(s_2,0)$	$\Delta(2)$		$(s_3,0)$	$\Delta(3)$	$(s_3,0)$	$(s_5,0)$	$\Delta(5)$
	a_{i25}	-	$(s_5,0)$	$(s_3,0)$	$\Delta(3)$	$(s_7,0)$	$(s_1,0)$	$\Delta(1)$	$(s_4,0)$	$(s_4,0)$	$\Delta(4)$
Quality	a_{i31}	-	$(s_3,0)$	$(s_5,0)$	$\Delta(5)$	$(s_1,0)$	$(s_7,0)$	$\Delta(7)$	$(s_7,0)$	$(s_1,0)$	$\Delta(1)$
	a_{i32}	-	$(s_3,0)$	$(s_5,0)$	$\Delta(5)$	$(s_1,0)$	$(s_7,0)$	$\Delta(7)$	$(s_6,0)$	$(s_2,0)$	$\Delta(2)$
	a_{i33}	-	$(s_3,0)$	$(s_5,0)$	$\Delta(5)$	$(s_1,0)$	$(s_7,0)$	$\Delta(7)$	$(s_4,0)$	$(s_4,0)$	$\Delta(4)$
	a_{i34}	-	$(s_4,0)$	$(s_4,0)$	$\Delta(4)$	$(s_1,0)$	$(s_7,0)$	$\Delta(7)$	$(s_4,0)$	$(s_4,0)$	$\Delta(4)$
Service	a_{i41}	+	$(s_7,0)$	$(s_7,0)$	$\Delta(7)$	$(s_7,0)$	$(s_7,0)$	$\Delta(7)$	$(s_2,0)$	$(s_2,0)$	$\Delta(2)$
	a_{i42}	+	$(s_7,0)$	$(s_7,0)$	$\Delta(7)$	$(s_7,0)$	$(s_7,0)$	$\Delta(7)$	$(s_3,0)$	$(s_3,0)$	$\Delta(3)$
	a_{i43}	+	$(s_6,0)$	$(s_6,0)$	$\Delta(6)$	$(s_4,0)$	$(s_4,0)$	$\Delta(4)$	$(s_2,0)$	$(s_2,0)$	$\Delta(2)$
	a_{i44}	-	$(s_6,0)$	$(s_2,0)$	$\Delta(2)$	$(s_4,0)$	$(s_4,0)$	$\Delta(4)$	$(s_2,0)$	$(s_6,0)$	$\Delta(6)$

continued on following page

Table 4. Continued

Response	a_{i51}	-	$(s_4,0)$	$(s_4,0)$	$\Delta(4)$	$(s_4,0)$	$(s_4,0)$	$\Delta(4)$	$(s_3,0)$	$(s_5,0)$	$\Delta(5)$
	a_{i52}	-	$(s_3,0)$	$(s_5,0)$	$\Delta(5)$	$(s_3,0)$	$(s_5,0)$	$\Delta(5)$	$(s_2,0)$	$(s_6,0)$	$\Delta(6)$
	a_{i53}	-	$(s_2,0)$	$(s_6,0)$	$\Delta(6)$	$(s_2,0)$	$(s_6,0)$	$\Delta(6)$	$(s_1,0)$	$(s_7,0)$	$\Delta(7)$
	a_{i54}	-	$(s_2,0)$	$(s_6,0)$	$\Delta(6)$	$(s_2,0)$	$(s_6,0)$	$\Delta(6)$	$(s_4,0)$	$(s_4,0)$	$\Delta(4)$
	a_{i55}	-	$(s_3,0)$	$(s_5,0)$	$\Delta(5)$	$(s_5,0)$	$(s_3,0)$	$\Delta(3)$	$(s_7,0)$	$(s_1,0)$	$\Delta(1)$

Table 5. Vector W and W^* on fitting strategies with four criteria

Strategy	Fuzzy linguistic quantifier	w_1	w_2	w_3	w_4	$Orness(W)$
		w_1^*	w_2^*	w_3^*	w_4^*	$Orness(W^*)$
Major	Most	0	0.4	0.5	0.1	0.4333
		0.1932	0.2269	0.2666	0.3133	0.4333
Critical	At least half	0.5	0.5	0	0	0.8333
		0.6478	0.2355	0.0856	0.0311	0.8333
Basic	As many as possible	0	0	0.5	0.5	0.1667
		0.0311	0.0856	0.2355	0.6478	0.1667

Table 6. Vector W and W^* on fitting strategies with five criteria

Strategy	Fuzzy linguistic quantifier	w_1	w_2	w_3	w_4	w_5	$Orness(W)$
		w_1^*	w_2^*	w_3^*	w_4^*	w_5^*	$Orness(W^*)$
Major	Most	0	0.2	0.4	0.4	0	0.4500
		0.1620	0.1791	0.1980	0.2189	0.2420	0.4500
Critical	At least half	0.4	0.4	0.2	0	0	0.8000
		0.5307	0.2565	0.1240	0.0599	0.0290	0.8000
Basic	As many as possible	0	0	0.2	0.4	0.4	0.2000
		0.0290	0.0599	0.1240	0.2565	0.5307	0.2000

Table 7. Evaluation results on each individual criterion and aggregation results (Maturity)

Candidates		Supplier A		Supplier B		Supplier C	
Criteria	Strategy	$\Delta(\theta)$	(s_t, α)	$\Delta(\theta)$	(s_t, α)	$\Delta(\theta)$	(s_t, α)
R&D	Basic	$\Delta(6.028)$	$(s_6, 0.028)$	$\Delta(5.178)$	$(s_5, 0.178)$	$\Delta(3.089)$	$(s_3, 0.089)$
Cost	Critical	$\Delta(2.318)$	$(s_2, 0.318)$	$\Delta(4.927)$	$(s_5, -0.073)$	$\Delta(6.669)$	$(s_7, -0.331)$
Quality	Major	$\Delta(4.687)$	$(s_5, -0.313)$	$\Delta(7.000)$	$(s_7, 0.000)$	$\Delta(2.527)$	$(s_3, -0.473)$
Service	Major	$\Delta(5.167)$	$(s_5, 0.167)$	$\Delta(5.260)$	$(s_5, 0.260)$	$\Delta(3.000)$	$(s_3, 0.000)$
Response	Major	$\Delta(5.10)$	$(s_5, 0.099)$	$\Delta(4.638)$	$(s_5, -0.362)$	$\Delta(4.316)$	$(s_4, 0.316)$
Entire supplier performance		$\Delta(4.498)$	$(s_4, 0.498)$	$\Delta(5.302)$	$(s_5, 0.302)$	$\Delta(3.787)$	$(s_4, -0.213)$

$$w_4^* = \frac{1}{\sum_{j=1}^{4} h^{4-j}} = \frac{1}{h^3 + h^2 + h + 1} = 0.3132640$$

Then the maximum entropy M-LOWA operator is introduced to aggregate 2-tuple linguistic information from intra-criteria (behaviors within individual criterion) to inter-criteria (entire criteria). Since the industry of notebook computer has reached the "Maturity" phase of its product life

Figure 3. Classification of the existing approaches for supplier performance assessment

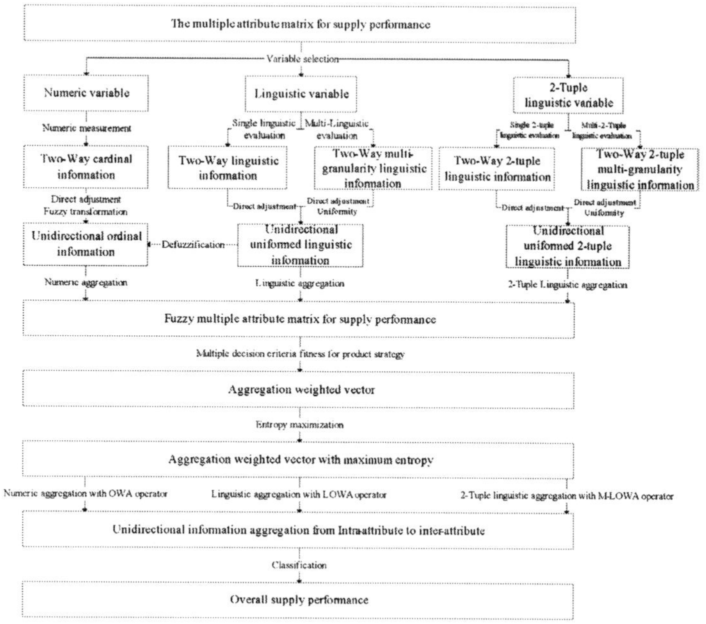

Table 8. Example of linguistic aggregating behaviors for criterion "Cost" of Supplier B

$$F_Q(\Delta(6),\Delta(1),\Delta(5),\Delta(3),\Delta(1)) =$$
$$[0.5307, 0.2565, 0.1240, 0.0599, 0.0290] \cdot$$
$$[\Delta(6),\Delta(5),\Delta(3),\Delta(1),\Delta(1)]^T$$

$$= C^5\{(0.5307,\Delta(6)),$$
$$(0.2565,\Delta(5)),$$
$$(0.1240,\Delta(3)),$$
$$(0.0599,\Delta(1)),$$
$$(0.0290,\Delta(1))\}$$

$$= 0.5307 \odot \Delta(5) \oplus (1-0.5465)$$
$$\odot C^3\{(0.5824,\Delta(3)),(0.2815,\Delta(1)),(0.1361,\Delta(1))\} \odot \Delta(6) \oplus (1-0.5307) \odot$$
$$C^4\{(0.5465,\Delta(5)),(0.2641,\Delta(3)),(0.1277,\Delta(1)),(0.0617,\Delta(1))\}$$

$$C^4\{(0.5465,\Delta(5)),(0.2641,\Delta(3)),(0.1277,\Delta(1)),(0.0617,\Delta(1))\}$$

$$= 0.5465 \odot \Delta(5) \oplus (1-0.5465) \odot C^3\{(0.5824,\Delta(3)),(0.2815,\Delta(1)),(0.1361,\Delta(1))\}$$

$$C^3\{(0.5824,\Delta(3)),(0.2815,\Delta(1)),(0.1361,\Delta(1))\}$$

$$= 0.5824 \odot \Delta(3) \oplus (1-0.5824) \odot C^2\{(0.6742,\Delta(1)),(0.3258,\Delta(1))\}$$

$$C^2\{(0.6742,\Delta(1)),(0.3258,\Delta(1))\} = 0.6742 \odot \Delta(1) \oplus (1-0.6742) \odot \Delta(1) = \Delta(\theta)$$

$$\theta = \min\{8, 1+[0.6742 \times (1-1)]\} = \min\{8,1\} = 1$$

$$C^2\{(0.6742,\Delta(1)),(0.3258,\Delta(1))\} = \Delta(1)$$

$$C^3\{(0.5824,\Delta(3)),(0.2815,\Delta(1)),(0.1361,\Delta(1))\} = 0.5824 \odot \Delta(3) \oplus (1-0.5824) \odot \Delta(1) = \Delta(\theta)$$

$$\theta = \min\{8, 1+[0.5824 \times (3-1)]\} = \min\{8, 2.1648\} = 2.1648$$

$$C^3\{(0.5824,\Delta(3)),(0.2815,\Delta(1)),(0.1361,\Delta(1))\} = \Delta(2.1648)$$

$$C^4\{(0.5465,\Delta(5)),(0.2641,\Delta(3)),(0.1277,\Delta(1)),(0.0617,\Delta(1))\}$$

$$= 0.5465 \odot \Delta(5) \oplus (1-0.5465) \odot \Delta(2.1648) = \Delta(\theta)$$

$$\theta = \min\{8, 2.1648+[0.5465 \times (5-2.1648)]\} = \min\{8, 3.7142\} = 3.7142$$

$$C^4\{(0.5465,\Delta(5)),(0.2641,\Delta(3)),(0.1277,\Delta(1)),(0.0617,\Delta(1))\} = \Delta(3.7142)$$

$$C^5\{(0.5307,\Delta(6)),(0.2565,\Delta(5)),(0.1240,\Delta(3)),(0.0599,\Delta(1)),(0.0290,\Delta(1))\} \quad \Delta(3.7142) = \Delta(\theta) \quad (1-0.5307)$$
$$\Delta(6) \quad = 0.5307 \quad \theta = \min\{8, 3.7142+[0.5307 \times (6-3.7142)]\} = \min\{8, 4.9272\} = 4.9272$$
$$C^5\{(0.5307,\Delta(6)),(0.2565,\Delta(5)),(0.1240,\Delta(3)),(0.0599,\Delta(1)),(0.0290,\Delta(1))\} = \Delta(4.927)$$

cycle, the "Cost" attribute is considered as a "Critical" factor, while "Quality", "Service" and "Response" of attributes are considered as "Major" factors. Since the mass production system is generally adopted for a mature industry, R&D advantage is no longer the focus of attention, as a result the "R&D" attribute is considered as a "Basic" factor. Table 7 lists the evaluation results along with each individual criterion and the aggregation results based on the maximum entropy M-LOWA operator using the fuzzy linguistic quantifier considering the strategy at the "Matu-

Table 9. The classification result for evaluation methods demonstrated in Figure 3

Data Form Category	Cardinal data (Numeric)	Ordinal data (Linguistic)	ordinal with cardinal data (2-Tuple linguistic)
Single scale	Chang et al. [5]	Wang et al. [30]	The proposed methodology
Multiple scales	Wang et al. [29]	Chang et al. [6]	Wang [28]

rity" phase of the product life cycle. An example of linguistic aggregating behaviors for criterion "Cost" of Supplier B is listed in Table 8.

The final aggregation results, which are based on the trade-off mechnism of the current product strategy, indicate that Supplier B has the best performance, followed by Supplier A and finally Supplier C.

DISCUSSION AND CONCLUSION

According to Herrera et al. (1996), single-tuple linguistic variable cannot be distinguished between the supplier performance of Suppliers A and C. Thus, adopting the 2-tuple linguistic variable and aggregated linguistic information by the M-LOWA operator can avoid a tied result and retain both the linguistic information and data. Hence, the proposed approach can help to enhance decision-making quality and sensitivity. Figure 3 summarized the existing approaches for supplier performance assessment. Chang et al. (2006) and Wang et al. (2009) employed the numeric approach to solve the issue. Chang et al. (2007) and Wang et al. (2007) manipulated the pure linguistic approach to proceed the assessment. Moreover, Wang (2008) and this study exercised 2-tuple linguistic approaches to perform the same task with multiple and single scales, respectively. A comprehensive framework for the assessment of supplier performance has also been accomplished in this study. The existing approaches of various data forms and scale categories are classified in Table 8. Furthermore, the proposed methodology, demonstrated in Figure 3, can be combined with other techniques such as quality function deployment (QFD) and analytical hierarchy process (AHP) which will be the topics for future research.

REFEREwNCES

Aitken, J., Childerhouse, P., & Towill, D. (2003). The impact of product life cycle on supply chain strategy. *International Journal of Production Economics, 85*(2), 127–140. doi:10.1016/S0925-5273(03)00105-1

Aouam, T., Chang, S. I., & Lee, E. S. (2003). Fuzzy MADM: An outranking method. *European Journal of Operational Research, 145*(2), 317–328. doi:10.1016/S0377-2217(02)00537-4

Carbonara, N., Giannoccaro, I., & Pontrandolfo, P. (2002). Supply chains within industrial districts: A theoretical framework. *International Journal of Production Economics, 76*(2), 159–176. doi:10.1016/S0925-5273(01)00159-1

Chan, F. T. S. (2003). Performance measurement in a supply chain. *International Journal of Advanced Manufacturing Technology, 21*(7), 534–548. doi:10.1007/s001700300063

Chang, S. L., Wang, R. C., & Wang, S. Y. (2006). Applying fuzzy linguistic quantifier to select supply chain partners at different phases of product life cycle. *International Journal of Production Economics, 100*(2), 348–359. doi:10.1016/j.ijpe.2005.01.002

Chang, S. L., Wang, R. C., & Wang, S. Y. (2007). Applying a direct multi-granularity linguistic and strategy-oriented aggregation approach on the assessment of supply performance. *European Journal of Operational Research, 177*(2), 1013–1025. doi:10.1016/j.ejor.2006.01.032

Chen, S. J., & Hwang, C. L. (1992). *Fuzzy multiple attribute decision making: Methods and applications*. Berlin, Germany: Springer-Verlag.

Choi, T. Y., & Hartley, J. L. (1996). An exploration of supplier selection practices across the supply chain. *Journal of Operations Management, 14*(4), 333–343. doi:10.1016/S0272-6963(96)00091-5

Delgado, M., Verdegay, J. L., & Vila, M. A. (1992). Linguistic decision making models. *International Journal of Intelligent Systems, 7*(5), 479–492. doi:10.1002/int.4550070507

Filev, D., & Yager, R. R. (1995). Analytic properties of maximum entropy OWA operators. *Information Sciences, 85*(1-3), 11–27. doi:10.1016/0020-0255(94)00109-O

Gunasekaran, A., Patel, C., & McGaughey, R. E. (2004). A framework for supply chain performance measurement. *International Journal of Production Economics, 87*(3), 333–347. doi:10.1016/j.ijpe.2003.08.003

Herrera, F., Herrera-Viedma, E., & Martinez, L. (2000). A fusion approach for managing multi-granularity linguistic term sets in decision making. *Fuzzy Sets and Systems, 114*(1), 43–58. doi:10.1016/S0165-0114(98)00093-1

Herrera, F., Herrera-Viedma, E., & Verdegay, J. L. (1995). A sequential selection process in group decision making with a linguistic assessment approach. *Information Sciences, 85*(4), 223–239. doi:10.1016/0020-0255(95)00025-K

Herrera, F., Herrera-Viedma, E., & Verdegay, J. L. (1996). Direct approach processes in group decision making using linguistic OWA operators. *Fuzzy Sets and Systems, 79*(2), 175–190. doi:10.1016/0165-0114(95)00162-X

Herrera, F., & Martinez, L. (2000). A 2-tuple fuzzy linguistic representation model for computing with words. *IEEE Transactions on Fuzzy Systems, 8*(6), 746–752. doi:10.1109/91.890332

Kacprzyk, J. (1986). Group decision making with a fuzzy liguistic majority. *Fuzzy Sets and Systems, 18*(2), 105–118. doi:10.1016/0165-0114(86)90014-X

Kannan, V. R., & Tan, K. C. (2002). Supplier selection and assessment: Their impact on business performance. *Journal of Supply Chain Management, 38*(4), 11–21. doi:10.1111/j.1745-493X.2002.tb00139.x

Krause, D. R., Pagell, M., & Curkovic, S. (2001). Toward a measure of competitive priorities for purchasing. *Journal of Operations Management, 19*(4), 497–512. doi:10.1016/S0272-6963(01)00047-X

Lee, H. L. (2003). Aligning supply chain strategies with product uncertainties. *IEEE Engineering Management Review, 31*(2), 26–34. doi:10.1109/EMR.2003.1207060

Muralidharan, C., Anantharaman, N., & Deshmukh, S. G. (2002). A multi-criteria group decisionmaking model for supplier rating. *Journal of Supply Chain Management, 38*(4), 22–33. doi:10.1111/j.1745-493X.2002.tb00140.x

Otto, A., & Kotzab, H. (2003). Does supply chain management really pay? Six perspectives to measure the performance of managing a supply chain. *European Journal of Operational Research, 144*(2), 306–320. doi:10.1016/S0377-2217(02)00396-X

Sarkis, J., & Talluri, S. (2002). A model for strategic supplier selection. *Journal of Supply Chain Management, 38*(1), 18–28. doi:10.1111/j.1745-493X.2002.tb00117.x

Sharland, A., Eltantawy, R. A., & Giunipero, L. C. (2003). The impact of cycle time on supplier selection and subsequent performance outcomes. *Journal of Supply Chain Management, 39*(3), 4–12. doi:10.1111/j.1745-493X.2003.tb00155.x

Talluri, S., & Narasimhan, R. (2004). A methodology for strategic sourcing. *European Journal of Operational Research, 154*(1), 236–250. doi:10.1016/S0377-2217(02)00649-5

Tracey, M., & Tan, C. L. (2001). Empirical analysis of supplier selection and involvement, customer satisfaction, and firm performance. *Supply Chain Management, 6*(3-4), 174–188.

Verma, R., & Pullman, M. E. (1998). An analysis of the supplier selection process. *Omega, 26*(6), 739–750. doi:10.1016/S0305-0483(98)00023-1

Vonderembse, M. A., & Tracey, M. (1999). The impact of supplier selection criteria and supplier involvement on manufacturing performance. *Journal of Supply Chain Management, 35*(3), 33–39. doi:10.1111/j.1745-493X.1999.tb00060.x

Wang, S. Y. (2008). Applying 2-tuple multi-granularity linguistic variables to determine the supply performance in dynamic environment based on product-oriented strategy. *IEEE Transactions on Fuzzy Systems, 16*(1), 29–39. doi:10.1109/TFUZZ.2007.903316

Wang, S. Y., Chang, S. L., & Wang, R. C. (2007). Applying a direct approach in linguistic assessment and aggregation on supply performance. *International Journal of Operations Research, 4*(4), 238–247.

Wang, S. Y., Chang, S. L., & Wang, R. C. (2009). Assessment of supplier performance based on product-development strategy by applying multi-granularity linguistic term sets. *Omega, 37*(1), 215–226. doi:10.1016/j.omega.2006.10.003

Yager, R. R. (1988). On ordered weighted averaging aggregation operators in multicriteria decision making. *IEEE Transactions on Systems, Man, and Cybernetics, 18*(1), 183–190. doi:10.1109/21.87068

Zhang, Q., Chen, J. C. H., He, Y. Q., Ma, J., & Zhou, D. N. (2003). Multiple attribute decision making: Approach integrating subjective and objective information. *International Journal of Manufacturing Technology and Management, 5*(4), 338–361. doi:10.1504/IJMTM.2003.003460

This work was previously published in the International Journal of Fuzzy System Applications, Volume 1, Issue 2, edited by Toly Chen, pp. 1-16, copyright 2010 by IGI Publishing (an imprint of IGI Global).

Compilation of References

Abarbanell, J. S., & Bushee, B. J. (1997). Fundamental analysis, future earnings, and stock prices. *Journal of Accounting Research, 35*, 1–24. doi:10.2307/2491464

Abdulghafour, M., Fellah, A., & Abidi, M. A. (1994). Fuzzy logic-based data integration: theory and application. In. *Proceedings of the IEEE International Conference in Multisensor Fusion and Integration for Intelligent Systems, 1*, 151–160.

Aggarwal, C. C., & Yu, P. S. (2001). Outlier detection for high dimensional data. In *Proceedings of the 2001 ACM SIGMOD international conference on Management of data* (pp. 37-46).

Aitken, J., Childerhouse, P., & Towill, D. (2003). The impact of product life cycle on supply chain strategy. *International Journal of Production Economics, 85*(2), 127–140. doi:10.1016/S0925-5273(03)00105-1

Akagi, H., Kanajawa, Y., & Nabae, A. (1984). Instantaneous reactive power compensator comprising switching devices without energy storage components. *IEEE Transactions on Industry Applications, 20*(3), 625–630. doi:10.1109/TIA.1984.4504460

Albus, J. S. (1972). *Theoretical and Experimental Aspects of a Cerebellar Model.* Unpublished PhD dissertation, University of Maryland.

Albus, J. S. (1975). A new approach to manipulator control: the cerebellar model articulation controller (CMAC). *Journal of Dynamics Systems, Measurement, and Control. Transactions of ASME, 97*(3), 220–227.

Al-Deek, H. M. (2007). Use of vessel freight data to forecast heavy truck movements at seaports. *Transportation Research Record: Journal of the Transportation Research Board, 1804*, 217–224. doi:10.3141/1804-29

Alpaydin, E. (2004). *Introduction to machine learning.* Cambridge, MA: MIT Press.

Altan, M. (2010). Reducing shrinkage in injection moldings via the Taguchi, ANOVA and neural network methods. *Materials & Design, 31*, 599–604. doi:10.1016/j.matdes.2009.06.049

Ang, K. K., & Quek, C. (2000). Improved MCMAC with momentum, neighbourhood, and averaged trapezoidal output. *IEEE Transactions on Systems. Man and Cybernetics: Part B, 30*(3), 491–500. doi:10.1109/3477.846237

Ang, K. K., & Quek, C. (2006). Stock trading using RSPOP: a novel rough set neuro-fuzzy approach. *IEEE Transactions on Neural Networks, 17*(5), 1301–1315. doi:10.1109/TNN.2006.875996

Anis Ibrahim, W. R., & Morcos, M. M. (2002). Artificial intelligence and advanced mathematical tools for power quality applications: A survey. *IEEE Transactions on Power Delivery, 17*(2), 668–673. doi:10.1109/61.997958

An, L., Zhikang, S., Wenji, Z., Ruixiang, F., & Chunming, T. (2009). Development of hybrid active power filter based on adaptive fuzzy dividing frequency-control method. *IEEE Transactions on Power Delivery, 24*(1), 424–432. doi:10.1109/TPWRD.2008.2005877

An, W., Angulo, C., & Sun, Y. (2008). Support vector regression with interval-input interval-output. *International Journal of Computational Intelligence Systems, 1*(4), 299–303. doi:10.2991/ijcis.2008.1.4.2

Aouam, T., Chang, S. I., & Lee, E. S. (2003). Fuzzy MADM: An outranking method. *European Journal of Operational Research, 145*(2), 317–328. doi:10.1016/S0377-2217(02)00537-4

Arabshahi, P., Marks, R. J., Oh, S., Caudell, T. P., Choi, J. J., & Song, B. G. (1997). Pointer adaptation and pruning of min-max fuzzy inference and estimation. *IEEE Transactions on Circuits and Systems-II: Analog and Digital Signal Processing*, *44*(9), 696–709. doi:10.1109/82.624992

Armstrong, E. *(1989). Principles for competitive semiconductor manufacturing. In* Proceedings of the IEEE/SEMI International Semiconductor Manufacturing Science Symposium

Arya, R. K. (2006). Approximations of Mamdani type multi-fuzzy sets fuzzy controller by simplest fuzzy controller. *International Journal of Computational Cognition*, *4*, 35–47.

Atanassov, K. T. (1986). Intuitionistic fuzzy sets. *Fuzzy Sets and Systems*, *50*, 87–96. doi:10.1016/S0165-0114(86)80034-3

Atanassov, K. T., & Gargov, G. (1989). Interval-valued intuitionistic fuzzy sets. *Fuzzy Sets and Systems*, *31*, 343–349. doi:10.1016/0165-0114(89)90205-4

Atanassov, K. T., & Georgiev, C. (1993). Intuitionistic fuzzy prolog. *Fuzzy Sets and Systems*, *53*, 121–128. doi:10.1016/0165-0114(93)90166-F

Atanassov, K. T., Pasi, G., & Yager, R. R. (2005). Intuitionistic fuzzy interpretations of multi-criteria multi-person and multi-measurement tool decision making. *International Journal of Systems Science*, *36*(14), 859–868. doi:10.1080/00207720500382365

Baba, N., Inoue, N., & Asakawa, H. (2000). *Utilization of neural networks & gas for constructing reliable decision support systems to deal stocks*. Osaka, Japan: Osaka-Kyoiku University.

Bárdossy, A. (1990). Note on fuzzy regression. *Fuzzy Sets and Systems*, *65*, 65–75. doi:10.1016/0165-0114(90)90064-D

Bazaraa, M. S. (2006). *Nonlinear programming-theory and algorithms*. Chichester, UK: John Wiley & Sons.

Beliakov, G., Pradera, A., & Calvo, T. (2007). *Aggregation functions: A guide for practitioners*. Heidelberg, Germany: Springer-Verlag.

Bendhia, S., Ramdani, M., & Sicard, E. (2005). *EMC of ICs: Techniques for low emission and susceptibility*. New York, NY: Springer.

Berkhin, P. (2002). *Survey of clustering data mining techniques*. Retrieved from http://www.accrue.com/products/researchpapers.html

Bertrand, P., & Goupil, F. (2000). Descriptive statistic for symbolic data. In Bock, H.-H., & Diday, E. (Eds.), *Analysis of symbolic data* (pp. 106–124). Heidelberg, Germany: Springer-Verlag.

Beyer, K., Goldstein, J., Ramakrishnan, R., & Shaft, U. (1999). When is nearest neighbor meaningful? In *ICDT '99: Proceedings of the 7th International Conference on Database Theory* (LNCS 1540, pp. 217-235).

Bezdek, J. C. (1981). *Pattern Recognition with Fuzzy Objective Function Algorithms*. New York: Plenum Press.

Bezine, H., Derbel, N., & Alimi, A. M. (2002). Fuzzy control of robotic manipulator: some issues on design and rule base size reduction. *Engineering Applications of Artificial Intelligence*, *15*, 401–416. doi:10.1016/S0952-1976(02)00075-1

Bien, Z., Park, K.-H., Bang, W.-C., & Stefanov, D. H. (2002). LARES: An intelligent sweet home for assisting the elderly and the handicapped. In *Proceedings of the 1st Cambridge Workshop on Universal Access and Assistive Technology* (pp.43-46).

Billard, L., & Diday, E. (2000). Regression analysis for interval-valued data. In *Proceedings of the 7th Conference of the International Federation of Classification Societies* (pp. 369-374).

Billard, L., & Diday, E. (2003). From the statistics of data to the statistics of knowledge: Symbolic data analysis. *Journal of the American Statistical Association*, *98*(462), 470–487. doi:10.1198/016214503000242

Bishop, C. M. (2006). *Pattern Recognition and Machine Learning*. Berlin: Springer Verlag.

Bock, H. H., & Diday, E. (2000). *Analysis of symbolic data*. Heidelberg, Germany: Springer-Verlag.

Bose, B. K. (1994). Expert systems, fuzzy logic, and neural network applications in power electronics and motion control. *Proceedings of the IEEE, 82*(8), 1303–1323. doi:10.1109/5.301690

Boulkroune, A., Tadjine, M., M'Saad, M., & Farza, M. (2010). Fuzzy adaptive controller for MIMO nonlinear systems with known and unknown control direction. *Fuzzy Sets and Systems, 161*, 797–820. doi:10.1016/j.fss.2009.04.011

Bourlard, H., & Wellekens, C. J. (1990). Links between Markov models and multilayer perceptrons. *IEEE Transactions on Pattern Analysis and Machine Intelligence, 12*(12), 1167–1178. doi:10.1109/34.62605

Bowerman, B. L., & O'Connell, R. T. (1993). *Forecasting and Time Series: An Applied Approach*. Pacific Grove, CA: Duxbury Press.

Boyd, S., & Vandenberghe, L. (1997). Semidefinite programming relaxations of non-convex problems in control and combinatorial optimization. In Paulraj, A., Roychowdhury, V., & Schaper, C. D. (Eds.), *Communications, computation, control and signal processing: A tribute to Thomas Kailath*. Amsterdam, The Netherlands: Kluwer Academic.

Boyd, S., & Vandenberghe, L. (2004). *Convex optimization*. Cambridge, UK: Cambridge University Press.

Braglia, M., Frosolini, M., & Montanari, R. (2003). Fuzzy TOPSIS approach for failure mode, effects and criticality analysis. *Quality and Reliability Engineering, 19*, 425–443. doi:10.1002/qre.528

Brogliato, B., & Lozano, R. (1994). Adaptive control of first-order nonlinear systems with reduced knowledge of the plant parameters. *IEEE Transactions on Automatic Control, 39*, 1764–1768. doi:10.1109/9.310070

Buckley, J. J., & Hayashi, Y. (1995). Neural nets for fuzzy systems. *Fuzzy Sets and Systems, 71*, 265–276. doi:10.1016/0165-0114(94)00282-C

Burillo, P., & Bustince, H. (1996). Vague sets are intuitionistic fuzzy sets. *Fuzzy Sets and Systems, 79*, 403–405. doi:10.1016/0165-0114(95)00154-9

Burns, R. S. (2001). *Advanced control engineering*. Oxford, UK: Butterworth-Heinemann.

Carbonara, N., Giannoccaro, I., & Pontrandolfo, P. (2002). Supply chains within industrial districts: A theoretical framework. *International Journal of Production Economics, 76*(2), 159–176. doi:10.1016/S0925-5273(01)00159-1

Carnes, R. *(1991). Long-term cost of ownership: Beyond purchase price. In* Proceedings of the IEEE/SEMI International Semiconductor Science Symposium *(pp. 39-43).*

Castro, J. L. (1995). Fuzzy logical controllers are universal approximators. *IEEE Transactions on Systems, Man, and Cybernetics. Part B, Cybernetics, 25*, 629–635.

Castro, J. R., Castillo, O., Melin, P., & Rodriguez-Diaz, A. (2009). A hybrid learning algorithm for a class of interval type-2 fuzzy neural networks. *Information Sciences, 179*, 2175–2193. doi:10.1016/j.ins.2008.10.016

Catelani, M., & Giraldi, S. (1998). Fault diagnosis of analog circuits with model-based techniques. In. *Proceedings of the IEEE Conference on Instrumentation and Measurement Technology, 1*, 501–504.

Chan, F. T. S. (2003). Performance measurement in a supply chain. *International Journal of Advanced Manufacturing Technology, 21*(7), 534–548. doi:10.1007/s001700300063

Chang, P. T., & Lee, E. S. (1994). Fuzzy linear regression with spreads unrestricted in sign. *Computers & Mathematics with Applications (Oxford, England), 28*, 61–70. doi:10.1016/0898-1221(94)00127-8

Chang, P.-C., & Hsieh, J.-C. (2003). A neural networks approach for due-date assignment in a wafer fabrication factory. *International Journal of Industrial Engineering, 10*(1), 55–61.

Chang, P.-C., Hsieh, J.-C., & Liao, T. W. (2005). Evolving fuzzy rules for due-date assignment problem in semiconductor manufacturing factory. *Journal of Intelligent Manufacturing, 16*, 549–557. doi:10.1007/s10845-005-1663-4

Chang, S. L., Wang, R. C., & Wang, S. Y. (2006). Applying fuzzy linguistic quantifier to select supply chain partners at different phases of product life cycle. *International Journal of Production Economics, 100*(2), 348–359. doi:10.1016/j.ijpe.2005.01.002

Chang, S. L., Wang, R. C., & Wang, S. Y. (2007). Applying a direct multi-granularity linguistic and strategy-oriented aggregation approach on the assessment of supply performance. *European Journal of Operational Research, 177*(2), 1013–1025. doi:10.1016/j.ejor.2006.01.032

Chang, Y. C. (2001). Adaptive fuzzy-based tracking control for nonlinear SISO systems via VSS and H^{∞} approaches. *IEEE Transactions on Fuzzy Systems, 9*, 278–292. doi:10.1109/91.919249

Chang, Y. H. O. (2001). Hybrid fuzzy least-squares regression analysis and its reliability measures. *Fuzzy Sets and Systems, 119*, 225–246. doi:10.1016/S0165-0114(99)00092-5

Chang, Y. H. O., & Ayyub, B. M. (2001). Fuzzy regression methods-a comparative assessment. *Fuzzy Sets and Systems, 119*, 187–203. doi:10.1016/S0165-0114(99)00091-3

Chen, T., & Wang, Y. C. (2010). A hybrid fuzzy and neural approach for forecasting the book-to-bill ratio in the semiconductor manufacturing industry. *International Journal of Advanced Manufacturing Technology.*

Chen, A.-S., Leung, M. T., & Daouk, H. (2003). Application of neural networks to an emerging financial market: forecasting and trading the Taiwan Stock Index. *Computers & Operations Research, 30*(6), 901–923. doi:10.1016/S0305-0548(02)00037-0

Chen, B. S., Lee, C. H., & Chang, Y. C. (1996). tracking design of uncertain nonlinear SISO systems: Adaptive fuzzy approach. *IEEE Transactions on Fuzzy Systems, 4*, 32–43.H^{∞}doi:10.1109/91.481843

Chen, C. T. (2000). Extension of the TOPSIS for group decision-making under fuzzy environment. *Fuzzy Sets and Systems, 114*, 1–9. doi:10.1016/S0165-0114(97)00377-1

Cheng, I. S., Tsujimura, Y., Gen, M., & Tozawa, T. (1995). An efficient approach for large scale project planning based on fuzzy Delphi method. *Fuzzy Sets and Systems, 76*, 277–288. doi:10.1016/0165-0114(94)00385-4

Chen, M. F., & Tzeng, G. H. (2004). Combining grey relation and TOPSIS concepts for selecting an expatriate host country. *Mathematical and Computer Modelling, 40*, 1473–1490. doi:10.1016/j.mcm.2005.01.006

Chen, S. J., & Hwang, C. L. (1992). *Fuzzy multiple attribute decision making: Methods and applications.* Berlin, Germany: Springer-Verlag.

Chen, S. M., & Tan, J. M. (1994). Handling multicriteria fuzzy decision-making problems based on vague set theory. *Fuzzy Sets and Systems, 67*, 163–172. doi:10.1016/0165-0114(94)90084-1

Chen, T. (2007). Evaluating the mid-term competitiveness of a product in a semiconductor fabrication factory with a systematic procedure. *Computers & Industrial Engineering, 53*, 499–513. doi:10.1016/j.cie.2007.05.008

Chen, T. (2008). A SOM-FBPN-ensemble approach with error feedback to adjust classification for wafer-lot completion time prediction. *International Journal of Advanced Manufacturing Technology, 37*(7-8), 782–792. doi:10.1007/s00170-007-1007-y

Chen, T. (2009). A FNP approach for evaluating and enhancing the long-term competitiveness of a semiconductor fabrication factory through yield learning modeling. *International Journal of Advanced Manufacturing Technology, 40*, 993–1003. doi:10.1007/s00170-008-1414-8

Chen, T. (2009). A fuzzy-neural knowledge-based system for job completion time prediction and internal due date assignment in a wafer fabrication plant. *International Journal of Systems Science, 40*(8), 889–902. doi:10.1080/00207720902974553

Chen, T. (2010). A FNP approach for establishing the optimal and efficient capacity re-allocation plans for enhancing the long-term competitiveness of a semiconductor product. *International Journal of Fuzzy Systems, 12*(2), 158–169.

Chen, T., & Lin, Y. C. (2008). A fuzzy-neural system incorporating unequally important expert opinions for semiconductor yield forecasting. *International Journal of Uncertainty, Fuzziness, and Knowledge-based Systems, 16*(1), 35–58. doi:10.1142/S0218488508005030

Chen, T., & Wang, Y. C. (2009). A fuzzy set approach for evaluating and enhancing the mid-term competitiveness of a semiconductor factory. *Fuzzy Sets and Systems, 160*, 569–585. doi:10.1016/j.fss.2008.06.006

Chen, T., Wang, Y. C., & Tsai, H. R. (2009a). Lot cycle time prediction in a ramping-up semiconductor manufacturing factory with a SOM-FBPN-ensemble approach with multiple buckets and partial normalization. *International Journal of Advanced Manufacturing Technology, 42*(11-12), 1206–1216. doi:10.1007/s00170-008-1665-4

Chen, T., Wang, Y. C., & Wu, H. C. (2009b). A fuzzy-neural approach for remaining cycle time estimation in a semiconductor manufacturing factory – a simulation study. *International Journal of Innovative Computing. Information and Control, 5*(8), 2125–2139.

Chen, T., Wu, H.-C., & Wang, Y.-C. (2009c). Fuzzy-neural approaches with example post-classification for estimating job cycle time in a wafer fab. *Applied Soft Computing*, 1225–1231. doi:10.1016/j.asoc.2009.03.006

Chen, W. C., Tai, P. H., Wang, M. W., Deng, W. J., & Chen, C. T. (2008). A neural network-based approach for dynamic quality prediction in a plastic injection molding process. *Expert Systems with Applications, 35*, 843–849. doi:10.1016/j.eswa.2007.07.037

Chen, W., & Zhang, Z. (2010). Globally stable adaptive backstepping fuzzy control for output-feedback systems with unknown high-frequency gain sign. *Fuzzy Sets and Systems, 161*, 821–836. doi:10.1016/j.fss.2009.10.026

Chiang, K. T. (2007). The optimal process conditions of an injection-molded thermoplastic part with a thin shell feature using grey-fuzzy logic: A case study on machining the PC/ABS cell phone shell. *Materials & Design, 28*, 1851–1860. doi:10.1016/j.matdes.2006.04.008

Chiang, W.-C., Urban, L., & Baldridge, G. W. (1996). A neural network approach to mutual fund asset value forecasting. *Omega. International Journal of Management Science, 24*(2), 205–215.

Chiu, S. (1994). Fuzzy model identification based on cluster estimation. *Journal of Intelligent and Fuzzy System, 2*, 267–278.

Choi, T. Y., & Hartley, J. L. (1996). An exploration of supplier selection practices across the supply chain. *Journal of Operations Management, 14*(4), 333–343. doi:10.1016/S0272-6963(96)00091-5

Chou, T. S.-C., Yang, C.-C., Chen, C.-H., & Lai, F. (1996). A rule-based neural stock trading decision support system. In *Proceedings of the IEEE/IAFE Conference on Computational Intelligence for Financial Engineering*, New York (pp.148-154).

Chu, T. C. (2002a). Facility location selection using fuzzy TOPSIS under group decisions. *International Journal of Uncertainty. Fuzziness and Knowledge-Based Systems, 10*, 687–701. doi:10.1142/S0218488502001739

Chu, T. C. (2002b). Selecting plant location via a fuzzy TOPSIS approach. *International Journal of Advanced Manufacturing Technology, 20*, 859–864. doi:10.1007/s001700200227

Chu, T. C., & Lin, Y. C. (2003). A fuzzy TOPSIS method for robot selection. *International Journal of Advanced Manufacturing Technology, 21*, 284–290. doi:10.1007/s001700300033

Ciliz, M. K. (2005). Rule base reduction for knowledge based fuzzy controller with application to vacuum cleaner. *Expert Systems with Applications, 28*, 175–184. doi:10.1016/j.eswa.2004.10.009

Coppi, R., D'Urso, P., Giordani, P., & Santoro, A. (2006). Least squares estimation of a linear regression model with LR fuzzy response. *Computational Statistics & Data Analysis, 51*(1), 267–286. doi:10.1016/j.csda.2006.04.036

Cordón, O., Gomide, F., Herrera, F., Hoffmann, F., & Magdalena, L. (2004). Ten years of genetic fuzzy systems: current framework and new trends. *Fuzzy Sets and Systems, 141*, 5–31. doi:10.1016/S0165-0114(03)00111-8

Crowder, B. C. *(1989). Manufacturing science and manufacturing competitiveness. In* Proceedings of the IEEE/SEMI International Semiconductor Manufacturing Science Symposium *(pp. 32-33).*

D'Urso, P. (2003). Linear regression analysis for fuzzy/crisp input and fuzzy/crisp output data. *Computational Statistics & Data Analysis, 42*, 47–72. doi:10.1016/S0167-9473(02)00117-2

D'Urso, P., & Giordani, P. (2006). A weighted fuzzy c-means clustering model for fuzzy data. *Computational Statistics & Data Analysis, 50*(6), 1496–1523. doi:10.1016/j.csda.2004.12.002

D'Urso, P., & Santoro, A. (2006). Goodness of fit and variable selection in the fuzzy multiple linear regression. *Fuzzy Sets and Systems*, *157*, 2627–2647. doi:10.1016/j.fss.2005.03.015

Dai, Y., & Xu, J. (1999). Analog circuit fault diagnosis based on noise measurement. *Microelectronics and Reliability*, *39*(8), 1293–1298. doi:10.1016/S0026-2714(99)00029-3

Davé, R. N. (1991). Characterization and detection of noise in clustering. *Pattern Recognition Letters*, *12*(11), 657–664. doi:10.1016/0167-8655(91)90002-4

De Oliveira, J. V., & Pdrycz, W. (2007). *Advances in fuzzy clustering and its applications*. New York, NY: John Wiley & Sons. doi:10.1002/9780470061190

Deffree, S. *(2007)*. Semiconductor winners and losers split by competitiveness. *Retrieved from*http://www.electronicsweekly.com/

Delgado, M., & Moral, S. (1987). On the concept of possibility-probability consistency. *Fuzzy Sets and Systems*, *21*, 311–318. doi:10.1016/0165-0114(87)90132-1

Delgado, M., Verdegay, J. L., & Vila, M. A. (1992). Linguistic decision making models. *International Journal of Intelligent Systems*, *7*(5), 479–492. doi:10.1002/int.4550070507

Deng, W. J., Chen, C. T., Sun, C. H., Chen, W. C., & Chen, C. P. (2008). An effective approach for process parameter optimization in injection molding of plastic housing components. *Polymer-Plastics Technology and Engineering*, *47*, 910–919. doi:10.1080/03602550802189142

Deschrijver, G., & Kerre, E. E. (2007). On the position of intuitionistic fuzzy set theory in the framework of theories modelling imprecision. *Information Sciences*, *177*, 1860–1866. doi:10.1016/j.ins.2006.11.005

Diamond, P. (1988). Fuzzy least squares. *Information Sciences*, *46*, 141–157. doi:10.1016/0020-0255(88)90047-3

Diamond, P. (1990). Least squares fitting of compact set-valued data. *Journal of Mathematical Analysis and Applications*, *14*, 531–544.

Diamond, P., & Kloeden, P. (1994). *Metric spaces of fuzzy sets*. Singapore: World Scientific.

Diamond, P., & Korner, R. (1997). Extended fuzzy linear models and least squares estimates. *Computers & Mathematics with Applications (Oxford, England)*, *33*(9), 15–32. doi:10.1016/S0898-1221(97)00063-1

Dickey, D. A., & Fuller, W. A. (1981). Likelihood ratio statistics for autoregressive time series with a unit. *Econometrica*, *49*(4), 1057–1072. doi:10.2307/1912517

Ding, C., & He, X. (2004a). *K*-means clustering via principal component analysis. In *Proceedings of the Twenty-first International Conference on Machine learning* (pp. 225-232).

Ding, C., & He, X. (2004b). Linearized cluster assignment via spectral ordering. In *Proceedings of the Twenty-first International Conference on Machine learning* (pp. 233-240).

Dixon, J. W., Contrado, J. M., & Morán, L. A. (1999). A fuzzy-controlled active front-end rectifier with current harmonic filtering characteristics and minimum sensing variables. *IEEE Transactions on Power Electronics*, *14*(4), 724–729. doi:10.1109/63.774211

Do, J.-H., Jang, H.-Y., Jung, S.-H., Jung, J.-W., & Bien, Z. (2005). Soft remote control system in the intelligent sweet home. In *Proceedings of the IEEE/RSJ International Conference on Intelligent Robots and Systems* (pp. 3984-3989).

Do, J.-H., Kim, J.-B., Park, K.-H., Bang, W.-C., & Bien, Z. Z. (2005). Soft remote control system in the intelligent sweet home. In *Proceedings of the IEEE/RSJ International Conference on Intelligent Robots and Systems*, Edmonton, AB, Canada (pp. 3984-3989).

Do, J.-H., Jang, H., Jung, S. H., Jung, J. W., & Bien, Z. (2002). Soft remote control system using hand pointing gesture. *International Journal of Human-friendly Welfare Robotic Systems*, *3*(1), 27–30.

Dong, W. H., Sun, X. X., & Lin, Y. (2007). Variable structure model reference adaptive control with unknown high frequency gain sign. *Acta Automatica Sinica*, *33*(4), 404–408.

Donoho, D. (2000, August). *High-dimensional Data Analysis: the Curses and Blessings of Dimensionality*. Paper presented at the conference "Math Challenges of the 21st Century" of the American Mathematics Society, Los Angeles.

Donoso, S., Marin, N., & Vila, M. A. *(2006). Quadratic programming models for fuzzy regression. In* Proceedings of the International Conference on Mathematical and Statistical Modeling in Honor of Enrique Castillo.

Dubois, D., Prade, H., & Sandri, S. (1991). On possibility/probability transformations. In *Proceedings of the Fourth IFSA Conference* (pp. 50-53).

Dubois, D., & Prade, H. (1982). On several representations of an uncertain body of evidence. In Gupta, M. M., & Sanchez, E. (Eds.), *Fuzzy information and decision processes* (pp. 167–181). Amsterdam, The Netherlands: North-Holland.

Dubois, D., & Prade, H. (1983). Unfair coins and necessary measures: A possible interpretation of histograms. *Fuzzy Sets and Systems, 10*, 15–20. doi:10.1016/S0165-0114(83)80099-2

Dubois, D., & Prade, H. (1996). What are fuzzy rules and how to use them. *Fuzzy Sets and Systems, 84*, 169–185. doi:10.1016/0165-0114(96)00066-8

Duda, R. O., Hart, P. E., & Stork, D. G. (2001). *Pattern classification* (2nd ed.). Hoboken, NJ: John Wiley & Sons.

Dunn, J. (1973). A fuzzy relative of the isodata process and its use in detecting compact well-separated clusters. *Cybernetics and Systems: An International Journal, 3*(3), 32–57. doi:10.1080/01969727308546046

Dunn, J. C. (1974). Well separated clusters and optimal fuzzy partitions. *Journal of Cybernetics, 4*(1), 95–104. doi:10.1080/01969727408546059

Eraslan, E. (2009). The estimation of product standard time by artificial neural networks in the molding industry. *Mathematical Problems in Engineering.*

Essounbouli, E., & Hamzaoui, A. (2006). Direct and indirect robust adaptive fuzzy controllers for a class of nonlinear systems. *International Journal of Control, Automation, and Systems, 4*(2), 146–154.

Filev, D., & Yager, R. R. (1995). Analytic properties of maximum entropy OWA operators. *Information Sciences, 85*(1-3), 11–27. doi:10.1016/0020-0255(94)00109-O

Fu, H.-C., Chang, H.-Y., Xu, Y. Y., & Pao, H.-T. (2000). User adaptive handwriting recognition by self-growing probabilistic decision-based neural networks. *IEEE Transactions on Neural Networks, 11*(6), 1373–1384. doi:10.1109/72.883451

Fukuyama, Y., & Sugeno, M. (1989). A new method of choosing the number of clusters for the fuzzy *c*-means method. In *Proceedings of the 5th Fuzzy System Symposium* (pp. 247-250).

Gabriella, P., Yager, R. R., & Atanassov, K. T. (2004). Intuitionistic fuzzy graph interpretations of multi-person multi-criteria decision making: Generalized net approach. In *Proceedings of the Second IEEE International Conference on Intelligent Systems* (pp. 434-439).

Gau, W. L., & Buehrer, D. J. (1993). Vague sets. *IEEE Transactions on Systems, Man, and Cybernetics, 23*, 610–614. doi:10.1109/21.229476

Ge, S. S., Hong, F., & Lee, T. H. (2004). Adaptive neural control of nonlinear time-delay systems with unknown virtual control coefficients. *IEEE Transactions on Systems, Man, and Cybernetics – Part B, 34*(1), 499-516.

Geer, J. F., & Klir, G. J. (1992). A mathematical analysis of information-preserving transformations between probabilistic and possibilistic formulations of uncertainty. *International Journal of General Systems, 20*, 143–176. doi:10.1080/03081079208945024

Ghazavi, S. N., & Liao, T. W. (2008). Medical data mining by fuzzy modeling with selected features. *Artificial Intelligence in Medicine, 43*, 195–206. doi:10.1016/j.artmed.2008.04.004

Gibbs, A. L., & Su, F. E. (2002). On choosing and bounding probability metrics. *International Statistical Review, 70*(3), 419–435. doi:10.2307/1403865

Gionis, A., Hinneburg, A., Papadimitriou, S., & Tsaparas, P. (2005). Dimension induced clustering. In *Proceedings of the Eleventh ACM SIGKDD International Conference on Knowledge Discovery in Data Mining* (pp. 51-60).

Glanz, F. H., Miller, W. T., & Kraft, L. G. (1991). An overview of the CMAC neural network. In *Proceedings of the IEEE Conference on Neural Networks for Ocean Engineering* (pp. 301-308).

Grady, W. M., Samotyj, M. J., & Noyola, A. H. (1990). Survey of active power line conditioning methodologies. *IEEE Transactions on Power Delivery*, *5*(3), 1536–1542. doi:10.1109/61.57998

Gruber, H. (1994). *Learning and strategic product innovation: Theory and evidence for the semiconductor industry*. Boston, MA: Elsevier.

Grzegorzewski, P., & Mrówka, E. (2005). Some notes on (Atanassov's) intuitionistic fuzzy sets. *Fuzzy Sets and Systems*, *156*(3), 492–495. doi:10.1016/j.fss.2005.06.002

Gunasekaran, A., Patel, C., & McGaughey, R. E. (2004). A framework for supply chain performance measurement. *International Journal of Production Economics*, *87*(3), 333–347. doi:10.1016/j.ijpe.2003.08.003

Guo, Z., Quek, C., & Maskell, D. L. (2006). FCMAC-AARS: a novel FNN architecture for stock market prediction and trading. In *Proceedings of the International Conference on Evolutionary Computation*, Vancouver, BC, Canada.

Habbi, A., & Zelmat, M. (2003). An improved self-tuning mechanism of fuzzy control by gradient descent method. In *Proceedings of the 17th European Simulation Multiconference*, Nottingham, UK.

Hagras, H. A. (2006). Comments on dynamical optimal training for interval type-2 fuzzy neural network (T2FNN). *IEEE Transactions on Systems, Man, and Cybernetics. Part B, Cybernetics*, *36*(5), 1206–1209. doi:10.1109/TCSI.2006.873184

Hagras, H. A. (2007). Type-2 FLCs: A new generation of fuzzy controllers. *IEEE Computational Intelligent Magazine*, *2*, 30–43. doi:10.1109/MCI.2007.357192

Hagras, H. A., & Hierarchical, A. A. (2004). Type-2 fuzzy logic control architecture for autonomous mobile robots. *IEEE Transactions on Fuzzy Systems*, *12*(4), 524–539. doi:10.1109/TFUZZ.2004.832538

Hampel, R., & Chaker, N. (1998). Minimizing the variable parameters for optimizing the fuzzy controller. *Fuzzy Sets and Systems*, *100*, 131–142. doi:10.1016/S0165-0114(97)00059-6

Han, J.-S. (2006). *New feature subset selection method and its application for EMG recognition*. Unpublished doctoral dissertation, Korean Advanced Institute of Science and Technology, Daejeon, Korea.

Hathaway, R. J., & Bezdek, J. C. (2001). Fuzzy *c*-means clustering of incomplete data. *IEEE Transactions on Systems, Man, and Cybernetics. Part B, Cybernetics*, *31*(5), 735–744. doi:10.1109/3477.956035

Hathaway, R. J., Bezdek, J. C., & Pedrycz, W. (1996). A parametric model for fusing heterogeneous fuzzy data. *IEEE Transactions on Fuzzy Systems*, *4*(3), 1277–1282. doi:10.1109/91.531770

Helms, R. (2001). Chips: The fastest downturn in history has its optimist. *Business Week*.

Herrera, F., Herrera-Viedma, E., & Martinez, L. (2000). A fusion approach for managing multi-granularity linguistic term sets in decision making. *Fuzzy Sets and Systems*, *114*(1), 43–58. doi:10.1016/S0165-0114(98)00093-1

Herrera, F., Herrera-Viedma, E., & Verdegay, J. L. (1995). A sequential selection process in group decision making with a linguistic assessment approach. *Information Sciences*, *85*(4), 223–239. doi:10.1016/0020-0255(95)00025-K

Herrera, F., Herrera-Viedma, E., & Verdegay, J. L. (1996). Direct approach processes in group decision making using linguistic OWA operators. *Fuzzy Sets and Systems*, *79*(2), 175–190. doi:10.1016/0165-0114(95)00162-X

Herrera, F., & Martinez, L. (2000). A 2-tuple fuzzy linguistic representation model for computing with words. *IEEE Transactions on Fuzzy Systems*, *8*(6), 746–752. doi:10.1109/91.890332

Herrera, F., Martinez, L., & Sanchez, P. J. (2005). Managing non-homogeneous information in group decision making. *European Journal of Operational Research*, *166*, 115–132. doi:10.1016/j.ejor.2003.11.031

Hinneburg, A., & Keim, D. A. (1999). Optimal grid-clustering: Towards breaking the curse of dimensionality in high-dimensional clustering. In *Proceedings of 25th International Conference on Very Large Data Bases* (pp. 506-517).

Honda, K., Ichihashi, H., Notsu, A., Masulli, F., & Rovetta, S. (2006). Several formulations for graded possibilistic approach to fuzzy clustering. In S. Greco, et al. (Eds.), *Rough Sets and Current Trends in Computing (RSCTC) 2006* (LNCS 4259, pp. 939-948). Berlin: Springer Verlag.

Honda, K., & Ichihashi, H. (2004). Linear fuzzy clustering techniques with missing values and their application to local principal component analysis. *IEEE Transactions on Fuzzy Systems, 12*(2), 183–193. doi:10.1109/TFUZZ.2004.825073

Honda, K., & Ichihashi, H. (2005). Regularized linear fuzzy clustering and probabilistic PCA mixture models. *IEEE Transactions on Fuzzy Systems, 13*(4), 508–516. doi:10.1109/TFUZZ.2004.840104

Honda, K., Ichihashi, H., Masulli, F., & Rovetta, S. (2007). Linear fuzzy clustering with selection of variables using graded possibilistic approach. *IEEE Transactions on Fuzzy Systems, 15*(5), 878–889. doi:10.1109/TFUZZ.2006.889946

Honda, K., Notsu, A., & Ichihashi, H. (2010). Fuzzy pca-guided robust *k*-means clustering. *IEEE Transactions on Fuzzy Systems, 18*(1), 67–79. doi:10.1109/TFUZZ.2009.2036603

Hong, P., Turk, M., & Huang, T. S. (2000). Gesture modeling and recognition using finite state machines. In *Proceedings of the 4th IEEE International Conference on Automatic Face and Gesture Recognition* (pp. 410-415).

Hong, D. H., & Hwang, C. (2005). Interval regression analysis using quadratic loss support vector machine. *IEEE Transactions on Fuzzy Systems, 13*, 229–237. doi:10.1109/TFUZZ.2004.840133

Hong, D. H., & Hwang, C. (2006). Support vector interval regression machine for crisp input and output data. *Fuzzy Sets and Systems, 157*, 1114–1125. doi:10.1016/j.fss.2005.09.008

Höppner, F., Klawonn, F., Kruse, R., & Runkler, T. (1999). *Fuzzy Cluster Analysis*. Chichester, UK: John Wiley & Sons.

Hsiao, F. H., Xu, S. D., Lin, C. Y., & Tsai, Z. R. (2008). Robustness design of fuzzy control for nonlinear multiple time-delay large-scale systems via neural-network-based approach. *IEEE Transactions on Systems, Man, and Cybernetics. Part B, Cybernetics, 38*(1), 244–251. doi:10.1109/TSMCB.2006.890304

Huang, M. S., & Lin, T. Y. (2008). An innovative regression model-based searching method for setting the robust injection molding parameters. *Journal of Materials Processing Technology, 198*, 436–444. doi:10.1016/j.jmatprotec.2007.07.022

Hung, W. L., & Yang, M. S. (2005). Fuzzy clustering on LR-type fuzzy numbers with an application in Taiwanese tea evaluation. *Fuzzy Sets and Systems, 150*(3), 561–577. doi:10.1016/j.fss.2004.04.007

Hwang, B. N., Chang, S. C., Yu, H. C., & Chang, C. W. (2008). Pioneering e-supply chain integration in semiconductor industry: A case study. *International Journal of Advanced Manufacturing Technology, 36*, 825–832. doi:10.1007/s00170-006-0886-7

Hwang, C. L., & Yoon, K. (1981). *Multiple attribute decision making: Methods and applications: A state of the art survey*. Berlin, Germany: Springer-Verlag.

IC-EMC. (2002). *61967-2: Methods of radiated emission – TEM cell method and wideband TEM cell method 150 KHz to 8 GHz*. Geneva, Switzerland: International Electrotechnical Commission.

IC-EMC. (2006). *62433-2: Models of integrated circuits for EMI behavioral simulation – ICEM-CE. ICEM conducted emission model*. Geneva, Switzerland: International Electrotechnical Commission.

IEEE. Computer Society. (1993). *Standard 519-1992: IEEE Recommended Practices and Requirements for Harmonic Control in Electrical Power Systems*. Washington, DC: Author.

Ioannou, P. A., & Sun, J. (1996). *Robust adaptive control*. Upper Saddle River, NJ: Prentice Hall.

Irpino, A., & Verde, R. (2008). Dynamic clustering for interval data using a Wasserstein-based distance. *Pattern Recognition Letters*, *29*(11), 1648–1658. doi:10.1016/j.patrec.2008.04.008

Ishibuchi, H., & Yamamoto, T. (2005). Rule weight specification in fuzzy rule-based classification systems. *IEEE Transactions on Fuzzy Systems*, *13*(4), 428–435. doi:10.1109/TFUZZ.2004.841738

Jain, S. K., Agrawal, P., & Gupta, H. O. (2002). Fuzzy logic controlled shunt active power filter for power quality improvement. *IEE Proceedings. Electric Power Applications*, *149*(5), 317–328. doi:10.1049/ip-epa:20020511

Jang, J.-S. R. (1996). Input selection for ANFIS learning. In *Proceedings of the IEEE International Conference on Fuzzy Systems* (pp. 1493-1499).

Jang, J. S., Sun, C. T., & Mizutani, E. (1997). *Neuro-fuzzy and soft computing: A computational approach to learning and machine intelligence*. Upper Saddle River, NJ: Prentice-Hall.

Jang, J.-S. R. (1993). ANFIS: Adaptive network-based fuzzy inference system. *IEEE Transactions on Systems, Man, and Cybernetics*, *23*(3), 665–685. doi:10.1109/21.256541

Janikow, C. Z. (1998). Fuzzy decision trees: Issues and methods. *IEEE Transactions on Systems, Man, and Cybernetics*, *28*(1), 1–14. doi:10.1109/3477.658573

Jeng, J. T., Chuang, C. C., & Su, S. F. (2003). Support vector interval regression networks for interval regression analysis. *Fuzzy Sets and Systems*, *138*, 283–300. doi:10.1016/S0165-0114(02)00570-5

Jenkins, T., Phail, F., & Sackman, S. (1990). Semiconductor competitiveness in the 1990s. *Proceedings of the Society of Automotive Engineers*, *223*, 249–255.

Jeon, M.-J. (2008). *Hand gesture recognition using multivariate fuzzy decision tree and user adaptation*. Unpublished master's thesis, Korean Advanced Institute of Science and Technology, Daejeon, Korea.

Jeyakumar, V., Rubinov, A. M., & Wu, Z. Y. (2007). Non-convex quadratic minimization problems with quadratic constraints: Global optimality conditions. *Mathematics Programming Series A*, *110*, 521–541. doi:10.1007/s10107-006-0012-5

Jung, S.-H. (2007). *Incremental user adaptation in Korean sign language recognition using motion similarity and prediction from adaptation history*. Unpublished master's thesis, Korea Advanced Institute of Science and Technology, Daejeon, Korea.

Jung, S.-H., Do, J.-H., & Bien, Z. (2006). Adaptive hand motion recognition in soft remote control system. In *Proceedings of the 7th International Workshop on Human-friendly Welfare Robotic Systems*.

Kacprzyk, J. (1986). Group decision making with a fuzzy liguistic majority. *Fuzzy Sets and Systems*, *18*(2), 105–118. doi:10.1016/0165-0114(86)90014-X

Kaloust, J., & Qu, Z. (1995). Continuous robust control design for nonlinear uncertain systems without a priori knowledge of control direction. *IEEE Transactions on Automatic Control*, *40*, 276–282. doi:10.1109/9.341792

Kaloust, J., & Qu, Z. (1997). Robust control design for nonlinear uncertain systems with an unknown time-varying control direction. *IEEE Transactions on Automatic Control*, *42*(3), 393–399. doi:10.1109/9.557583

Kamoun, A., Jaziri, M., & Chaabouni, M. (2009). The use of the simplex method and its derivatives to the on-line optimization of the parameters of an injection moulding process. *Chemometrics and Intelligent Laboratory Systems*, *96*, 117–122. doi:10.1016/j.chemolab.2008.04.010

Kannan, V. R., & Tan, K. C. (2002). Supplier selection and assessment: Their impact on business performance. *Journal of Supply Chain Management*, *38*(4), 11–21. doi:10.1111/j.1745-493X.2002.tb00139.x

Karnik, N. N., & Mendel, J. M. (1999). Applications of type-2 fuzzy logic systems to forecasting of time-series. *Information Science*, *120*, 89–111. doi:10.1016/S0020-0255(99)00067-5

Karuppanan, P., & Mahapatra, K. K. (2011). PLL with fuzzy logic controller based shunt active power filter for harmonic and reactive power compensation. In *Proceedings of the International Conference on Power Electronics*, New Delhi, India (pp. 1-6).

Kaufmann, A., & Gupta, M. M. (1998). *Fuzzy Mathematical Models in Engineering and Management Science*. Amsterdam, The Netherlands: North-Holland.

Keller, A. (2000). Fuzzy clustering with outliers. In *Proceedings of the 19th International Conference of the North American Fuzzy Information Processing Society* (pp. 143-147).

Ker, J., Hsu, C., Kuo, Y., & Liu, B. (1997). A fuzzy CMAC model for color reproduction. *Fuzzy Sets and Systems*, *91*(1), 53–68. doi:10.1016/S0165-0114(96)00083-8

Kim, B. S. (2003). Rehabilitation and assistive devices. In *Proceedings of the 8th International Conference on Rehabilitation Robotics*, Daejeon, Korea.

Kim, D.-J. (2004). *Image-based personalized facial expression recognition system using fuzzy neural networks.* Unpublished doctoral dissertation, Korea Advanced Institute of Science and Technology, Daejeon, Korea.

Kim, H., & Lin, C. S. (1992). Use of adaptive resolution for better CMAC learning. In *Proceedings of the International Joint Conference on Neural Networks* (pp. 517-522).

Kim, D.-W., Lee, K. H., & Lee, D. (2003). Fuzzy cluster validation index based on inter-cluster proximity. *Pattern Recognition Letters*, *24*(15), 2561–2574. doi:10.1016/S0167-8655(03)00101-6

Klement, E. P., Mesiar, R., & Pap, E. (2000). *Triangular norms*. Dordrecht, The Netherlands: Kluwer Academic.

Klir, G. J. (1989). Probability-possibility conversion. In *Proceedings of the Third IFSA Congress* (pp. 408-411).

Klir, G. J. (1990). A principle of uncertainty and information invariance. *International Journal of General Systems*, *17*, 249–275. doi:10.1080/03081079008935110

Klir, G. J. (2006). *Uncertainty and information*. New York, NY: John Wiley & Sons.

Klir, G. J., & Parviz, B. (1992). Probability-possibility transformations: A comparison. *International Journal of General Systems*, *21*, 291–310. doi:10.1080/03081079208945083

Krause, D. R., Pagell, M., & Curkovic, S. (2001). Toward a measure of competitive priorities for purchasing. *Journal of Operations Management*, *19*(4), 497–512. doi:10.1016/S0272-6963(01)00047-X

Krishnapuram, R., & Keller, J. M. (1993). A possibilistic approach to clustering. *IEEE Transactions on Fuzzy Systems*, *1*(2), 98–110. doi:10.1109/91.227387

Krstic, M., Kanellapoulos, I., & Kokotovic, P. (1995). *Nonlinear and adaptive control design*. New York, NY: Wiley Interscience.

Kruse, R., Döring, C., & Lesot, M.-J. (2007). Advances in fuzzy clustering and its applications. In *Fundamentals of Fuzzy Clustering* (pp. 3–30). London: John Wiley & Sons.

Kumar, P., & Mahajan, A. (2009). Soft computing techniques for the control of an active power filter. *IEEE Transactions on Power Delivery*, *24*(1), 452–461. doi:10.1109/TPWRD.2008.2005881

Labiod, S., & Boucherit, M. S. (2006). Indirect fuzzy adaptive control of a class of SISO nonlinear systems. *Arabian Journal for Science and Engineering*, *31*(1B), 61–74.

Labiod, S., Boucherit, M. S., & Guerra, T. M. (2005). Adaptive fuzzy control of a class of MIMO nonlinear systems. *Fuzzy Sets and Systems*, *151*(1), 59–77. doi:10.1016/j.fss.2004.10.009

Labiod, S., & Guerra, T. M. (2007). Direct adaptive fuzzy control for a class of MIMO nonlinear systems. *International Journal of Systems Science*, *38*(8), 665–675. doi:10.1080/00207720701500583

Labiod, S., & Guerra, T. M. (2010). Indirect adaptive fuzzy control for a class of nonaffine nonlinear systems with unknown control directions. *International Journal of Control, Automation, and Systems*, *8*(4), 903–907. doi:10.1007/s12555-010-0425-z

Labussiere-Dorgan, C., Bendhia, S., Sicard, E., Junwu, T., Quaresma, H. J., & Lochot, C. (2008). Modeling the electromagnetic emission of a microcontroller using a single model. *IEEE Transactions on Electromagnetic Compatibility*, *50*, 22–34. doi:10.1109/TEMC.2007.911918

Lam, M. (2004). Neural network techniques for financial performance prediction: integrating fundamental and technical analysis. *Decision Support Systems*, *37*(4), 567–581. doi:10.1016/S0167-9236(03)00088-5

Lawley, D. N., & Maxwell, A. E. (1963). *Factor Analysis as a Statistical Method.* London: Butterworth.

Leachman, R. C. (2002). *Competitive semiconductor manufacturing: Final report on findings from benchmarking eight-inch, sub-350nm wafer fabrication lines*. Berkley, CA: University of California.

Lee, H. L. (2003). Aligning supply chain strategies with product uncertainties. *IEEE Engineering Management Review*, *31*(2), 26–34. doi:10.1109/EMR.2003.1207060

Leigh, W., Purvis, R., & Ragusa, J. M. (2002). Forecasting the NYSE composite index with technical analysis, pattern recognizer, neural network, and genetic algorithm: a case study in romantic decision support. *Decision Support Systems*, *32*(4), 361–377. doi:10.1016/S0167-9236(01)00121-X

Leontaris, I. J., & Billings, S. A. (1987). Model selection and validation methods for nonlinear systems. *International Journal of Control*, *45*, 311–341. doi:10.1080/00207178708933730

Li, M. (2004). *Test frequency selection for analog circuits based on bode diagrams and equivalent fault grouping*. Unpublished master's thesis, University of Cincinnati, OH.

Li, Q., Xu, M., Zhang, H., & Liu, F. (2006). Neural network based flow forecast and diagnosis. In *Computational Intelligence and Security* (LNCS 3802, pp. 542-547).

Liang, Q., & Mendel, J. M. (2000). Equalization of nonlinear time-varying channels using type-2 fuzzy adaptive filters. *IEEE Transactions on Fuzzy Systems*, *8*, 551–563. doi:10.1109/91.873578

Liang, Q., & Mendel, J. M. (2000). Interval type-2 logic systems: Theory and design. *IEEE Transactions on Fuzzy Systems*, *8*, 535–550. doi:10.1109/91.873577

Liao, S. H., & Hu, T. C. (2007). Knowledge transfer and competitive advantage on environmental uncertainty: An empirical study of the Taiwan semiconductor industry. *Technovation*, *27*(6-7), 402–411. doi:10.1016/j.technovation.2007.02.005

Liao, T. W. (2004). Fuzzy reasoning based automatic inspection of radiographic welds: weld recognition. *Journal of Intelligent Manufacturing*, *15*, 69–85. doi:10.1023/B:JIMS.0000010076.56537.07

Liao, T. W. (2006). Mining Human Interpretable Knowledge Using Automatic Data-Driven Fuzzy Modeling Methods – A Review. In Triantaphyllou, E., & Felici, G. (Eds.), *Data Mining and Knowledge Discovery Approaches Based on Rule Induction Techniques* (pp. 495–550). New York: Springer. doi:10.1007/0-387-34296-6_15

Liao, T. W., Celmins, A. K., & Hammell, R. J. Jr. (2003). A fuzzy c-means variant for the generation of fuzzy term sets. *Fuzzy Sets and Systems*, *135*, 241–257. doi:10.1016/S0165-0114(02)00136-7

Liao, T. W., Li, D.-M., & Li, Y.-M. (1999). Detection of welding flaws from radiographic images using fuzzy clustering methods. *Fuzzy Sets and Systems*, *108*, 145–158. doi:10.1016/S0165-0114(97)00307-2

Liao, T. W., Li, D.-M., & Li, Y.-M. (2000). Extraction of welds from radiographic images using fuzzy classifiers. *Information Sciences*, *126*, 21–40. doi:10.1016/S0020-0255(00)00016-5

Li, D. F. (2003). *Fuzzy multiobjective many-person decision makings and games*. Beijing, China: National Defense Industry Press.

Li, D. F. (2004). Some measures of dissimilarity in intuitionistic fuzzy structures. *Journal of Computer and System Sciences*, *68*, 115–122. doi:10.1016/j.jcss.2003.07.006

Li, D. F. (2005). Multiattribute decision making models and methods using intuitionistic fuzzy sets. *Journal of Computer and System Sciences*, *70*, 73–85. doi:10.1016/j.jcss.2004.06.002

Li, D. F. (2010a). TOPSIS-based nonlinear-programming methodology for multiattribute decision making with interval-valued intuitionistic fuzzy sets. *IEEE Transactions on Fuzzy Systems*, *18*(2), 299–311.

Li, D. F. (2010b). Linear programming method for MADM with interval-valued intuitionistic fuzzy sets. *Expert Systems with Applications*, *37*(8), 5939–5945. doi:10.1016/j.eswa.2010.02.011

Li, D. F., Chen, G. H., & Huang, Z. G. (2010). Linear programming method for multiattribute group decision making using IF sets. *Information Sciences*, *180*(9), 1591–1609. doi:10.1016/j.ins.2010.01.017

Li, D. F., & Cheng, C. T. (2002). New similarity measures of intuitionistic fuzzy sets and application to pattern recognitions. *Pattern Recognition Letters*, *23*, 221–225. doi:10.1016/S0167-8655(01)00110-6

Li, D. F., Wang, Y. C., Liu, S., & Shan, F. (2009). Fractional programming methodology for multi-attribute group decision-making using IFS. *Applied Soft Computing*, *9*, 219–225. doi:10.1016/j.asoc.2008.04.006

Lima Neto, E. A., & de Carvalho, F. A. T. (2008). Centre and range method for fitting a linear regression model to symbolic interval data. *Computational Statistics & Data Analysis*, *52*, 1500–1515. doi:10.1016/j.csda.2007.04.014

Lin, T. C., Chang, Y. M., & Kuo, M. J. (2008). Circuit extraction for ICEM model based on fuzzy logic system. In *Proceedings of the EMC Technology and Practice Symposium*.

Lin, T. C., Chen, Y. C., & Kuo, M. J. (2007). Analog circuits fault diagnosis under parameter variations based on fuzzy logic system. In *Proceedings of the VLSI Design/CAD Symposium*, Hualien, Taiwan.

Lin, T. C., Kuo, M. J., & Chen, Y. C. (2007). Frequency domain analog circuit fault diagnosis based on radial basis function neural network. In *Proceedings of the VLSI Design/CAD Symposium*, Hualien, Taiwan.

Lin, Y., Hsu, L., Costa, R. R., & Lizarralde, F. (2003). Variable structure model reference adaptive control for systems with unknown high frequency gain. In *Proceedings of the 42nd IEEE Conference on Decision and Control* (pp. 3525-3530).

Lin, L., Yuan, X. H., & Xia, Z. Q. (2007). Multicriteria fuzzy decision-making methods based on intuitionistic fuzzy sets. *Journal of Computer and System Sciences*, *73*, 84–88. doi:10.1016/j.jcss.2006.03.004

Lin, T. C. (2010a). Analog circuits fault diagnosis under parameter variations based on type-2 fuzzy logic systems. *International Journal of Innovative Computing. Information and Control*, *6*(5), 2137–2158.

Lin, T. C. (2010b). Observer-based robust adaptive interval type-2 fuzzy tracking control of multivariable nonlinear systems. *Engineering Applications of Artificial Intelligence*, *23*(3), 386–399. doi:10.1016/j.engappai.2009.11.007

Lin, T. C., Kuo, M. J., & Hsu, C. H. (2010). Robust adaptive tracking control of multivariable nonlinear systems based on interval type-2 fuzzy approach. *International Journal of Innovative Computing. Information and Control*, *6*(3), 941–961.

Lin, T. C., Liu, H. L., & Kuo, M. J. (2009). Direct adaptive interval type-2 fuzzy control of multivariable nonlinear systems. *Engineering Applications of Artificial Intelligence*, *22*, 420–430. doi:10.1016/j.engappai.2008.10.024

Liu, H. W., & Wang, G. J. (2007). Multi-criteria decision-making methods based on intuitionistic fuzzy sets. *European Journal of Operational Research*, *179*, 220–233. doi:10.1016/j.ejor.2006.04.009

Liu, L., & Huang, J. (2006). Global robust stabilization of cascade-connected systems with dynamic uncertainties without knowing the control direction. *IEEE Transactions on Automatic Control*, *51*, 1693–1699. doi:10.1109/TAC.2006.883023

Liu, L., & Huang, J. (2008). Global robust output regulation of lower triangular systems with unknown control direction. *Automatica*, *44*, 1278–1284. doi:10.1016/j.automatica.2007.09.014

Liu, R.-W. (1991). *Testing and diagnosis of analog circuits and systems*. New York, NY: Van Nostrand Reinhold.

Liu, X. (2007). Parameterized defuzzification with maximum entropy weighting function - another view of the weighting function expectation method. *Mathematical and Computer Modelling*, *45*, 177–188. doi:10.1016/j.mcm.2006.04.014

MacQueen, J. B. (1967). Some methods of classification and analysis of multivariate observations. In *Proceedings of the Fifth Berkeley Symposium on Mathematical Statistics and Probability* (Vol. 1, pp. 281-297).

Maia, A. L. S., de Carvalho, F. A. T., & Ludermir, T. B. (2008). Forecasting models for interval-valued time series. *Neurocomputing*, *71*, 3344–3352. doi:10.1016/j.neucom.2008.02.022

Ma, M., Friedman, M., & Kandel, A. (1997). General fuzzy least squares. *Fuzzy Sets and Systems*, *88*, 107–118. doi:10.1016/S0165-0114(96)00051-6

Mamdani, E. H., & Assilina, S. (1975). An experiment in linguistic synthesis with a fuzzy logic controller. *International Journal of Man-Machine Studies*, *7*(1), 1–13. doi:10.1016/S0020-7373(75)80002-2

Mantaras, R. L. (1990). *Approximate Reasoning Models*. Chichester, UK: Ellis Horwood Limited.

Marks, R. J., II, Oh, S., Arabshahi, P., Caudell, T. P., Choi, J. J., & Song, B. G. (1992). Steepest descent adaptation of min-max fuzzy if-then rules. In *Proceedings of the IEEE/INNS International Joint Conference on Neural Networks*, Beijing, China.

Martinez, R., Castillo, O., & Aguilar, L. T. (2009). Optimization of interval type-2 fuzzy logic controllers for a perturbed autonomous wheeled mobile robot using genetic algorithm. *Information Sciences*, *179*, 2157–2174. doi:10.1016/j.ins.2008.12.028

Masulli, F., & Rovetta, S. (2006). Soft transition from probabilistic to possibilistic fuzzy clustering. *IEEE Transactions on Fuzzy Systems*, *14*(4), 516–527. doi:10.1109/TFUZZ.2006.876740

Ma, X. J., & Sun, Z. Q. (2000). Output tracking and regulation of nonlinear system based on Takagi-Sugeno fuzzy model. *IEEE Transactions on Systems, Man, and Cybernetics. Part B, Cybernetics*, *30*, 47–59. doi:10.1109/3477.826946

Mendel, J. M. (1995). Fuzzy logic systems for engineering: A tutorial. *Proceedings of the IEEE*, *83*(3), 345–377. doi:10.1109/5.364485

Mendel, J. M. (2004). Computing derivatives in interval type-2 fuzzy logic systems. *IEEE Transactions on Fuzzy Systems*, *12*(1), 84–98. doi:10.1109/TFUZZ.2003.822681

Merz, C. J., & Murphy, P. M. (1996). *UCI repository for machine learning data-bases*. Retrieved from http://archive.ics.uci.edu/ml/

Miller, W. T., Glanz, F. H., & Kraft, L. G. (1990). CMAC: an associative neural network alternative to backpropagation. *Proceedings of the IEEE*, *78*, 1561–1567. doi:10.1109/5.58338

Mishra, S., & Bhende, C. N. (2007). Bacterial foraging technique-based optimized active power filter for load compensation. *IEEE Transactions on Power Delivery*, *22*(1), 457–465. doi:10.1109/TPWRD.2006.876651

Mitchell, H. B. (2005). Pattern recognition using type-II fuzzy sets. *Information Sciences*, *170*, 409–418. doi:10.1016/j.ins.2004.02.027

Miyamoto, S., Ichihashi, H., & Honda, K. (2008). *Algorithms for Fuzzy Clustering*. Berlin: Springer Verlag.

Mohan, N., Undeland, T., & Robbins, W. P. (2003). *Power Electronics-Converters, Applications and Design*. New York, NY: John Wiley & Sons.

Moody, J. (1989). Fast learning in multi-resolution hierarchies. *Advances in Neural Information Processing Systems*, *1*, 29–38.

Moody, J., & Saffell, M. (2001). Learning to trade via direct reinforcement. *IEEE Transactions on Neural Networks*, *12*(4), 875–889. doi:10.1109/72.935097

Moody, J., Wu, L., Liao, Y., & Saffell, M. (1998). Performance functions and reinforcement learning for trading systems and portfolios. *Journal of Forecasting*, *17*(5-6), 441–470. doi:10.1002/(SICI)1099-131X(1998090)17:5/6<441::AID-FOR707>3.0.CO;2-#

Moor, R. E. (1979). *Methods and applications of interval analysis*. Philadelphia, PA: SIAM.

Moser, B., & Navarata, M. (2002). Fuzzy controllers with conditionally firing rules. *IEEE Transactions on Fuzzy Systems*, *10*(3), 340–349. doi:10.1109/TFUZZ.2002.1006437

Muralidharan, C., Anantharaman, N., & Deshmukh, S. G. (2002). A multi-criteria group decisionmaking model for supplier rating. *Journal of Supply Chain Management*, *38*(4), 22–33. doi:10.1111/j.1745-493X.2002.tb00140.x

Mutnury, B. (2005). *Macromodeling of nonlinear driver and receiver circuits*. Unpublished doctoral dissertation, Georgia Institute of Technology, Atlanta, GA.

Nakanishi, H., Turksen, I. B., & Sugeno, M. (1993). A review and comparison of six reasoning methods. *Fuzzy Sets and Systems*, *57*, 257–294. doi:10.1016/0165-0114(93)90024-C

Nam, Y., & Wohn, K. (1996). Recognition of. space-time hand-gesture using hidden Markov model. In *Proceedings of the ACM Symposium on Virtual Reality Software and Technology* (pp. 51-58).

Narendra, K. S., & Parthasarathy, K. (1990). Identification and control of dynamical systems using neural networks. *IEEE Transactions on Neural Networks*, *1*, 4–27. doi:10.1109/72.80202

Näther, W. (2000). On random fuzzy variables of second order and their application to linear statistical inference with fuzzy data. *Metrika*, *51*, 201–221. doi:10.1007/s001840000047

Nesterov, Y. E., & Todd, M. J. (1998). Primal-dual interior-point methods for self-scaled cones. *SIAM Journal on Optimization*, *8*(2), 324–364. doi:10.1137/S1052623495290209

Nesterov, Y. E., Wolkowicz, H., & Ye, Y. (2000). Semidefinite programming relaxations of nonconvex quadratic optimization. In Wolkowicz, H., Saigal, R., & Vandenberghe, L. (Eds.), *Handbook of semidefinite programming*. Amsterdam, The Netherlands: Kluwer Academic.

Nhut, M., Shi, D., & Quek, C. (2006). FCMAC-BYY: fuzzy CMAC using Bayesian Ying-Yang learning. *IEEE Transactions on Systems, Man and Cybernetics. Part B*, *36*(5), 1180–1190.

Nomura, H., Hayashi, I., & Wakami, N. (1992). A learning method of fuzzy inference rules by descent method. In *Proceedings of the IEEE International Conference on Fuzzy Systems*, San Diego, CA (pp. 203-210).

Nussbaum, R. D. (1983). Some remarks on the conjecture in parameter adaptive control. *Systems & Control Letters*, *3*, 243–246. doi:10.1016/0167-6911(83)90021-X

Oktem, H., Erzurumlu, T., & Uzman, I. (2007). Application of Taguchi optimization technique in determining plastic injection molding process parameters for a thin-shell part. *Materials & Design*, *28*, 1271–1278. doi:10.1016/j.matdes.2005.12.013

Ong, N. S., & Koh, Y. H. (2005). Experimental investigation into micro injection molding of plastic parts. *Materials and Manufacturing Processes*, *20*, 245–253. doi:10.1081/AMP-200042004

Ordonez, R., & Passino, K. M. (1999). Stable multi-input multi-output adaptive fuzzy/neural control. *IEEE Transactions on Fuzzy Systems*, *7*(3), 3453–3453. doi:10.1109/91.771089

Otto, A., & Kotzab, H. (2003). Does supply chain management really pay? Six perspectives to measure the performance of managing a supply chain. *European Journal of Operational Research*, *144*(2), 306–320. doi:10.1016/S0377-2217(02)00396-X

Pal, N. R., Pal, K., Keller, J. M., & Bezdek, J. C. (2005). A possibilistic fuzzy *c*-means clustering algorithm. *IEEE Transactions on Fuzzy Systems*, *13*(4), 508–516. doi:10.1109/TFUZZ.2004.840099

Pankowska, A., & Wygralak, M. (2006). General IF-sets with triangular norms and their applications to group decision making. *Information Sciences*, *176*, 2713–2754. doi:10.1016/j.ins.2005.11.011

Park, J. H., Kim, S. H., & Moon, C. J. (2006). Adaptive fuzzy controller for the nonlinear system with unknown sign of the input gain. *International Journal of Control, Automation, and Systems*, *4*(2), 178–186.

Passino, K. M., & Yurkovich, S. (1998). *Fuzzy control*. Reading, MA: Addison-Wesley.

Pedrycz, W. (2001). Fuzzy equalization in the construction of fuzzy sets. *Fuzzy Sets and Systems*, *119*, 329–335. doi:10.1016/S0165-0114(99)00135-9

Pedrycz, W., Bezdek, J. C., Hathaway, R. J., & Rogers, G. W. (1998). Two nonparametric models for fusing heterogeneous fuzzy data. *IEEE Transactions on Fuzzy Systems*, *6*(3), 411–425. doi:10.1109/91.705509

Peng, C. Y., & Chien, C. F. (2003). *Data value development to enhance competitive advantage: A retrospective study of EDA systems for semiconductor fabrication.* International Journal of Services. *Technology and Management*, *4*(4-6), 365–383.

Peng, F. Z., Akagi, H., & Nabae, A. (1990). Study of active power filters using quad series voltage source PWM converters for harmonic compensation. *IEEE Transactions on Power Electronics*, *5*(1), 9–15. doi:10.1109/63.45994

Peters, G. (1994). Fuzzy linear regression with fuzzy intervals. *Fuzzy Sets and Systems*, *63*, 45–55. doi:10.1016/0165-0114(94)90144-9

Piramuthu, S. (1991). Theory and methodology – financial credit-risk evaluation with neural and neuralfuzzy systems. *European Journal of Operational Research*, *112*, 310–321. doi:10.1016/S0377-2217(97)00398-6

Polik, I., & Terlaky, T. A. (2007). Survey of the S-Lemma. *SIAM, 49*(3), 371-418.

Porter , M. E. (1979). How competitive forces shape strategy. Harvard Business Review.

Pring, M. J. (2002). *Technical Analysis Explained: The Successful Investor's Guide to Spotting Investment Trends and Turning Points*. New York: McGraw-Hill.

Quek, C., & Singh, A. (2005). POP-Yager: a novel self-organising fuzzy neural network based on the Yager inference. *Expert Systems with Applications, 29*(1), 229–242. doi:10.1016/j.eswa.2005.03.001

Quek, C., & Zhou, R. W. (2006). Structure and learning algorithms of a nonsingleton input fuzzy neural network based on the approximate analogical reasoning schema. *Fuzzy Sets and Systems, 157*(13), 1814–1831. doi:10.1016/j.fss.2005.12.010

Rabiner, L. R. (1989). A tutorial on hidden Markov models and selected applications in speech recognition. *Proceedings of the IEEE, 77*(2), 257–285. doi:10.1109/5.18626

Redden, D. T., & Woodall, W. H. (1994). Properties of certain fuzzy linear regression methods. *Fuzzy Sets and Systems, 64*, 361–375. doi:10.1016/0165-0114(94)90159-7

Rovithakis, G. A., & Christodoulou, M. A. (1994). Adaptive control of unknown plants using dynamical neural networks. *IEEE Transactions on Systems, Man, and Cybernetics. Part B, Cybernetics, 24*, 400–412.

Runkler, T. A. (2007). Pareto optimality of cluster objective and validity functions. In *Proceedings of 2007 IEEE International Conference on Fuzzy Systems* (pp. 79-84).

Saad, E. W., Prokhorov, D. V., & Wunsch, D. C. II. (1998). Comparative study of stock trend prediction using time delay, recurrent and probabilistic neural networks. *IEEE Transactions on Neural Networks, 9*(6), 1456–1470. doi:10.1109/72.728395

Sakawa, M., & Yano, H. (1992). Fuzzy linear regression analysis for fuzzy input-output data. *Information Sciences, 63*, 191–206. doi:10.1016/0020-0255(92)90069-K

Sambhoos, K., Guiffrida, A. L., & Llinas, J. (2005). *Research on possibility-probability transformations in support of innovative approaches to fusion 2*. Buffalo, NY: University of Buffalo.

Sarkis, J., & Talluri, S. (2002). A model for strategic supplier selection. *Journal of Supply Chain Management, 38*(1), 18–28. doi:10.1111/j.1745-493X.2002.tb00117.x

Schon, S., & Kutterer, H. (2005). Using zonotopes for overestimation-free interval least-squares -some geodetic applications. *Reliable Computing, 11*, 137–155. doi:10.1007/s11155-005-3034-4

Scott, G. M., & Ray, W. H. (1993). Creating efficient nonlinear network process models that allow model interpretation. *Journal of Process Control, 3*(3), 163–178. doi:10.1016/0959-1524(93)80022-4

Sepulveda, R., Castillo, O., Melin, P., Rodriguez-Diaz, A., & Montiel, O. (2009). Experimental study of intelligent controller under uncertainty using type-1 and type-2 fuzzy logic. *Information Sciences, 177*, 2023–2048. doi:10.1016/j.ins.2006.10.004

Sharland, A., Eltantawy, R. A., & Giunipero, L. C. (2003). The impact of cycle time on supplier selection and subsequent performance outcomes. *Journal of Supply Chain Management, 39*(3), 4–12. doi:10.1111/j.1745-493X.2003.tb00155.x

Sicard, E., & Boyer, A. (2009). *IC-EMC – user's manual*. Retrieved from http://www.ic-emc.org

Sim, J., Tung, W. L., & Quek, C. (2006). FCMAC-Yager: A novel Yager inference scheme based fuzzy CMAC. *IEEE Transactions on Neural Networks, 17*(6), 1394–1410. doi:10.1109/TNN.2006.880362

Simoes, M. G., Bose, B. K., & Spiegel, R. J. (1997). Fuzzy logic based intelligent control of a variable speed cage machine wind generation system. *IEEE Transactions on Power Electronics, 12*(1), 87–95. doi:10.1109/63.554173

Singh, B., Al-Haddad, K., & Chandra, A. (1999). A review of active power filters for power quality improvement. *IEEE Transactions on Industrial Electronics, 46*(5), 960–969. doi:10.1109/41.793345

Singh, B., Chandra, A., & Al-Haddad, K. (1999). Computer aided modeling and simulation of active power filters. *Journal of Electric Machines & Power Systems, 27*, 1227–1241. doi:10.1080/073135699268687

Singh, G. K., Singh, A. K., & Mitra, R. (2007). A simple fuzzy logic based robust active power filter for harmonic minimization under random load variation. *Electric Power Systems Research*, *77*, 1101–1111. doi:10.1016/j.epsr.2006.09.006

Sjoberg, J., Zhang, Q., Ljung, L., Benveniste, A., Delyon, B., & Glorennec, P. (1995). Nonlinear black-box modeling in system identification: a unified overview. *Automatica*, *31*, 1691–1724. doi:10.1016/0005-1098(95)00120-8

Slotine, J. E., & Li, W. (1991). *Applied nonlinear control*. Upper Saddle River, NJ: Prentice Hall.

Socha, K., & Dorigo, M. (2008). Ant colony optimization for continuous domains. *European Journal of Operational Research*, *185*, 1155–1173. doi:10.1016/j.ejor.2006.06.046

Song, M. C., Liu, Z., Wang, M. J., Yu, T. M., & Zhao, D. Y. (2007). Research on effects of injection process parameters on the molding process for ultra-thin wall plastic parts. *Journal of Materials Processing Technology*, *187-188*, 668–671. doi:10.1016/j.jmatprotec.2006.11.103

Spooner, J. T., Maggiore, M., Ordonez, R., & Passino, K. M. (2002). *Stable adaptive control and estimation for nonlinear systems*. Chichester, UK: John Wiley & Sons. doi:10.1002/0471221139

Spooner, J. T., & Passino, K. M. (1996). Stable adaptive control using fuzzy systems and neural networks. *IEEE Transactions on Fuzzy Systems*, *4*, 339–359. doi:10.1109/91.531775

Stefanov, D. H., Bien, Z. Z., & Bang, W.-C. (2004). The smart house for older persons and persons with physical disabilities: Structure, technology arrangements, and perspectives. *IEEE Transactions on Neural Systems and Rehabilitation Engineering*, *12*(2), 228–250. doi:10.1109/TNSRE.2004.828423

Steinbach, M., Ertös, L., & Kumar, V. (2004). The challenges of clustering high dimensional data. In Wille, L. T. (Ed.), *New Directions in Statistical Physics: Econophysics, Bioinformatics, and Pattern Recognition* (pp. 273–307). Berlin: Springer.

Stephen, H. L., David, A. H., & Jack, J. G. (1992). Theory and development of higher-order CMAC neural networks. *IEEE Control Systems*, *12*(2), 23–30. doi:10.1109/37.126849

Sugeno, M. (1974). *Theory of fuzzy integrals and its application*. Unpublished doctoral dissertation, Tokyo Institute of Technology, Tokyo, Japan.

Sugeno, M. (1977). Fuzzy measures and fuzzy integrals: A survey. In Gupta, M. M., Saridis, G. N., & Gaines, B. R. (Eds.), *Fuzzy automata and decision process* (pp. 89–102). Amsterdam, The Netherlands: North-Holland.

Su, M. C. (2000). A fuzzy rule-based approach to spatio-temporal hand gesture recognition. *IEEE Transactions on Systems, Man, and Cybernetics. Part C*, *30*(2), 276–281.

Szmidt, E., & Kacprzyk, J. (2004). A concept of similarity for intuitionistic fuzzy sets and its use in group decision making. In *Proceedings of the International Joint Conference on Neural Networks and the IEEE International Conference on Fuzzy Systems* (pp. 25-29).

Szmidt, E., & Kacprzyk, J. (2001). Distances between intuitionistic fuzzy sets. *Fuzzy Sets and Systems*, *114*, 505–518. doi:10.1016/S0165-0114(98)00244-9

Szmidt, E., & Kacprzyk, J. (2002). Using intuitionistic fuzzy sets in group decision making. *Control and Cybernetics*, *31*, 1037–1053.

Takagi, T., & Sugeno, M. (1985). Fuzzy identification of systems and its applications to modeling and control. *IEEE Transactions on Systems, Man, and Cybernetics*, *15*, 116–132.

Talluri, S., & Narasimhan, R. (2004). A methodology for strategic sourcing. *European Journal of Operational Research*, *154*(1), 236–250. doi:10.1016/S0377-2217(02)00649-5

Tanaka, H. (1987). Fuzzy data analysis by possibility linear model. *Fuzzy Sets and Systems*, *24*, 363–375. doi:10.1016/0165-0114(87)90033-9

Tanaka, H., & Lee, H. (1998). Interval regression analysis by quadratic programming approach. *IEEE Transactions on Fuzzy Systems*, *6*(4), 473–481. doi:10.1109/91.728436

Tanaka, H., Uejima, S., & Asai, K. (1982). Fuzzy limear regression model. *IEEE Transactions on Systems, Man, and Cybernetics*, *12*, 903–907. doi:10.1109/TSMC.1982.4308925

Tanaka, H., Uejima, S., & Asia, K. (1982). Linear regression analysis with fuzzy model. *IEEE Transactions on Systems, Man, and Cybernetics*, *12*, 903–907. doi:10.1109/TSMC.1982.4308925

Tanaka, H., & Watada, J. (1988). Possibilistic linear systems and their application to the linear regression model. *Fuzzy Sets and Systems*, *272*, 275–289. doi:10.1016/0165-0114(88)90054-1

Tang, S. H., Tan, Y. J., Sapuan, S. M., Sulaiman, S., Ismail, N., & Samin, R. (2007). The use of Taguchi method in the design of plastic injection mould for reducing warpage. *Journal of Materials Processing Technology*, *182*, 418–426. doi:10.1016/j.jmatprotec.2006.08.025

Teng, T. K., Shieh, J. S., & Chen, C. S. (2003). Genetic algorithms applied in online autotuning PID parameters of a liquid-level control system. *Transactions of the Institute of Measurement and Control*, *25*(5), 433–450. doi:10.1191/0142331203tm0098oa

Thammano, A. (1999). Neuro-fuzzy model for stock market prediction. In *Proceedings of the Artificial Neural Networks in Engineering Conference (ANNIE'99)* (pp. 587-591).

Tong, S. C., Tang, J., & Wang, T. (2000). Fuzzy adaptive control of multivariable nonlinear systems. *Fuzzy Sets and Systems*, *111*, 153–167. doi:10.1016/S0165-0114(98)00058-X

Tracey, M., & Tan, C. L. (2001). Empirical analysis of supplier selection and involvement, customer satisfaction, and firm performance. *Supply Chain Management*, *6*(3-4), 174–188.

Tran, L., & Duckstein, L. (2002). Comparison of fuzzy numbers using a fuzzy distance measure. *Fuzzy Sets and Systems*, *130*, 331–341. doi:10.1016/S0165-0114(01)00195-6

Triantaphyllou, E., & Lin, C. T. (1996). Development and evaluation of five fuzzy multiattribute decision-making methods. *International Journal of Approximate Reasoning*, *14*, 281–310. doi:10.1016/0888-613X(95)00119-2

Trippi, R. R., & Turban, E. (1993). *Neural Networks in Finance and Investing: Using Artificial Intelligence to Improve Real-World Performance*. Chicago: Probus Publishing Company.

Tseng, F. M., Tzeng, G. H., Yu, H. C., & Yuan, B. J. C. (2001). Fuzzy ARIMA model for forecasting the foreign exchange market. *Fuzzy Sets and Systems*, *118*, 9–19. doi:10.1016/S0165-0114(98)00286-3

Tung, W. L., & Quek, C. (2002). GenSoFNN: a generic self-organizing fuzzy neural network. *IEEE Transactions on Neural Networks*, *13*(5), 1075–1086. doi:10.1109/TNN.2002.1031940

Tung, W. L., Quek, C., & Cheng, P. (2004). GenSo-EWS: a novel neural-fuzzy based early warning system for predicting bank failures. *Neural Networks*, *17*(4), 567–587. doi:10.1016/j.neunet.2003.11.006

Turksen, I. B., & Zhong, Z. (1990). An approximate analogical reasoning schema based on similarity measures and interval-valued fuzzy sets. *Fuzzy Sets and Systems*, *34*, 223–346. doi:10.1016/0165-0114(90)90218-U

Vandenberghe, L., & Boyd, S. (1996). Semidefinite programming. *SIAM Review*, *38*, 49–95. doi:10.1137/1038003

Verleysen, M. (2003). Learning high-dimensional data. In Ablameyko, S., Gori, M., Goras, L., & Piuri, V. (Eds.), *Limitations and future trends in neural computation* (pp. 141–162). IOS Press.

Verma, R., & Pullman, M. E. (1998). An analysis of the supplier selection process. *Omega*, *26*(6), 739–750. doi:10.1016/S0305-0483(98)00023-1

Vonderembse, M. A., & Tracey, M. (1999). The impact of supplier selection criteria and supplier involvement on manufacturing performance. *Journal of Supply Chain Management*, *35*(3), 33–39. doi:10.1111/j.1745-493X.1999.tb00060.x

Walsh, S. T., Boylan, R. L., McDermott, C., & Paulson, A. (2005). The semiconductor silicon industry roadmap: Epochs driven by the dynamics between disruptive technologies and core competencies. *Technological Forecasting and Social Change*, *72*, 213–236. doi:10.1016/S0040-1625(03)00066-0

Wang, C. H., Cheng, C., & Lee, T. (2004). Dynamical optimal training for interval type-2 fuzzy neural network (T2FNN). *IEEE Transactions on Systems, Man, and Cybernetics. Part B, Cybernetics*, *34*(3), 1472–1477. doi:10.1109/TSMCB.2004.825927

Wang, C. H., Liu, H. L., & Lin, C. T. (2001). Dynamic learning rate optimalization of the back propagation algorithm. *IEEE Transactions on Systems, Man, and Cybernetics. Part B, Cybernetics*, *31*, 669–677.

Wang, L. X. (1993). Stable adaptive fuzzy control of nonlinear systems. *IEEE Transactions on Fuzzy Systems*, *1*, 146–155. doi:10.1109/91.227383

Wang, L. X. (1994). *Adaptive fuzzy systems and control: Design and stability analysis*. Upper Saddle River, NJ: Prentice-Hall.

Wang, L.-X., & Mendel, J. M. (1992). Generating fuzzy rules by learning from examples. *IEEE Transactions on Systems, Man, and Cybernetics*, *22*(6), 1414–1427. doi:10.1109/21.199466

Wang, S. Y. (2008). Applying 2-tuple multi-granularity linguistic variables to determine the supply performance in dynamic environment based on product-oriented strategy. *IEEE Transactions on Fuzzy Systems*, *16*(1), 29–39. doi:10.1109/TFUZZ.2007.903316

Wang, S. Y., Chang, S. L., & Wang, R. C. (2007). Applying a direct approach in linguistic assessment and aggregation on supply performance. *International Journal of Operations Research*, *4*(4), 238–247.

Wang, S. Y., Chang, S. L., & Wang, R. C. (2009). Assessment of supplier performance based on product-development strategy by applying multi-granularity linguistic term sets. *Omega*, *37*(1), 215–226. doi:10.1016/j.omega.2006.10.003

Wang, Y. (2003). Mining stock price using fuzzy rough set system. *Expert Systems with Applications*, *24*, 13–23. doi:10.1016/S0957-4174(02)00079-9

Weber, S. (1983). A general concept of fuzzy connectives, negations and implications based on t-norms. *Fuzzy Sets and Systems*, *11*, 115–134. doi:10.1016/S0165-0114(83)80073-6

White, H. (1988). Economic prediction using neural networks: the case of IBM daily stock returns. In *Proceedings of the Second Annual IEEE Conference on Neural Networks* (pp. 451-458).

Wilson, C. L. (1994). Self-organizing neural network system for trading common stocks. In *Proceedings of the IEEE International Conference on Neural Networks* (pp. 3651-3654).

Wu, H.-C. (2003). Linear regression analysis for fuzzy input and output data using the extension principle. *Computers & Mathematics with Applications (Oxford, England)*, *45*(12), 1849–1859. doi:10.1016/S0898-1221(03)90006-X

Wu, K. C. (1996). Fuzzy interval control of mobile robots. *Computers & Electrical Engineering*, *22*(3), 211–229. doi:10.1016/0045-7906(95)00038-0

Xie, X. L., & Beni, G. (1987). A validity measure or fuzzy clustering. *IEEE Transactions on Pattern Analysis and Machine Intelligence*, *13*(8), 841–847. doi:10.1109/34.85677

Xu, Z. S. (2007). Intuitionistic preference relations and their application in group decision making. *Information Sciences*, *177*, 2363–2379. doi:10.1016/j.ins.2006.12.019

Yabuuchi, Y., & Watada, J. (1997). Fuzzy principal component analysis and its application. *Biomedical Fuzzy and Human Sciences*, *3*, 83–92.

Yager, R. R. (1985). On the question of credibility of evidence for expert systems. In *Proceedings of the Conference on the Analysis of Decision Problems within an Uncertain and Imprecise Environment* (pp. 1-15).

Yager, R. R. (1980). Fuzzy subsets of type II in decisions. *Journal of Cybernetics*, *10*, 137–159. doi:10.1080/01969728008927629

Yager, R. R. (1986). The entailment principle for Dempster-Shafer granules. *International Journal of Intelligent Systems*, *1*, 247–262. doi:10.1002/int.4550010403

Yager, R. R. (1988). On ordered weighted averaging aggregation operators in multicriteria decision making. *IEEE Transactions on Systems, Man, and Cybernetics*, *18*(1), 183–190. doi:10.1109/21.87068

Yager, R. R. (1993). Element selection from a fuzzy subset using the fuzzy integral. *IEEE Transactions on Systems, Man, and Cybernetics*, *23*, 467–477. doi:10.1109/21.229459

Yager, R. R. (1998). On measures of specificity. In Kaynak, O., Zadeh, L. A., Turksen, B., & Rudas, I. J. (Eds.), *Computational intelligence: Soft computing and fuzzy-neuro integration with applications* (pp. 94–113). Berlin, Germany: Springer-Verlag. doi:10.1007/978-3-642-58930-0_6

Yager, R. R., & Filev, D. P. (1994). *Essentials of fuzzy modeling and control*. New York, NY: John Wiley & Sons.

Yager, R. R., & Kreinovich, V. (2007). Entropy conserving transforms and the entailment principle. *Fuzzy Sets and Systems*, *158*, 1397–1405. doi:10.1016/j.fss.2007.01.019

Yaghoobi, M. A., Jones, D. F., & Tamiz, M. (2008). Weighted additive models for solving fuzzy goal programming problems. *Asia-Pacific Journal of Operational Research*, *25*(5), 715–733. doi:10.1142/S0217595908001973

Yamato, J., Ohya, J., & Ishii, K. (1992). Recognizing human action in time-sequential images using hidden Markov model. In *Proceedings of the IEEE Conference on Computer Vision and Pattern Recognition* (pp. 375-385).

Yang, S.-E. (2007). *Gesture spotting using fuzzy garbage model and user adaptation.* Unpublished master's thesis, Korea Advanced Institute of Science and Technology, Daejeon, Korea.

Yang, M. S., Hwang, P. Y., & Chen, D. H. (2004). Fuzzy clustering algorithms for mixed feature variables. *Fuzzy Sets and Systems*, *141*(2), 301–317. doi:10.1016/S0165-0114(03)00072-1

Yang, M. S., & Ko, C. H. (1996). On a class of fuzzy c-numbers clustering procedures for fuzzy data. *Fuzzy Sets and Systems*, *84*(1), 4960. doi:10.1016/0165-0114(95)00308-8

Yang, M. S., & Liu, H. H. (1999). Fuzzy clustering procedures for conical fuzzy vector data. *Fuzzy Sets and Systems*, *106*(2), 189–200. doi:10.1016/S0165-0114(97)00277-7

Yang, Y., & Gao, F. (2006). Injection molding product weight: Online prediction and control based on a nonlinear principal component regression model. *Polymer Engineering and Science*, *46*(4), 540–548. doi:10.1002/pen.20522

Ye, X., & Jiang, J. (1998). Adaptive nonlinear design without a priori knowledge of control directions. *IEEE Transactions on Automatic Control*, *43*, 1617–1621. doi:10.1109/9.728882

Yildiz, O. T., & Alpaydin, E. (2001). Omnivariate decision trees. *IEEE Transactions on Neural Networks*, *12*(6), 1539–1546. doi:10.1109/72.963795

Ying, H. (2000). *Fuzzy Control and Modeling: Analytical foundations and applications*. New York, NY: Wiley-IEEE Press.

Zadeh, L. A. (1965). Fuzzy sets. *Information and Control*, *8*, 338–353. doi:10.1016/S0019-9958(65)90241-X

Zadeh, L. A. (1975). Calculus of fuzzy restrictions. In Zadeh, L. A.,(Eds.), *A. Fuzzy Sets and Their Applications to Cognitive and Decision Processes* (pp. 1–39). New York: Academic Press.

Zadeh, L. A. (1979). A theory of approximate reasoning. In Hayes, J., Michie, D., & Mikulich, L. I. (Eds.), *Machine intelligence 9* (pp. 149–194). New York, NY: Halstead Press.

Zadeh, L. A. (1996). Fuzzy logic = computing with words. *IEEE Transactions on Fuzzy Systems*, *4*, 103–111. doi:10.1109/91.493904

Zadeh, L. A. (1999). Outline of a computational theory of perceptions based on computing with words. In Sinha, N. K., & Gupta, M. M. (Eds.), *Soft computing and intelligent systems* (pp. 3–22). Boston, MA: Academic Press.

Zadeh, L. A. (2005). Toward a generalized theory of uncertainty (GTU)-An outline. *Information Sciences*, *172*, 1–40. doi:10.1016/j.ins.2005.01.017

Zeng, X., & Singh, M. G. (1995). Approximation theory of fuzzy systems-MIMO case. *IEEE Transactions on Fuzzy Systems*, *3*(2), 219–235. doi:10.1109/91.388175

Zhang, Q., Chen, J. C. H., He, Y. Q., Ma, J., & Zhou, D. N. (2003). Multiple attribute decision making: Approach integrating subjective and objective information. *International Journal of Manufacturing Technology and Management*, *5*(4), 338–361. doi:10.1504/IJMTM.2003.003460

Zhang, T. P., & Ge, S. S. (2007). Adaptive neural control of MIMO nonlinear state time-varying delay systems with unknown dead-zones and gain signs. *Automatica*, *43*, 1021–1033. doi:10.1016/j.automatica.2006.12.014

Zhao, M. Y., Cheng, C. T., Chau, K. W., & Li, G. (2006). Multiple criteria data envelopment analysis for full ranking units associated to environment impact assessment. *International Journal of Environment and Pollution*, *28*(3-4), 448–464. doi:10.1504/IJEP.2006.011222

Zhou, R. W., & Quek, C. (1999). POPFNN-AAR(S): A pseudo outer-product based fuzzy neural network. *IEEE Transactions on Systems, Man, and Cybernetics*, *29*(6), 859–870. doi:10.1109/3477.809038

Zimmermann, H. J. (2001). *Fuzzy set theory and its applications*. Amsterdam, The Netherlands: Kluwer Academic. doi:10.1007/978-94-010-0646-0

About the Contributors

Tin-Chih Toly Chen received the Ph. D. degree in Industrial Engineering from National Tsin Hua University. He is now a Professor in the Department of Industrial Engineering and Systems Management of Feng Chia University. His research interests include fuzzy and neural computing, competitiveness analysis, operations research, semiconductor manufacturing, and global warming issues. Dr. Chen has published more than one hundred papers in refereed journals, and is the recipient of several research and paper awards. Dr. Chen has been the guest editor of SCI journals including *Fuzzy Sets and Systems* and *International Journal of Advanced Manufacturing Technology*, and is the founding Editor-in-Chief of *International Journal of Fuzzy System Applications.*

* * *

Abbas Al-Refaie is currently working as an assistant professor in the Department of Industrial Engineering, University of Jordan, Amman. His research interests include: Data Envelopment Analysis, Robust Design, Statistical Quality Control, Design of Experiments, Taguchi Methods, Operation Research and Optimization, and Quality Management.

Rakesh K. Arya is a senior resource scientist in the remote sensing application center, M.P. Council of Science and Technology, Bhopal, India. He has obtained his Ph. D. and Master degree from Indian Institute of Technology, Roorkee, India in 2006 and 2000 respectively, and Bachelor degree from Government Engineering College, Sagar, Madhya Pradesh, India in 1994. His research interests include nonlinear control theory, fuzzy control theory, optimization techniques, GIS and image processing.

Zeungnam Bien (S'72-M'75-SM'91-F'07) received the BS degree in electronics engineering from Seoul National University, Korea, in 1969,and the MS and PhD degrees in electrical engineering from the University of Iowa, in1972 and 1975, respectively. During 1976-1977, he was an assistant professor in the Department of Electrical Engineering, University of Iowa. He, thereafter, joined the Computer Applications and Software Engineering (CASE) Center, Syracuse University, New York. During 1987-1988, he was a visiting professor in the Department of Control Engineering, Tokyo Institute of Technology, Japan. Since 1977, he had been with the Department of Electrical Engineering and Computer Science, Korea Advanced Institute of Science and Technology (KAIST), Daejeon, until2008, and became professor emeritus. In 2009, he joined the School of Electrical and Computer Engineering, Ulsan National Institute of Science and Technology (UNIST), Korea, where he is currently a chaired professor. His current research interests include intelligent automation and learning control methods, soft computing techniques with emphasis on fuzzy logic systems, and service robotics and rehabilitation engineering systems. He is a fellow of the IEEE and the IFSA (2007).He was a distinguished lecturer of the IEEE Robotics and Automation Society during 2004-2005. He was the president of IFSA during 2003-2005. He has been the director of the Human-Friendly Welfare Robot System Engineering Research Center, KAIST.

Hamid Boubertakh received the Engineer diploma in automatic control from the University of Annaba, Algeria in 1998, and the M. Sc. degree in automatic control from the Military Polytechnic School of Algiers (EMP), Algeria in 2001. He received his Ph. D. degree in automatic control from the National Polytechnic School of Algiers (ENP), Algeria in 2009. He is currently working as an associate professor in the Automatic Control Department at the University of Jijel, Algeria, where he is pursuing his research and teaching activities. His area of interest is intelligent control and optimization.

Alexandre Boyer received his engineering diploma, Masters and PhD degrees in electronics from the National Institute of Applied Sciences (INSA) in Toulouse, France, in 2004 and in 2007 respectively. He is currently an Assistant Professor in the Department of Electrical and Computer Engineering at INSA, Toulouse. His current research interests include IC susceptibility modeling, reliability of ICs and computer aided design (CAD) tool development for electromagnetic compatibility (EMC).

Mark Burgin received his Ph.D. in mathematics from Moscow State University and Doctor of Science in logic and philosophy from the National Academy of Sciences of Ukraine. He was a Professor at the Institute of Education, Kiev; at International Solomon University, Kiev; at Kiev State University, Ukraine; and Head of the Assessment Laboratory in the Research Center of Science at the National Academy of Sciences of Ukraine. Now he is a Visiting Scholar at UCLA, Los Angeles, California, USA. Dr. Burgin is a member of New York Academy of Sciences and an Honorary Professor of the Aerospace Academy of Ukraine. He is a Chief Editor of the journal Integration and Associate Editor of the International Journal on Computers and their Applications. He was a member of organizing and program committees of more than 30 conferences. Dr. Burgin is doing research, has publications, and taught courses in mathematics, computer science, information sciences, system theory, artificial intelligence, software engineering, logic, psychology, education, social sciences, and methodology of science. Dr. Burgin is the author or co-author of more than 500 published papers and 17 published books, including "Theory of Information" (2010), "Neoclassical Analysis: Calculus Closer to the Real World" (2008), "Super-recursive Algorithms" (2005), "On the Nature and Essence of Mathematics" (1998), "Intellectual Components of Creativity" (1998), "Fundamental Structures of Knowledge and Information" (1997), "Introduction to the Modern Exact Methodology of Science" (1994).

Ming-Hsien Caleb Li is a Professor in the Department of Industrial Engineering and Systems Management at Feng Chia University, Taiwan. His interests are Six Sigma Management, Quality Engineering, Taguchi method, Design of Experiment, and Quality Control. Li is a member of Chinese IIE and Chinese Society for Quality.

Sheng-Lin Chang received his Ph.D. degree in Quantitative Management Science from University of Houston, USA. He was with the Department of Industrial Management, National Taiwan University of Science & Technology, until 2006. He is currently a Professor with the Department of Industrial Engineering & Management at China University of Science & Technology, Taiwan. His main research interests include operations strategy and management.

Yi-Ming Chang was born in Taichung, Taiwan, in 1985. He received the B.S. degree and M.S. degree in electronic engineering from the Feng-Chia University, Taichung, Taiwan in 2007 and 2009, respectively. He is currently working toward the Ph.D. degree in the Department of Electrical Engineering, National Cheng Kung University, Chungli, Taiwan. His research interests are in the area of adaptive control and fuzzy-neural-neural.

Oktay Duman received his Ph.D. from the University of Ankara in 2003. Now he is a professor of mathematics at the TOBB Economics and Technology University (TOBB-ETU). His research includes summability theory, approximation theory and fuzzy mathematics.

Thierry Marie Guerra was born in Mulhouse, France in 1963. He is currently professor at the University of Valenciennes et du Hainaut-Cambrésis (UVHC), France. He received his Ph.D degree in automatic control from the UVHC in 1991 and the HDR in 1999. He is head of the Laboratory of Industrial and Human Automation, Mechanics and Computer Science (LAMIH CNRS FRE 3304) (86 researchers and staff, 69 PhD students and post-docs). He is vice-chair of the Technical Committee 3.2 "Computational Intelligence in Control" for IFAC (International Federation of Automatic Control), member of the IFAC TC 7.1 "Automotive Control", Area Editor of the international journal Fuzzy Sets & Systems and member of IEEE Vehicle Power and Propulsion Committee. His major research fields and topics of interest are, wine, hard rock, chess, nonlinear control, LPV, quasi-LPV (Takagi-Sugeno) models control and observation, LMI constraints, Non quadratic Lyapunov functions. Applications to powertrain systems (IC engine, electrical motors, hybrid vehicles, fuel cells).

Zaiyi Guo received the B.Eng. degree (2003) in Computer Engineering and is currently pursuing her the Ph.D. degree in iartificial immune system from the School of Computer Engineering, Nanyang Technological University, Singapore. Her research interests include intelligent control, intelligent architectures, artificial immune systems, neural networks, fuzzy neural systems, neurocognitive informatics, and genetic algorithms.

Katsuhiro Honda received his BE, ME and DEng in Industrial Engineering from Osaka Prefecture University, Osaka, Japan, in 1997, 1999 and 2004, respectively. He is currently an Associate Professor in the Department of Computer Sciences and Intelligent Systems, Osaka Prefecture University. His research interests include hybrid techniques of fuzzy clustering and multivariate analysis, data mining with fuzzy data analysis and neural networks.

Hidetomo Ichihashi received his BE and DEng in Industrial Engineering from Osaka Prefecture University, Osaka, Japan, in 1971 and 1986, respectively. From 1971 to 1981, he was with the Information System Center of Matsushita Electric Industrial Co., Ltd.. From 1981 to 1993, he was a Research Associate, Assistant Professor and Associate Professor at Osaka Prefecture University, where he is currently a Professor in the Department of Computer Sciences and Intelligent Systems. His fields of interest are in adaptive modeling of GMDH-type neural networks, fuzzy c-means clustering and classifier, data mining with fuzzy data analysis, human-machine interface and cognitive engineering.

Moon-Jin Jeon received the B.S. degree in electrical engineering and computer science from Hanyang University, Korea, in 2006 and the M.S. degree in electrical engineering from Korea Advanced Institute of Science and Technology (KAIST), Korea, in 2008. He is currently a researcher of Korea Aerospace Research Institute (KARI). In the company, he participates in development of a low earth orbit satellite. His current research interests include spacecraft mission simulation, soft computing technique for estimating satellite state of heart data.

Frank Klawonn received his M.Sc. in mathematics and his Ph.D. in computer science from the University of Braunschweig in 1988 and 1992, respectively. He is currently the head of the Laboratory for Data Analysis and Pattern Recognition at Ostfalia University of Applied Sciences in Wolfenbuettel (Germany) and the head of the Bioinformatics and Statistics Group at Helmholtz Centre of Infection Research. His main research interests focus on techniques for intelligent data analysis, especially clustering and classification. He is a member of the editorial boards of various journals, for example Data Mining, Modelling & Management and Knowledge Engineering & Soft Data Paradigms.

Rudolf Kruse received his M.Sc. and his Ph.D. in mathematics from the University of Braunschweig in 1979 and 1980, respectively. He is currently a professor at the Department of Computer Science of the Otto-von-Guericke University of Magdeburg (Germany) where he is leading the computational intelligence research group. He has carried out research and projects in statistics, artificial intelligence, expert systems, fuzzy control, fuzzy data analysis, computational intelligence, and data mining. He is a fellow of the International Fuzzy Systems Association (IFSA), fellow of the European Coordinating Committee for Artificial Intelligence (ECCAI) and fellow of the Institute of Electrical and Electronics Engineers (IEEE).

Ming-Jen Kuo was born in Taipei, Taiwan, in 1982. He received the B.S. degree in electronic engineering from the Fu Jen Catholic University, Taipei, Taiwan, and M.S. degree in electronic engineering from the Feng Chia University, Taichung, Taiwan in 2005 and 2007, respectively. He is currently working toward the Ph.D. degree in the Department of Electrical Engineering, National Central University, Chungli, Taiwan. His research interests include electromagnetic compatibility, fuzzy and neural control.

Salim Labiod is currently an associate professor at the Department of Automatic Control at the University of Jijel, Algeria. He received his Engineer diploma, his Ph. D. degree and his HDR in automatic control from the National Polytechnic School of Algiers in 1995, 2005 and 2007, respectively. He has authored and co-authored more than 50 scientific papers in refereed journals and conferences. He has served as reviewer in many international journals and conferences and as member of the program committee of many conferences. His research interests include fuzzy control, nonlinear control and adaptive control.

Sang Wan Lee received his B.S. degree in electrical engineering from Yonsei University, Korea, in 2003 and the M.S. and Ph.D. degrees in electrical engineering from Korea Advanced Institute of Science and Technology (KAIST), Korea, in 2005 and 2009, respectively. He is currently a Postdoctoral Associate in Professor Poggio's lab at Massachusetts Institute of Technology (MIT). His current research interests include machine learning and computational neuroscience with emphasison attention model in the human visual system, hypothetical associative learning, and integrated information in discrete dynamic systems.

Tun-Yuan Lee was born in Taichung, Taiwan, in 1988. He received the B.S. degree in Electronic Engineering from the Feng-Chia University, Taichung, Taiwan in 2010. He is currently working toward the M.S. degree in Electronic Engineering, Electronic Engineering University, Taichung, Taiwan. His research interests are in the area of adaptive control and fuzzy-neural-neural.

Minghuang Li received the M.S. degree in Applied Mathematics from Beijing Normal University in 2009. His research focuses on fuzzy information processing, programming, data mining and intelligent computing.

Deng-Feng Li was born in 1965. He received the B.Sc. and M.Sc. degrees in applied mathematics from the National University of Defense Technology, Changsha, China, in 1987 and 1990, respectively, and the Ph.D. degree in system science and optimization from the Dalian University of Technology, Dalian, China, in 1995. From 2003 to 2004, he was a Visiting Scholar with the School of Management, University of Manchester Institute of Science and Technology, Manchester, U. K. He is currently a Professor with the School of Management, Fuzhou University, Fuzhou, China. He has authored or coauthored more than 180 journal papers and three monographs. He has coedited one proceeding of the international conference. His current research interests include fuzzy decision analysis, group decision making, fuzzy game theory, supply chain, fuzzy sets and system analysis, fuzzy optimization, and differential game.

T. Warren Liao is currently a Professor with Department of Mechanical and Industrial Engineering at Louisiana State University. His research interests include soft computing, data mining, metaheustics, advanced materials and manufacturing processes, and supply chain. He has more than 70 refereed journal publications included in the Web of Science with more than 900 citations and is durrently serving as an Associate Editor for Applied Soft Computing.

Tsung-Chih, Lin was born in Changhua, Taiwan, in 1961. He received the B.S. degree in electrical engineering from the Feng Chia University, Taichung, Taiwan in 1984, the M.S. in control engineering from the National Chiao Tung University in 1986 and the Ph.D. degree in the School of Microelectronic Engineering, Griffith University, Brisbane, Australia in 2002. He is currently Associate Professor in the Department of Electronic Engineering, Feng Chia University, Taichung, Taiwan. His current research interests and publications are in the areas of adaptive control, fuzzy-neural-network, robust control and analog circuit testing.

Douglas L. Maskell received the B.E. (Hons.), MEngSc and Ph.D. degrees in electronic and computer engineering from James Cook University (JCU), Townsville, Australia, in 1980, 1984, and 1996, respectively. He is currently an Associate Professor in the School of Computer Engineering, Nanyang Technological University, Singapore. He is the Director of the Centre for High Performance Embedded Systems (CHiPES) and is an active member of the Centre for Computational Intelligence. His research interests are in the areas of embedded systems, reconfigurable computing, intelligent systems, and algorithm acceleration for hybrid high performance computing (HPC) systems.

Tomohiro Matsui received his BE and MEng in Computer Sciences and Intelligent Systems from Osaka Prefecture University, Osaka, Japan, in 2008 and 2010, respectively. He had been a graduate student in the Graduate School of Engineering, Osaka Prefecture University, and is with the Kansai Electric Power Co., Inc..

Jiang-Xia Nan was born in 1978. She received the B.Sc. and M.Sc. degrees in applied mathematics from the Shanxi Normal University and Guangxi Normal University, Taiyuan and Guilin, China, in 2001 and 2004, respectively, and the Ph.D. degree in the school of mathematical sciences from the Dalian University of Technology, Dalian, China, in 2010. She is currently a Lecturer in the school of mathematics and computing sciences, Guilin University of Electronic Technology, Guilin, China. She has authored or coauthored more than 10 journal papers. Her current research interests include fuzzy decision analysis, group decision making, fuzzy game theory and fuzzy optimization.

Akira Notsu received his BE, MI and D. Informatics from Kyoto University in 2000, 2002 and 2005, respectively. He is currently an Assistant Professor in the Department of Computer Sciences and Intelligent Systems, Osaka Prefecture University. His research interests include agent-based social simulation, communication networks, game theory, human-machine interface and cognitive engineering.

Kwang-Hyun Park received the BS, MS and PhD degrees in electrical engineering and computer science from Korea Advanced Institute of Science and Technology (KAIST), Daejeon, Korea, in 1994, 1997, and 2001, respectively. During 2001 to 2003, he was a research associate at the Human-friendly Welfare Robot System Engineering Research Center at KAIST. He was a research professor in the Department of Electrical Engineering and Computer Science, KAIST, from 2003 to 2004, a visiting researcher at the College of Computing, Georgia Institute of Technology, Atlanta, from 2004 to 2005, and a visiting professor in the Department of Electrical Engineering and Computer Science, KAIST, from 2005 to 2008. He is currently an assistant professor in the Department of Information and Control Engineering, Kwangwoon University since 2008. His research interests include service robot, assistive technology, human-robot interaction, machine learning, and pattern recognition.

Chai Quek (SM) received the B.Sc. degree (1986) in electrical and electronics engineering and the Ph.D. degree (1990) in intelligent control from Heriot Watt University, Edinburgh, Scotland. He is an Associate Professor and a Member of the Centre of Computational Intelligence, formerly the Intelligent Systems Laboratory, School of Computer Engineering, Nanyang Technological University. His research interests include intelligent control, intelligent architectures, AI in education, neural networks, fuzzy neural systems, neurocognitive informatics, and genetic algorithms. Dr. Quek is a senior member of IEEE and a member of the IEEE Technical Committee on Computational Finance.

Zahra S. Razaee was born in Tehran, Iran in 1985. She received the B.S degree in Applied Mathematics from University of Tehran in 2007 and the M.S degree in Industrial Engineering from Amirkabir University of Technology (Tehran Polytechnic) in 2009 under the supervision of Professor Fazel Zarandi, with the thesis entitled "A fuzzy case based reasoning system for value engineering". Her research interests are pattern recognition, information retrieval, image processing, Natural language processing and expert systems.

Rambir Singh is a research scholar in the Electrical Engineering Department, MNNIT, Allahabad, India. He has served the maintenance branch of Indian Air Force for more than 17 years. He has obtained his M.Tech degree from MNNIT, Allahabad, India in 2004, and Bachelor degree from IEI, Kolkata, India in 2001, both in Electrical Engineering. His research interests include power quality, active power filters, fuzzy logic control, artificial intelligence and evolutionary algorithms.

Asheesh K. Singh is an associate professor in the Electrical Engineering Department, MNNIT, Allahabad, India. He received the Ph.D. degree from the Indian Institute of Technology, Roorkee, India, the M.Tech. degree from the REC, Kurukshetra, India, and the B.Tech. degree from HBTI, Kanpur, India, in 2007, 1994 and 1991 respectively. Since 1995, he has been on the academic staff of MNNIT, Allahabad, India. His research interests include application of soft computing techniques in power systems, distributed generation, power quality and reliability etc.

Reay-Chen Wang received a Ph.D. degree in Industrial Management from National Taiwan University of Science and Technology, Taiwan. In the past, he is a Professor with the Department of Industrial Management at National Taiwan University of Science & Technology until 2004. He is currently a Professor with the Department of Industrial Engineering & Management at Tungnan University, Taiwan. His main research interests include operations management, quality management, and the applications of fuzzy theory in industrial management.

Shih-Yuan Wang received a Bachelor degree in Industrial Management from National Yun-Lin institute of Technology, Taiwan, in 1996. He has also received an MBA and Ph.D. degree in Industrial Management from National Taiwan University of Science and Technology, Taiwan, in 1998 and 2005, respectively. He was with the Department of Industrial Engineering & Management at Hsiuping Institute of Technology, Taiwan, until 2009. He is currently an Associate Professor with the Department of Information Management at Jinwen University of Science & Technology, Taiwan. His main research interests include the applications of fuzzy theory in industrial management.

Yi-Chi Wang is an Associate Professor in the Department of Industrial Engineering and Systems Management at Feng Chia University. He received his B.S. in Mechanical Engineering from Tatung Institute of Technology, Taiwan, 1993, M.S. in Manufacturing Engineering from Syracuse University, New York, and his Ph.D. degree in Industrial Engineering from Mississippi State University in 2003. His major research interests are in areas of agent-based manufacturing systems, joint replenishment problems, supply chain system simulation, and metal cutting optimization. He has conducted several research projects funded by National Science Council of Taiwan and has published over 40 articles in refereed journals and conference proceedings. Dr. Wang was the co-founder of the Society of Lean Enterprise Systems of Taiwan (SLEST). He currently serves as the secretary of SLEST. In 2011, he co-chaired the 21st International Conference on Flexible Automation and Intelligent Manufacturing, hosted in Taiwan. He is a member of SME, IIE, and CIIE.

Roland Winkler received his M.SC. in computer science from the Otto-von-Guericke University Magdeburg in 2002. He is currently with the Department of Air Transportation of the German Aerospace Centre in Braunschweig, Germany. His research area focus on data analysis for air traffic management.

Ronald R. Yager has worked in the area of machine intelligence for over twenty-five years. He has published over 500 papers and fifteen books in areas related to fuzzy sets, decision making under uncertainty and the fusion of information. He is among the world's top 1% most highly cited researchers with over 7000 citations. He was the recipient of the IEEE Computational Intelligence Society Pioneer award in Fuzzy Systems. Dr. Yager is a fellow of the IEEE, the New York Academy of Sciences and the Fuzzy Systems Association. He was given a lifetime achievement award by the Polish Academy of Sciences for his contributions. He served at the National Science Foundation as program director in the Information Sciences program. He was a NASA/Stanford visiting fellow and a research associate at the University of California, Berkeley. He has been a lecturer at NATO Advanced Study Institutes. He is a distinguished honorary professor at the Aalborg University Denmark. He is an affiliated distinguished researcher at the European Centre for Soft Computing. He received his undergraduate degree from the City College of New York and his Ph. D. from the Polytechnic University of New York. Currently, he is Director of the Machine Intelligence Institute and Professor of Information Systems at Iona College. He is editor and chief of the International Journal of Intelligent Systems. He serves on the editorial board of numerous technology journals.

Seung-Eun Yang received his B.S. degree in electronic engineering from Sogang University, Korea, in 2005 and the M.S. degree in electrical engineering from Korea Advanced Institute of Science and Technology (KAIST), Korea, in 2007. He is a researcher of Korea Aerospace Research Institute (KARI). In the company, he participates in development of a low earth orbit satellite. He currently researches the autonomous fault management and ground command execution system for the satellite.

Fusheng Yu is a professor in the School of Mathematics Sciences, Beijing Normal University, Beijing, China. He received the M.S. degree and PhD degree in Applied Mathematics from Beijing Normal University in 1989 and in 1998 respectively. From 2002 to 2004, he was with the Department of Electrical and Computer Engineering, University of Alberta, Edmonton, Alberta, Canada as a visiting scholar. He is pursuing research in computational intelligence, fuzzy systems and fuzzy modeling, knowledge discovery and data mining, fuzzy neural networks, expert systems and fault diagnosis, and knowledge representation.

Fazel Zarandi is currently a Professor of Industrial Engineering in Amirkabir University of Technology (Tehran Polytechnic).

Index

H

I

L

M

N

O

P

Q

R

S

T

U

V

W

X

CPSIA information can be obtained at www.ICGtesting.com
Printed in the USA
BVOW030035250512

290858BV00007B/4/P

9 781466 618701